Lecture Notes in Artificial Intelligence 8765

Subseries of Lecture Notes in Computer Science

LNAI Series Editors

Randy Goebel
University of Alberta, Edmonton, Canada
Yuzuru Tanaka
Hokkaido University, Sapporo, Japan
Wolfgang Wahlster
DFKI and Saarland University, Saarbrücken, Germany

LNAI Founding Series Editor

Joerg Siekmann
DFKI and Saarland University, Saarbrücken, Germany

Luc Lamontagne   Enric Plaza (Eds.)

Case-Based Reasoning Research and Development

22nd International Conference, ICCBR 2014
Cork, Ireland, September 29 - October 1, 2014
Proceedings

 Springer

Volume Editors

Luc Lamontagne
Université Laval
Department of Computer Science and Software Engineering
Québec, Canada
E-mail: luc.lamontagne@ift.ulaval.ca

Enric Plaza
IIIA, Artificial Intelligence Research Institute CSIC
Spanish National Research Council
Bellaterra, Catalonia, Spain
E-mail: enric@iiia.csic.es

ISSN 0302-9743   e-ISSN 1611-3349
ISBN 978-3-319-11208-4   e-ISBN 978-3-319-11209-1
DOI 10.1007/978-3-319-11209-1
Springer Cham Heidelberg New York Dordrecht London

Library of Congress Control Number: 2014948036

LNCS Sublibrary: SL 7 – Artificial Intelligence

© Springer International Publishing Switzerland 2014
This work is subject to copyright. All rights are reserved by the Publisher, whether the whole or part of
the material is concerned, specifically the rights of translation, reprinting, reuse of illustrations, recitation,
broadcasting, reproduction on microfilms or in any other physical way, and transmission or information
storage and retrieval, electronic adaptation, computer software, or by similar or dissimilar methodology
now known or hereafter developed. Exempted from this legal reservation are brief excerpts in connection
with reviews or scholarly analysis or material supplied specifically for the purpose of being entered and
executed on a computer system, for exclusive use by the purchaser of the work. Duplication of this publication
or parts thereof is permitted only under the provisions of the Copyright Law of the Publisher's location,
in ist current version, and permission for use must always be obtained from Springer. Permissions for use
may be obtained through RightsLink at the Copyright Clearance Center. Violations are liable to prosecution
under the respective Copyright Law.
The use of general descriptive names, registered names, trademarks, service marks, etc. in this publication
does not imply, even in the absence of a specific statement, that such names are exempt from the relevant
protective laws and regulations and therefore free for general use.
While the advice and information in this book are believed to be true and accurate at the date of publication,
neither the authors nor the editors nor the publisher can accept any legal responsibility for any errors or
omissions that may be made. The publisher makes no warranty, express or implied, with respect to the
material contained herein.

Typesetting: Camera-ready by author, data conversion by Scientific Publishing Services, Chennai, India

Printed on acid-free paper

Springer is part of Springer Science+Business Media (www.springer.com)

Preface

This volume comprises the papers presented at ICCBR 2014: the 22nd International Conference on Case-Based Reasoning (http://www.iccbr.org/iccbr14/) which took place from September 29 through October 1, 2014 at the Imperial Hotel in Cork, Ireland. There were 49 submissions from 17 countries spanning North America, Europe, Africa, Asia and the Pacific. Each one was reviewed by three Program Committee members using the criteria of relevance, significance, originality, technical quality, and presentation. The committee accepted 19 papers for oral presentation and 16 papers for poster presentation at the conference.

The International Conference on Case-Based Reasoning (ICCBR) is the preeminent international meeting on case-based reasoning (CBR). Previous ICCBR conferences have been held in Sesimbra, Portugal (1995), Providence, USA (1997), Seeon Monastery, Germany (1999), Vancouver, Canada (2001), Trondheim, Norway (2003), Chicago, USA (2005), Belfast, UK (2007), Seattle, USA (2009), Alessandria, Italy (2010), London, UK (2011), Lyon, France (2012) and most recently in Saratoga Springs, USA.

Day 1 of ICCBR featured topical workshops on current and interesting aspects of CBR including case-based agents, reasoning with social media content, reasoning about time in CBR and synergies between CBR and Data Mining. The Doctoral Consortium involved presentations by ten graduate students in collaboration with their respective senior CBR research mentors. Day 1 also hosted the Computer Cooking Contest, the aim of which is to promote the use of AI technologies such as case-based reasoning, information extraction, information retrieval, and semantic technologies.

Days 2 and 3 consisted of scientific paper presentations on theoretical and applied CBR research as well as invited talks from two distinguished scholars: Tony Veale, lecturer at the School of Computer Science at University College Dublin, Ireland, and Frode Sørmo, Chief Technology Officer at Verdande Technology, Trondheim, Norway. Tony Veale gave a keynote address on creative reuse of past solutions and how cut-up techniques can be applied to compositional reuse. Frode Sørmo presented case studies explaining how Verdande Technology applies case-based reasoning techniques to oil drilling operations, financial services and healthcare monitoring. This volume includes contributions describing the main ideas presented in these keynote presentations.

The presentations and posters covered a wide range of CBR topics of interest both to researchers and practitioners including case retrieval and adaptation, case base maintenance, case-based planning, textual CBR, and applications to Web technologies, traffic management, product recommendation, facial recognition, home automation, and fault diagnosis.

Many people participated in making ICCBR 2014 a success. Derek Bridge, University College Cork, Ireland, served as the conference chair with Luc Lamontagne, Laval University, Canada and Enric Plaza, IIIA-CSIC, Spain, as program co-chairs. We would like to thank David Leake, Indiana University, USA, and Jean Lieber, LORIA, France, who acted as workshop chairs. Our thanks also go to Rosina Weber, Drexel University, USA, and Nirmalie Wiratunga, Robert Gordon University, UK, for organizing the Doctoral Consortium. We thank Mirjam Minor, Goethe University, Germany and Emmanuel Nauer, LORIA, France, who were responsible for the Computer Cooking Competition. We are also very grateful to all our sponsors, which at the time of printing included the Artificial Intelligence Journal, the Department of Computer Science at University College Cork, Empolis, Fáilte Ireland, Knexus Research Corporation and Verdande Technology. We also want to acknowledge the support of AAAI and the Cork Convention Bureau.

We thank the Program Committee and the additional reviewers for their timely and thorough participation in the reviewing process. We appreciate the time and effort put in by the local organizers. Finally, we acknowledge the support of EasyChair in the submission, review, and proceedings creation processes, and we thank Springer for its continued support in publishing the proceedings of ICCBR.

July 2014   Luc Lamontagne
                                                             Enric Plaza

Organization

Program Committee

Agnar Aamodt	NTNU, Norway
David Aha	Naval Research Laboratory, USA
Klaus-Dieter Althoff	DFKI/University of Hildesheim, Germany
Ralph Bergmann	University of Trier, Germany
Isabelle Bichindaritz	State University of New York at Oswego, USA
Alan Black	Drexel University, USA
Derek Bridge	University College Cork, Ireland
William Cheetham	CDPHP, USA
Alexandra Coman	Northern Ohio University, USA
Amélie Cordier	LIRIS-CNRS, France
Susan Craw	The Robert Gordon University, UK
Sarah Jane Delany	Dublin Institute of Technology, Ireland
Belen Diaz-Agudo	Complutense University of Madrid, Spain
Michael Floyd	Carleton University, Canada
Ashok Goel	Georgia Institute of Technology, USA
Mehmet Goker	Salesforce, USA
Pedro González Calero	Complutense University of Madrid, Spain
Luc Lamontagne	Laval University, Canada
David Leake	Indiana University, USA
Jean Lieber	LORIA - Inria Lorraine, France
Ramon López de Mántaras	IIIA-CSIC, Spain
Cindy Marling	Ohio University, USA
Lorraine McGinty	University College Dublin, Ireland
David McSherry	University of Ulster, UK
Alain Mille	LIRIS-CNRS, France
Mirjam Minor	Goethe Universitaet Frankfurt, Germany
Stefania Montani	University Piemonte Orientale, Italy
Hector Munoz-Avila	Lehigh University, USA
Santiago Ontañón	Drexel University, USA
Miltos Petridis	Brighton University, UK
Enric Plaza	IIIA - CSIC, Spain
Luigi Portinale	University of Piemonte Orientale, Italy
Ashwin Ram	PARC, USA
Juan Recio-Garcia	Complutense University of Madrid, Spain
Thomas Roth-Berghofer	University of West London, UK
Jonathan Rubin	PARC, USA

Barry Smyth	University College Dublin, Ireland
Antonio Sánchez-Ruiz	Complutense University of Madrid, Spain
Ian Watson	University of Auckland, New Zealand
Rosina Weber	Drexel University, USA
David Wilson	University of North Carolina at Charlotte, USA
Nirmalie Wiratunga	The Robert Gordon University, UK

Additional Reviewers

Black, Alan	McGreggor, Keith
Canensi, Luca	Mueller, Gilbert
Cox, Michael	Quijano Sanchez, Lara
Cunningham, Padraig	Reuss, Pascal
Dufour-Lussier, Valmi	Sahay, Saurav
Fitzgerald, Tesca	Sani, Sadiq
Goerg, Sebastian	Sauer, Christian Severin
Leonardi,Giorgio	Schulte-Zurhausen, Eric
Kendall-Morwick, Joseph	Schumacher, Pol
Massie, Stewart	Vattam, Swaroop

Table of Contents

Invited Talks

Research Papers

What I Talk about When I Talk about CBR

Frode Sørmo

Verdande Technology, Trondheim, Norway
frode@verdandetechnology.com

Abstract. In the last two years, Verdande Technology has branched out from our first application in monitoring drilling operations of oil and gas wells to look at other use-cases of case-based reasoning (CBR). In particular, we have looked at monitoring applications that use real-time sensor data and require a person in the decision making loop.

This talk will briefly present our pre-existing use-case in oil & gas, as well as two use cases from financial services and healthcare that were taken to the prototype stage. In exploring these use cases as well as others that were not taken as far, we have spoken to customers, end users, data scientists, industry analysts and managers in order to understand if CBR is or could be on their radar. We present what we learned from these interactions, as well as how we learned to talk about CBR to these different groups. Finally, we will ask what, if anything, holds CBR back from the type of large scale industrial adoption we see happening with rule-based systems and "analytics" technologies.

The oil & gas industry is going through a demographic shift in its workforce that is known in the industry as the "Great Crew Change." Due to an aging workforce, a majority of the workers in many oil & gas firms are due to retire over the next 5 to 10 years. At the same time, the number of wells and difficulty of drilling are increasing as the easy oil has already been extracted. Today, monitoring of drilling operations requires a highly skilled staff to monitor the real-time sensor data from the rig in real-time, and the industry is seeking ways to improve consistency and efficiency. Verdande Technology's DrillEdge product, which provides decision support for real-time monitoring of drilling operations, has been available in this market since 2009, and is now live on over 60 rigs across the world at any one time.

Ten years ago, a cardiovascular surgeon and an oil & gas manager met and found that their jobs had more in common than what was obviously apparent. The flow loops used to pump drilling fluid to circulate out cuttings use many of the same flow models as the vascular system in the body, and these days small drill strings are even used by surgeons to clear out arteries. Based on this insight, the Pumps & Pipes conference was born as a technology sharing forum for the oil & gas and healthcare industries. Verdande has worked with one of the pioneers behind this conference, Dr. Alan Lumsden, to examine if case-based reasoning can be used to monitor patients before, during and after open heart surgery.

In financial trading, the potential revenue for those who react quickest is great. Some years ago, it was found that it was possible to shave a few milliseconds off the time it took a data signal to go from New York to Chicago by digging a trench as straight as possible between the two cities. This would be used

L. Lamontagne and E. Plaza (Eds.): ICCBR 2014, LNCS 8765, pp. 1–2, 2014.
© Springer International Publishing Switzerland 2014

to run a direct fiber connection, thus cutting out switches and circuitous routes and improving the latency slightly. However, the revenues gained through speed are often at the expense of normal traders, such as day traders or pension funds. To counter this trend, a small group of industry insiders decided to start a new exchange, IEX, which would provide a fairer environment to traders. Verdande Technology has worked with IEX to look at how to monitor an IT infrastructure so sensitive to latency that shaving off a few milliseconds is worth a $300M investment.

In exploring these use cases as well as others that were not taken as far, we have spoken to customers, end users, data scientists, industry analysts and managers across the healthcare, financial services and oil & gas industries. Typically, the first question we discussed with them was when it makes sense to consider using case-based reasoning. In particular, we have often been asked to compare how CBR relates to traditional rule-based systems and "analytics" technologies. Rule-based systems have seen wide adoption, especially in the financial services industry, and as such it is reasonably understood what they do well and do less well. "Analytics" has become a buzz word for methods from machine learning and statistics used to build predictive or descriptive models from data. Tools and services around analytics are available from big vendors such as Oracle, IBM and EMC, and they are actively marketing them as the solution for what firms should do with the "Big Data" they have collected. We have found knowledge of CBR among industry analysts and big IT organizations to be quite limited, even among technical people with a machine learning or statistics background. That said, we have not found it difficult to explain why case-based reasoning is an appropriate technology for many use cases. In particular, the advantage CBR has is how easy it is to understand the concept and the reasoning process. We have seen so many potential use cases first hand that we believe that there are uncountable potential applications out there for CBR. Although there are a decent amount of CBR applications in use, many of the applications out there now are created by companies with specialized expertise in CBR or by internal teams in larger organizations that have a similar specialized competence. In contrast, rule-based systems have now come to the point where the internal IT organizations of large corporations routinely use them in their own projects, or with help from big system integrators such as Accenture or IBM, and "analytics" technologies are taking that step now. What would it take for CBR to get this kind of industry adoption?

Running with Scissors:
Cut-Ups, Boundary Friction and Creative Reuse

Tony Veale*

School of Computer Science and Informatics, University College Dublin,
Belfield, Dublin D4, Ireland
Tony.Veale@UCD.ie
http://Afflatus.UCD.ie

Abstract. Our experience of past problems can offer valuable insights into the solution of current problems, though since novel problems are not merely re-occurrences of those we have seen before, their solutions require us to integrate multiple sources of inspiration into a single, composite whole. The degree to which the seams of these patchwork solutions are evident to an end-user offers an inverse measure of the practical success of the reuse process: the less visible its joins, the more natural a solution is likely to seem. However, since creativity is neither an objective nor an intrinsic property of a solution, but a subjective label ascribed by a community, the more perceptible the tensions between parts, and the more evident the wit that one must employ to ameliorate these tensions, then the more likely we are to label a solution as creative. We explore here the conceit that creative reuse is more than practical problem-solving: it is reuse that draws attention to itself, by reveling in *boundary friction*.

Keywords: Boundary friction, incongruity resolution, semantic tension, computational creativity, creative reuse.

1   Introduction

Robert Altman's movie *The Player* opens with a glorious tracking shot that swoops through the offices of a fictitious Hollywood film studio. Joining the camera on its sweeping arcs, the audience is allowed to eavesdrop on insider conversations that illuminate the Hollywood philosophy of filmmaking. In one office we hear a group of technocrats discussing the longest and most impressive tracking shots in the history of cinema, while in another we hear a writer pitching a new idea to a studio executive. The writer frames his pitch as a high-concept blend; this new film, he suggests, is best imagined as a cross between *Ghost* and *The Manchurian Candidate*. Altman wants us to laugh at these scenes, to set the tone for the sharp-edged parody to follow. For we can only marvel at why a writer should want to marry elements from a then-recent blockbuster romance about

---

* This work forms part of the EC-funded WHIM project (*The What-If Machine*). The author would like to acknowledge the valuable contribution of his WHIM collaborator Alessandro Valitutti to the work described in this paper.

L. Lamontagne and E. Plaza (Eds.): ICCBR 2014, LNCS 8765, pp. 3–16, 2014.
© Springer International Publishing Switzerland 2014

a lovelorn ghost with aspects of a grim political thriller about brainwashing and McCarthyism. Altman seems to be suggesting that Hollywood's love of the high-concept pitch, combined with a ruthless determination to recycle past successes, has led to a cynical dumbing-down of its creative processes.

Hollywood has long sought to create familiar surprises by recombining past successes in high-concept way ([1]). Consider Akira Kurosawa's 1954 Japanese film *The Seven Samurai*, which became the 1960 American Western *The Magnificent Seven* when sword-wielding samurai were re-imagined as heroic gun-slingers. Kurosawa's own *Ran* can also be seen as a Japanese reworking of Shakespeare's *King Lear*, though Shakespeare was hardly the first to repackage the Celtic legend of Lir and his daughters. The 1961 musical *West-Side Story* is a Hollywood reworking of Shakespeare's 1594 play *Romeo and Juliet*, in which rival Verona families the Montagues and the Capulets become the rival street-gangs of the Jets and the Sharks. Baz Luhrman's movie *Romeo + Juliet* is also a modern musical set amongst crime families in Verona beach, with most of its dialogue taken directly from Shakespeare's play. Surprisingly, the re-combination not only works but excels, and big risks often produce the biggest rewards. Though cinematic titans like Orson Welles had struggled and failed to film a worthy treatment of Joseph Conrad's novel *Heart of Darkness*, the director Francis Ford Coppola succeeded with his 1979 movie *Apocalypse Now*, which transposed the action from the Belgian Congo to Vietnam and Cambodia so as to critique the U.S. role in the Vietnam war. In its own way, the 1994 film *Forrest Gump* also found success by making Vietnam the centerpiece of its rather loose re-combination of elements from Voltaire's 1759 novel *Candide*. The Coen brothers depression-era comedy *O Brother, Where Art Thou* was also a rather loosely-based reworking its classical source, Homer's *Odyssey*, but the comedic re-combination works so well that the film won an Academy Award for best adapted screenplay.

When stripped of their commercial motivation, the reuse mechanisms favored by Hollywood studios bear a surprising similarity to the recombinant techniques pioneered by the surrealist school of experimental art. The surrealists sought more than a conscious disavowal of cliché; they sought a radical means of escaping the deep-hewn ruts that unconsciously inhibit our spontaneity and creativity. As William Burroughs put it, one *"cannot will spontaneity into being"*, but one can "introduce the unpredictable spontaneous factor with a pair of scissors". The scissors here alludes to the cut-up technique pioneered by Burroughs and artist/writer Brion Gysin, in which a linear text is sliced, diced and randomly re-spliced to form new texts that give rise to new and unexpected interpretations ([2]). The purpose of the cut-up technique is actually two-fold: not only does it aim to create new combinations from old, much like the Hollywood system, it also consciously aims to disrupt the mind's attempts to automatically group commonly co-occurring words and ideas into familiar gestalts. Unlike the Hollywood approach, the technique embraces uncertainty and incongruity, and aims to challenge rather than to comfort, to unsettle rather than to placate.

The cut-up technique was originally inspired by the collage movement in visual art, in which fragments of images and texts are re-combined to form a novel patchwork whole. When the cut-up technique is applied to a linear text, the text is segmented into short strands of contiguous words that do not necessarily respect either phrase or sentence boundaries. These strands are then randomly recombined, to form a new text that uses the same words in different linear juxtapositions, to facilitate – if one charitably overlooks the inevitable bad grammar and illogical punctuation – very different global interpretations. Gysin originally applied the technique to layers of newsprint, which he sliced into linguistic chunks with a razor, and Burroughs later extended the technique to audio tapes. In principle, any linear source of information, from text to audio to video and even DNA, can be sliced and re-spliced using the cut-up technique to deliberately subvert familiar patterns and spontaneously suggest new meanings. Note, however, that the cut-up technique does not actually create new meanings, and is "*merely generative*" in the purest sense. Rather, the goal of the cut-up technique – and of related techniques developed by Burroughs, such as the *fold-in* and the *drop-in* – is to generate candidates for interpretation that an artist then evaluates, filters and ranks according to their creative potential. By automatizing the production process in a way that frees it from the often baleful effects of cliché, such techniques also free a creator to step back, observe, and focus instead on the crucial acts of selection and evaluation. In other words, these techniques are the arts-and-crafts equivalent of a generative algorithm

1.1   Boundary Friction

The cut-up technique has intriguing resonances with automatic approaches to language processing, which also aim to understand complex texts by carving them into smaller and more manageable chunks. For instance, *boundary friction* is a problematic phenomenon observed in a case-based approach to machine translation called "*Example-Based Machine Translation*" (or EBMT; see [3], [4], [5]). In a typical EBMT system, a target-language translation of a text is stitched together from a combination of pre-translated linguistic chunks that were previously extracted from a parallel bilingual corpus. While these chunks are internally coherent, the recombination of chunks can often give rise to cross-chunk incoherencies. For example, a system may translate a novel source-language sentence by combining a target-language chunk for its subject noun-phrase with a target-language chunk for its verb-phrase. But if these NP and VP chunks, derived from different translation contexts, now disagree on the grammatical number of their head noun and head verb, this friction must be ameliorated with an appropriate word-level change to achieve native fluency. Commercial EBMT systems succeed or fail on their ability to reduce boundary frictions and thus produce a fluent, natural translation. In contrast, the creation of boundary friction is the very *raison d'être* of the cut-up technique, whose explicit goal is the generation of texts whose juxtapositions are both challenging and surprising. It is rare that the cut-up technique generates fluent texts straight out of the gate; though such outputs are possible, albeit with careful curation, the goal is not simply to replace one natural text with another.

EBMT is most successfully applied in domains where there is a logical progression from past to present to future texts, as in the evolving documentation of a maturing software product. Since present texts are known to the system and exist in multiple translations, an EBMT system can carve out a collection of reusable translation chunks by aligning parallel translations of the same source texts. Though future texts are unknown and unseen, their subject matter is easy to predict, so in an important sense the present allows the future to be predicted. The general principle at work here – which sees the future as a functional re-arrangement of the present – also applies to artistic cut-ups. By cutting and re-arranging a text on a given subject, one is likely to produce another text on a similar subject, albeit one with some provocative juxtapositions. Occasionally, it is even possible that a cut-up produces a text that is not just a logical progression of the original, but a testable prediction of some future state of affairs. Burroughs describes one such experiment with cut-ups, in which a text about the oil billionaire John Paul Getty was randomly re-arranged to produce the sentence *"It's a bad thing to sue your own father."* A year later, Burroughs claims, the billionaire was indeed sued by one of his own sons. This is a hand-picked example, one that shows the necessity of carefully filtering the generative products of the cut-up technique, but it does show the capacity of cut-ups to bring into focus the latent possibilities that implicitly reside in a text. Burroughs described the relation between combinatorial generation and prediction thusly: *"when you cut into the present, the future leaks out.'* In truth, cut-ups are no more predictive of the future than fortune cookies or horoscopes, yet if creativity is our goal, spontaneity and unpredictability are often the better parts of generation when placed in the hands of a selective creator.

1.2   Forgiven, Not Forgotten

An EBMT system that fails to resolve all of the frictions that arise at the boundaries of a patchwork translation is effectively creating more work for a human post-editor. Though such frictions are primarily grammatical in nature, they undermine a reader's faith in the competence of a translation even if the gist of its meaning is still apparent. In contrast, the cut-technique often produces piecemeal texts that ask for a suspension of superficial grammatical rigour from their readers, by posing an implicit bargain that promises conceptual rewards at a deeper level of meaning. Whereas an EBMT system strives to produce a target-language translation that would not look out of place in its original corpus of target-language texts, the cut-up technique strives to produce novel composites that reside comfortably in *none* of its contributing corpora. EBMT systems thus seek to obliterate the joins in its composite outputs, while instances of the cut-up technique wear these joins proudly, as signifiers of creativity.

A cut-up text is a text of divided references, for it simultaneously evokes different viewpoints on the world as captured in different contributory fragments. This rift may prompt a subtle convergence of views that goes unnoticed as a creative act, or it may force an unsubtle divergence that calls attention to itself as a creative provocation. As Howard Pollio puts it, *"split reference yields humour*

*if the joined items (or the act joining them) emphasize the boundary or line sep-
arating them; split reference yields metaphor if the boundary between the joined
items (or the act joining them) is obliterated and the two items fuse to form a
single entity"* [6]. While metaphor is so pervasive in language that speakers are
often unaware of just how much they rely on common figures of speech, the spell
of metaphor is easily broken, and with jarring effect, when metaphors are inju-
diciously mixed so as to clash at ill-fitting boundaries. Creative combinatorial
reuse resolves the tensions that arise between the components that contribute
to a solution, so that the resulting whole can achieve its intended goals, but it
does not obliterate all traces of these tensions. Rather, though the whole works,
it continues to surprise, by drawing attention to itself as an unnatural – albeit
useful – marriage of strange bedfellows. In humorous wit, as in ostentatious creativity more generally, one must not fail to address the internal tensions in a
juxtaposition of forms and ideas, but neither must one resolve them so utterly
that an audience does not recognize their influence, nor appreciate the means
of their resolution ([7], [8]). Such tensions are a vital part of how creativity
announces itself to an audience

1.3 Structure of This Paper

EBMT systems suffer from *superficial* boundary friction, which can diminish our
faith in the quality of a translation if not fully resolved. Creative applications
of the cut-up technique revel in *deep* boundary friction, at the semantic and
pragmatic levels of meaning rather than at the surface levels of language. In
this paper we will computationally explore the latter, deeper source of friction,
to generate cut-ups of our knowledge-representations that are non-obvious and
challenging, yet ultimately meaningful and occasionally even humorous, insight-
ful and profound.

  In section two we introduce the notions of a *knowledge cut-up* and a *conceptual
mash-up*, to demonstrate that the surrealist cut-up technique can be applied as
much to our logical representations of knowledge as to our linguistic realizations
of this knowledge in text. To a computer, after all, a symbolic knowledge representation is merely another kind of symbolic text, albeit one that is unambiguous
and logically-privileged. In section three we describe a novel, inferential use of
these new techniques, and describe, in section four, how inferential cut-ups form
the basis of a creative, fully-automated Twitterbot. Section five then uses the
outputs of this system to obtain some initial experimental findings regarding the
kinds of boundary friction that are most satisfying to human audiences. The paper concludes in section six with some closing observations about the creativity
of compositional reuse.

2 Knowledge Cut-Ups and Conceptual Mash-Ups

The mechanics of the cut-up technique are easily automated, though superficial
boundary friction abounds when the technique is randomly applied to raw texts.

The more informative the mark-up that can be extracted from a text, the more well-formed the outputs can be. Consider a twitterbot called *Two Headlines* by bot-designer Darius Kazemi (Twitter handle *@twoheadlines*), which does exactly what its name suggests. Two random headlines from today's news, as provided by a popular news aggregator, are mashed together to yield a headline that is well-formed yet factually provocative. Because the source headlines contain additional mark-up that identifies the named-entities in the text – whether *Justin Bieber* or *Barack Obama* or *Google* or *Italy* – it is a simple matter to mix-and-match the entities from two headlines into one without introducing any boundary friction at the grammatical level. The goal, of course, is to introduce friction at the conceptual level by, for example, associating Justin Bieber with momentous international political decisions or scientific discoveries, or Barack Obama with the laughable or criminal indiscretions of pampered celebrities. The resulting headline often produces what Koestler ([9]) terms a *"bisociation"*, a clash of two overlapping but conflicting frames of reference that is jarring but meaningful.

The more mark-up present in the input texts, the more insight that can be gleaned about their structure and their meaning, and so the more informed the decisions that a computational cut-up agent can make when creating a new text. The most informative texts are those that explicitly define the knowledge possessed by a computer for a particular domain: that is, the computer's own symbolic knowledge representations. For example, these representations are commonly structured around the notion of a semantic triple, in which a logical subject is linked to a logical object by an explicit semantic relation. These triples carve up the world at its joints, but it is an easy matter to further cut-up these triples at their joints, to mix-and-match subjects, relations and objects from different triples into provocative new hypotheses, in much the same way that *@twoheadlines* serves up a bisociative mix of today's news headlines. The triples in question can come from any generic triple-store on the Semantic Web, or they may form part of the bespoke knowledge-base for an AI/NLP reasoning system. Consider the triples harvested by Veale and Li ([10], [11]) from the texts of the Web. These authors glean completions to common why do questions from the Google search engine, such as *"why do dogs chase cats"*, '*why do philosophers ask questions"* and '*why do poets use rhyme"*. They then extract the implicit consensus presupposition at the core of each question (such as e.g. that most poets use rhyme, imagery and metaphor) and automatically convert each presupposition into an explicit semantic triple (such as $< poet, use, metaphor >$).

These triples are grist for a metaphor interpretation and generation system named *Metaphor Eyes*. Suppose *Metaphor Eyes* were to consider the metaphorical potential of viewing philosophers as poets. The system generates knowledge-level cut-ups from the triples it possesses for *poet* by replacing the subject *poet* with the subject *philosopher* in each *poet* triple. This simple, generative action is then followed by a validation phase that automatically evaluates each of the newly-produced cut-up triples, by looking to the Web for evidence that the new triple captures a relationship that is attested in one or more Web texts. Thus, the cut-up that *"philosophers use metaphors"* is attested by a Web search that finds

hundreds of supporting documents. But *Metaphor Eyes* does more than generate
and test cut-ups for a given pairing of concepts: it suggests interesting pairings
for itself, by examining how words and ideas are clustered in a large text corpus
such as the Google n-grams ([12]). Thus, since *poets* and *philosophers* are seen
in sufficiently frequent proximity to each other, and each is found in proximity
to ideas that are related to the other (such as *metaphor*, or *idea*, or *argument*),
Metaphor Eyes actively suggests the pairing of poets and philosophers to the
user, and proceeds to generate a Web-validated knowledge cut-up as its output.
Metaphor Eyes fluidly plays with the boundaries between knowledge and text,
moving between one and the other to extract maximum reuse value from each.

Veale and Li also explore the potential of metaphorical cut-ups to plug gaps
in a system's knowledge representation of the world. Suppose a system possesses
little or no knowledge about philosophers, but possesses knowledge about concepts that are not so far removed, such as scholars, theologians, scientists and
poets. *Metaphor Eyes* combines triples from neighboring concepts to mash together a proxy representation for *philosopher*, by combining the Web-validated
triples for the cut-ups *philosopher as scholar* (each accumulates and applies
knowledge, each performs research), *philosopher as theologian* (each spreads doc-
trines), *philosopher as scientist* (each develops and explores ideas) and *philosopher as poet* (each nurtures ideals). The resulting set of cut-up triples from diverse sources is called a *"conceptual mash-up"*, and Veale and Li rank the triples
in a mash-up according to how much Web evidence can be found for each. They
demonstrate empirically that the top-ranked triples in a mash-up have a high
probability of being attested in real Web texts, and show that a mash-up is a
good substitute for human-engineered knowledge in an AI system. Like human
users of the cut-up technique, they explore the space of ideas with a scissors.

3   Inferential Cut-Ups Lead to Surprising Conclusions

Creative producers are masters of ambiguity. They make the most of the ambiguity in their inputs, and induce ambiguity in their outputs to foster indeterminism
and the emergence of new, unexpected meanings. The cut-up technique is designed to unleash the latent ambiguity in an otherwise business-as-usual text,
such as a news story or a well-thumbed novel. In a knowledge-based computational setting, ambiguity allows a system to transcend one of its most vexing
limitations: the knowledge that comprises most AI systems is *safe* knowledge,
facts and rules that most informed users would consider to be true if they ever
gave the matter any thought. So how does one go from safe premises to conclusions that can genuinely surprise us, and that force us to view the world
anew from a very different perspective? The ancient sophists were masters of
using familiar knowledge to reach desirable ends, and even Socrates, who openly
disdained the sophists, practiced similar techniques for nobler ends: by linking
familiar assumptions in the right order, a sophisticated thinker can lead an audience from their safe moorings in received wisdom into new, uncharted territories.

Suppose we ponder the dystopian scenario of a *"world without beauty"*? Since
Beauty is an unqualified positive in most world-views, and most knowledge-bases

that address the matter at all will axiomatize it using other positive qualities such as *Art* and *Love*, most of us (and our systems) will initially conclude that a world without beauty would be a much-diminished place to live. Yet a good repository of common-sense knowledge will also capture the relationship between *Beauty* and *Jealousy*, noting that the former often causes the latter. *Jealousy*, in turn, promotes *Hate* and *Conflict*, with the latter sometimes escalating to the level of *War*. A good sophist might thus conclude – and show how others might reasonably reach this conclusion for themselves – that a world without beauty might be a world with less jealousy, less hate, and less war. A world without beauty might, in fact, be a desirable place to live.

The elements of this causal chain are unassuming semantic triples of a knowledge base, and none have been subjected to the dislocations of the cut-up technique. Yet the chain as a whole can be viewed as a cut-up in its own right, akin to a newspaper cut-up that reuses whole sentences as its building blocks rather than the phrasal sub-components of these sentences. The guiding principle of such a cut-up is to produce a surprising conclusion from the most unsurprising premises, much as Socrates and his followers, and their rivals the sophists, did in ancient Athens. Such a process can also be modeled on a computer, as computers excel at pursuing inferential chains to their logical ends. Each successive link in a chain may exhibit little or no boundary friction with the link that goes before or the link that comes after, but the cumulative effect will be to produce a very evident friction between the first and last links in the chain.

The potential for semantic and pragmatic friction between the end-points of a chain is increased if one is willing to use a sophist's disregard for logical rigor when adding new links to the chain. Consider $< dictator, suppress, critic >$, a triple which captures the widespread belief that dictators censure their critics, or worse. A system that also believes that $< critic, criticize, artist >$ and $< artist, produce, art >$ may well construct an inferential chain from dictators to art via critics and artists, to infer that dictators indirectly promote art by thwarting the critics that impede its producers. Such a chain embodies a surprising claim, that more dictators lead to more art, but it is predicated on several acts of sophistry. Firstly, critics come in multiple guises, and the political critics that decry a regime are not typically the same as those that criticize art. Secondly, to criticize is not always to hinder or deter. Thirdly, art critics do more than criticize, and often encourage artists too. Lastly, even if the inferential chain can be taken at face value, it says nothing about whether dictators promote more *good* art. These may be considered the weak points of the causal argument, but they may also be seen as the resolution of the semantic friction – between bad dictators and good art – that gives the argument its shock value. One man's sophistry may well be another's *semantic slippage* ([13]). In either case, inference chains that exhibit deep friction often prove to be surprising in ways that make us think about what we know.

This notion of deep friction in an inferential chain may be operationalized in different ways, but the simplest and most effective employs the affective profile of the concepts concerned. *Art* and *Beauty* are positive concepts, and are denoted by words with a strong positive sentiment. *Dictators* and *Critics* are negative concepts, and are denoted by words with a strong negative sentiment. It follows that more art, more beauty and fewer critics should be considered positive outcomes, just as less art, less beauty and more dictators should be considered a negative outcome. An inferential chain that shows how a positive outcome can be derived from a negative cause (such more art from more dictators), or how a negative outcome can be derived from a positive cause (such as more war from more beauty) thus exhibits a deep friction between its initial premises and its final conclusion. This friction poses a challenge to an audience – *how can this be so?* – that must be resolved meaningfully, either by accepting the conclusion at face value or by identifying the sophistry at its heart. In either case, the audience is aware of both the friction and its resolution; indeed, the resolution actively draws our attention to the friction, and draws us into its worldview.

4   Creative Twitterbots

Twitterbots are fully-automated, generative systems whose outputs are designed to be distributed via the micro-blogging service *Twitter*. Though easily dismissed as trivial, Twitterbots embody all of the anarchic spirit of the cut-up technique and of the conceptual artists that created it. Consider again the *@twoheadlines* bot, which uses the cut-up technique just as Burroughs and Gysin imagined it, to *"introduce the unpredictable spontaneous factor"* into our readings of the daily news. However, as a simple, uncurated system, *@twoheadlines* captures only the generative part of the cut-up process, and omits the vital reflection phase in which candidate outputs are ranked, evaluated and then chosen or discarded by an intelligent agent with creative intent. Though a productive system, *@twoheadlines* is ultimately a *"merely generative"* system that lacks any self-appreciation, and cannot thus be considered a computationally creative agent.

In contrast, *@MetaphorMagnet* is a Twitterbot that employs the knowledge cut-ups of section 2 (via the computational service *Metaphor Eyes*) and the inferential cut-ups of section 3 to produce metaphorical insights that are carefully evaluated and filtered by the system itself. *@MetaphorMagnet* is not merely a generative system; rather, it is a system that only tweets metaphors that it considers to be well-formed and thought-provoking, the latter quality arising out of its deliberate use of resolvable, causal boundary friction. The bot generates a new metaphorical insight every hour, based on its explorations of its own sizable knowledge-base of semantic triples. Because these explorations allow it to detect deep frictions between its own knowledge of the world – knowledge that it believes most humans will also possess – *@MetaphorMagnet* has a sharp eye for hypocrisy and disappointment that is well-suited to ironic commentary. Consider these representative tweets:

> #Irony: When the initiates that learn about engaging mysteries participate in boring rituals. #EngagingOrBoring #Initiate

> Writers write metaphors. Dictators suppress the critics that criticize the authors that write metaphors. Who is better? #DictatorOrWriter

Slavery imposes the shackles that impose bonds. Marriage creates the attachments that create bonds. Take your pick. #MarriageOrSlavery?

Truth provides the knowledge that underpins arts. Illusion enhances the beauty that underpins arts. Take your pick. #TruthOrIllusion?

#Irony: When the tests that are conducted in tidy laboratories are conducted by rumpled professors. #TidyOrRumpled #Test

Twitter's 140-character size limitation per tweet imposes obvious limitations on the length of the inferential chains that can be entertained by the system, though its biggest challenge is not the size of its search horizon, nor the size of its knowledge-base, but the breadth of its repertoire for expressing new insights in suggestive forms. The best Twitterbots exhibit an identifiable aesthetic and an identifiable world-view, in much the same way that human Twitter users exude an identifiable personality. *@MetaphorMagnet* exudes a distinctly hard-boiled personality through its causal linking of positive and negative concepts. As shown in the *#MarriageOrSlavery* and *#TruthOrIllusion* tweets above, *@Metaphor-Magnet* is a sophist that assumes moral equivalence wherever it can demonstrate an apparent causal equivalence. We leave it to other Twitterbots to present a more positive and uplifting view of the world.

5   Half-Baked vs. Deep-Fried: Comparing Friction Types

We have hypothesized that certain forms of boundary friction are inherently valuable, insofar as they signal not so much a lapse in combinatorial finesse by a computational system, but a conceptual challenge that draws an audience into the creative act itself. Superficial friction typically arises either as a loose-end – a flaw overlooked by a careless or inept creator that detracts from his or her achievement – or as a weak attempt at generating a mere frisson of semantic tension. Deep friction, like a gnashing of conceptual gears, is both surprising yet appropriate, a piquant insight emerging from a plate of bland generalizations. We test this claim in the context of inferential cut-ups by using a crowd-sourcing platform to elicit the feedback of human judges. But we do not present judges with the polished outputs of *@MetaphorMagnet*, as this twitterbot carefully packages its tweets in a variety of rhetorical guises that may influence a judge's perception of the underlying reasoning. Rather, judges are presented with inferential chains in unpolished, quantitative forms, so e.g. the chain *dictators suppress the critics that criticize the artists that produce art* is presented simply as *more dictators → fewer critics → more artists → more art.*

Inferential chains are randomly generated from the underlying knowledge-base to support three different test conditions. In the first condition, the *random* condition, inferential chains of connected generalizations are produced without regard to whether they contain any kind of friction at all. These chains are logically well-formed, but no interestingness criterion is applied as a filter. In the second condition, the *surface* condition, chains of connected generalizations are produced so that the first and last concepts in the chain exhibit a polar opposition in sentiment, such as *love vs. war* or *art vs. death*. Any friction in these chains is most likely the product of superficial differences in sentiment between connected concepts, rather than of a surprising twist of causality. In the third condition, the *deep* condition, inference chains are produced that exhibit precisely this kind of causal friction, so that a positive idea (such as *more beauty*) is shown to indirectly have negative consequences (such as *more war*), or a negative idea (such as *more war*) is shown to indirectly have positive consequences (such as *more patriotism* or *more prosperity*)

A pool of 80 inferential chains was randomly generated for each condition, 30 of which (per condition) were manually annotated as a gold standard to detect scammers, and 50 of which (per condition) were finally annotated by independent judges. The crowd-sourcing platform *CrowdFlower* was used to recruit a panel of 70 human judges to estimate, for each of these 3x50 inference chains, the degree of surprise exhibited by each. The full inferential pathway was presented in each case, so that judges could see not only its conceptual end-points, but the coarse logic at work in each link of the chain. The gold standard paths were used to detect unengaged scammers, resulting in 2.5% of judgments overall being discarded. Ultimately, 50 chains for each condition were judged by 15 or more judges, producing 765 judgments for the *random* condition, 751 for the *surface* condition, and 750 for the *deep* condition. Each elicited judgment provided a measure of surprise on a scale from 0 (no surprise) to 3 (very surprising) for a given inference chain.

The mean surprise value for the 765 judgments of the *random* condition is 1.06, that for the 751 judgments of the *surface* condition is 0.96, and that for the 750 judgments for the *deep* condition is 1.44. There is little here to distinguish the chains of the random condition from those of the surface condition, suggesting that combinations that rely only on superficial differences in form, rather than on deep differences in causality, fail to reliably elicit any interest. However, as we have hypothesized, there is a statistically significant difference in surprisingness between, on one hand, the chains of the *surface* and *random* conditions, and on the other, those of the *deep* condition. A one-sided Wilcoxon rank-sum test verifies that the increase in mean surprisingness from the *surface and random* conditions to the *deep* condition is significant at the $p < .001$ level. To be surprised one must have prior expectations that are thwarted in some way. Surprise turns to fascination when initial dissonance gives way, after some consideration, to a deeper resonance of ideas. The inventors of the cut-up technique did not want to generate jarring combinations for their own sake, but to produce meaningful bisociations that break with banality. So to be appreciated as *"creative"*, combinatorial reuse systems must do more than seek out combinations that work; they must seek out combinations that work in spite of themselves.

6 Conclusions

Creative thinkers must constantly question received wisdom, and look beyond the superficial, isolated meaning of consensus beliefs. Creative thinkers actively seek out boundary friction whenever different texts, ideas, viewpoints or rules are combined, because this friction allows one to see wherever convention has fixed the acceptable boundaries of everyday categories. By identifying the boundaries of the conventional mindset, a creative thinker can *break set* and search for untapped value beyond those boundaries. Though certain forms of boundary friction are vexing issues to be resolved away, others offer a map to what is novel, unexpected and interesting.

The recombinant elements of past solutions rarely click together as cleanly as Lego bricks, but for that we should be thankful. Combinatorial reuse is a well-proven strategy for problem-solving that offers many opportunities for creative friction, at various levels of resolution. Reusable solution elements can range from individual words or concepts to heftier chunks of meaning or text to physical artefacts that exhibit different functionalities in different contexts. The artist provocateur Marcel Duchamp introduced the notion of a creative readymade to art when, in 1917, when he displayed a signed urinal (christened *"Fountain"*) at a Dadaist art exhibition in New York. Duchamp shifted the emphasis of artistic production from the act of original generation to the act of aesthetic selection, arguing that the artifacts at the core of the artistic process need not be constructed by the artists themselves ([14]). Duchamp used physical objects, constructed by skilled artisans, as his *"readymades"*, but Burroughs and Gysin were to show that any object, physical or textual or conceptual, could be creatively re-purposed and re-used as a readymade in a new context. By viewing the phrasal elements of a text database such as the Google n-grams ([12]) as a database of reusable elements that gain new meanings in new contexts, algorithms can create and exploit linguistic readymades of their own ([11], [15]).

Reusable elements may themselves by complex software components that strive for creativity in their own right. The creative friction that spurs innovation from antagonistic partners, of a kind that is well-attested by famous creative human partnerships, may also be evident in the qualities that arise from productive mash-ups of independent Web-services. A service oriented architecture ([16]) supports combinatorial reuse at multiple levels of resolution, and allows boundary friction to subtly influence the creativity of the end-result at each of these levels. For instance, other creative systems may use the outputs of the *@MetaphorMagnet* twitterbot as inputs to their own generative processes, or they may call directly upon the corresponding *Metaphor Magnet* Web-service to ask for specific kinds of metaphors for specific topics ([17], [18]). Web-services allow us to convert the generative components of past successes into stand-alone, modular, reusable, discoverable, generative engines in their own right. These services will exploit their own internal forms of boundary friction, much as the *Metaphor Magnet* Web-service exploits its own, intra-service notion of semantic tension to generate exciting new metaphors. But they will also give rise to higher-level, inter-service frictions that are harder to predict but just as useful to exploit. In a thriving ecosystem of competing Web services, creativity will arise not just from the carefully planned interactions of these services – whether for metaphor generation, language translation, story-telling, affective filtering, multimodal conversion, and so on – but from the unplanned frictions that emerge from the less-than-seamless integration of these services. In such complex systems of imperfectly described, underspecified modules, the potential for boundary friction abounds at every level, offering opportunities for creativity all the way down. William Burroughs viewed his scissors as a deliberate instigator of boundary friction. We too can harness boundary friction to *"introduce the unpredictable spontaneous factor"* into our combinatorial reuse systems to make them more genuinely creative.

References

1. Veale, T.: Creativity as pastiche: A computational treatment of metaphoric blends, with special reference to cinematic *"borrowing"*. In: Proceedings of the Mind II: Computational Models of Creative Cognition, Dublin, Ireland (1997)
2. Robinson, E.S.: Shift Linguals: Cut-Up Narratives from William S. Burroughs to the Present. Rudolphi, Amsterdam, The Netherlands (2011)
3. Sumita, E., Iida, H., Kohyama, H.: Translating With Examples: A new approach to Machine Translation. In: Proceedings of the 3rd International Conference on Theoretical and Methodological Issues in Machine Translation of Natural Language, Austin, Texax, USA (1990)
4. Veale, T., Way, A.: Gaijin: A template-driven bootstrapping approach to example-based machine translation. In: Proc. NeMNLP Conference on New Methods in Natural Language Processessing, Bulgaria (1997)
5. Way, A., Gough, N.: wEBMT: developing and validating an example-based machine translation system using the world wide web. Computational Linguistics 29(3), 421–457 (2003)
6. Pollio, H.R.: Boundaries in humor and metaphor. In: Mio, J.S., Katz, A.N. (eds.) Metaphor, Implications and Applications, pp. 231–253. Lawrence Erlbaum Associates, Mahwah (1996)
7. Raskin, V.: Semantic Mechanisms of Humor. Reidel, Dordrecht (1985)
8. Attardo, S., Hempelmann, C.F., Di Maio, S.: Script oppositions and logical mechanisms: Modeling incongruities and their resolutions. Humor: International Journal of Humor Research 15(1), 3–46 (2002)
9. Koestler, A.: The Act of Creation. Hutchinsons, London (1964)
10. Veale, T., Li, G.: Creative Introspection and Knowledge Acquisition: Learning about the world through introspective questions and exploratory metaphors. In: Burgard, W., Roth, D. (eds.) Proc. of 25th AAAI International Conference of the Association for the Advancement of AI, San Francisco, California (2011)
11. Veale, T.: Exploding the Creativity Myth: The computational foundations of linguistic creativity. Bloomsbury Academic, London (2012)
12. Brants, T., Franz, A.: Web 1T 5-gram Ver. 1. Linguistic Data Consortium (2006)
13. Hofstadter, D.R.: Fluid Concepts and Creative Analogies: Computer Models of the Fundamental Mechanisms of Thought. Basic Books, New York (1995)
14. Taylor, M.R.: Marcel Duchamp: Étant donnés (Philadelphia Museum of Art). Yale University Press (2009)
15. Veale, T.: Linguistic Readymades and Creative Reuse. Transactions of the SDPS: Journal of Integrated Design and Process Science 17(4), 37–51 (2013)
16. Erl, T.: SOA: Principles of Service Design. Prentice Hall (2008)

17. Veale, T.: A Service-Oriented Architecture for Computational Creativity. Journal of Computing Science and Engineering 7(3), 159–167 (2013)
18. Veale, T.: A Service-Oriented Architecture for Metaphor Processing. In: Proc. of the 2nd ACL Workshop on Metaphor in NLP, at ACL 2014, the 52nd Annual Meeting of the Association for Computational Linguistics, Baltimore, USA (2014)

Appendix: Online Resources

Metaphor Magnet exists as both a Twitterbot, which pushes metaphors of its own invention, and as a public Web-Service, which generates specific families of metaphors for specific topics, on demand, via a HTML and an XML Web interface. It also provides other creative products, such as poems and blends, for these metaphors. This service can be accessed at: `http://boundinanutshell.com/metaphor-magnet-acl`

Learning Solution Similarity
in Preference-Based CBR

Amira Abdel-Aziz[1], Marc Strickert[1], and Eyke Hüllermeier[2]

[1] Department of Mathematics and Computer Science
University of Marburg, Germany
{amira,strickert}@mathematik.uni-marburg.de
[2] Department of Computer Science
University of Paderborn, Germany
eyke@upb.de

Abstract. This paper is a continuation of our recent work on preference-based CBR, or Pref-CBR for short. The latter is conceived as a case-based reasoning methodology in which problem solving experience is represented in the form of contextualized preferences, namely preferences for candidate solutions in the context of a target problem to be solved. In our Pref-CBR framework, case-based problem solving is formalized as a preference-guided search process in the space of candidate solutions, which is equipped with a similarity (or, equivalently, a distance) measure. Since the efficacy of Pref-CBR is influenced by the adequacy of this measure, we propose a learning method for adapting solution similarity on the basis of experience gathered by the CBR system in the course of time. More specifically, our method makes use of an underlying probabilistic model and realizes adaptation as Bayesian inference. The effectiveness of this method is illustrated in a case study that deals with the case-based recommendation of red wines.

Keywords: Preferences, distance, similarity learning, probabilistic modeling, Bayesian inference.

1 Introduction

Building on recent research on preference handling in artificial intelligence, we recently started to develop a coherent and generic methodological framework for case-based reasoning (CBR) on the basis of formal concepts and methods for knowledge representation and reasoning with *preferences*, referred to as Pref-CBR [6,8,5]. A preference-based approach to CBR appears to be appealing for several reasons, notably because case-based experiences naturally lend themselves to representations in terms of preferences over candidate solutions. Moreover, the flexibility and expressiveness of a preference-based formalism well accommodate the uncertain and approximate nature of case-based problem solving.

Like many other CBR approaches, Pref-CBR proceeds from a formal framework consisting of a problem space $\mathbb{X}$ and a solution space $\mathbb{Y}$. Yet, somewhat less common, it assumes a similarity (or distance) measure to be defined not

L. Lamontagne and E. Plaza (Eds.): ICCBR 2014, LNCS 8765, pp. 17–31, 2014.
© Springer International Publishing Switzerland 2014

only on $\mathbb{X}$ but also on $\mathbb{Y}$. Moreover, it assumes a strong connection between the notions of *preference* and *similarity*. More specifically, for each problem $x \in \mathbb{X}$, it assumes the existence of a theoretically *ideal* solution[1] $y^* \in \mathbb{Y}$, and the less another solution y differs from y^* in the sense of a distance measure Δ_Y, the more this solution is preferred.

As a consequence, the performance and effectiveness of Pref-CBR is strongly influenced by the distance measure Δ_Y: The better this measure captures the true differences between solutions, the more effective Pref-CBR will be. In this paper, we therefore extend our framework through the integration of a *distance learning* module. Thus, the idea is to make use of the experience collected in a problem solving episode, not only to extend the case base through memorization of preferences, but also to adapt the distance measure Δ_Y.

The rest of the paper is organized as follows. By way of background, and to assure a certain level of self-containedness, we recall the essentials of Pref-CBR in the next section. Our approach to learning solution similarity in Pref-CBR is then presented in Section 3 and illustrated by means of two simulation studies in Section 4. The paper ends with some concluding remarks and an outlook on future work in Section 5.

2   Preference-Based CBR

Preference-based CBR replaces experiences of the form "solution y (optimally) solves problem x", as commonly used in CBR, by weaker information of the form "y is better (more preferred) than z as a solution for x", that is, by a preference between two solutions *contextualized* by a problem x. More specifically, the basic "chunk of information" we consider is symbolized in the form $y \succeq_x z$ and suggests that, for the problem x, the solution y is supposedly at least as good as z.

As argued in [5], this type of knowledge representation overcomes several problems of more common approaches to CBR. In particular, the representation of experience is less demanding: As soon as two candidate solutions y and z have been tried as solutions for a problem x, these two alternatives can be compared and, correspondingly, a strict preference in favor of one of them (or an indifference) can be expressed. To this end, it is neither required that one of these solutions is optimal, nor that their suitability is quantified in terms of a numerical utility.

2.1   A Formal Framework

In the following, we assume the problem space $\mathbb{X}$ to be equipped with a similarity measure $S_X : \mathbb{X} \times \mathbb{X} \to \mathbb{R}_+$ or, equivalently, with a (reciprocal) distance measure $\Delta_X : \mathbb{X} \times \mathbb{X} \to \mathbb{R}_+$. Thus, for any pair of problems $x, x' \in \mathbb{X}$, their similarity is denoted by $S_X(x, x')$ and their distance by $\Delta_X(x, x')$. Likewise, we assume the

---

[1] As will be explained in more detail in Section 2, this solution might be fictitious and perhaps impossible to be materialized.

solution space $\mathbb{Y}$ to be equipped with a similarity measure S_Y or, equivalently, with a (reciprocal) distance measure Δ_Y. In general, $\Delta_Y(\boldsymbol{y}, \boldsymbol{y}')$ can be thought of as a kind of adaptation cost, i.e., the (minimum) cost that needs to be invested to transform the solution $\boldsymbol{y}$ into $\boldsymbol{y}'$. As will become clear later on, our framework suggests a natural connection between distance and similarity, which involves a parameter $\beta \geq 0$ and is of the following form:

$$S_Y(\boldsymbol{y}, \boldsymbol{y}') = \exp\big(-\beta \cdot \Delta_Y(\boldsymbol{y}, \boldsymbol{y}')\big) \in (0, 1] \qquad (1)$$

In preference-based CBR, problems $\boldsymbol{x} \in \mathbb{X}$ are not associated with single solutions but rather with preferences over solutions, that is, with elements from a class of preference structures $\mathfrak{P}(\mathbb{Y})$ over the solution space $\mathbb{Y}$. Here, we make the assumption that $\mathfrak{P}(\mathbb{Y})$ is given by the class of all weak order relations $\succeq$ on $\mathbb{Y}$, and we denote the relation associated with a problem $\boldsymbol{x}$ by $\succeq_{\boldsymbol{x}}$. More precisely, we assume that $\succeq_{\boldsymbol{x}}$ has a specific form, which is defined by an *ideal* solution $\boldsymbol{y}^* = \boldsymbol{y}^*(\boldsymbol{x}) \in \mathbb{Y}$ and the distance measure Δ_Y: the closer a solution $\boldsymbol{y}$ to $\boldsymbol{y}^*$, the more it is preferred. Thus, $\boldsymbol{y} \succeq_{\boldsymbol{x}} \boldsymbol{z}$ iff $\Delta_Y(\boldsymbol{y}, \boldsymbol{y}^*) \leq \Delta_Y(\boldsymbol{z}, \boldsymbol{y}^*)$, for which we shall also use the notation $[\boldsymbol{y} \succeq \boldsymbol{z} \,|\, \boldsymbol{y}^*]$. In conjunction with the regularity assumption that is commonly made in CBR, namely that similar problems tend to have similar (ideal) solutions, this property legitimates a preference-based version of this assumption: *Similar problems are likely to induce similar preferences over solutions.*

We like to mention that the solution $\boldsymbol{y}^*$ could be purely fictitious and does not necessarily exist in practice. For example, anticipating one of our case studies in Section 4 as an illustration, a solution could be a red wine characterized by a certain number of chemical properties. One could then imagine that, for a specific person, there is an ideal (most preferred) wine in a space of all theoretically conceivable (i.e., chemically feasible) red wines, even if this wine is not a practically possible solution, either because it is not produced or not offered by the wine seller—finding another wine that closely resembles it is then the best one can do. More formally, this means that we assume a set of theoretically conceivable solutions $\mathbb{Y}^*$ that contains the space of practically reachable solutions $\mathbb{Y}$ as a subset.

2.2 Case-Based Inference

The key idea of preference-based CBR is to exploit experience in the form of previously observed preferences, deemed relevant for the problem at hand, in order to support the current problem solving episode; like in standard CBR, the *relevance* of a preference will typically be decided on the basis of problem similarity, i.e., those preferences will be deemed relevant that pertain to similar problems. An important question that needs to be answered in this connection is the following: Given a set of observed preferences on solutions, considered representative for a problem $\boldsymbol{x}_0$, what is the underlying preference structure $\succeq_{\boldsymbol{x}_0}$ or, equivalently, what is most likely the ideal solution $\boldsymbol{y}^*$ for $\boldsymbol{x}_0$?

We approach this problem from a statistical perspective, considering the true preference model $\succeq_{\boldsymbol{x}_0} \in \mathfrak{P}(\mathbb{Y})$ associated with the query $\boldsymbol{x}_0$ as a random variable with distribution $\mathbf{P}(\cdot \mid \boldsymbol{x}_0)$, where $\mathbf{P}(\cdot \mid \boldsymbol{x}_0)$ is a distribution $\mathbf{P}_\theta(\cdot)$ parametrized by $\theta = \theta(\boldsymbol{x}_0) \in \Theta$. The problem is then to estimate this distribution or, equivalently, the parameter θ on the basis of the information available. This information consists of a set

$$\mathcal{D} = \left\{ \boldsymbol{y}^{(i)} \succ \boldsymbol{z}^{(i)} \right\}_{i=1}^N \tag{2}$$

of preferences of the form $\boldsymbol{y} \succ \boldsymbol{z}$ between solutions.

The basic assumption underlying nearest neighbor estimation is that the conditional probability distribution of the output given the input is (approximately) locally constant, that is, $\mathbf{P}(\cdot \mid \boldsymbol{x}_0) \approx \mathbf{P}(\cdot \mid \boldsymbol{x})$ for $\boldsymbol{x}$ close to $\boldsymbol{x}_0$. Thus, if the preferences (2) are coming from problems $\boldsymbol{x}$ similar to $\boldsymbol{x}_0$ (namely from the nearest neighbors of $\boldsymbol{x}_0$ in the case base), then this assumption justifies considering $\mathcal{D}$ as a representative sample of $\mathbf{P}_\theta(\cdot)$ and, hence, estimating θ via maximum likelihood (ML) inference by

$$\theta^{ML} = \underset{\theta \in \Theta}{\operatorname{argmax}} \mathbf{P}_\theta(\mathcal{D}) . \tag{3}$$

An important prerequisite for putting this approach into practice is a suitable data generating process, i.e., a process generating preferences in a stochastic way.

Our data generating process is based on the idea of a discrete choice model as used in choice and decision theory [11]. Recall that the (absolute) preference for a solution $\boldsymbol{y} \in \mathbb{Y}$ supposedly depends on its distance $\Delta_Y(\boldsymbol{y}, \boldsymbol{y}^*) \geq 0$ to an ideal solution $\boldsymbol{y}^*$, where $\Delta_Y(\boldsymbol{y}, \boldsymbol{y}^*)$ can be seen as a "degree of suboptimality" of $\boldsymbol{y}$. As explained in [10], more specific assumptions on an underlying (latent) utility function on solutions justify the following model of discrete choice, which can be seen as a generalization of the well-known Bradley-Terry model:

$$\mathbf{P}(\boldsymbol{y} \succ \boldsymbol{z}) = \mathbf{P}(\boldsymbol{y} \succ \boldsymbol{z} \mid \boldsymbol{y}^*) = \frac{S_Y(\boldsymbol{y}, \boldsymbol{y}^*)}{S_Y(\boldsymbol{y}, \boldsymbol{y}^*) + S_Y(\boldsymbol{z}, \boldsymbol{y}^*)} , \tag{4}$$

where $S_Y(\boldsymbol{y}, \boldsymbol{y}^*)$ is defined as

$$S_Y(\boldsymbol{y}, \boldsymbol{y}^*) = \exp\left(-\beta \cdot \Delta_Y(\boldsymbol{y}, \boldsymbol{y}^*) \right)$$

according to (1). This similarity can be seen as the degree to which $\boldsymbol{y}$ resembles the ideal solution $\boldsymbol{y}^*$; likewise, $S_Y(\boldsymbol{z}, \boldsymbol{y}^*)$ is the degree to which $\boldsymbol{z}$ is close to ideal. Thus, the probability of observing the (revealed) preference $\boldsymbol{y} \succ \boldsymbol{z}$ depends on the degree of optimality of $\boldsymbol{y}$ and $\boldsymbol{z}$, namely their respective closeness to the ideal solution: The less optimal $\boldsymbol{z}$ in comparison to $\boldsymbol{y}$, the larger the probability to observe $\boldsymbol{y} \succ \boldsymbol{z}$; if $\Delta_Y(\boldsymbol{z}, \boldsymbol{y}^*) = \Delta_Y(\boldsymbol{y}, \boldsymbol{y}^*)$, then $\mathbf{P}(\boldsymbol{y} \succ \boldsymbol{z}) = 1/2$.

The coefficient β can be seen as a measure of precision of the preference feedback. For large β, $\mathbf{P}(\boldsymbol{y} \succ \boldsymbol{z})$ converges to 0 if $\Delta_Y(\boldsymbol{z}, \boldsymbol{y}^*) < \Delta_Y(\boldsymbol{y}, \boldsymbol{y}^*)$ and to 1 if $\Delta_Y(\boldsymbol{z}, \boldsymbol{y}^*) > \Delta_Y(\boldsymbol{y}, \boldsymbol{y}^*)$; this corresponds to a deterministic (error-free) information source. The other extreme case, namely $\beta = 0$, models a completely unreliable source reporting preferences at random.

The probabilistic model outlined above is specified by two parameters: the ideal solution $\boldsymbol{y}^*$ and the (true) precision parameter $\beta^* \in \mathbb{R}_+$. Depending on the context in which these parameters are sought, the ideal solution might be unrestricted (i.e., any element of $\mathbb{Y}$ is an eligible candidate), or it might be restricted to a certain subset $\mathbb{Y}_0 \subseteq \mathbb{Y}$ of candidates.

As mentioned before, the parameter vector $\theta^* = (\boldsymbol{y}^*, \beta^*) \in \mathbb{Y}_0 \times \mathbb{R}^*$ can be estimated from a given set (2) of observed preferences using the maximum likelihood principle. Assuming independence of the preferences, the likelihood of $\theta = (\boldsymbol{y}, \beta)$ is given by

$$\ell(\theta) = \prod_{i=1}^{N} \mathbf{P}\left(\boldsymbol{y}^{(i)} \succ \boldsymbol{z}^{(i)} \mid \theta\right) \tag{5}$$

The ML estimation $\theta_{ML} = (\boldsymbol{y}^{ML}, \beta^{ML})$ of θ^* is given by the maximizer of (5):

$$\theta_{ML} = \left(\boldsymbol{y}^{ML}, \beta^{ML}\right) = \operatorname*{argmax}_{\boldsymbol{y} \in \mathbb{Y}_0, \, \beta \in \mathbb{R}_+} \ell(\boldsymbol{y}, \beta) \tag{6}$$

The problem of finding this estimation in an efficient way is discussed in [10].

2.3   CBR as Preference-Guided Search

Case-based inference as outlined above realizes a "one-shot prediction" of a promising solution for a query problem, given preferences in the context of similar problems encountered in the past. In a case-based problem solving process, this prediction may thus serve as an initial solution, which is then adapted step by step. An adaptation process of that kind can be formalized as a search process, namely a traversal of a suitable space of candidate solutions that is guided by preference information collected in previous problem solving episodes.

To this end, we assume the solution space $\mathbb{Y}$ to be equipped with a topology that is defined through a *neighborhood structure*: For each $\boldsymbol{y} \in \mathbb{Y}$, we denote by $\mathcal{N}(\boldsymbol{y}) \subseteq \mathbb{Y}$ the neighborhood of this candidate solution. The neighborhood is thought of as those solutions that can be produced through a single modification of $\boldsymbol{y}$, i.e., by applying one of the available adaptation operators to $\boldsymbol{y}$.

Our case base **CB** stores problems $\boldsymbol{x}_i$ together with a set of preferences $\mathcal{P}(\boldsymbol{x}_i)$ that have been observed for these problems. Thus, each $\mathcal{P}(\boldsymbol{x}_i)$ is a set of preferences of the form $\boldsymbol{y} \succ_{\boldsymbol{x}_i} \boldsymbol{z}$, which are collected while searching for a good solution to $\boldsymbol{x}_i$.

We conceive preference-based CBR as an iterative process in which problems are solved one by one; our current implementation of this process is described in pseudo-code in Algorithm 1. In each problem solving episode, a good solution for a new query problem is sought, and new experiences in the form of preferences are collected. In what follows, we give a high-level description of a single problem solving episode:

(i) Given a new query problem $\boldsymbol{x}_0$, the K nearest neighbors $\boldsymbol{x}_1, \ldots, \boldsymbol{x}_K$ of this problem (i.e., those with smallest distance in the sense of Δ_X) are retrieved from the case base **CB**, together with their preference information $\mathcal{P}(\boldsymbol{x}_1), \ldots, \mathcal{P}(\boldsymbol{x}_K)$.

(ii) This information is collected in a single set of preferences $\mathcal{P}$, which is considered representative for the problem $\boldsymbol{x}_0$ and used to guide the search process.

(iii) The search for a solution starts with an initial candidate $\boldsymbol{y}^\bullet \in \mathbb{Y}$, namely the "one-shot prediction" (6) based on $\mathcal{P}$, and iterates L times. Restricting the number of iterations by an upper bound L reflects our assumption that an evaluation of a candidate solution is costly.

(iv) In each iteration, a new candidate $\boldsymbol{y}^{query}$ is determined, again based on (6), and given as a query to an underlying oracle. The oracle is a (possibly imperfect) information source that compares $\boldsymbol{y}^{query}$ with the current best solution $\boldsymbol{y}^\bullet$. The preference reported by the oracle is memorized by adding it to the preference set $\mathcal{P}_0 = \mathcal{P}(\boldsymbol{x}_0)$ associated with $\boldsymbol{x}_0$, as well as to the set $\mathcal{P}$ of preferences used for guiding the search process. Moreover, the better solution is retained as the current best candidate.

(v) When the search stops, the current best solution $\boldsymbol{y}^\bullet$ is returned, and the case $(\boldsymbol{x}_0, \mathcal{P}_0)$ is added to the case base.

The preference-based guidance of the search process is realized in (iii) and (iv). Here, our case-based inference method is used to find the most promising candidate among the neighborhood of the current solution $\boldsymbol{y}^\bullet$, based on the preferences collected in the problem solving episode so far. By providing information about which of these candidates will most likely constitute a good solution for $\boldsymbol{x}_0$, it (hopefully) points the search into the most promising direction.

3   Distance Learning in Pref-CBR

This section is devoted to the main extension of our Pref-CBR framework, namely the distance adaptation component in line 24 of Algorithm 1. In our framework, we assume preference information to be produced according to the probabilistic model (4). Therefore, it is natural to approach the distance learning problem from a probabilistic point of view. Correspondingly, we shall propose a Bayesian method to tackle this problem.

3.1   A Local-Global Representation of Distance

We begin with a simplifying assumption on the structure of the distance measure Δ_Y, namely that it adheres to the *local-global principle* [2] and takes the form

$$\Delta_Y(\boldsymbol{y}, \boldsymbol{y}^*) = \sum_{i=1}^{k} \alpha_i \cdot \Delta_i(\boldsymbol{y}, \boldsymbol{y}^*) \ , \tag{7}$$

where $\Delta_1, \ldots, \Delta_k$ are local distances pertaining to different properties of solutions, and $\boldsymbol{\alpha} = (\alpha_1, \ldots, \alpha_k)$ is a partition of unity (i.e., the coefficients α_i are non-negative and sum up to 1). We assume the Δ_i to be known, whereas the α_i, which are modeling the importance of the local distances, are supposed to be unknown. Learning the distance measure (7) is thus equivalent to learning these parameters.

Algorithm 1. Pref-CBR Search(K, L, J)

Require: K = number of nearest neighbors collected in the case base
$\phantom{\textbf{Require:} }L$ = number of queries to the oracle
$\phantom{\textbf{Require:} }J$ = number of preferences used to guide the search process

1. $\mathbb{X}_0 \leftarrow$ list of problems to be solved ▷ a subset of $\boldsymbol{X}$
2. $Q \leftarrow [\cdot]$ ▷ empty list of performance degrees
3. $\mathbf{CB} \leftarrow \emptyset$ ▷ initialize empty case base
4. **while** $\mathbb{X}_0$ not empty **do**
5. $\boldsymbol{x}_0 \leftarrow$ pop first element from $\mathbb{X}_0$ ▷ new problem to be solved
6. $\{\boldsymbol{x}_1, \ldots, \boldsymbol{x}_K\} \leftarrow$ nearest neighbors of $\boldsymbol{x}_0$ in $\mathbf{CB}$ (according to $\varDelta_X$)
7. $\{\mathcal{P}(\boldsymbol{x}_1), \ldots, \mathcal{P}(\boldsymbol{x}_K)\} \leftarrow$ preferences associated with nearest neighbors
8. $\mathcal{P} \leftarrow \mathcal{P}(\boldsymbol{x}_1) \cup \mathcal{P}(\boldsymbol{x}_2) \cup \ldots \cup \mathcal{P}(\boldsymbol{x}_k)$ ▷ combine neighbor preferences
9. $\boldsymbol{y}^{\bullet} \leftarrow \text{CBI}(\mathcal{P}, \mathbb{Y})$ ▷ select an initial candidate solution
10. $\mathbb{Y}^{vis} \leftarrow \{\boldsymbol{y}^{\bullet}\}$ ▷ candidates already visited
11. $\mathcal{P}_0 \leftarrow \emptyset$ ▷ initialize new preferences
12. **for** $i = 1$ to L **do**
13. $\mathcal{P}^{nn} = \{\boldsymbol{y}^{(j)} \succ \boldsymbol{z}^{(j)}\}_{j=1}^{J} \leftarrow J$ preferences in $\mathcal{P} \cup \mathcal{P}_0$ closest to $\boldsymbol{y}^{\bullet}$
14. $\mathbb{Y}^{nn} \leftarrow$ neighborhood $\mathcal{N}(\boldsymbol{y}^{\bullet})$ of $\boldsymbol{y}^{\bullet}$ in $\mathbb{Y} \setminus \mathbb{Y}^{vis}$
15. $\boldsymbol{y}^{query} \leftarrow \text{CBI}(\mathcal{P}^{nn}, \mathbb{Y}^{nn})$ ▷ find next candidate
16. $[\boldsymbol{y} \succ \boldsymbol{z}] \leftarrow \text{Oracle}(\boldsymbol{x}_0, \boldsymbol{y}^{query}, \boldsymbol{y}^{\bullet})$ ▷ check if new candidate is better
17. $\mathcal{P}_0 \leftarrow \mathcal{P}_0 \cup \{\boldsymbol{y} \succ \boldsymbol{z}\}$ ▷ memorize preference
18. $\boldsymbol{y}^{\bullet} \leftarrow \boldsymbol{y}$ ▷ adopt the current best solution
19. $\mathbb{Y}^{vis} \leftarrow \mathbb{Y}^{vis} \cup \{\boldsymbol{y}^{query}\}$
20. **end for**
21. $q \leftarrow$ performance of solution $\boldsymbol{y}^{\bullet}$ for problem $\boldsymbol{x}_0$
22. $Q \leftarrow [Q, q]$ ▷ store the performance
23. $\mathbf{CB} \leftarrow \mathbf{CB} \cup \{(\boldsymbol{x}_0, \mathcal{P}_0)\}$ ▷ memorize new experience
24. Adapt distance measure $\varDelta_Y$
25. **end while**
26. return list Q of performance degrees

3.2 Bayesian Distance Learning

Adopting the above representation of the distance measure $\varDelta_Y$, our choice model (4) is now given by

$$\mathbf{P}(\boldsymbol{y} \succ \boldsymbol{z}) = \frac{S_Y(\boldsymbol{y}, \boldsymbol{y}^*)}{S_Y(\boldsymbol{y}, \boldsymbol{y}^*) + S_Y(\boldsymbol{z}, \boldsymbol{y}^*)} \tag{8}$$

with

$$S_Y(\boldsymbol{y}, \boldsymbol{y}^*) = \exp\left(-\sum_{i=1}^{k} \gamma_i \cdot \varDelta_i(\boldsymbol{y}, \boldsymbol{y}^*)\right) \tag{9}$$

and $\gamma_i = \beta \cdot \alpha_i \geq 0$. Thus, learning $\boldsymbol{\gamma} = (\gamma_1, \ldots, \gamma_k)$ means learning β and $\boldsymbol{\alpha}$ simultaneously. In fact, these parameters can be recovered from $\boldsymbol{\gamma}$ as follows:

$$\beta = \gamma_1 + \gamma_2 + \ldots + \gamma_k$$
$$\alpha_i = \gamma_i / \beta$$

For simplicity, suppose that $\boldsymbol{\gamma} = (\gamma_1, \ldots, \gamma_k)$ only assumes values in a finite (or at least countable) set $\Gamma \subset \mathbb{R}_+^k$; this allows us to work with probability distributions instead of density functions. For example, Γ could be a suitable discretization of a continuous domain, such as a grid on a hypercube.

Since the true $\boldsymbol{\gamma}$ (used by the oracle) is assumed to be unknown, we model our belief about the parameters γ_i in (9) in the form of a probability distribution

$$\mathbf{P} : \Gamma \to [0, 1] \ ,$$

i.e., for each vector $\boldsymbol{\gamma} \in \Gamma$, $\mathbf{P}(\boldsymbol{\gamma})$ denotes the prior probability of that vector. Unless specific (prior) knowledge is available, this probability can be initialized by the uniform distribution over Γ.

Now, suppose a preference $p = [\boldsymbol{y} \succ \boldsymbol{z} \,|\, \boldsymbol{y}^*]$ to be revealed by the oracle. Since the oracle is supposed to generate preferences according to (8), this observation provides a hint at the true value of $\boldsymbol{\gamma}$. More specifically, it can be used for performing a Bayesian inference step to update our belief about $\boldsymbol{\gamma}$:

$$\mathbf{P}(\boldsymbol{\gamma} \,|\, p) = \frac{\mathbf{P}(p \,|\, \gamma)\mathbf{P}(\gamma)}{\mathbf{P}(p)} \ , \tag{10}$$

where $\mathbf{P}(p \,|\, \boldsymbol{\gamma})$ is given by (8). Concretely, this means realizing the following update for each $\boldsymbol{\gamma} \in \Gamma$:

$$\mathbf{P}(\boldsymbol{\gamma}) \leftarrow \frac{1}{C} \cdot \mathbf{P}(\boldsymbol{\gamma}) \cdot \mathbf{P}(p \,|\, \boldsymbol{\gamma}) \ , \tag{11}$$

where $\mathbf{P}(p \,|\, \boldsymbol{\gamma}) = \mathbf{P}(\boldsymbol{y} \succ \boldsymbol{z})$ is given by (8) and C is a normalizing constant assuring that the (posterior) probability degrees sum up to 1. To the best of our knowledge, there is no parameterized family of distributions that is conjugate with (8), so that the posterior (10) needs to be computed numerically.

3.3   Integration with Pref-CBR Search

As mentioned before, the adaptation of Δ_Y as outlined above is integrated in our Pref-CBR search procedure in line 24 of Algorithm 1. Thus, the idea is to update the belief about $\boldsymbol{\gamma}$ (and hence about Δ_Y, which is uniquely determined by this parameter) after each problem solving episode, making use of the newly observed preferences. Here is a summary of the main steps:

– Suppose our current belief about $\boldsymbol{\gamma}$ to be specified in the form of a probability $\mathbf{P}$ on Γ; in the beginning, this could be the uniform distribution, for example.

- In a single problem solving episode (lines 5–23 of Algorithm 1), Pref-CBR Search is used to solve a new problem x_0. This requires a concrete distance Δ_Y, and therefore a concrete parameter vector γ, which is used to "mimic" the (ground-truth) similarity measure of the oracle. To this end, we can reasonably choose the expectation according to our current distribution, which is considered as our current "best guess" of the true vector:[2]

$$\widehat{\gamma} = \frac{1}{|\Gamma|} \sum_{\gamma \in \Gamma} \gamma \cdot \mathbf{P}(\gamma) \qquad (12)$$

Using the distance measure (7) and choice model (8) parameterized by $\widehat{\gamma}$ or, more specifically, the induced parameters

$$\widehat{\beta} = \widehat{\gamma}_1 + \ldots + \widehat{\gamma}_k \ , \qquad (13)$$

$$\widehat{\alpha}_i = \widehat{\gamma}_i / \widehat{\beta} \ , \qquad (14)$$

the Pref-CBR search procedure is performed as usual.
- Upon termination of a problem solving episode (line 20 of Algorithm 1), Pref-CBR Search yields a solution $y^\bullet$, which is not necessarily the truly ideal solution y^* but at least an approximation thereof. Moreover, Pref-CBR Search returns a set of preferences $\mathcal{P}_0$ that have been collected during the search process. These preferences can now be used for updating our belief about γ.[3] To this end, the adaptation (10) is carried out for each of the preferences in $\mathcal{P}_0$. More specifically, for each preference $y \succ z$ observed in the last episode, a learning step is carried out with $[y \succ z \,|\, y^\bullet]$.

It is important to note that contextualizing an observed preference $y \succ z$ by $y^\bullet$ instead of y^* makes our distance learning method *approximate* and may possibly affect its efficacy. In fact, since preferences of the form $\widehat{p} = [y \succ z \,|\, y^\bullet]$ can be seen as "noisy" versions of the true preferences $p = [y \succ z \,|\, y^*]$, our method is actually learning from noisy data. We shall return to this issue in the experimental section further below.

3.4 Related Work

The learning and adaptation of similarity or distance measures has been studied intensively in the literature, not only in CBR but also in related fields like machine learning. Yet, our approach has a number of properties that distinguish it from most others: similarity is learned in the solution space, not in the problem space; training information is purely qualitative and based on paired comparisons; learning is done within the framework of Bayesian inference, making use of a probabilistic model.

---

[2] An alternative would be to choose the mode of the distribution instead of the mean.
[3] In principle, this set of preferences can be enriched further, assuming that each solution adopted in a later stage of the search process is preferred to each earlier solution.

Similarity learning in CBR has almost exclusively focused on learning similarity in the problem space. This is also true for the work of Stahl [12,16,13,15], which nevertheless shares a number of commonalities with our approach. In particular, he also considers the learning of weights in a linear combination of local similarity functions [12,14], albeit based on different types of training information and using other learning techniques. Our own previous work [3] is related, too, as it learns from qualitative feedback in the form of preferences. Essentially, this is what Stahl in [13] refers to as *relative case utility feedback*.

Appropriate metrics are also essential for the performance of distance-based methods such as nearest neighbor estimation, which are used for classification, regression, and related problems. Metric learning has therefore been studied quite intensively in machine learning and pattern recognition. While Mahalanobis distance metric learning has received specific attention in this regard, more involved problems such as nonlinear metric learning, local metric learning, semi-supervised metric learning, and metric learning for structured data have been tackled more recently. We refer to [1] for a comprehensive and up-to-date survey of the metric learning literature.

A Bayesian approach to distance metric learning has been proposed in [17]. Here, the authors estimates a posterior distribution for the distance metric from labeled pairwise constraints, namely equivalence constraints (pairs of similar objects) and inequivalence constraints (pairs of dissimilar objects). Worth mentioning is also the Bayesian approach to preference elicitation by [9]. Although it is concerned with utility instead of distance learning, the authors proceed from training information in the form of paired comparisons, and assume preferences to be generated by the Bradley-Terry model. Their model is still a bit simpler than our model (4) and permits the derivation of closed-form Bayesian updates (using a suitable family of conjugate priors).

4 Experiments

4.1 Synthetic Data

To illustrate our Bayesian approach to distance learning, independently of its use within the Pref-CBR framework, we conducted some very simple experiments for the case $\mathbb{Y} = [0,1]^2$, $\Delta_1(\boldsymbol{y}, \boldsymbol{y}') = |y_1 - y_1'|$, $\Delta_2(\boldsymbol{y}, \boldsymbol{y}') = |y_2 - y_2'|$, and $\boldsymbol{\alpha} = (\alpha_1, \alpha_2)$. For simplicity, we also assumed β to be known and only learned $\boldsymbol{\alpha}$.

To this end, we generated triplets $(\boldsymbol{y}, \boldsymbol{z}, \boldsymbol{y}^*) \subset \mathbb{Y}$ uniformly at random and derived exemplary preferences $[\boldsymbol{y} \succ \boldsymbol{z} \,|\, \boldsymbol{y}^*]$ or $[\boldsymbol{z} \succ \boldsymbol{y} \,|\, \boldsymbol{y}^*]$ according to our probabilistic model (4). Starting with a uniform prior on the simplex $\{(\alpha_1, \alpha_2) \,|\, \alpha_1, \alpha_2 \geq 0, \alpha_1 + \alpha_2 = 1\}$, N updates (11) were realized based on N random preferences of that kind.

Figure 1 shows typical examples of the marginal distributions for α_1 after $N = 50$, $N = 200$ and $N = 500$ examples. As expected, the distributions are fluctuating around the true value of α_1 (here taken as 0.3) and become more and more peaked with increasing N. Moreover, comparing the distributions on the left and the right panel, it can be seen that learning becomes easier for

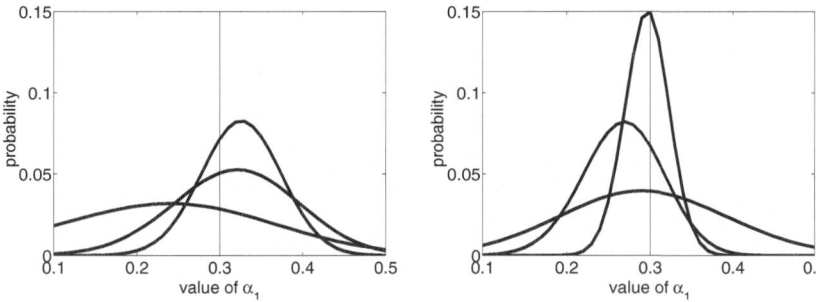

Fig. 1. Probability distributions for the parameter α_1 after $N = 50$, $N = 200$ and $N = 500$ examples, with $\beta = 5$ (left) and $\beta = 10$ (right)

larger values of β: for $\beta = 5$, the distributions are less peaked than for $\beta = 10$. Since β reflects the reliability of the oracle, this is again in agreement with our expectation. The same effect can also be observed in Figure 2 (left), which shows the boxplots for the mean value estimator (14); 100 of such estimators were derived from distributions for $N = 100$ and different values of β. Again, the larger β, the more precise the estimate of α_1 (and correspondingly of α_2).

As explained above, our Pref-CBR framework only produces an estimate $\boldsymbol{y}^{\bullet}$ of the ideal solution $\boldsymbol{y}^*$. Therefore, our distance learning method is based on preferences $[\boldsymbol{y} \succ \boldsymbol{z} \,|\, \boldsymbol{y}^{\bullet}]$ that can be seen as "noisy" versions of the true preferences $[\boldsymbol{y} \succ \boldsymbol{z} \,|\, \boldsymbol{y}^*]$. To simulate this property, we generated triplets $(\boldsymbol{y}, \boldsymbol{z}, \boldsymbol{y}^*) \subset \mathbb{Y}$ as above and set $\boldsymbol{y}^{\bullet} = \boldsymbol{y}^* + (\epsilon_1, \epsilon_2)^{\top}$, where ϵ_1 and ϵ_2 are (independent) normally distributed random variables with mean 0 and standard deviation σ. The observed preference (either $\boldsymbol{y} \succ \boldsymbol{z}$ or $\boldsymbol{z} \succ \boldsymbol{y}$) was then generated with our model (4) using the true $\boldsymbol{y}^*$, while distance learning was done using this model with the estimate $\boldsymbol{y}^{\bullet}$.

The effect of learning from noisy examples can be seen in Figure 2 (right), where we again show boxplots for the mean value estimator (14) based on 100 repetitions of the learning procedure with $N = 100$. As can be seen, the noise level σ does not seem to have a strong influence on the *variance* of the estimation. What is notable, however, is an apparent *bias* of the estimate: The larger σ, the more the estimates of α_1 are moving away from the true value 0.3 toward 0.5. Although this result cannot easily be generalized beyond the specific setting of our experiment, a tendency toward uniform weights of the α-coefficients (i.e., $\alpha_1 = \alpha_2 = 0.5$ in our case) is plausible: The more $\boldsymbol{y}^{\bullet}$ deviates from $\boldsymbol{y}^*$, the more noisy the examples will be for our distance learner—in the limit, they will become purely random, and on average, all local distances Δ_i will seemingly have the same influence then.

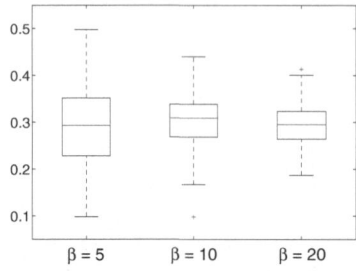 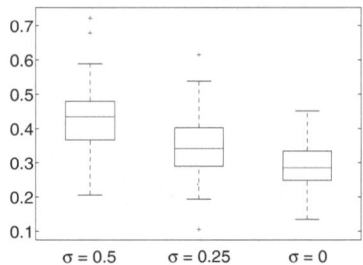

Fig. 2. Boxplots for the mean value estimate of α_1. Left: Different values of the precision parameter β. Right: Different levels of noise in the estimate $y^\bullet$.

4.2   Red Wine Recommendation

In this case study, we applied Pref-CBR to the problem of red wine recommendation. The scenario is as follows: A wine merchant tries to find his best offer for a customer, i.e., that wine in his cellar the customer likes the most. For an average customer, it will be much easier to (qualitatively) compare two wines instead of rating an individual wine—this nicely fits the assumption of our framework. Thus, the merchant can offer different candidate wines to the customer (who plays the role of our oracle), which are always compared to the current favorite. For obvious reason, however, the number of such comparisons needs to be limited.

To simulate this scenario, we made use of the red wine data set from the UCI machine learning repository [7]. This data describes 1599 red wines in terms of different chemical properties; here, we only used three of them, namely sulphates (y_1), pH (y_2), and total sulfur dioxide (y_3), which were found to have the strongest influence on preference [4]. We randomly extracted 500 wines to constitute the wines in the cellar, while the remaining 1500 were used as queries. Thus, a query is a wine that is thought of as the ideal solution of a customer (in this example, problem space and solution space therefore have the same structure).

We defined the distance measures in terms of (7) with the local distances given as $\Delta_i(y, y^*) = |y_i - y_i^*|$, $i = 1, 2, 3$. Moreover, assuming that the different chemical properties have a different influence on taste, we defined the *ground truth* distances $\Delta_X = \Delta_Y$ by setting $\alpha_1 = 0.1$, $\alpha_2 = 0.6$, $\alpha_3 = 0.3$. However, we assume this ground truth measure, which is the target of our similarity learning method, to be unknown. Instead, the measure used in the solution space as an initial measure subject to adaptation (and in the problem space without adaptation) is the *default measure* with uniform weights $\alpha_1 = \alpha_2 = \alpha_3 = 1/3$.

We used Pref-CBR Search (Algorithm 1) for sequential problem solving, starting with an empty case base. Then, we applied the algorithm to the 1500 query cases one by one and monitored its performance. To measure the quality q of a proposed solution $y^\bullet$ for a problem x_0 (line 22), we computed the position of this solution in the complete list of $|\mathbb{Y}| = 500$ wines in the store ranked by

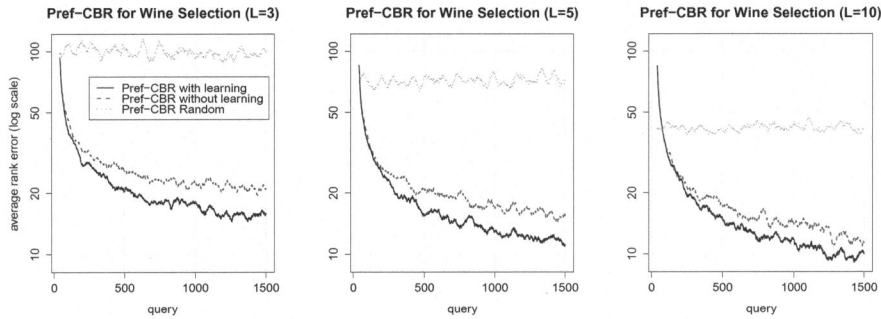

Fig. 3. Evolution of the average rank error in sequential problem solving (each query gives rise to one problem solving episode) for $L = 3, 5$ and 10 queries

(ground truth) similarity to the query $x_0 = y^*$ (i.e., 1 would be the optimal performance). To stabilize the results and make trends more visible, the corresponding sequence of performance degrees produced by a single run of Pref-CBR Search was averaged over 100 such runs.

We compared two versions of Pref-CBR Search, namely with and without (solution) similarity adaptation. Moreover, as a baseline we also used a search strategy in which the preference-guided selection of the next candidate solution in line 15 of Algorithm 1 is replaced by a random selection (i.e., an element from $\mathbb{Y}^{nn}$ is selected uniformly at random). Although this is a very simple strategy, it is suitable to isolate the effect of guiding the search behavior on the basis of preference information.

We applied our Algorithm 1 with $K = 5$, $L \in \{3, 5. 10\}$, $J = 25$; since the solution space is quite small, we used a global neighborhood structure, i.e., we defined the neighborhood of a solution y as $\mathcal{N}(y) = \mathbb{Y} \setminus \{y\}$. As can be seen from the results in Figure 3, our preference-based CBR approach shows a clear trend toward improvement from episode to episode, as opposed to the random variant of the search algorithm.

More importantly, however, similarity adaptation is clearly beneficial: Making use of the preference information gathered in the first episodes, Pref-CBR Search succeeds in learning the ground truth similarity measure, which in turn leads to better search performance and solution quality. The variant without similarity adaptation finds reasonably good solutions, too, because even the suboptimal (default) measure is guiding the search in a right direction—yet, with similarity adaptation enabled, the search becomes more effective, and the smaller the number of queries (L), the more pronounced the relative improvement.

5   Summary and Outlook

In this paper, we extended our framework for preference-based CBR by a Bayesian learning method for adapting the similarity (distance) measure in the solution space. As already explained, this measure equips this space with a topology, which in turn influences the search-based problem solving strategy in Pref-CBR. Consequently, a data-driven adaptation of this measure in cases where it does not seem to be optimally predefined can be seen as an important prerequisite for the effectiveness of Pref-CBR. In fact, first experimental studies presented in this paper have shown that an adaptation of similarity can lead to an improved overall performance of the CBR system.

Needless to say, there is still much scope to improve our method. For example, our implementation of Bayesian inference based on a fixed discretization Γ of the parameter space is not very efficient, especially if the vector γ to be learned is high-dimensional. Therefore, this approach should be replaced by more sophisticated methods for approximate Bayesian inference.

While the last problem is purely computational, the learning bias that is due to the use of estimated instead of truly ideal solutions is of a more conceptual nature. One way to overcome this problem is to make reasonable statistical assumptions about the relationship between $y^\bullet$ and y^* (like normality of the difference in our simulation study), and to extend our probabilistic model of the data-generating process correspondingly. Realizing Bayesian learning on the basis of this extended model would then explicitly account for the potential sub-optimality of $y^\bullet$.

A nice feature of Bayesian learning is the possibility to incorporate prior knowledge in a simple and elegant way. In our case, this would concern knowledge about the importance of the local distance measures (parameter α) and/or knowledge about the reliability of the information source (parameter β). Exploring the usefulness of this option is another point on our agenda.

Finally, instead of only learning the solution similarity S_Y, it would of course make sense to adapt the problem similarity S_X, too. This could be done, for example, on the basis of the preferences collected for different problems x_i. In fact, our neighborhood-based approach to case retrieval assumes that similar problems lead to similar preferences on solutions, or, more concretely: if two cases x_i and x_j are similar in the sense of S_X, the respective sets of preferences $\mathcal{P}(x_i)$ and $\mathcal{P}(x_j)$ are similar (in a sense to be specified), too. The idea, then, would be to adapt S_X in such a way as to make this assumption as valid as possible.

Acknowledgments. This work has been supported by the German Research Foundation (DFG).

References

1. Bellet, A., Habrard, A., Sebban, M.: A survey on metric learning for feature vectors and structured data. CoRR, abs/1306.6709 (2013)
2. Burkhard, H.-D., Richter, M.M.: On the notion of similarity in case based reasoning and fuzzy theory. In: Soft Computing in Case Based Reasoning, pp. 29–45. Springer (2001)

3. Cheng, W., Hüllermeier, E.: Learning similarity functions from qualitative feedback. In: Althoff, K.-D., Bergmann, R., Minor, M., Hanft, A. (eds.) ECCBR 2008. LNCS (LNAI), vol. 5239, pp. 120–134. Springer, Heidelberg (2008)
4. Cortez, P., Cerdeira, A., Almeida, F., Matos, T., Reis, J.: Modeling wine preferences by data mining from physicochemical properties. Decision Support Systems 47(4), 547–553 (2009)
5. Domshlak, C., Hüllermeier, E., Kaci, S., Prade, H.: Preferences in AI: An overview. Artificial Intelligence (2011)
6. Doyle, J.: Prospects for preferences. Comput. Intell. 20(2), 111–136 (2004)
7. Frank, A., Asuncion, A.: UCI machine learning repository (2010)
8. Goldsmith, J., Junker, U.: Special issue on preference handling for Artificial Intelligence. Computational Intelligence 29(4) (2008)
9. Guo, S., Sanner, S.: Real-time multiattribute Bayesian preference elicitation with pairwise comparison queries (2010)
10. Hüllermeier, E., Schlegel, P.: Preference-based CBR: First steps toward a methodological framework. In: Ram, A., Wiratunga, N. (eds.) ICCBR 2011. LNCS, vol. 6880, pp. 77–91. Springer, Heidelberg (2011)
11. Peterson, M.: An Introduction to Decision Theory. Cambridge Univ. Press (2009)
12. Stahl, A.: Learning feature weights from case order feedback. In: Aha, D.W., Watson, I. (eds.) ICCBR 2001. LNCS (LNAI), vol. 2080, pp. 502–516. Springer, Heidelberg (2001)
13. Stahl, A.: Learning similarity measures: A formal view based on a generalized CBR model. In: Muñoz-Ávila, H., Ricci, F. (eds.) ICCBR 2005. LNCS (LNAI), vol. 3620, pp. 507–521. Springer, Heidelberg (2005)
14. Stahl, A., Gabel, T.: Using evolution programs to learn local similarity measures. In: Ashley, K.D., Bridge, D.G. (eds.) ICCBR 2003. LNCS, vol. 2689, pp. 537–551. Springer, Heidelberg (2003)
15. Stahl, A., Gabel, T.: Optimizing similarity assessment in case-based reasoning. In: Proceedings of the 21st National Conference on Artificial Intelligence, AAAI (2006)
16. Stahl, A., Schmitt, S.: Optimizing retrieval in CBR by introducing solution similarity. In: Proceedings of the International Conference on Artificial Intelligence, IC-AI, Las Vegas, USA (2002)
17. Yang, L., Jin, R., Sukthankar, R.: Bayesian active distance metric learning. In: Proc. UAI, Uncertainty in Artificial Intelligence (2007)

Case-Based Parameter Selection for Plans: Coordinating Autonomous Vehicle Teams

Bryan Auslander[1], Tom Apker[2], and David W. Aha[3]

[1] Knexus Research Corporation, Springfield VA, USA
[2] Exelis Inc., Alexandria VA, USA
[3] Navy Center for Applied Research in Artificial Intelligence,
Naval Research Laboratory (Code 5514), Washington, DC, USA
bryan.auslander@knexusresearch.com,
{thomas.apker.ctr,david.aha}@nrl.navy.mil

Abstract. Executing complex plans for coordinating the behaviors of multiple heterogeneous agents often requires setting several parameters. For example, we are developing a decision aid for deploying a set of autonomous vehicles to perform situation assessment in a disaster relief operation. Our system, the *Situated Decision Process* (SDP), uses parameterized plans to coordinate these vehicles. However, no model exists for setting the values of these parameters. We describe a case-based reasoning solution for this problem and report on its utility in simulated scenarios, given a case library that represents only a small percentage of the problem space. We found that our agents, when executing plans generated using our case-based algorithm on problems with high uncertainty, performed significantly better than when executing plans using baseline approaches.

Keywords: Case-based reasoning, parameter selection, robotic control.

1 Introduction

Real-world plans can be complex; they may require many parameters for coordinating multiple agents, resources, and decision points. Furthermore, multiple performance metrics may be used to assess a plan's execution, and may involve trade offs. For example, we consider the problem of how to deploy a team of autonomous unmanned vehicles (AUVs), managed by a human operator, to conduct situation assessment in preparation for a Humanitarian Assistance/Disaster Relief (HADR) mission. The team includes a heterogeneous set of robotic platforms that vary in their capabilities. If we want to minimize vehicle energy consumption while maximizing coverage of the area surveyed, how many vehicles (and of what type) should we deploy, and how should they behave? If we're not conservative, we may expend too many resources (vehicles or energy), reducing our capability to respond to other emergencies in the near-term. Likewise, we run the risk of poor performance if too few resources are deployed, which may have serious consequences to the effected civilians.

L. Lamontagne and E. Plaza (Eds.): ICCBR 2014, LNCS 8765, pp. 32–47, 2014.
© Springer International Publishing Switzerland 2014

This problem is not unique to military mission planning, as similar decisions must be made for many other types of resource-bounded tasks such as emergency first response, sports team financial management, and local government budget planning. In each case, parameterized plans may exist to solve a complex problem, but how should their parameters be set? In some situations (such as ours), this problem can be compounded when the model for mapping parameter settings to expected performance is incomplete. While such models may be acquired, this requires access to problem-solving expertise or a massive database that records problem attributes, parameter settings, and the resulting performance metrics. Unfortunately, we lack both for our task.

We describe a case-based approach to solve this problem where our parameters include the number and types of AUVs to send along with algorithm specific options for surveying an area, and report on its utility in an initial empirical study. Our algorithm uses a similarity-weighted vote to set parameter values, and generates a single score for a set of performance metrics. We found that our approach generates plans that performed well in simulation studies versus baseline approaches, particularly when state uncertainty is high.

We describe related work in Section 2, then propose how AUVs can support HADR missions in Section 3. In Section 4, we overview the Situated Decision Process (SDP), which defines a role for case-based parameter selection. We present our case-based algorithm in Section 5, report its application in simulated scenarios in Section 6, and discuss the results in Section 7 before concluding.

2   Related Work

We focus on the problem of setting the values for multiple parameters, which are then used by a goal reasoning (GR) module to assist with multi-agent plan generation. CBR has previously been used for parameter setting. For example, Wayland [16] is a deployed case-based system used to select the parameter values for an aluminum die-casting machine. It uses a type-constrained and feature-weighted 1-nearest neighbor rule for retrieval and a rule-based approach for adaptation. Unlike our approach, Wayland's cases do not include specific performance outcome data, nor use them for case reuse.

Several CBR systems set parameter values while interacting with other reasoning modules, though to our knowledge ours is the first to supply them to a GR module for controlling multiple AUVs in coordinated plans. Weber et al. [19] uses artificial neural networks to model software programs and biological systems, resulting in a case representation of problem, solution, and outcome similar to ours. Genetic algorithms are used to learn the initial cases, which are then clustered using their solutions and given to a discriminant analysis technique to find select problem features. While they are concerned with one outcome metric we focus on a multi-objective problem, and using GAs may be infeasible in our domain due to long run times. Jin and Zhu [5] use CBR to set the initial parameter values for an injection molding process. These settings are repeatedly tested and modified by a fuzzy reasoner until all defects are eliminated, at which time a new case is stored. While not emphasized in this paper, we use CBR to repeatedly, rather than only initially, recommend parameter settings throughout plan execution. Montani [11] uses CBR to set the parameters for multiple systems, including a rule-based reasoner used to modify therapies for diabetes patients. We focus on multi-agent planning rather than rule revision, and we focus on the control of AUV teams rather than a health sciences application. Finally, Pavon et al. [15] use CBR to revise Bayesian network models for setting the control parameters of a genetic algorithm that performs root identification for geometric problems. In contrast, our CBR algorithm uses a performance-based voting procedure rather than abduction to set parameters.

Jaidee et al. [3] also integrated CBR with a GR module for team coordination planning. However, they use CBR to formulate goals while we use it to set parameter values, and we focus on robotic control rather than video games.

Other have studied CBR for robotics applications. For example, Likhachev et al. [7] use CBR to learn parameter settings for the behavior-based control of a ground robot in environments that change over time. Their approach learns and adapts cases, effectively searching the behavior parameter space, until good performance is obtained. While their work focuses on motion control for a single robot, we instead focus on the high-level control of robot teams. Karol et al. [6] also focus on robot team coordination (for RoboCup soccer). They use CBR to select actions that are transformed to motion control parameters. Ros et al. [18] also focus on action selection for RoboCup soccer, and use a sophisticated representation and reasoning method. Of interest to us is that each agent can behave independently and abort the plan, which is also essential in our domain because unexpected state changes could have catastrophic consequences. However, this body of research focuses on motion planning for relatively short-term behaviors, whereas we focus on longer duration plans that are monitored by a GR module.

Finally, CBR has been studied for several military applications, including those involving disaster response. For example, Abi-Zeid et al. [1] studied incident prosecution, including real time support for situation assessment in search and rescue missions. Their prototype system, ASISA, uses CBR to select hierarchical information-gathering plans for situation assessment. Muñoz-Avila et al.'s [12] HICAP instead uses a conversational CBR system to assist operators with refining tasks in support of noncombatant evacuation operations. SiN [13] extends their work, integrating a planner to automatically decompose tasks where feasible. However, while these systems use planning modules to support rescue operations, they do not set parameters for multiagent plans, nor focus on coordinating robot team behaviors.

3   Domain: Military HADR Operations

HADR operations [14] are performed by several countries in response to events such as Hurricane Katrina (August 2005), the Haiti earthquake (2010), and Typhoon Haiyan (November 2013). Before support operations can arrive an initial wave of responders must gather required information about the impact zone (e.g., infrastructure status, suggested ingress and evacuation routes, and survivor locations). Current operations employ remotely controlled drones and human-piloted helicopters to gather this information. We instead propose deploying a heterogeneous team of AUVs with appropriate sensor platforms to automate much of this process, so as to reduce time and cost. We claim that this should enable responders to perform critical tasks more quickly for HADR operations.

In this paper we focus a module of the SDP, which we are developing to assist with HADR operation. Under a human operator's guidance, the SDP will deploy a team of heterogeneous AUVs, coordinated by a GR module to identify which goals need to be accomplished, and deploy the AUVs to best achieve them [17]. To perform this task the GR module needs to generate, compare, schedule, and dispatch plans for achieving these goals. Plans vary in their performance metrics based on resource allocation (e.g., of AUVs and their search algorithm), and selecting among them requires the GR to deliberate about their predicted performance (e.g., to maximize search area coverage and minimize energy consumption). To do this, we use a case-based algorithm to select parameter settings for the GR module, where cases associate problems with solutions (i.e., parameter settings) and their performance metrics. This enables the SDP to propose multiple solutions for a goal by optimizing on different metrics.

4   Simulating the Situated Decision Process (SDP)

To provide context for our work on case-based parameter selection, we briefly describe the SDP's modules, the simulated AUVs that it will control, their sensing models and search algorithms, and scenario performance metrics.

4.1   SDP Modules

Figure 1 highlights SDP's primary modules. It will support a Forward Air Controller (FAC) in surveying and assessing Areas of Interest (AoIs) of a HADR mission. The FAC will provide mission objectives and constraints using a Human-Robot Interface, whose GUI will permit highlighting of AoIs, important assets, and related information. These will be provided to a GR module [3], which will help to decompose the given objectives. This will result in selecting goals and conditions for synthesizing a finite state automaton to control the motions of the AUVs. Our Goal Reasoner depends on a Case-Based Parameter Selector to recommend parameter settings given geographical regions and constraints.

4.2   Simulation Platforms

HADR missions begin with generating a new road map, as a common feature of disasters is the collapse of buildings or washing out of roads. We propose to use three AUV types (Figure 2), which vary in their motion properties, to perform infrastructure assessment. The number of each type corresponds to a parameter that needs to be set in HADR mission plans.

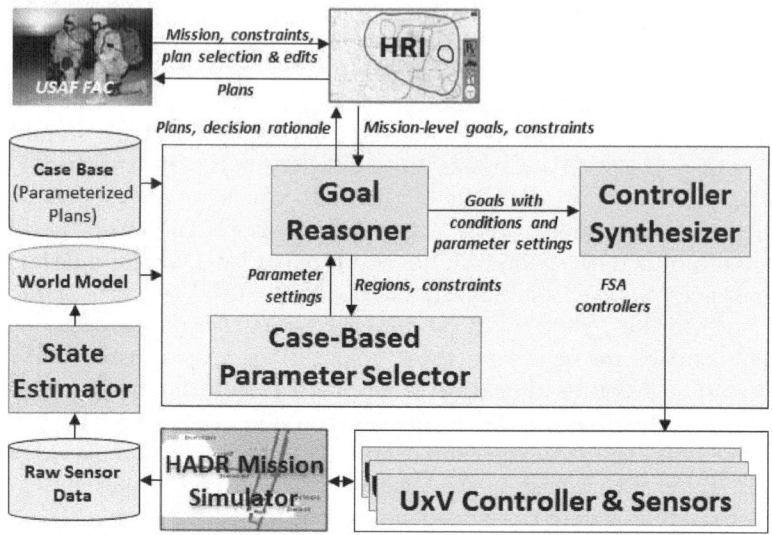

Fig. 1. The Situated Decision Process (SDP)

(a) STUAS   (b) UGV   (c) MAV

Fig. 2. Platforms Simulated in our Study (Acknowledgements: (A) US Marines, Shannon Arledge; (B) US Navy, Elizabeth R. Allen; (C) NASA, Sean Smith)

From the air, the roadways that are still intact are visible to downward looking cameras, which we assume are carried at relatively high altitude (> 1000m) by small tactical unmanned aircraft systems (STUAS) and at low altitudes ($< 100m$) by multirotor air vehicles (MAVs). From the ground, the best roadmapping sensor is a 3D omnidirectional lidar mounted on unmanned ground vehicles (UGVs).

We model AUVs as nonholomonic physicomimetic agents [10]. This allows us to run them simultaneously in the MASON [9] multi-agent simulator to compare the performance of teams that vary in their numbers of each AUV type.

4.3 Sensing Model

Downward facing cameras and scanning lidars take measurements of the region surrounding an agent as shown in Figure 3a. For flying agents, the width d of the sampled area depends on the agent's altitude h and the field of view of the camera α. For UGVs, d is the maximum effective range of their lidar. Both of

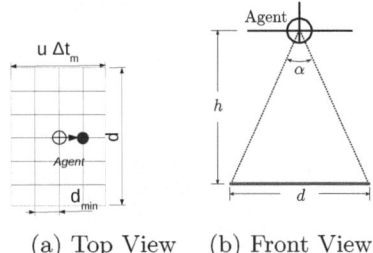

(a) Top View   (b) Front View

Fig. 3. Schematic of Downward-Facing Camera Image Segmented into Discrete Regions

these sensors are data-rich and information-poor, and so rather than modeling transmission of whole images, we assume they segment their sensed area into discrete grid cells d_{min} meters across, where d_{min} is half the expected width of the target signal (e.g., roads). The agent's speed u and measurement interval Δt_m determine the minimum depth of the simulated sensor grid. We assume a square grid of $d = 60$m and $d_{min} = 6$m for all agents in our simulator.

At each time step, where $\Delta t_m = .1$s, our simulated sensors return a Boolean grid that classifies each cell based on whether it contains the target signal. Each sensor updates an estimate of the percentage of coverage of a set of global grid cells using a one-dimensional Kalman filter, which fuses data from multiple platforms (whose sensor views overlap). A Kalman filter allows us to control the rate at which cell uncertainty can grow or be reduced with guaranteed and predictable rates of convergence of each cell to its mean value.

4.4 Area Coverage Search Algorithm

We use three parameters that can affect how the AUVs search a geographical area during HADR missions, where we assume they all begin a scenario near a single, randomly selected point and follow a grazing area coverage algorithm [8]. Our first parameter determines how a simulated AUV searches. *Greedy* search applies a physicomimetics force to drive each agent towards the nearest "food", here simulated by global grid cell covariance. In contrast, *Segmented* search, depicted in Figure 4, guides each agent towards the nearest "food" in its own Voronoi cell. Our second parameter, *On Demand Recharging*, is true iff the AUVs can automatically recharge themselves. Finally, if the third parameter *Mobile Charger* is true, then a mobile base station follows the AUVs.

4.5 Scenario Metrics

Time is critical in HADR missions, both for reaching survivors and collecting information to ensure that aid is properly distributed when it arrives. However, the infrastructure damage that needs to be detected often makes resources such as electricity and fuel more precious. Thus, we define the following performance metrics to assess agent performance in our simulation studies:

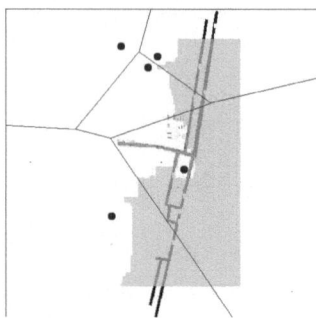

Fig. 4. Voronoi Partition of a Search Area among Multiple MAV Agents

1. *Coverage*: Percentage of the AoIs searched
2. *Energy Consumption* (in Joules): Total energy consumed

We measure Coverage over a HADR scenario's entire duration, which we set to 30 minutes in our empirical study because we expect HADR personnel to arrive within 30 minutes after the AUVs are deployed. Energy Consumption is calculated uniformly across fuel and battery types used to power the AUVs.

In our evaluation (Section 6), we also use the compound *Efficiency* metric, defined as the ratio of Coverage to Energy Consumption. We found that it is a better heuristic (than the other two metrics) for guiding parameter selection.

5   Case-Based Parameter Selection

We use a case-based algorithm in the SDP to select parameter settings for HADR scenarios. We describe our case representation, including a case's recommended parameter settings, in Section 5.1. Our algorithm employs case retrieval and reuse, as described in Sections 5.2 and 5.3, but does not perform case revision or retention, which we leave for future work.

5.1   Case Representation

We represent a case $C = \langle P, S, O \rangle$ as a vector of attributes denoting a problem P, its multi-dimensional solution S, and the multi-dimensional performance outcome O recorded when executing a HADR mission plan with parameters S to the scenario summarized by P. Table 1 details this representation.

The attributes of P characterize the HADR scenario and are used to define case similarity (see Section 5.2). P's parameters are predominately focused on geometric features given the task of surveying an area. These include the number of disjoint Areas of Interest (AoIs) in the scenario, the distance between opposite corners of the (rectangular) Area of Operations (AO), the total area sizes of the AoIs, and Coverage Decay, a binary parameter indicating whether the sensor information for a grid cell decays over time and becomes obsolete (i.e., uncertainty increases). This allows us to model changes to the environment, such as when

Table 1. Case Representation used for SDP Parameter Selection

Case Component	Attribute Name	Value Range
Problem (*P*)	# Disjoint AoIs	[1, 10]
	AO Diagonal Distance	[424, 1980]
	Total Area Size of AoIs	[500, 1,568,000] m^2
	Coverage Decay	{1=true, 0=false}
Solution (*S*)	# STUASs	[0, 3]
	# UGVs	[0, 3]
	# MAVs	[0, 10]
	Search algorithm	{g=greedy, s=segmented}
	On Demand Recharging	{true, false}
	Mobile Charger	{true, false}
Outcome (*O*)	Coverage	[0%, 95%]
	Energy Consumption	[0, 8,244,000] Joules

a fire spreads across grid cells. (Evaluating the effectiveness of these features versus others is left for future research.)

A case's solution S contains the settings for six parameters that define how to apply the SDP to P. These include the number of each type of unmanned vehicle to deploy (i.e., STUASs, UGVs, and MAVs), the type of area coverage search algorithm to use (i.e., Greedy or Segmented), whether to use On Demand Recharging, and whether to deploy a Mobile Charger (for UGVs). Finally, the outcome attributes O are the scenario metrics described in Section 4.5.

In our empirical study, our case library includes multiple cases with the same problem, but differ in their solutions, because our simulator is non-deterministic.

5.2 Case Retrieval

Our parameter selection algorithm retrieves cases in two phases. In the first phase it computes the similarity of a given *query* (whose attributes are those listed for Problems in Table 1) with the distinct problems among cases in a library L, and retrieves those cases $L' \in L$ whose similarity exceeds a threshold t, where the similarity of a query q and a case's problem $c.P$ is defined as follows:

$$sim(q, c.P) = \frac{1}{|P|} \sum_{a \in P} 1 - \frac{|q_a - c.P_a|}{a_{max}}, \tag{1}$$

where a_{max} is the largest possible absolute difference among two values of an attribute a.

In the second phase, L' is filtered to remove cases whose performance metrics are low. This phase is needed to remove cases whose problems may have had a poor sampling bias. Filtering is done using a function that scores the outcomes of each case $c \in L'$ relative to every other case in L'. We score each case c by

---

Algorithm 1. Case Reuse Algorithm for Setting Parameter Values

Inputs: Query q, Cases N_q
Returns: $Solution[]$ // A vector of parameter settings for q
Legend:
  N_q // q's k-Neighborhood of cases
  s // A parameter among those in q's predicted solution S
  $Votes[]$ // Summed weighted scores, indexed by parameter
  $SumWeights[]$ // Summed weights, indexed by parameter

SetParameters$(q, N_q) =$
foreach $s \in S$ **do**
  $\lfloor$ $Votes[s] = SumWeights[s] = 0$;

foreach $c \in N_q$ **do**
  foreach $s \in c.S$ **do**
    $weight = \mathbf{sim}(q, c.P)$;
    $Votes[s] \mathrel{+}= weight \times \mathbf{score}(c, N_q)$;
    $SumWeights[s] \mathrel{+}= weight$;

foreach $s \in c.S$ **do**
  // Compute value for parameter s with the highest weighted vote
  $\lfloor$ $Solution[s] = \mathbf{maxParamValue}(Votes[s]/SumWeights[s])$;

return $Solution[]$;

---

calculating a Student t-statistic[1] for its Efficiency relative to some set of cases C as follows:

$$score(c, C) = \frac{c_e - \overline{C}_e}{s}, \tag{2}$$

where c_e is the efficiency of case c, $\overline{C}_e$ is the mean efficiency of all cases in C, and s is the sample deviation of C.

For each case $c \in L'$, we compute its $score(c, L')$. For each subset of cases $L'_p \subset L'$ with problem p, we identify its $n\%$ most Efficient cases and compute their mean, denoted by $mean(L', p, n)$. We then rank these mean values across all problems p represented in L', identify the problems P' with the k highest values of $mean(L', p, n)$, and return all cases with a problem $p \in P'$. This yields, for query q, a neighborhood of retrieved (and filtered) cases N_q.

5.3   Case Reuse

SDP uses Algorithm 1 for case reuse. It computes a similarity-weighted vote among the retrieved cases N_q, where a vote of a case $c \in N_q$ is the product of its problem's similarity to q and its $score(c, N_q)$ as defined in Equation 2.

---

[1] We use the Student t-statistic because the population statistics of our non-deterministic scenarios are unknown, and we have verified that Coverage and Energy Consumption are both normally distributed.

Our HADR sceanrio simulator is non-deterministic. Thus, the metrics computed for a given ⟨problem,solution⟩ pair can vary each time a scenario is executed. To account for this we compute these metrics using their mean values across a set of scenario executions.

Given a query q and a set of cases N_q, Algorithm 1 computes a score for each case $c \in N_q$ (the k-Neighborhood of q) and then computes the weighted vote of each parameter in q's solution vector S. Our algorithm allows for all cases in N_q to contribute votes to the adapted solution, and assumes that the parameters are independent. It weights each vote by the similarity of $c.P$ to q, giving more (normalized) weight to cases that are more similar to query q. Finally, it returns a solution (i.e., a vector of parameter settings).

6   Empirical Study

We empirically tested four research hypotheses:

H1 Solutions obtained using problem knowledge can outperform random solutions.
H2 No single solution performs optimally on all problems.
H3 Using a case-based approach to set parameters yields solutions that perform well in comparison to a known good solution.
H4 Case adaptation increases performance metrics.

In the following sections we describe the metrics, simulation data, empirical method, algorithms tested, the results, and their analysis.

6.1   Metrics

We initially planned to use the (raw) outcome metrics in O listed in Table 1, namely Coverage and Energy Consumption. However, neither metric alone provides comprehensive insights for the results. This motivated us to introduce Efficiency (Section 4.5), which we use for results analysis.

6.2   Simulation Data

We conducted our study using MASON [9], a discrete-event multiagent simulator that models all physics behaviors and physicomimetics control. A problem scenario in MASON is defined by the problem attributes P (see Table 1), the locations and sizes of the AoIs, and the AUVs' starting locations. (We did not include these latter attributes in P due, in part, to their high variance.) Running MASON scenarios also requires setting the parameters in S (e.g., the number of AUVs per platform type, the type of area search algorithm to use). Running a parameterized MASON scenario yields the set of outcomes in O. MASON is non-deterministic; these outcomes can vary each time it runs a parameterized scenario because it executes the actions of multiple agents in a random order.

Table 2. Test Scenario Problems

Problem #	Disjoint AoIs	AO Diagonal Distance	AoIs' Area	Coverage Decay
1	1	1621	259972	
2	7	1553	396580	No
3	6	1400	375275	
4	6	1001	69279	
5	4	1332	130684	
6	10	1451	285652	
7	5	636	14452	Yes
8	3	1773	197925	
9	10	1222	165494	
10	3	1833	194512	

To generate the case library L for our experiments, we created a problem scenario generator that outputs random scenarios (according to a uniform distribution) over the attributes for problems and solutions, using the ranges shown in Table 1. We used it to generate 20 problem scenarios and paired each with 100 solution vectors to yield the P and S values for 2000 cases. We then executed each ⟨Scenario,Solution⟩ pair in MASON 10 times, and recorded the average values of the outcome metrics (O) with each case.

6.3   Empirical Method

We tested our CBR algorithms (Section 6.4) using a partial cross-validation strategy. In particular, we selected the first 10 problem scenarios (see Table 2) among the 20 generated, and performed leave-one-out testing. That is, when we tested a problem scenario p_i, we temporarily removed the 100 cases in L with problem p_i for testing, leaving 1900 cases in L. We repeated this cross-validation 10 times to generate the data for our analyses.

The time required to generate L was four days, even after parallelizing the 10 trials per problem and dividing the problems among two multi-cpu machines. Thus, time constraints prevented us from running a complete leave-one-out test on all 20 problems, which we leave for future work.

6.4   Algorithms Tested

We tested two CBR algorithms that vary in whether they perform case adaptation/reuse. They both apply the case retrieval algorithm described in Section 5.2, where we set $k = 3$, $n = 20\%$ and $t = 75\%$. (Values were not optimized).

CBR$_N$ (Non-adapted) This retrieves N_q, selects a case $c \in N_q$ that has the maximum score(c,N_q), and executes MASON using the solution vector $c.S$.
CBR$_A$ (Adapted) After case retrieval, this applies Algorithm 1 to q and N_q. It outputs solution vector S, which is given to MASON to execute.

We also included the following three baselines in our experiments:

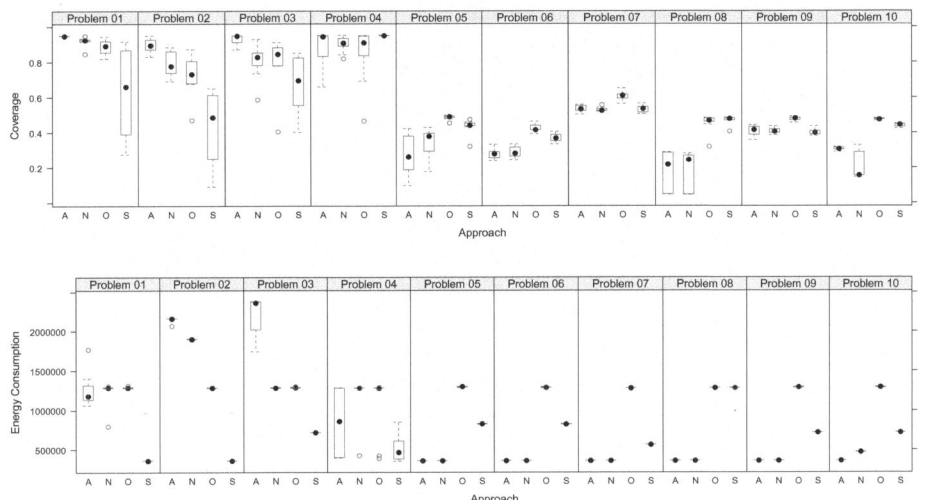

Fig. 5. Performance of CBR_A (A), CBR_N (N), **O**, and **S** on Coverage, where higher values are preferred, and Energy Consumption, where lower values are preferred

Random Mean (R). This does not require simulation. Instead, for a given scenario, it locates the 100 cases in L whose problem P matches the scenario, and yields the mean value of their outcome metrics.

Best Overall (O). This evaluates how well a single good solution performs across all problems. It finds the case $c \in L$ whose solution $c.S$ has the highest score(c,L) and applies it to every problem in the test set.

Best Sampled (S). This is designed to produce a good known solution for any query. It locates the 100 cases $L' \subset L$ whose problem matches the given scenario, finds the case $c \in L'$ with highest score(c, L'), and returns its outcome metrics $c.O$.

6.5 Results and Analysis

H1: Figure 5 displays plots for Coverage and Energy Consumption for the test problems. Using a one-tailed t-test, we found that **R** consistently recorded significantly lower Efficiency (not shown for space reasons) than the other algorithms, which supports **H1**. For Coverage, **S** performed significantly better on 6 scenarios, **O** 9 times, CBR_N 4 times, and CBR_A 6 times. For Energy Consumption, these algorithms significantly outperformed **R** for all scenarios (except problem 4 for CBR_N and **O**). Given this, we will ignore **R** for the rest of this section.

H2: We expected that **O** would be consistently outperformed by, or perform simmilarly to, the remaining algorithms for the 9 test problems (**O**'s solution came from a case with Problem 1, which we excluded from testing). Comparing Efficiency, **S** significantly outperformed **O** on 8 test problems, while CBR_N (CBR_A) significantly outperformed **O** on 4 of the 6 problems in which Coverage

Table 3. Solutions generated by algorithms **S**, CBR_N (N), and CBR_A (A) for problems 1-10 (where t for Greedy indicates that Greedy search was used, *Recharging* means On Demand Recharging, and *Mobile C* means a Mobile Charger was used)

	#MAVs			#UGVs			#STUASs			Greedy			Recharging			Mobile C		
P	S	N	A	S	N	A	S	N	A	S	N	A	S	N	A	S	N	A
1	0	2	2	0	0	3	1	3	3	t	f	f	t	t	t	f	f	f
2	0	1	0	0	1	3	1	3	3	f	t	t	t	t	t	f	t	f
3	0	2	2	0	0	3	2	3	3	f	f	f	t	t	t	f	f	f
4	1	2	2	1	0	0	3	3	3	t	f	f	t	t	t	t	f	f
5	1	0	0	0	0	0	2	1	1	f	t	t	t	t	t	f	f	f
6	1	0	0	0	0	0	2	1	1	f	t	f	t	t	t	f	f	f
7	2	0	0	0	0	0	1	1	1	t	t	f	t	t	t	f	f	f
8	2	0	0	0	0	0	3	1	1	f	t	t	t	t	t	f	f	f
9	0	0	0	0	0	0	1	1	1	t	t	t	f	t	t	t	f	f
10	0	1	0	0	0	0	2	1	1	f	f	f	t	t	t	f	f	f

Decay exists (primarily due to lower Energy Consumption). These results support this hypothesis (i.e., that one solution does not perform well across all test problems and tailoring of solutions is preferable), particularly for when Coverage Decay (i.e., higher state uncertainty) exists.

H3: The results support this hypothesis, especially for Coverage Decay problems. For example, when comparing Efficiency CBR_A significantly outperformed **S** on problems {6,7,9,10} while CBR_N outperformed **S** on {5,6,7,9}. We discuss this further in Section 7.

H4: We analyzed whether CBR_A outperformed CBR_N, which seems to be indicated by Figure 5. For Coverage, CBR_A significantly outperforms CBR_N on 4 problems, though sometimes at the expense of Energy Consumption (e.g., for problem 3). However, while the graphs suggest some advantages to case reuse, the significance tests are inconclusive.

7 Discussion

The results are challenging to analyze because this is a multi-objective problem. Although Efficiency combines them, it treats both outcome metrics equally, independently of whether this is preferable. The CBR algorithms performed well on Coverage, but were mediocre on Energy Consumption compared to **S** on problems 1-4, which are not Coverage Decay problems. However, while **S** is supposed to be a good solution, the plans generated by the CBR algorithms perform comparably without cheating (i.e., training on test problems).

Table 3 displays the solutions generated by **S**, CBR_N, and CBR_A for each problem. (Baseline **O** always chose 2 MAVs, 0 UGVs, 3 STUASs, Segmented search, On Demand Recharging, and no Mobile Charger.) The large variance in Coverage (Figure 5) is partially caused by solutions that use few STUAS (Table 3). This is due in part to a physics model that permits STUAS, in some conditions, to depart (i.e., fly out of) an AoI. However, Coverage variations are much smaller for problems {6,7,9} than problems {5,8}, even though they all use only one STUAS. This is because the latter problems have fewer disjoint AoIs, which may reduce the frequency with which STUAS depart the map. We will address this issue in future work.

For Coverage Decay problems, 95% Coverage becomes nearly impossible to obtain, and the CBR algorithms converge to using one STUAS to increase Efficiency (i.e., low Energy Consumption and large Coverage). Operating multiple AUVs yields diminishing benefits (i.e., marginally larger Coverage at a higher Energy Consumption), although an operator may prefer higher Coverage. We will further explore what metrics to use in future work.

8   Conclusion

To our knowledge, this is the first study that uses CBR to set parameters that control the behavior of a heterogeneous AUV team (here, to perform situation assessment for HADR missions). Our simulation study showed that our case-based algorithm can perform well on this task (i.e., outperforms a random approach as well as the parameter values found to perform best for any scenario, and performs comparably to the best known solution for a given problem scenario). While case adaptation tends to improve performance, the increases are not significant.

Our simulation models only some of the real-world environment's characteristics (e.g., it ignores wind, ground cover, and vehicle overhead costs), which could impact our results. For example, substantial ground cover could degrade the data collected by a STUAS, and increase reliance on the other two platforms. Adding these parameters would make pre-calculation of solutions for the entire problem space infeasible, and strongly motivate the need for case retention.

Our case reuse algorithm makes a strong independence assumption and is sensitive to small training sets. We will study other approaches for adapting cases such as using stretched neighborhoods for each dimension [4]. We will also study methods for seeding the case base with good solutions learned from multi-objective optimization functions. Although this type of optimization has been used previously (e.g., [2]), we propose to use it to seed the case base rather than use cases to seed further multi-objective searches.

Finally, we will test the SDP for its ability to help an operator guide a team of AUVs in outdoor field tests under controlled conditions. We have scheduled such tests, with increasingly challenging scenarios, during the next three years.

Acknowledgements. The research described in this paper was funded by OSD ASD (R&E). We also thank Michael Floyd and the reviewers for their comments on an earlier draft.

References

1. Abi-Zeid, I., Yang, Q., Lamontagne, L.: Is CBR applicable to the coordination of search and rescue operations? A feasibility study. In: Althoff, K.-D., Bergmann, R., Branting, L.K. (eds.) ICCBR 1999. LNCS (LNAI), vol. 1650, pp. 358–371. Springer, Heidelberg (1999)
2. Cobb, C., Zhang, Y., Agogino, A.: Mems design synthesis: Integrating case-based reasoning and multi-objective genetic algorithms. In: Proceedings of SPIE, vol. 6414 (2006)
3. Jaidee, U., Muñoz-Avila, H., Aha, D.W.: Case-based goal-driven coordination of multiple learning agents. In: Delany, S.J., Ontañón, S. (eds.) ICCBR 2013. LNCS, vol. 7969, pp. 164–178. Springer, Heidelberg (2013)
4. Jalali, V., Leake, D.: An ensemble approach to instance-based regression using stretched neighborhoods. In: Proceedings of the Twenty-Sixth International Florida Artificial Intelligence Research Society Conference (2013)
5. Jin, X., Zhu, X.: Process parameter setting using case-based and fuzzy reasoning for injection molding. In: Proceedings of the Third World Congress on Intelligent Control and Automation, pp. 335–340. IEEE (2000)
6. Karol, A., Nebel, B., Stanton, C., Williams, M.-A.: Case based game play in the roboCup four-legged league: Part I The theoretical model. In: Polani, D., Browning, B., Bonarini, A., Yoshida, K. (eds.) RoboCup 2003. LNCS (LNAI), vol. 3020, pp. 739–747. Springer, Heidelberg (2004)
7. Likhachev, M., Kaess, M., Arkin, R.: Learning behavioral parameterization using spatio-temporal case-based reasoning. In: Proceedings of the International Conference on Robotics and Automation, vol. 2, pp. 1282–1289. IEEE (2002)
8. Liu, S.-Y., Hedrick, J.: The application of domain of danger in autonomous agent team and its effect on exploration efficiency. In: Proceedings of the 2011 IEEE American Control Conference, San Francisco, CA, pp. 4111–4116 (2011)
9. Luke, S., Cioffi-Revilla, C., Panait, L., Sullivan, K., Balan, G.: Mason: A multiagent simulation environment. Simulation 81(7), 517–527 (2005)
10. Martinson, E., Apker, T., Bugajska, M.: Optimizing a reconfigurable robotic microphone array. In: International Conference on Intelligent Robots and Systems, pp. 125–130. IEEE (2011)
11. Montani, S.: Exploring new roles for case-based reasoning in heterogeneous ai systems for medical decision support. Applied Intelligence 28, 275–285 (2008)
12. Muñoz-Avila, H., Aha, D.W., Breslow, L., Nau, D.: HICAP: An interactive case-based planning architecture and its application to noncombatant evacuation operations. In: Proceedings of the Ninth National Conference on Innovative Applications of Artificial Intelligence, pp. 879–885. AAAI Press (1999)
13. Muñoz-Avila, H., Aha, D.W., Nau, D., Weber, R., Breslow, L., Yaman, F.: SiN: Integrating case-based reasoning with task decomposition. In: Proceedings of the Seventeenth International Joint Conference on Artificial Intelligence, pp. 999–1004. Morgan Kaufmann (2001)
14. O'Connor, C.: Foreign humanitarian assistance and disaster-relief operations: Lessons learned and best practices. Naval War College Review 65 (2012)

15. Pavón, R., Díaz, F., Laza, R., Luzón, V.: Automatic parameter tuning with a Bayesian case-based reasoning system: A case study. Expert Systems with Applications 36, 3407–3420 (2009)
16. Price, C.J., Pegler, I.S.: Deciding parameter values with case-based reasoning. In: Watson, I.D. (ed.) UK CBR 1995. LNCS, vol. 1020, pp. 119–133. Springer, Heidelberg (1995)
17. Roberts, M., Vattam, S., Alford, R., Auslander, B., Karneeb, J., Molineaux, M., Apker, T., Wilson, M., McMahon, J., Aha, D.W.: Iterative goal refinement for robotics. In: ICAPS Workshop on Planning and Robotics (2014)
18. Ros, R., Arcos, J., Lopez de Mantaras, R., Veloso, M.: A case-based approach for coordinated action selection in robot soccer. Artificial Intelligence 173, 1014–1039 (2009)
19. Weber, R., Proctor, J.M., Waldstein, I., Kriete, A.: CBR for modeling complex systems. In: Muñoz-Ávila, H., Ricci, F. (eds.) ICCBR 2005. LNCS (LNAI), vol. 3620, pp. 625–639. Springer, Heidelberg (2005)

Automatic Case Capturing
for Problematic Drilling Situations

Kerstin Bach[1], Odd Erik Gundersen[1,2],
Christian Knappskog[2], and Pinar Öztürk[2]

[1] Verdande Technology AS
Trondheim, Norway
http://www.verdandetechnology.com
[2] Department of Computer and Information Science
Norwegian University of Science and Technology, Trondheim, Norway
http://www.ntnu.no

Abstract. Building a high quality case base for knowledge intensive Case-Based Reasoning (CBR) applications is expensive and time consuming, especially when it requires manual work from experienced knowledge engineers. This paper presents a clustering-based method for capturing cases in time series data within the oil well drilling domain. We present a novel method for automatically detecting and capturing predictive cases originally created by domain experts. The research presented is evaluated within Verdande's *DrillEdge*, in which until today case capturing is an experience-driven and thoroughly manual process. Our findings show that this process can be partially automated and customizing an individual CBR application in a complex domain can be further developed.

Keywords: Case-Based Reasoning, Clustering, Semi-Automatic Case Acquisition, Data Streams, Knowledge Mining, Spatio-temporal Data.

1 Introduction

Learning from experience is one of a Case-Based Reasoning (CBR) system's key success factors. However, in some domains case acquisition is a knowledge intensive task that requires a high level of domain expertise. This affects the ability to grow the case base, as the cost of adding new cases is high. The domain expertise that is required to build competent case bases in domains such as health care, financial services and oil and gas is scarce. A high level of expertise in these domains typically makes the domain experts sought after, which again make them both busy and costly and therefore unavailable for experience transfer. High quality cases require domain experts to understand case-based reasoning as well as the use case. This is another obstacle that also adds to the cost. As cases in these domains can help prevent extremely costly outcomes, the return on the case acquisition investment can be high, and thus the high cost is accepted. However, because of the high cost of building cases in these domains, measures that can be taken to support the case acquisition process are welcome.

L. Lamontagne and E. Plaza (Eds.): ICCBR 2014, LNCS 8765, pp. 48–62, 2014.
© Springer International Publishing Switzerland 2014

The cases we are aiming at within the scope of this paper are predictive cases derived from experience. Predictive cases in a real-time setting run the retrieve phase of the CBR cycle [1] regularly in order to warn of problematic situations. A problematic situation is characterized by single problematic symptoms clustered over time until a point of a critical situation is reached. Incoming data is analyzed for reoccurring events, which are subsequently compared to an existing case base. Once similar cases are found, they are presented to the user to take actions.

Our focus in this paper is oil well drilling and the application *DrillEdge*, which is a commercial streaming data case-based reasoning decision support system that is widely deployed (see Gundersen [18] for an introduction to the system). Building cases for *DrillEdge* is a time-consuming and knowledge intensive task performed by highly skilled domain experts, but the savings that are possible by using the application make the effort required worth it. Building one case is estimated to take 40 to 60 hours of focused work by a domain expert, which means that the cost of one case is between USD 6,000 to USD 9,000.

We present research that seeks to automate parts of the highly manual case acquisition process, to aid the domain expert's effort and reduce the time required for building cases. The proposed solution combines the automatic symptom detection in real-time drilling logs offered by *DrillEdge* with domain knowledge and clustering algorithms to propose cases to the domain expert. In this way, the domain expert can focus on validating cases suggested by the system rather than manually analyzing data covering months of real-time drilling data and textual reports.

The rest of the paper is structured as follows: In section 2 we will discuss related work to our approach before we describe the application domain of real-time data in oil well drilling in section 3. The following two sections will focus on the case capturing method: section 4 discusses the assumptions and requirements while section 5 describes the introduced method in detail using a real-life example. Section 6 contains the description of the experiments we conducted and presents the results along with a discussion of the evaluation results. The final section draws conclusions and gives an outlook on future work.

2   Related Research

CBR research has been targeting the importance of case base maintenance in the past. There are different goals the automatic case creation approaches pursue: Analyzing an existing case base and restructuring (adding, deleting or revising) it for a competence gain or on the other hand analyzing raw data in order to build new cases fitting a given case structure.

Smyth and McKenna [27] introduce a model of case competence that assesses the performance of a case base by calculating the coverage of clusters of cases (group coverage) in order to determine the global competence of a case base. They apply this model to their case authoring software, visually showing the current competence of the case base and how the case builder's new case will affect it. In [26], the same authors use deletion of cases as a method to maintain a case base. They apply the same competence model as in [28] and use it to tackle the utility problem without degrading competence. Finally Zhy, Yang and Columbia [33] use these deletion policies in reverse to add cases to a case base to improve its competence. In recent works, Cummins and Bridge [7,8] picked up further case maintenance or editing algorithms in order to follow an ensemble approach for optimizing case bases by combining various maintenance strategies.

Case mining as a combination of data mining and case building has been exercised in various settings where raw data is provided in natural language. Bach et al. [4] introduce a CBR system based on a database of customer support reports written in free text format in the manufacturing industry. In order to create cases from the customer reports, a knowledge model is built using natural language processing and refined in discussions with domain experts. Each customer support report is a candidate for a case – based on a fixed case structure, information is extracted and mapped to a case. In a similar fashion, Farley [12] explored automatically building cases for IDS (Integrated Diagnostic System), in which IDS monitors messages sent from air planes to the ground. The system initially had 70 manually created cases built from either automatic malfunction reports recorded by the system or text reports submitted by pilots. Farley used domain experts and natural language processing to extract information from the messages in order to build cases. Similarly, Dufour-Lussier et al. [10] apply simplified natural language processing techniques for extracting rich case representations for processes from texts. The authors use an approach derived from detecting grammatical structures to detect work flows described in texts.

Manzoor et al. suggests a method of generating cases using genetic algorithms [20]. In their study they apply their approach to the problem of assigning exam dates. This is a combinatorial problem where certain assignments are more preferable than others, which is represented in a fitness function. The initial case base is created by randomly generating a case and then using the genetic algorithm to improve it before it is submitted. Any successful application of a case is run through the same algorithm as generated to improve and retain the case.

In comparison to our approach previously mentioned case acquisition tasks either dealt with maintaining an existing case base or creating cases from existing raw data. In our domain we have neither, because the drilling logs do not necessarily contain cases. Instead we have to identify whether a case can be build. Similarly, Ram and Santamaria [22,23] used CBR with a reactive control system for the navigation of an autonomous robot. The goal of the system is to avoid collisions. Each case represents a series of navigation steps over time and the system learns by either changing old cases, reinforcing successfully used cases or creating a new case. The authors use genetic algorithms for detecting problems and setting is similar to how we implemented our algorithm for detecting a problem and then set the windows for it.

Flinter and Keane [13] as well as Floyd and Esfandiari [14] on the other hand follow an approach of reproducing an expert capturing a case for particular situations. However, both approaches are made for real-time strategy games, which produce a huge amount rather than complex cases as we are aiming at.

On the other hand, cluster algorithms in CBR have mainly been used during the retrieval step rather than the creation of cases. Applying clustering algorithms, such as k-Means and Self-Organizing Maps for organizing case bases has been investigated mostly for improving the performance of large case bases [15,30]. The goal of these approaches was providing an unsupervised alternative to indexing large amounts of cases. The authors cluster cases by their features and derive a structure that can be reused during the retrieval. Furthermore, Smyth [25] leaned how to index already existing case bases and Patterson [21] extended it by using clustering for a regression-based numerical value prediction.

Further on, Fornells et al. [16], for instance, provide cluster-based retrieval for jCOLIBRI that provides the WEKA tool set as in memory organization of the case base. Therewith they provide different retrieval strategies depending on the chosen case base clustering algorithm.

In medical applications and biomedicine where cases have a huge amount of attributes, clustering has been widely applied. Arshadi et al. [3], for instance, use clustering for enhancing the accuracy of CBR systems. In their work they describe how to utilize clusters for selecting features from data sets with large records.

3   Acquiring Cases from Real-Time Drilling Data

Building cases for *DrillEdge* requires comprehensive knowledge of the drilling domain along with experience in CBR when selecting which information is relevant to describe a problematic situation. Cases in *DrillEdge* are built based on multidimensional abstractions (see [24]) of real-time logs: incoming drilling data is analyzed, filtered, combined, and eventually represented as events. In *DrillEdge*, events are created by Pattern Recognition agents, which identify complex events based on different input streams (such as rotation of the drill string, mud circulating in a well, load and pressure on tools) and mapped to a drilling log. The creation of the events follow the idea of Complex Event Processing (CEP) systems as they are described by Luckham [19].

Case matching is carried out based on the events with regards to time and depth. Both dimensions as well as the type of events are weighted within a case's similarity measure which makes case matching in the given domain a complex task. Details on how the similarity measures are designed have been described by Gundersen [17]. When building cases manually, the expert reviews the time and depth view of a log and the associated events in order to determine the window that describes a problematic situation. In a following step, the expert weights these factors with regard to their utility for the case [5] and verifies the findings with descriptions in drilling reports.

Figure 1 shows an example drilling log in the time and depth view. The log contains the measurement of the hole depth and bit position during the entire drilling process. Monitoring hole depth and bit depth shows the progress of drilling: the top left corner of the figure marks the starting point of the drilling operation and as long as hole and bit depth proceed in parallel, the well is being drilled. The lower right corner is the end of the log while the x-axis contains the time and the y-axis the hole depth. Once bit and hole depth differ over a period of time the drilling process does not gain depth. Such interruption could be caused by the change of the tools used or problematic situations.

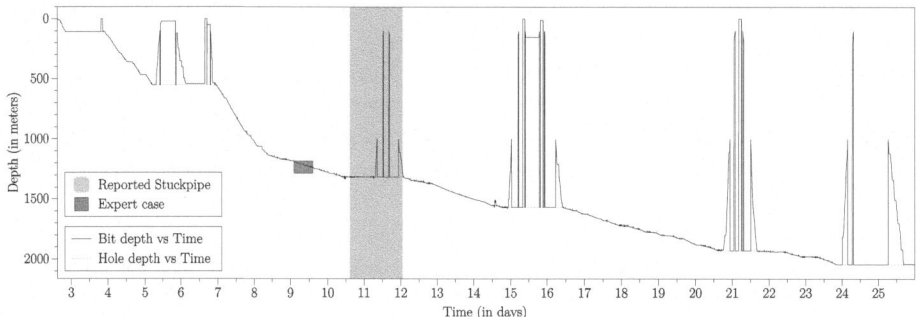

Fig. 1. Overview of a drilling log including a case captured by an expert and a reported mechanical stuck pipe situation

The drilling profile given in figure 1 contains numerous occasions when hole and bit depth differ as well as one problematic situation (marked gray). At $time = 12$ there is a reported mechanical stuck pipe. A mechanical stuck pipe is an incident when gravel filled up around the drill bit until it is not possible to rotate properly and drill the well further. Such an incident causes non-productive time (NPE), because the drilling engineers have to take actions to remove the gravel and free the drill bit in order to continue. This incident is usually succeeded by proceeding problems (events):

- Increased pressure of the mud fluids running through the well – because the fluids cannot be pushed through the gravel easily
- The drill string is hard to move – because gravel closes the hole around it

These problems occur in different variations and intensities during the drilling operation. A single occurrence usually does not cause problematic situations. However, a frequent occurrence of such an event can be seen as a signature for an upcoming problem. In the given example, one can see that after the reported mechanical stuck pipe there is a NPE of about 18 hours before drilling continues.

In order to avoid such a problematic situation the case to be built should contain relevant events over a certain amount of time in a certain depth. In the given example, we have a case captured by an expert about 30 hours before the incident. Until today, the identification of a case is manual work and in the remaining of this paper, we will introduce how this can be automated.

4   Assumptions and Requirements

The approach presented focuses on one particular problem, the mechanical stuck pipe. We have analyzed the causes of a mechanical stuck pipe with identifying relevant events and unique signatures, which resulted in a set of events as well as a flat spot incident as significant for a mechanical stuck pipe situation.

4.1   Assumptions

We assume that a mechanical stuck pipe situation is always introduced by a so called flat spot. A flat spot is described as a situation where the drill bit is moving or lifted, but no depth is gained. In the example given in figure 1 flat spots are at $time = 5 - 6$, $time = 11$ or $time = 15 - 16$. The hole depth at these periods stays constant over time while the bit depth changes frequently. These changes mark the drill bit being lifted and lowered to remove the gravel around it. Since a flat spot is caused by a stuck pipe, we assume that the mechanical stuck pipe occurred at the beginning of the flat spot. Furthermore, the events produced by the Pattern Recognition agents are filtered before they are fed into the clustering algorithm, so they only contain events relevant for a mechanical stuck pipe. This assumption has been made in collaboration with Verdande's domain experts based on their lessons learned when building cases.

4.2   Requirements

Choosing the appropriate clustering algorithm for a data set mainly relies on being able to model the data appropriately.

   We experimented with clustering using all of the different attributes, and with some of the more well-known algorithms: K-Means [29], Affinity Propagation (AP) [31], Mean Shift [6], DBSCAN [11] and OPTICS [2]. K-Means, AP and Mean Shift are complete as they do not allow noise. K-Means also requires the number of clusters to be known beforehand, which is not given in our scenario. OPTICS produced clusters that were too fine without matching assumptions the experts had suggested. Out of this experimentation we selected DBSCAN, as it is density based, handles noise and we were able to achieve clusters that matched well with the ones suggested by domain experts.

   Density-based spatial clustering of applications with noise (DBSCAN) takes two parameters, ϵ and *minimum points*. For each point it checks to see if there are more than *minimum points* within ϵ distance of itself, if true it is considered a *core point*. Core points within ϵ distance of each other merge to form a cluster. Points within ϵ distance that are not core points will still be part of the cluster. Points that do not meet these requirements are marked as noise.

   In order to provide an entirely unsupervised approach, the two parameters either have to be set as constants across all logs, or calculated for each. By defining two extreme cases, one where we did not want the cluster to be any bigger, and one where we did not want it to be smaller, we found the optimal *minimum points* value at 10. The event distribution density varies across logs, so we applied a method suggested by Zhou et al [32] to calculate an ϵ value on each log. This approach sets ϵ to be the average distance between all the samples. More advanced ways of calculating ϵ are available, but we found that this approach gave satisfactory results.

5   Automatic Case Capturing Method

In order to build a case for *DrillEdge*, we need a time and depth window containing relevant events for a specific problem. We suggest that if we are able to locate an area indicative of a drilling problem, the cluster of events preceding the location in time and depth will contain the symptoms for this problem. An overview of our approach is shown in figure 2. The input is time series data and events on which the DBSCAN cluster algorithm is applied for two different dimensions: time and depth. Further, the log profile of the time series data is analyzed to identify so called flat spots where the bit does not move. Whether the bit is stuck or not is not analyzed. Both the flat spots and the event clusters are merged in order to build and mark an area in the log where a case is located. In the following, we build an intersection set of the time and depth clusters. This window is then narrowed down by using the given flat spot and subsequently each cluster's medoids. Finally events are weighted based on their occurrence in the remaining case window. The event types are filtered so that we only work on the four relevant event types.

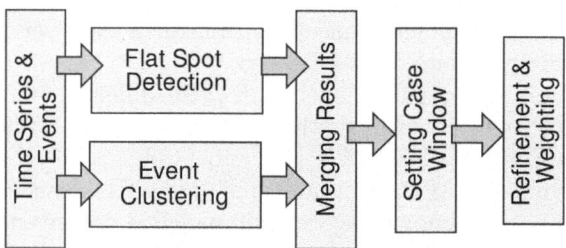

Fig. 2. Automatic case capturing process

In the following we explain the automatic case capturing method in detail using an anonymized, but real-life drilling log. Figure 3 shows the detailed actions taken by the case capturing method. Plot (a) shows how an expert would capture a case for the given drilling log as well as the reported stuck pipe. The following plots (b) to (f) on the other hand shows the our proposed method finds a case that warns of a mechanical stuck pipe.

5.1   Flat Spot Detection

Our hypothesis is that in the logs an area of no drilling activity is an interesting area. These are recognizable in the time versus depth-view as *flat spots*, areas

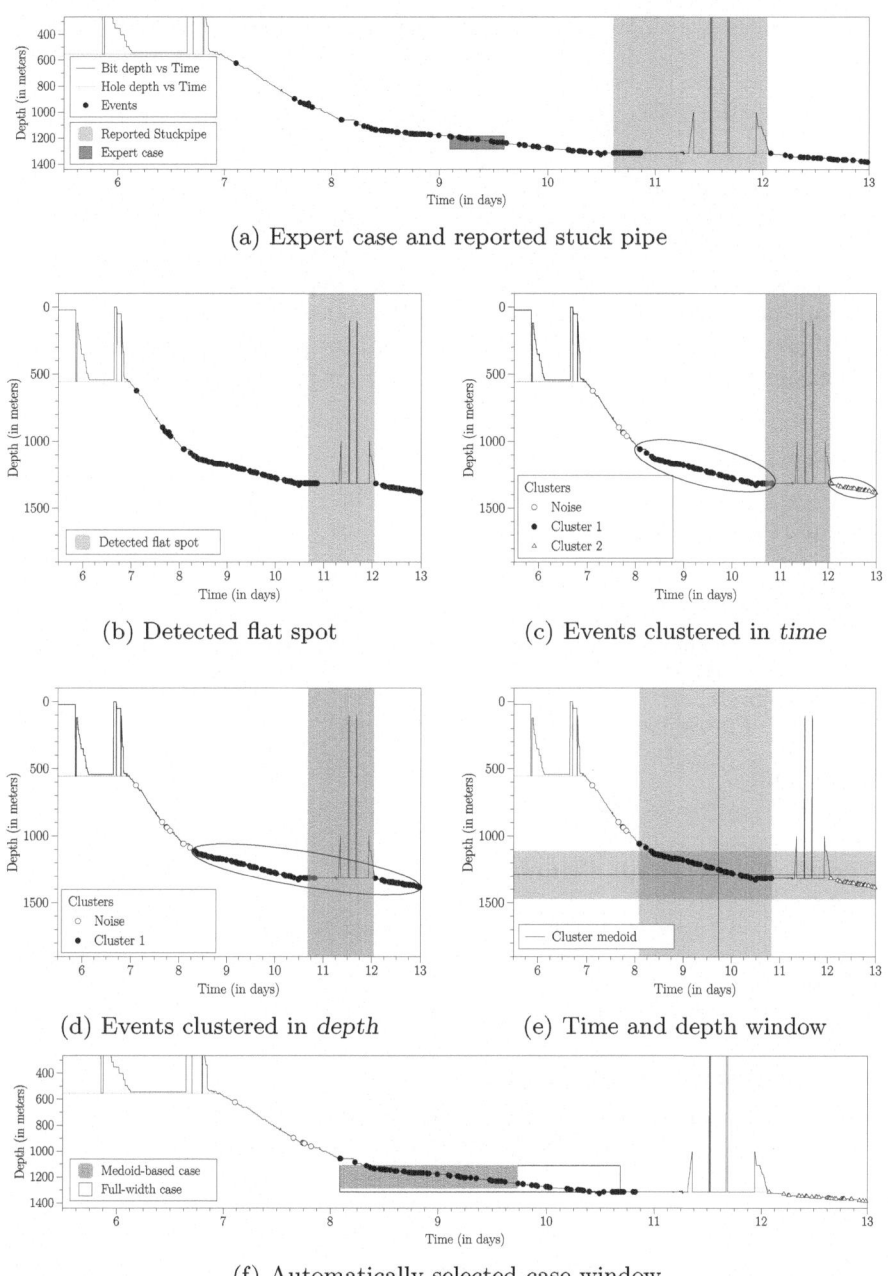

(a) Expert case and reported stuck pipe

(b) Detected flat spot

(c) Events clustered in *time*

(d) Events clustered in *depth*

(e) Time and depth window

(f) Automatically selected case window

Fig. 3. Detecting and capturing a case that prevents a mechanical stuck pipe situation

where the hole depth is constant for a noticeable period of time. Not all of these areas are caused by drilling problems. Pulling the equipment up to the surface to make changes, waiting for bad weather to pass are just some of the reasons a flat-spot might be logged. In (b) of Figure 3 a possible flat spot detected by the algorithm given in Algorithm 1 is marked.

When the drill pipe gets stuck, remedial actions are performed, and thus no drilling occurs and no hole depth is gained. Hence, our assumption is that the areas of the log where no hole depth is gained are interesting candidates for us to analyze. Each log consists of data points with values for different measurements. Detecting a flat spot requires only the hole depth for which the algorithm is comparing the difference over 100 data points. The logs we are working on contain a data point every 5 to 15 seconds, so 100 data points represents a window of 10-20 minutes. Considering that connecting new pipes or pulling the drill bit up can easily take 5 minutes this time frame is long enough to see a flat spot.

```
DetectFlatSpots ()
    input  : A log of dataPoints of size l
    output: Flat spots including stuck bits
    for index ← 0 to l do
        lookAhead ← index + 100;
        startIndex = index; endIndex = −1;
        while dataPoints[i].holeDepth == dataPoints[lookAhead].holeDepth do
            endIndex = lookAhead;
            if endIndex != -1 then
                FlatSpots ← newFlatSpot[start, end];
            end
        end
    end
    foreach flatSpot within FlatSpots do
        i ← flatSpot.start;
        while i < flatSpot.end&BitAndHoleDepthEqual(dataPoints[i]) do
            i + +;
        end
        FlatSpotWithStuckBits ← [flatSpot.start, i]
    end

    BitAndHoleDepthEqual (dataPoint)

    return (abs(dataPoint.holeDepth − dataPoint.bitDepth) < 5)
```
Algorithm 1: Locating a Flat Spot

Further, since there are more reasons that can cause a flat spot, we are also looking for a *potential stuck bit*. This is characterized by a subset of the flat spots where the drill bit is not immediately pulled out of the hole. This is seen as an indicator that the operators have issues with pulling the drill string out of the hole.

Within each flat spot we check if the bit depth and hole depth were the same, within a tolerance of five meters, for at least a period of five minutes. This filters out the situations where the flat spot is caused by the operators pulling out the drill string. It also allows for some movement of the bit, as this is common when trying to free a mechanically stuck pipe. The value of five meters is set based on observations of reported mechanically stuck pipe.

5.2   Event Clustering

DrillEdge performs matching of event sequences in time and depth separately. This allows problems that occur at specific depths (location in the vertical hole) to be identified, while also identifying problems that occur as a series of actions over time. By applying the DBSCAN clustering approach on depth and time separately, just a single dimension, we found that the clusters more closely matched those suggested by domain experts. Plots (c) and (d) of Figure 3 show the cluster results for both dimensions. Applying DBSCAN separately reduces the problem of clustering to a much simpler one. In the given example the clustering in time (c) results in two clusters – one before and one after the detected start of a flat spot. Events clustered by depth (plot (d)) results in one cluster for the given example section, which can be explained that after the flat spot the gain of depth happened gradually. Hence the difference is not large enough to enforce a second cluster.

5.3   Merging Results and Setting Case Window

After creating the clusters, they are merged based on their time and depth range to built an initial case window. Intersection sets are built wherever they occur and subsequently mapped to identified flat spots. If there are intersection sets without flat spots they are not considered as a case window.

Plot (e) of Figure 3 shows the result as a time and depth window. In this initial step it purely contains the ranges set by the clusters. The flat spots are then used as the end point for a potential case, so if the window ends after the flat spot, the case is cut at the flat spot. This is the *full width* cases, using the entire cluster width if possible.

As the result shows, the case window can be fairly large compared to the expert case in (a). Hence we narrowed the size of a case window by using the cluster medoid. The medoid is computed based on the time and depth dimension independently and hence reduces the case window in those dimensions. Eventually narrowing down the case window increases the *warning time* for the case. The warning time in *DrillEdge* is the time between a case alerting in the system and the predicted problem occurring. It can be seen in (f) that the end of the case window and the beginning of the flat spot incident increases by re-scaling the case window.

5.4   Refinement and Weighting of Events

After setting the case window, the containing features have to be weighted in order to perform accordingly on logs. Weights can be set on different feature types such as event types, operational properties like drilling fluid, well properties such as well geometry or geology, tools or type of activity, etc. In our case we only set feature weights on the event types and assign weights based on their occurrence. In the current implementation we only weighted the event types we used for the clustering, because we aim to test the clustering method and not further feature weights.

We compared the results with the weighting done by expert, we saw that the calculated weights matched or were very close to those suggested by the automatic approach. Our approach is however not able to differentiate between the relative importance of the depth or time sequences. Where a case builder would weight only one of the sequences, we weight both similarly.

6   Experiments and Evaluation

For evaluating our approach we decided to compare the performance of automatically captured cases to cases created by experts. Verdande has a general case base containing cases preventing problematic situations when drilling. Cases from the generic case library have been exercised and refined over years and they contain the most expert knowledge. From these cases we selected a subset of twelve cases as expert cases for the experiments, which target mechanical stuck pipe and cases with similar symptoms.

Further, we have selected eight different drilling log files that evidently contain mechanical stuck pipe problems. We have run the automatic case capturing method on all eight logs and it detected 15 potential mechanical stuck cases.

To compare the reference cases with the automatically generated cases, we ran them pairwise on the eight drilling logs and compared their similarity scores, then computed the true and false positives as follows:

- True Positive: expert case > 50% AND automatically captured case > 50%
- True Negative: expert case > 50% AND automatically captured case < 50%

In *DrillEdge* we call a case with a similarity score > 50% a case on the radar. Similarly to Delany [9], we measure the precision of a case by comparing whether the generated cases are on the radar at the same time as expert cases.

6.1   Results

Figure 4 shows the results of our experiments. The plots contain the precision values of the automatically captured mechanical stuck pipes in comparison to expert cases.

Plot (a) contains the results captured for the so called full-width cases and (b) contains the results for cases with a reduced window based on the cluster medoids (see section 5.3).

Each box plot contains the average precision achieved by all 15 automatically captured cases on each of the 8 logs compared to the expert cases. A box contains the median and stretches the second and third quantiles while the whiskers are

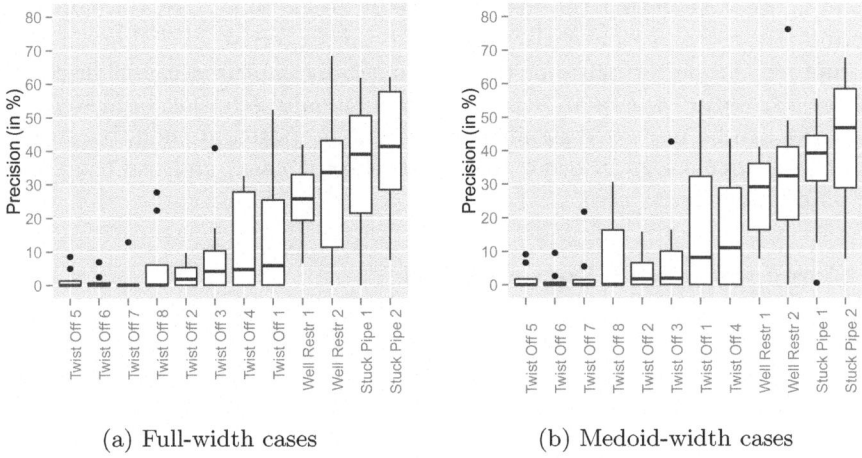

(a) Full-width cases   (b) Medoid-width cases

Fig. 4. Precision for automatically captured cases compared to expert cases on test logs (boxplots with whiskers' maximum of 1.5IQR and ordered by median)

the first and fourth quantile accordingly. The dots show the outliers, if any. Long boxes and whiskers show that there is a higher variance in the precision scores. For example, the comparison against the *stuck pipe 1* expert case reduces the variance as we reduce the width of the cases, while the median stays almost the same. This means that reducing the window produces fewer outliers in the precision scores. Further, the median of (a) and (b) differs as it increases for stuck pipe and well restriction cases and decreases for twist off cases.

Overall, stuck pipe cases have the highest precision, followed by well restriction cases and twist off cases in both scenarios (a) and (b). Also the tendencies for well restriction cases and twist off cases are the same. Reducing the window size has a clear effect on the comparison of the automatically captured stuck pipe cases to the twist off cases: while in (a) their common time on the radar is shorter than in (b). For most of the cases the interquartile range reduces since there's less variance among the cases. Also the maximum values and medians rise slightly while the minimum stays constant.

6.2 Evaluation and Discussion

Furthermore the precision values show that automatically captured stuck pipe cases and expert captured stuck pipe cases share the most time on the radar. A detailed comparison of the similarity score for an automatically captured and an expert case on the logs indicated that the similarity curve – the behavior of the case on the radar – are following a similar signature, but the expert case scores higher similarities and is a shorter time on the radar. This can be explained by the fact that the expert case usually stretches over a smaller window as it can be seen in plot (b) in contrast to plot (f) of figure 3. For this reason, we decided to reduce the size of the window by using the cluster medoids. The plots in figure 4 show that the cases, especially stuck pipe cases, develop in this direction. However, we have also seen that the expertise of a domain expert will improve the performance of the automatically captured cases. In contrast to today's situation where an expert has to manually scan logs and iteratively build cases can be improved significantly by the proposed method, because now we can pre-process logs in order to find stuck-pipe cases and use the expert's time on refining these cases.

7   Conclusion and Future Work

In this paper we have shown how the acquisition of cases in a knowledge intensive application can be supported by event detection and clustering. We introduced a method that automatically scans oil well drilling logs for problematic situations and marks a case window on time series data. We evaluated our approach against domain experts doing the same task and found out that our approach reliably chooses correct areas for a mechanical stuck pipe that can be refined by experts. These findings, however, can reduce the work of knowledge engineers considerably.

As mentioned before, the cases we were mainly looking at shared the types of events the clustering is based on: Stuck pipe cases are identified by the four event types. Twist off cases share these four and used three more while well restriction cases share three common event types and use three different ones. This explains why the captured cases react on those situations as well. In fact, a mechanical stuck pipe, well restriction, and twist off cases share a number of common indicators, but the the additional events included in their detection clearly differentiate them from each other. Overall, the overlap of cases shows that we are indicating the right types of problems.

In the future we will work on generalizing and extending the approach in order to work on various problematic situations as well as providing more elaborate feature weightings. Moreover, our finding should be included in the part of *DrillEdge* that supports knowledge engineers capturing cases.

References

1. Aamodt, A., Plaza, E.: Case-based reasoning: Foundational issues, methodological variations, and system approaches. Artificial Intelligence Communications 7(1), 39–59 (1994)
2. Ankerst, M., Breunig, M.M., Kriegel, H.P., Sander, J.: OPTICS: ordering points to identify the clustering structure. ACM SIGMOD Record 28(2), 49–60 (1999)
3. Arshadi, N., Jurisica, I.: Data mining for case-based reasoning in high-dimensional biological domains. IEEE Trans. on Knowledge and Data Engineering 17(8), 1127–1137 (2005)
4. Bach, K., Althoff, K.-D., Newo, R., Stahl, A.: A case-based reasoning approach for providing machine diagnosis from service reports. In: Ram, A., Wiratunga, N. (eds.) ICCBR 2011. LNCS, vol. 6880, pp. 363–377. Springer, Heidelberg (2011)

5. Bergmann, R., Richter, M.M., Schmitt, S., Stahl, A., Vollrath, I.: Utility-oriented matching: A new research direction for case-based reasoning. In: Schnurr, H.P., Staab, S., Studer, R., Stumme, G., Sure, Y. (eds.) Professionel Knowledge Management (Proc. of the 9th German Workshop on Case-Based Reasoning, GWCBR 2001), pp. 264–274. Shaker-Verlag, Aachen (2001)
6. Cheng, Y.: Mean shift, mode seeking, and clustering. IEEE Transactions on Pattern Analysis and Machine Intelligence 17(8), 790–799 (1995)
7. Cummins, L., Bridge, D.: Maintenance by a committee of experts: The mace approach to case-base maintenance. In: McGinty, L., Wilson, D.C. (eds.) ICCBR 2009. LNCS (LNAI), vol. 5650, pp. 120–134. Springer, Heidelberg (2009)
8. Cummins, L., Bridge, D.: On dataset complexity for case base maintenance. In: Ram, A., Wiratunga, N. (eds.) ICCBR 2011. LNCS (LNAI), vol. 6880, pp. 47–61. Springer, Heidelberg (2011)
9. Delany, S.J.: The good, the bad and the incorrectly classified: Profiling cases for case-base editing. In: McGinty, L., Wilson, D.C. (eds.) ICCBR 2009. LNCS, vol. 5650, pp. 135–149. Springer, Heidelberg (2009)
10. Dufour-Lussier, V., Le Ber, F., Lieber, J., Nauer, E.: Automatic case acquisition from texts for process-oriented case-based reasoning. Information Systems 40 (2014)
11. Ester, M., Kriegel, H.P., Sander, J., Xu, X.: A density-based algorithm for discovering clusters in large spatial databases with noise. In: Second International Conference on Knowledge Discovery and Data Mining, pp. 226–231 (1996)
12. Farley, B.: From free-text repair action messages to automated case generation. In: Proceedings of AAAI 1999 Spring Symposium: AI in Equipment Maintenance and Support (1999)
13. Flinter, S., Keane, M.T.: On the automatic generation of case libraries by chunking chess games. In: Veloso, M., Aamodt, A. (eds.) ICCBR 1995. LNCS, vol. 1010, pp. 421–430. Springer, Heidelberg (1995)
14. Floyd, M.W., Esfandiari, B.: An active approach to automatic case generation. In: McGinty, L., Wilson, D.C. (eds.) ICCBR 2009. LNCS, vol. 5650, pp. 150–164. Springer, Heidelberg (2009)
15. Fornells, A., Armengol, E., Golobardes, E.: Explanation of a clustered case memory organization. In: Proceedings of the 2007 Conference on Artificial Intelligence Research and Development, pp. 153–160. IOS Press, Amsterdam (2007)
16. Fornells, A., Recio-García, J.A., Díaz-Agudo, B., Golobardes, E., Fornells, E.: Integration of a methodology for cluster-based retrieval in jColibri. In: McGinty, L., Wilson, D.C. (eds.) ICCBR 2009. LNCS, vol. 5650, pp. 418–433. Springer, Heidelberg (2009)
17. Gundersen, O.E.: Toward measuring the similarity of complex event sequences in real-time. In: Díaz Adudo, B., Watson, I. (eds.) ICCBR 2012. LNCS, vol. 7466, pp. 107–121. Springer, Heidelberg (2012)
18. Gundersen, O.E., Sørmo, F., Aamodt, A., Skalle, P.: A real-time decision support system for high cost oil-well drilling operations. In: Twenty-Fourth IAAI Conference. AAAI Publications (2012)
19. Luckham, D.C.: The Power of Events: An Introduction to Complex Event Processing in Distributed Enterprise Systems. Addison-Wesley Longman Publishing Co., Inc., Boston (2001)
20. Manzoor, J., Asif, S., Masud, M., Khan, M.J.: Automatic case generation for case-based reasoning systems using genetic algorithms. In: 2012 Third Global Congress on Intelligent Systems, pp. 311–314 (2012)

21. Patterson, D., Rooney, N., Galushka, M., Anand, S.S.: Towards dynamic mainte-nance of retrieval knowledge in cbr. In: Proceedings of the Fifteenth International Florida Artificial Intelligence Research Society Conference, pp. 126–131. AAAI Press (2002)
22. Ram, A.: Continuous case-based reasoning. Artificial Intelligence 90(1-2), 25–77 (1997)
23. Ram, A., Santamaria, J.C.: Multistrategy learning in reactive control systems for autonomous robotic navigation. Informatica 17, 347–369 (1993)
24. Roddick, J.F., Spiliopoulou, M.: A survey of temporal knowledge discovery paradigms and methods. IEEE Transactions on Knowlegde and Data Engineer-ing 14(4), 750–767 (2002)
25. Smyth, B., Bonzano, A., Cunningham, P.: Using introspective learning to improve retrieval in CBR: A case study in air traffic control. In: Leake, D.B., Plaza, E. (eds.) ICCBR 1997. LNCS, vol. 1266, pp. 291–302. Springer, Heidelberg (1997)
26. Smyth, B., Keane, M.T.: Remembering To Forget A Competence-Preserving Case Deletion Policy for Case-Based Reasoning Systems, pp. 377–383. Springer, Heidel-berg (1995)
27. Smyth, B., McKenna, E.: Modelling the Competence of Case-Bases. In: Smyth, B., Cunningham, P. (eds.) EWCBR 1998. LNCS (LNAI), vol. 1488, pp. 208–220. Springer, Heidelberg (1998)
28. Smyth, B., McKenna, E.: Building Compact Competent Case-Bases. In: Althoff, K.-D., Bergmann, R., Branting, L.K. (eds.) ICCBR 1999. LNCS (LNAI), vol. 1650, pp. 329–342. Springer, Heidelberg (1999)
29. Steinhaus, H.: Sur la division des corps matériels en parties. Bull. Acad. Pol. Sci., Cl. III 4, 801–804 (1957)
30. Vernet, D., Golobardes, E.: An unsupervised learning approach for case-based clas-sifier systems. Expert Update. The Specialist Group on Artificial Intelligence 6(2), 37–42 (2003)
31. Zhang, H., Song, K.: Research and experiment on Affinity Propagation clustering algorithm. In: 2011 Second International Conference on Mechanic Automation and Control Engineering, pp. 5996–5999 (2011)
32. Zhou, H., Wang, P., Li, H.: Research on Adaptive Parameters Determination in DBSCAN Algorithm. Journal of Information and Computational Science 7, 1967–1973 (2012)
33. Zhu, J., Yang, Q., Columbia, B.: Remembering to Add: Competence-preserving Case-Addition Policies for Case-Base Maintenance Case-deletion Policies, pp. 234–241. Morgan Kaufmann Publishers Inc., San Francisco (1999)

Algorithm for Adapting Cases Represented in a Tractable Description Logic

Liang Chang[1,2], Uli Sattler[2], and Tianlong Gu[1]

[1] Guangxi Key Laboratory of Trusted Software,
Guilin University of Electronic Technology, Guilin, 541004, China
[2] School of Computer Science,
The University of Manchester, Manchester, M13 9PL, UK
changl.guet@gmail.com,sattler@cs.manchester.ac.uk,cctlgu@guet.edu.cn

Abstract. Case-based reasoning (CBR) based on description logics (DLs) has gained a lot of attention lately. Adaptation is a basic task in CBR that can be modeled as a knowledge base revision problem which has been solved in propositional logic. However, in DLs, adaptation is still a challenge problem since existing revision operators only work well for DLs of the *DL-Lite* family. It is difficult to design revision algorithms that are syntax-independent and fine-grained. In this paper, we propose a new method for adaptation based on the tractable DL $\mathcal{EL}_\perp$. Following the idea of adaptation as revision, we firstly extend the logical basis for describing cases from propositional logic to DL and present a formalism for adaptation based on $\mathcal{EL}_\perp$. Then we show that existing revision operators and algorithms in DLs can not be used for this formalism. Finally we present our adaptation algorithm. Our algorithm is syntax-independent and fine-grained, and satisfies the requirements on revision operators.

Keywords: description logic, case-based reasoning, adaptation, knowledge base revision, $\mathcal{EL}_\perp$.

1 Introduction

Description logic (DL) is a family of logics for representing and reasoning about knowledge of static application domains [2]. It is playing a central role in the Semantic Web, serving as the basis of the W3C-recommended Web ontology language OWL [11]. The main strength of DLs is that they offer considerable expressive power often going far beyond propositional logic, while reasoning is still decidable. Furthermore, DLs have well-defined semantics and are supported by many efficient reasoners.

In the last few years, there has been a growing interest in bringing the power and character of DLs into case-based reasoning (CBR) [9,10,18]. CBR is a type of analogical reasoning in which a new problem is solved by reusing past experiences called *source cases*. There are two basic tasks in the CBR inference: retrieval and adaptation. *Retrieval* aims at selecting a source case that is similar to the new problem according to some similarity criterion. *Adaptation* aims at

L. Lamontagne and E. Plaza (Eds.): ICCBR 2014, LNCS 8765, pp. 63–78, 2014.
© Springer International Publishing Switzerland 2014

generating a solution for the new problem by adapting the solution contained in the source case. At present, most research is concerned with the retrieval task when introducing DLs into CBR [9,18].

In comparison to retrieval, adaptation is often considered to be the more difficult task. One approach for this task is to model the adaptation process as the *revision* problem of a knowledge base (KB) [7,17]; it is hoped that an adaptation algorithm satisfies the AGM postulates on revision operators [1,13]. In propositional logic, there are many revision operators which satisfy the AGM postulates and can be applied to complete the adaptation task [17]. However, in DLs, it is difficult to design revision operators and algorithms that satisfy the AGM postulates [4]. Especially, it is a great challenge to design revision algorithms that are independent of the syntactical forms of KBs and are fine-grained for the minimal change principle.

According to the semantics adopted for defining "minimal change", existing revision operators and algorithms for DLs can be divided into two groups: *model-based approaches* (MBAs) [14] and *formula-based approaches* (FBAs) [4,16,20]. In MBAs, the semantics of minimal change is defined by measuring the distance between models. MBAs are syntax-independent and fine-grained, but at present they only work for DLs of the *DL-Lite* family [14]. In FBAs, the semantics of minimal change is reflected in the minimality of formulas removed by the revision process. There are two FBAs in the literature. One is based on the deductive closure of a KB [4,16]; it is syntax-independent and fine-grained, but again only works for *DL-Lite*. Another is based on justifications [20]; although it is applicable to DLs such as $\mathcal{SHOIN}$, it is syntax-dependent and not fine-grained.

DLs of the $\mathcal{EL}$ family are popular for building large-scale ontologies [3]. Some important medical ontologies and life science ontologies are built in $\mathcal{EL}$, such as the SNOMED CT [19] and the Gene Ontology [8]. A feature of this family of DLs is that they allow for reasoning in polynomial time, while being able to describe "relational structures". They are promising DLs for CBR since they are, on the one hand, of interesting expressive power and, on the other hand, restricted enough so that we can hope for a practical adaptation approach.

In the literature, some good results on combining DLs of the $\mathcal{EL}$ family with the retrieval task of CBR have been presented [18]; many algorithms for measuring the similarity of concepts in these DLs have also been proposed [15]. However, adaptation based on these DLs is still an open problem. One reason is that existing revision operators applicable to these DLs, to the best of our knowledge, are syntax-dependent and not fine-grained.

In this paper we present a new method for adaptation in the DL $\mathcal{EL}_\perp$ of the $\mathcal{EL}$ family. Our contributions regard three aspects. Firstly, we extend the logical basis for describing cases from propositional logic to the DL $\mathcal{EL}_\perp$, with a powerful way of describing cases as ABoxes in DL. Secondly, we extend the "adaptation as KB revision" view from [17] to the above setting and get a formalism for adaptation based on $\mathcal{EL}_\perp$. Finally, for the adaptation setting we provide an adaptation algorithm which is syntax-independent and fine-grained. The proofs of all of our technical results are given in the accompanying technical report [5].

2   The Description Logic $\mathcal{EL}_\perp$

The DL $\mathcal{EL}_\perp$ extends $\mathcal{EL}$ with bottom concept (and consequently disjointness statements) [3]. Let N_C, N_R and N_I be disjoint sets of *concept names*, *role names* and *individual names*, respectively. $\mathcal{EL}_\perp$-*concepts* are built according to the following syntax rule $C ::= \top \mid \perp \mid A \mid C \sqcap D \mid \exists r.C$, where $A \in N_C$, $r \in N_R$, and C, D range over $\mathcal{EL}_\perp$-concepts.

A *TBox* $\mathcal{T}$ is a finite set of *general concept inclusions* (GCIs) of the form $C \sqsubseteq D$, where C and D are concepts. An *ABox* $\mathcal{A}$ is a finite set of *concept assertions* of the form $C(a)$ and *role assertions* of the form $r(a, b)$, where $a, b \in N_I$, $r \in N_R$, and C is a concept. A *knowledge base* (KB) is a pair $\mathcal{K} = \langle \mathcal{T}, \mathcal{A} \rangle$.

Example 1. Consider the example on breast cancer treatment discussed in [17]. We add some background knowledge to it and describe the knowledge by the following GCIs in a TBox $\mathcal{T}$:

$Tamoxifen \sqsubseteq Anti\text{-}oestrogen,$ $Anti\text{-}aromatases \sqsubseteq Anti\text{-}oestrogen,$

$Tamoxifen \sqsubseteq \exists metabolizedTo.(Compounds \sqcap \exists bindto.OestrogenReceptor),$

$(\exists hasGene.CYP2D6) \sqcap (\exists TreatBy.Tamoxifen) \sqsubseteq \perp.$

These GCIs state that both tamoxifen and anti-aromatases are anti-oestrogen; tamoxifen can be metabolized into compounds which will bind to the oestrogen receptor; and tamoxifen is contraindicated for people with some gene of CYP2D6.

Suppose $Mary$ is a patient with a gene of the class CYP2D6 and with some symptom captured by a concept $Symp$. Then we can describe these information by an ABox $\mathcal{N}_{pb} = \{Symp(Mary), \exists hasGene.CYP2D6(Mary)\}$.   □

The semantics of $\mathcal{EL}_\perp$ is defined by an *interpretation* $\mathcal{I} = (\Delta^\mathcal{I}, \cdot^\mathcal{I})$, where the *interpretation domain* $\Delta^\mathcal{I}$ is a non-empty set composed of individuals, and $\cdot^\mathcal{I}$ is a function which maps each concept name $A \in N_C$ to a set $A^\mathcal{I} \subseteq \Delta^\mathcal{I}$, maps each role name $r \in N_R$ to a binary relation $r^\mathcal{I} \subseteq \Delta^\mathcal{I} \times \Delta^\mathcal{I}$, and maps each individual name $a \in N_I$ to an individual $a^\mathcal{I} \in \Delta^\mathcal{I}$. The function $\cdot^\mathcal{I}$ is inductively extended to arbitrary concepts as follows: $\top^\mathcal{I} := \Delta^\mathcal{I}$, $\perp^\mathcal{I} := \emptyset$, $(C \sqcap D)^\mathcal{I} := C^\mathcal{I} \cap D^\mathcal{I}$, and $(\exists r.C)^\mathcal{I} := \{x \in \Delta^\mathcal{I} \mid$ there exists $y \in \Delta^\mathcal{I}$ such that $(x, y) \in r^\mathcal{I}$ and $y \in C^\mathcal{I}\}$.

The *satisfaction relation* "$\models$" between any interpretation $\mathcal{I}$ and any GCI $C \sqsubseteq D$, concept assertion $C(a)$, role assertion $r(a, b)$, TBox $\mathcal{T}$ or ABox $\mathcal{A}$ is defined inductively as follows: $\mathcal{I} \models C \sqsubseteq D$ iff $C^\mathcal{I} \subseteq D^\mathcal{I}$; $\mathcal{I} \models C(a)$ iff $a^\mathcal{I} \in C^\mathcal{I}$; $\mathcal{I} \models r(a, b)$ iff $(a^\mathcal{I}, b^\mathcal{I}) \in r^\mathcal{I}$; $\mathcal{I} \models \mathcal{T}$ iff $\mathcal{I} \models X$ for every $X \in \mathcal{T}$; and $\mathcal{I} \models \mathcal{A}$ iff $\mathcal{I} \models X$ for every $X \in \mathcal{A}$.

$\mathcal{I}$ is a *model* of a KB $\mathcal{K} = \langle \mathcal{T}, \mathcal{A} \rangle$ if $\mathcal{I} \models \mathcal{T}$ and $\mathcal{I} \models \mathcal{A}$. We use $mod(\mathcal{K})$ to denote the set of models of a KB $\mathcal{K}$. Two KBs $\mathcal{K}_1$ and $\mathcal{K}_2$ are *equivalent* (written $\mathcal{K}_1 \equiv \mathcal{K}_2$) iff $mod(\mathcal{K}_1) = mod(\mathcal{K}_2)$.

There are many inference problems on DLs. Here we only introduce *consistency* and *entailment*. A KB $\mathcal{K} = \langle \mathcal{T}, \mathcal{A} \rangle$ is *consistent* (or $\mathcal{A}$ is *consistent* w.r.t. $\mathcal{T}$) if $mod(\mathcal{K}) \neq \emptyset$. $\mathcal{K}$ *entails* a GCI, assertion or ABox X (written $\mathcal{K} \models X$) if $\mathcal{I} \models X$ for every $\mathcal{I} \in mod(\mathcal{K})$. For these two inference problems the following results hold: a KB $\mathcal{K}$ is inconsistent iff $\mathcal{K} \models \top \sqsubseteq \bot$ iff $\mathcal{K} \models \bot(a)$ for some individual name a occurring in $\mathcal{K}$. If $\mathcal{K}$ is inconsistent, then we say that $\mathcal{K}$ *entails a clash*.

Example 2. Consider the TBox $\mathcal{T}$ and ABox $\mathcal{N}_{pb}$ presented in Example 1. Suppose there is an ABox $\mathcal{A}_{sol} = \{TreatBy(Mary, y), Tamoxifen(y)\}$ and a KB $\mathcal{K} = \langle \mathcal{T}, \mathcal{A}_{sol} \cup \mathcal{N}_{pb} \rangle$. It is obvious that $\mathcal{K}$ is inconsistent and entails a clash. More precisely, we have that $\mathcal{K} \models \top \sqsubseteq \bot$ and $\mathcal{K} \models \bot(Mary)$.   □

For any TBox, ABox or KB X, we use N_C^X (resp., N_R^X, N_I^X) to denote the set of concept names (resp., role names, individual names) occurring in X, and define the *signature* of X as $sig(X) = N_C^X \cup N_R^X \cup N_I^X$.

For any concept C, the *role depth* $rd(C)$ is the maximal nesting depth of "$\exists$" in C. For any TBox or ABox X, let $sub(X) = \bigcup_{C \sqsubseteq D \in X} \{C, D\}$ if X is a TBox, and $sub(X) = \{C \mid C(a) \in X\}$ if X is an ABox, then the *depth* of X is defined as $depth(X) = \max\{rd(C) \mid C \in sub(X)\}$.

3   Formalization of Adaptation Based on $\mathcal{EL}_\bot$

In this section we present a formalism for adaptation based on $\mathcal{EL}_\bot$. There are many different approaches for the formalization of adaptation in CBR. Here we follow the approach presented in [17] to formulate adaptation as knowledge base revision, with the difference that our formalism is based on the DL $\mathcal{EL}_\bot$ instead of propositional logic.

The basic idea of CBR is to solve similar problems with similar solutions. The new problem that needs to be solved is called *target problem*. The problems which have been solved and stored are called *source problems*. Each source problem *pb* has a *solution sol*, and the pair (*pb, sol*) is called a *source case*. A finite set of source cases forms a *case base*. Given a target problem *tgt*, the retrieval step of CBR will pick out a source case (*pb, sol*) according to the similarity between target problem and source problems, then the adaptation step will generate a solution *sol_{tgt}* for *tgt* by adapting *sol*.

In [17], the adaptation process is modeled as KB revision in propositional logic. More precisely, let two formulas $kb_1 = pb \wedge sol$ and $kb_2 = tgt$, then the solution sol_{tgt} is generated by calculating $(dk \wedge kb_1) \circ (dk \wedge kb_2)$, where dk is a formula describing the domain knowledge, and $\circ$ is a revision operator that satisfies the AGM postulates in propositional logic.

In our paper, after introducing the DL $\mathcal{EL}_\bot$ into CBR, knowledge in a CBR system is composed of three parts:

- the *domain or background knowledge* which is represented as a TBox $\mathcal{T}$;
- the knowledge about *case base* in which each *source case* is described by an ABox $\mathcal{A} = \mathcal{A}_{pb} \cup \mathcal{A}_{sol}$, where $\mathcal{A}_{pb}$ describes the *source problem* and $\mathcal{A}_{sol}$ describes the *solution*;

– the knowledge about *target problem* described by an ABox $\mathcal{N}_{pb}$.

With such a framework, given a target problem $\mathcal{N}_{pb}$, we can make use of similarity-measuring algorithms presented in the literature [18,15] to select a source case $\mathcal{A} = \mathcal{A}_{pb} \cup \mathcal{A}_{sol}$ such that, by treating individual names occurring in *srce* as variables, there exists a substitution σ such that $\sigma(\mathcal{A}_{pb})$ and $\mathcal{N}_{pb}$ has the maximum similarity. The retrieval algorithm will applies σ on $\mathcal{A}_{sol}$ and return $\sigma(\mathcal{A}_{sol})$ as a *possible solution* for the target problem.

Since retrieval algorithm is not the topic of this paper, we do not discuss it in detail here. In the rest of this paper, we will omit the notation σ and use $\mathcal{A}_{sol}$ directly to denote the possible solution returned by the retrieval algorithm.

Now suppose a possible solution has been returned by the retrieval algorithm, we define adaptation setting as follows.

Definition 1. *An* adaptation setting *based on* $\mathcal{EL}_\perp$ *is a triple* $\mathcal{AS} = (\mathcal{T}, \mathcal{A}_{sol}, \mathcal{N}_{pb})$, *where* $\mathcal{T}$ *is a TBox describing the domain knowledge of the CBR system,* $\mathcal{N}_{pb}$ *is an ABox describing the target problem, and* $\mathcal{A}_{sol}$ *is an ABox describing the possible solution returned by the retrieval algorithm.*

An ABox $\mathcal{A}'$ *is a* solution case *for an adaptation setting* $\mathcal{AS} = (\mathcal{T}, \mathcal{A}_{sol}, \mathcal{N}_{pb})$ *if* $sig(\mathcal{A}') \subseteq sig(\mathcal{T}) \cup sig(\mathcal{A}_{sol}) \cup sig(\mathcal{N}_{pb})$ *and the following statements hold:*

(R1) $\langle \mathcal{T}, \mathcal{A}' \rangle \models \mathcal{N}_{pb}$;
(R2) $\mathcal{A}' = \mathcal{A}_{sol} \cup \mathcal{N}_{pb}$ *if* $\mathcal{A}_{sol} \cup \mathcal{N}_{pb}$ *is consistent w.r.t.* $\mathcal{T}$*; and*
(R3) *if* $\mathcal{N}_{pb}$ *is consistent w.r.t.* $\mathcal{T}$ *then* $\mathcal{A}'$ *is also consistent w.r.t.* $\mathcal{T}$.

There may be more than one solution cases for an adaptation setting. From these solution cases, the user will select the best one and get a solution for the target problem.

The adaptation setting defined above is similar to the *instance-level revision* based on DLs [4]; **R1-R3** are just the basic requirements specified by the AGM postulates on revision operators [1,13]. More precisely, **R1** specifies that a revision result must entail the new information $\mathcal{N}_{pb}$; **R2** states that the revision operator should not change the KB $\langle \mathcal{T}, \mathcal{A}_{sol} \cup \mathcal{N}_{pb} \rangle$ if there is no conflict; **R3** states that the revision operator must preserve the consistency of KBs.

From the point of view of adaptation, these requirements on solutions are explained as follows [17]. If **R1** is violated, then it means that the adaptation process failed to solve the target problem. **R2** states that if the possible solution does not contradict the target problem w.r.t. the background knowledge, then it can be applied directly to the target problem. **R3** states that whenever the description of the target problem is consistent w.r.t. the domain knowledge, the adaptation process provides satisfiable result.

In the literature, there exist many revision operators and algorithms that can generate revision results satisfying the above requirements [4,14,16,20]. However, in practice, besides the necessary requirements specified by the definition, we hope that the adaptation algorithm satisfies two more requirements.

Firstly, the adaptation algorithm should be syntax-independent. Especially, if two target problems are logically equivalent w.r.t. the domain knowledge, then they should have the same solution. This requirement is formalized as follows:

(R4) Let $\mathcal{AS}_1 = (\mathcal{T}, \mathcal{A}_{sol_1}, \mathcal{N}_{pb_1})$ and $\mathcal{AS}_2 = (\mathcal{T}, \mathcal{A}_{sol_2}, \mathcal{N}_{pb_2})$ be two adaptation settings with $\langle \mathcal{T}, \mathcal{A}_{sol_1} \rangle \equiv \langle \mathcal{T}, \mathcal{A}_{sol_2} \rangle$ and $\langle \mathcal{T}, \mathcal{N}_{pb_1} \rangle \equiv \langle \mathcal{T}, \mathcal{N}_{pb_2} \rangle$. If $\mathcal{A}'_1$ is a solution case for $\mathcal{AS}_1$, then there must be a solution case $\mathcal{A}'_2$ for $\mathcal{AS}_2$ such that $\langle \mathcal{T}, \mathcal{A}'_1 \rangle \equiv \langle \mathcal{T}, \mathcal{A}'_2 \rangle$.

Secondly, the adaptation algorithm should guarantee a minimal change so that the experience contained in the solution of source cases is preserved as much as possible. Without such a requirement, given an adaptation setting $\mathcal{AS} = (\mathcal{T}, \mathcal{A}_{sol}, \mathcal{N}_{pb})$, if $\mathcal{A}_{sol} \cup \mathcal{N}_{pb}$ is inconsistent w.r.t. $\mathcal{T}$, then the ABox $\mathcal{N}_{pb}$ itself is a solution case by the definition. However, it is obvious that $\mathcal{N}_{pb}$ does not contain any information on solution, and all the experiences contained in the solution $\mathcal{A}_{sol}$ of the source case are completely lost.

We hope to specify the requirement on minimal change formally. However, it is non-trivial to do it in a framework based on DLs. Furthermore, it is well-accepted that there is no general notion of minimality that will "do the right thing" under all circumstances [4]. Therefore, under the framework of adaptation setting, we only specify this requirement informally as follows:

(R5) If $\mathcal{A}'$ is a solution case for the adaptation setting $\mathcal{AS} = (\mathcal{T}, \mathcal{A}_{sol}, \mathcal{N}_{pb})$, then the change from the KB $\langle \mathcal{T}, \mathcal{A}_{sol} \rangle$ to the KB $\langle \mathcal{T}, \mathcal{A}' \rangle$ is minimal.

To sum up, given an adaptation setting, we hope to generate solution cases which not only satisfy **R1-R3** specified by Definition 1, but also satisfy **R4** and "some reading" of **R5**.

Before the end of this section, we look an example of adaptation setting.

Example 3. Consider the TBox $\mathcal{T}$ and ABox $\mathcal{N}_{pb}$ presented in Example 1. Suppose $\mathcal{N}_{pb}$ is a description of the target problem. Suppose many successful treatment cases have been recorded in the case base, and from them a possible solution $\mathcal{A}_{sol} = \{TreatBy(Mary, y), Tamoxifen(y)\}$ is returned by the retrieval algorithm. Then we get an adaptation setting $\mathcal{AS} = (\mathcal{T}, \mathcal{A}_{sol}, \mathcal{N}_{pb})$. □

4   Existing Approaches to Instance-Level Revision

As we mentioned in Section 1, there are two groups of revision operators and algorithms for DLs in the literature. In this section, we show that they either do not support the DL $\mathcal{EL}_\perp$ or do not satisfy **R4** and **R5**.

4.1   Model-based Approaches

MBAs define revision operators over the distance between interpretations [14]. Under the framework of adaptation setting, suppose $\mathcal{A}'_i$ $(1 \leq i \leq n)$ are all the solution cases for $\mathcal{AS} = (\mathcal{T}, \mathcal{A}_{sol}, \mathcal{N}_{pb})$, and let $\mathcal{M} = \bigcup_{1 \leq i \leq n} mod(\langle \mathcal{T}, \mathcal{A}'_i \rangle)$, then, with MBAs, $\mathcal{M}$ should satisfy the following equation:

$$\mathcal{M} = \{\mathcal{J} \in mod(\langle \mathcal{T}, \mathcal{N}_{pb} \rangle) \mid there \ exists \ \mathcal{I} \in mod(\langle \mathcal{T}, \mathcal{A}_{sol} \rangle) \ with \ dist(\mathcal{I}, \mathcal{J}) = \min\{dist(\mathcal{I}', \mathcal{J}') \mid \mathcal{I}' \in mod(\langle \mathcal{T}, \mathcal{A}_{sol} \rangle), \mathcal{J}' \in mod(\langle \mathcal{T}, \mathcal{N}_{pb} \rangle)\} \}.$$

With MBAs, we can firstly calculate the set $\mathcal{M}$ by the above equation, and then construct the ABoxes $\mathcal{A}'_i$ $(1 \leq i \leq n)$ correspondingly.

In propositional logic, since each interpretation is only a truth assignment on propositional symbols, it is not difficult to measure the distance between interpretations and to calculate the set of models according to the distance [17]. However, it becomes very complex for DL, since basic symbols of DL are concept names, role names and individual names which are interpreted with set theory.

Let Σ be the set of concept names and role names occurring in $\mathcal{AS}$. There are four different approaches for measuring the distance $dist(\mathcal{I}, \mathcal{J})$:

- $dist^s_{\#}(\mathcal{I}, \mathcal{J}) = \#\{X \in \Sigma \mid X^{\mathcal{I}} \neq X^{\mathcal{J}}\}$,
- $dist^s_{\subseteq}(\mathcal{I}, \mathcal{J}) = \{X \in \Sigma \mid X^{\mathcal{I}} \neq X^{\mathcal{J}}\}$,
- $dist^a_{\#}(\mathcal{I}, \mathcal{J}) = \underset{X \in \Sigma}{sum} \, \#(X^{\mathcal{I}} \ominus X^{\mathcal{J}})$,
- $dist^a_{\subseteq}(\mathcal{I}, \mathcal{J}, X) = X^{\mathcal{I}} \ominus X^{\mathcal{J}}$ for every $X \in \Sigma$,

where $X^{\mathcal{I}} \ominus X^{\mathcal{J}} = (X^{\mathcal{I}} \setminus X^{\mathcal{J}}) \cup (X^{\mathcal{J}} \setminus X^{\mathcal{I}})$. Distances under $dist^s_{\#}$ and $dist^a_{\#}$ are natural numbers and are compared in the standard way. Distances under $dist^s_{\subseteq}$ are sets and are compared by set inclusion. Distances under $dist^a_{\subseteq}$ are compared as follows: $dist^a_{\subseteq}(\mathcal{I}_1, \mathcal{J}_1) \leq dist^a_{\subseteq}(\mathcal{I}_2, \mathcal{J}_2)$ iff $dist^a_{\subseteq}(\mathcal{I}_1, \mathcal{J}_1, X) \subseteq dist^a_{\subseteq}(\mathcal{I}_2, \mathcal{J}_2, X)$ for every $X \in \Sigma$. It is assumed that all models have the same interpretation domain and the same interpretation on individual names. In [14], the above four different semantics for MBAs are denoted as $\mathcal{G}^s_{\#}$, $\mathcal{G}^s_{\subseteq}$, $\mathcal{G}^a_{\#}$, and $\mathcal{G}^a_{\subseteq}$ respectively.

The limitation of applying these MBAs in our adaptation setting is shown by the following example.

Example 4. Consider an adaptation setting $\mathcal{AS}_1 = (\mathcal{T}_1, \mathcal{A}_1, \mathcal{N}_1)$, where

$$\mathcal{T}_1 = \{A \sqsubseteq \exists R.A, \; A \sqsubseteq C, \; E \sqcap \exists R.A \sqsubseteq \bot\}, \quad \mathcal{A}_1 = \{A(a)\}, \quad \mathcal{N}_1 = \{E(a)\}.$$

Firstly, by applying MBAs with the semantics $\mathcal{G}^s_{\subseteq}$ and $\mathcal{G}^s_{\#}$, we can calculate the set $\mathcal{M}$ and get the following result:

$$\mathcal{M} = \{\mathcal{J} \in mod(\langle \mathcal{T}_1, \mathcal{N}_1 \rangle) \mid \; there \; exists \; \mathcal{I} \in mod(\langle \mathcal{T}_1, \mathcal{A}_1 \rangle) \; such \; that$$
$$C^{\mathcal{I}} = C^{\mathcal{J}} \; and \; R^{\mathcal{I}} = R^{\mathcal{J}}\}.$$

We can show that there does not exist a finite number of ABoxes $\mathcal{A}'_i$ $(1 \leq i \leq n)$ such that $sig(\mathcal{A}'_i) \subseteq sig(\mathcal{T}_1) \cup sig(\mathcal{A}_1) \cup sig(\mathcal{N}_1)$ and $\underset{1 \leq i \leq n}{\bigcup} mod(\langle \mathcal{T}_1, \mathcal{A}'_i \rangle)$ $= \mathcal{M}$. Interested readers can refer to [6] for further details. Therefore, MBAs under the semantics $\mathcal{G}^s_{\subseteq}$ and $\mathcal{G}^s_{\#}$ suffer from inexpressibility.

Secondly, by applying MBAs with the semantics $\mathcal{G}^a_{\subseteq}$ and $\mathcal{G}^a_{\#}$, we will get

$$\mathcal{M} = \{\mathcal{J} \in mod(\langle \mathcal{T}_1, \mathcal{N}_1 \rangle) \mid there \; exists \; \mathcal{I} \in mod(\langle \mathcal{T}_1, \mathcal{A}_1 \rangle) \; such \; that$$
$$A^{\mathcal{I}} \ominus A^{\mathcal{J}} = E^{\mathcal{I}} \ominus E^{\mathcal{J}} = \{a^{\mathcal{I}}\}, C^{\mathcal{I}} = C^{\mathcal{J}} \; and \; R^{\mathcal{I}} = R^{\mathcal{J}}\}.$$

From $\mathcal{M}$ we can construct an ABox $\mathcal{A}'_1 = \{E(a), C(a), R(a, a)\}$ which satisfies $mod(\langle \mathcal{T}_1, \mathcal{A}'_1 \rangle) = \mathcal{M}$. Therefore, under the semantics $\mathcal{G}^a_{\subseteq}$ and $\mathcal{G}^a_{\#}$, $\mathcal{A}'_1$ is the only solution case. This result is very strange, since during the adaptation process there seems to be no "good" reason to enforce the assertion $R(a, a)$ to hold, and to exclude other possible assertions such as $\exists R.A(a)$. $\qquad \square$

To sum up, there are four notions of computing models in existing MBAs. For the adaptation based on $\mathcal{EL}_\perp$, two notions suffer from inexpressibility and the other two notions only generate solution cases which are counterintuitive.

4.2   Formula-Based Approaches

In the literature there are two typical formula-based approaches for instance-level revision in DLs.

The first one is based on deductive closures [4,16]. With this approach, given an adaptation setting $\mathcal{AS} = (\mathcal{T}, \mathcal{A}, \mathcal{N})$, we will firstly calculate the deductive closure of $\mathcal{A}$ w.r.t. $\mathcal{T}$ (denoted $cl_\mathcal{T}(\mathcal{A})$), then find a maximal subset $\mathcal{A}_m$ of $cl_\mathcal{T}(\mathcal{A})$ that does not conflict with $\mathcal{N}$ and $\mathcal{T}$, and finally return $\mathcal{A}_m \cup \mathcal{N}$ as a solution case. This approach works for restricted forms of *DL-Lite*, by assuming that $\mathcal{A}$ only contains assertions of the form $A(a)$, $\exists R(a)$ and $R(a,b)$, with A and R concept names or role names, and therefore $cl_\mathcal{T}(\mathcal{A})$ is finite and can be calculated effectively. However, it does not work for $\mathcal{EL}_\perp$, since in our adaptation setting any $\mathcal{EL}_\perp$ concept can be used for describing assertions in the ABox $\mathcal{A}$.

The second FBA is based on justifications (also known as MinAs or kernel) [20]. With this approach, given an adaptation setting $\mathcal{AS} = (\mathcal{T}, \mathcal{A}, \mathcal{N})$, we will firstly construct a KB $\mathcal{K}_0 = \langle \mathcal{T}, \mathcal{A} \cup \mathcal{N} \rangle$ and find all the minimal subsets of $\mathcal{K}_0$ that entail a clash (i.e., all justifications for clashes); then we will compute a minimal set $\mathcal{R} \subseteq \mathcal{A}$ that contains at least one element from each justification (such a set is also called a *repair*); finally we will return $(\mathcal{A} \setminus \mathcal{R}) \cup \mathcal{N}$ as a solution case. This approach is applicable to DLs such as $\mathcal{SHOIN}$ and therefore can deal with $\mathcal{EL}_\perp$. However, as shown by the following examples, it is syntax-dependent and not fine-grained, and therefore does not satisfy **R4** and **R5**.

Example 5. Consider the adaptation setting $\mathcal{AS}_1 = (\mathcal{T}_1, \mathcal{A}_1, \mathcal{N}_1)$ described in the previous example. It is obvious that $\langle \mathcal{T}_1, \mathcal{A}_1 \cup \mathcal{N}_1 \rangle \models \perp(a)$ and for which there is only one justification $\mathcal{J} = \{A \sqsubseteq \exists R.A, E \sqcap \exists R.A \sqsubseteq \perp, A(a), E(a)\}$. Therefore the only repair is $\mathcal{R} = \{A(a)\}$ and the only solution case is $\mathcal{A}'_1 = (\mathcal{A}_1 \setminus \mathcal{R}) \cup \mathcal{N}_1 = \mathcal{N}_1 = \{E(a)\}$.

This result is not good, since $\mathcal{A}'_1$ only contains the description of target problem and does not contain any information on solution. All the experiences provided by the solution $\mathcal{A}_1$ of source case are completely lost, therefore this approach is not fine-grained and does not satisfy **R5**. $\square$

Example 6. Consider another adaptation setting $\mathcal{AS}_2 = (\mathcal{T}_2, \mathcal{A}_2, \mathcal{N}_2)$, where $\mathcal{T}_2 = \mathcal{T}_1$, $\mathcal{N}_2 = \mathcal{N}_1$ and $\mathcal{A}_2 = \{A(a), C(a), \exists R.C(a)\}$. Apply the FBA based on justifications again, we will get a solution case $\mathcal{A}'_2 = \{E(a), C(a), \exists R.C(a)\}$.

It is obvious that $\langle \mathcal{T}_2, \mathcal{A}_2 \rangle \equiv \langle \mathcal{T}_1, \mathcal{A}_1 \rangle$ but $\langle \mathcal{T}_2, \mathcal{A}'_2 \rangle \not\equiv \langle \mathcal{T}_1, \mathcal{A}'_1 \rangle$. In other words, for the same target problem, we get two totally different solutions from two descriptions of experiences that are syntactically different but logically equivalent. The reason is that this approach does not satisfy **R4**. $\square$

To sum up, for the adaptation based on $\mathcal{EL}_\perp$, existing FBAs either can not be applied directly, or can be applied but is syntax-dependent and not fine-grained.

5   Our Approach for Adaptation Based on $\mathcal{EL}_\bot$

In this section we present an algorithm for adaptation based on $\mathcal{EL}_\bot$. Given an adaptation setting $\mathcal{AS} = (\mathcal{T}, \mathcal{A}, \mathcal{N})$, our algorithm will firstly construct a non-redundant depth-bounded model for the KB $\langle \mathcal{T}, \mathcal{A} \rangle$; then a revision process based on justifications will be carried out on this model by treating a model as a set of assertions; finally the resulting model will be mapped back to an ABox which will be returned as a solution case.

Our algorithm is based on a structure named revision graph, which is close to the completion graph used in classical tableau decision algorithms of DLs [12]. We firstly introduce some notions and operations on this structure and then present the algorithm.

5.1   Notions and Operations on Revision Graph

First, we consider a set N_V of *variables*, and extend interpretations $\mathcal{I}$ of $\mathcal{EL}_\bot$ to interpret these variables just like individual names.

A *revision graph* for $\mathcal{EL}_\bot$ is a directed graph $\mathcal{G} = (V, E, \mathcal{L})$, where

- V is a finite set of nodes composed of individual names and variables;
- $E \subseteq V \times V$ is a set of edges satisfying:
    - there is no edge from variables to individual names, and
    - for each variable $y \in V$, there is at most one node x with $\langle x, y \rangle \in E$;
- each node $x \in V$ is labelled with a set of concepts $\mathcal{L}(x)$; and
- each edge $\langle x, y \rangle \in E$ is labelled with a set of role names $\mathcal{L}(\langle x, y \rangle)$; furthermore, if y is a variable then $\sharp \mathcal{L}(\langle x, y \rangle) = 1$.

For each edge $\langle x, y \rangle \in E$, we call y a *successor* of x and x a *predecessor* of y. *Descendant* is the transitive closure of successor.

For any node $x \in V$, we use $level(x)$ to denote the level of x in the graph, and define it inductively as follows: $level(x) = 0$ if x is an individual name, $level(x) = level(y) + 1$ if x is a variable with a predecessor y, and $level(x) = +\infty$ if x is a variable without predecessor.

A graph $\mathcal{B} = (V', E', \mathcal{L}')$ is a *branch* of $\mathcal{G}$ if $\mathcal{B}$ is a tree and a subgraph of $\mathcal{G}$.

A branch $\mathcal{B}_1 = (V_1, E_1, \mathcal{L}_1)$ is *subsumed* by another branch $\mathcal{B}_2 = (V_2, E_2, \mathcal{L}_2)$ if $\mathcal{B}_1$ and $\mathcal{B}_2$ have the same root node, $\sharp(V_1 \cap V_2) = 1$, and there is a function $f : V_1 \to V_2$ such that: $f(x) = x$ if x is the root node, $\mathcal{L}_1(x) \subseteq \mathcal{L}_2(f(x))$ for every node $x \in V_1$, $\langle f(x), f(y) \rangle \in E_2$ for every edge $\langle x, y \rangle \in E_1$, and $\mathcal{L}_1(\langle x, y \rangle) \subseteq \mathcal{L}_2(\langle f(x), f(y) \rangle)$ for every edge $\langle x, y \rangle \in E_1$.

A branch $\mathcal{B}$ is *redundant* in $\mathcal{G}$ if $\mathcal{B}$ is subsumed by another branch in $\mathcal{G}$, and every node in $\mathcal{B}$ except the root is a variable.

Revision graphs can be seen as ABoxes with variables. Given a revision graph $\mathcal{G} = (V, E, \mathcal{L})$, we call $\mathcal{A}_\mathcal{G} = \bigcup_{x \in V} \{C(x) \mid C \in \mathcal{L}(x)\} \cup \bigcup_{\langle x, y \rangle \in E} \{R(x, y) \mid R \in \mathcal{L}(\langle x, y \rangle)\}$ as the *ABox representation* of $\mathcal{G}$, and call $\mathcal{G}$ as the *revision-graph representation* of $\mathcal{A}_\mathcal{G}$.

Procedure B-MW($\mathcal{K}$, k)

Input: a KB $\mathcal{K} = \langle \mathcal{T}, \mathcal{A} \rangle$ and a non-negative integer k.
Output: a revision graph $\mathcal{G} = (V, E, \mathcal{L})$.
1 Initialize the revision graph $\mathcal{G} = (V, E, \mathcal{L})$ as
   - $V := N_I^{\mathcal{K}}$,
   - $\mathcal{L}(a) := \{C \mid C(a) \in \mathcal{A}\}$ for each node $a \in V$,
   - $E := \{\langle a, b \rangle \mid$ *there is some* R *with* $R(a, b) \in \mathcal{A}\}$,
   - $\mathcal{L}(\langle a, b \rangle) := \{R \mid R(a, b) \in \mathcal{A}\}$ for each edge $\langle a, b \rangle \in E$.

2 **while** *there exists an expansion rule in Fig. 1 that is applicable to $\mathcal{G}$* **do**
   $\lfloor$ expand $\mathcal{G}$ by applying this rule.

3 **for** *each node $x \in V$* **do**
   $\lfloor$ $\mathcal{L}(x) := \{C \mid C \in \mathcal{L}(x)$ and C is a concept name $\}$.

4 **while** *there exists a redundant branch $\mathcal{B} = (V_{\mathcal{B}}, E_{\mathcal{B}}, \mathcal{L}_{\mathcal{B}})$ in $\mathcal{G}$* **do**
   $E := E \setminus E_{\mathcal{B}}$;
   $V := V \setminus (V_{\mathcal{B}} \setminus \{x_{\mathcal{B}}\})$, where $x_{\mathcal{B}}$ is the root of $\mathcal{B}$.

5 Return $\mathcal{G} = (V, E, \mathcal{L})$.

Given a KB $\mathcal{K} = \langle \mathcal{T}, \mathcal{A} \rangle$ and a non-negative integer k, we use the procedure B-MW($\mathcal{K}$, k) to construct a revision graph for them, and call this revision graph a *k-role-depth bounded minimal witness* for $\mathcal{K}$.

Example 7. Consider the adaptation setting $\mathcal{AS}_1$ described in Example 4 and call the procedure B-MW($\langle \mathcal{T}_1, \mathcal{A}_1 \rangle, 1$). During the execution of this procedure, two variables x_1 and x_2 will be introduced by the $\exists$-rule. Let $\mathcal{G}_1$ be the revision graph returned by this procedure, then its ABox representation is $\mathcal{A}_{\mathcal{G}_1} = \{A(a), C(a), R(a, x_1), A(x_1), C(x_1), R(x_1, x_2), A(x_2), C(x_2)\}$.

Consider the adaptation setting $\mathcal{AS}_2$ described in Example 6 and call the procedure B-MW($\langle \mathcal{T}_2, \mathcal{A}_2 \rangle, 1$). During the execution of step **2**, three variables y_1, z_1 and z_2 will be introduced by the $\exists$-rule. Let $\mathcal{G}_2'$ be the revision graph after the execution of step **3**, then its ABox representation is $\mathcal{A}_{\mathcal{G}_2'} = \{A(a), C(a), R(a, y_1), C(y_1), R(a, z_1), A(z_1), C(z_1), R(z_1, z_2), A(z_2), C(z_2)\}$. In $\mathcal{G}_2'$ there exists a redundant branch which will be removed by step **4**. Finally, let $\mathcal{G}_2$ be the revision graph returned by the procedure, then its ABox representation is $\mathcal{A}_{\mathcal{G}_2} = \{A(a), C(a), R(a, z_1), A(z_1), C(z_1), R(z_1, z_2), A(z_2), C(z_2)\}$.   □

GCI$_I$-rule: if $x \in N_I^{\mathcal{K}}$, $C \sqsubseteq D \in \mathcal{T}$, $D \notin \mathcal{L}(x)$, and $\langle \mathcal{T}, \mathcal{A} \rangle \models C(x)$,
then set $\mathcal{L}(x) := \mathcal{L}(x) \cup \{D\}$.

GCI$_V$-rule: if $x \notin N_I^{\mathcal{K}}$, $C \sqsubseteq D \in \mathcal{T}$, $D \notin \mathcal{L}(x)$, and $\langle \mathcal{T}, \{E(x) \mid E \in \mathcal{L}(x)\} \rangle \models C(x)$,
then set $\mathcal{L}(x) := \mathcal{L}(x) \cup \{D\}$.

⊓-rule: if $C_1 \sqcap C_2 \in \mathcal{L}(x)$, and $\{C_1, C_2\} \nsubseteq \mathcal{L}(x)$,
then set $\mathcal{L}(x) := \mathcal{L}(x) \cup \{C_1, C_2\}$.

∃-rule: if $\exists R.C \in \mathcal{L}(x)$, $level(x) \leq k$, and x has no successor z with $C \in \mathcal{L}(z)$,
then introduce a new variable z, set $V := V \cup \{z\}$, $E := E \cup \{\langle x, z \rangle\}$,
$\mathcal{L}(z) := \{C\}$, and $\mathcal{L}(\langle x, z \rangle) := \{R\}$.

Fig. 1. Expansion rules used by the procedure B-MW($\mathcal{K}$, k)

Given a TBox $\mathcal{T}$ and an ABox representation $\mathcal{A}_{\mathcal{G}}$ of some revision graph $\mathcal{G}$, we use the procedure Rolling($\mathcal{A}_{\mathcal{G}}$, $\mathcal{T}$) to roll up variables contained in $\mathcal{A}_{\mathcal{G}}$.

Procedure Rolling($\mathcal{A}_{\mathcal{G}}$, $\mathcal{T}$)

Input: an ABox $\mathcal{A}_{\mathcal{G}}$ that may contain variables, and a TBox $\mathcal{T}$.
Output: an ABox $\mathcal{A}$ without variables.

1 Transform $\mathcal{A}_{\mathcal{G}}$ into its revision-graph representation $\mathcal{G} = (V, E, \mathcal{L})$.
2 Delete from V the variables which are not descendants of any individual name.
3 **while** *there exists variable in V* **do**
  select a variable $y \in V$ that has no successor;
  $x := $ the predecessor of y;
  if $\mathcal{L}(y) \neq \emptyset$ **then** $C_y := \bigsqcap_{C \in \mathcal{L}(y)} C$ **else** $C_y := \top$;
  $R := $ the role name contained in $\mathcal{L}(\langle x, y \rangle)$;
  if $\langle \mathcal{T}, \{D(x) \mid D \in \mathcal{L}(x)\} \rangle \not\models (\exists R.C_y)(x)$ **then** $\mathcal{L}(x) := \mathcal{L}(x) \cup \{\exists R.C_y\}$;
  $E := E \setminus \{\langle x, y \rangle\}$;
  $V := V \setminus \{y\}$.
4 Return $\mathcal{A} := \bigcup_{x \in V} \{C(x) \mid C \in \mathcal{L}(x)\} \cup \bigcup_{\langle x, y \rangle \in E} \{R(x, y) \mid R \in \mathcal{L}(\langle x, y \rangle)\}$.

Example 8. Let us continue Example 7. Both the procedure Rolling($\mathcal{A}_{\mathcal{G}_1}$, $\mathcal{T}_1$) and the procedure Rolling($\mathcal{A}_{\mathcal{G}_2}$, $\mathcal{T}_2$) return the same ABox $\{A(a), C(a)\}$.

Let $\mathcal{A}'_{\mathcal{G}_1} = \mathcal{A}_{\mathcal{G}_1} \setminus \{A(a), A(x_1)\} = \{C(a), R(a, x_1), C(x_1), R(x_1, x_2), A(x_2), C(x_2)\}$, and let $\mathcal{A}''_{\mathcal{G}_1} = \mathcal{A}_{\mathcal{G}_1} \setminus \{A(a), R(a, x_1)\} = \{C(a), A(x_1), C(x_1), R(x_1, x_2), A(x_2), C(x_2)\}$. Then the procedure Rolling($\mathcal{A}'_{\mathcal{G}_1}$, $\mathcal{T}_1$) returns the ABox $\{C(a), \exists R.(C \sqcap \exists R.(A \sqcap C))(a)\}$, and Rolling($\mathcal{A}''_{\mathcal{G}_1}$, $\mathcal{T}_1$) returns the ABox $\{C(a)\}$. □

5.2 The Adaptation Algorithm

Let $\mathcal{T}$ be a TBox, and let $\mathcal{A}$, $\mathcal{N}$ be two ABoxes. If $\langle \mathcal{T}, \mathcal{A} \cup \mathcal{N} \rangle \models \top \sqsubseteq \bot$, then:

– a set $\mathcal{J} \subseteq \mathcal{A}$ is a $(\mathcal{A}, \mathcal{N})$-*justification* for a clash w.r.t. $\mathcal{T}$ if $\langle \mathcal{T}, \mathcal{J} \cup \mathcal{N} \rangle \models \top \sqsubseteq \bot$ and $\langle \mathcal{T}, \mathcal{J}' \cup \mathcal{N} \rangle \not\models \top \sqsubseteq \bot$ for every $\mathcal{J}' \subset \mathcal{J}$;

– a set $\mathcal{R} \subseteq \mathcal{A}$ is a $(\mathcal{A}, \mathcal{N})$-*repair* for clashes w.r.t. $\mathcal{T}$ if $\mathcal{R} \cap \mathcal{J} \neq \emptyset$ for every $(\mathcal{A}, \mathcal{N})$-justification $\mathcal{J}$, and for every $\mathcal{R}_i \subset \mathcal{R}$ there must be some $(\mathcal{A}, \mathcal{N})$-justification $\mathcal{J}_i$ such that $\mathcal{R}_i \cap \mathcal{J}_i = \emptyset$.

Now we are ready to present our algorithm. Given an adaptation setting $\mathcal{AS} = (\mathcal{T}, \mathcal{A}, \mathcal{N})$ and any integer k, where k is greater than the role depths of all the concepts occurring in $\mathcal{AS}$, the algorithm Adaptation($\mathcal{AS}$, k) operates as follows. Firstly, a revision graph $\mathcal{G}$ will be constructed for the KB $\langle \mathcal{T}, \mathcal{A} \rangle$ and the integer k. Secondly, a revision process based on justifications will be carried out on the ABox representation $\mathcal{A}_\mathcal{G}$ of $\mathcal{G}$. Thirdly, for each maximal subset $\mathcal{A}_i$ of $\mathcal{A}_\mathcal{G}$ that does not conflict with $\mathcal{N}$ and $\mathcal{T}$, the procedure Rolling($\mathcal{A}_i$, $\mathcal{T}$) will be used to roll up variables and get an ABox $\mathcal{A}'_i$. Finally, the ABox $\mathcal{A}'_i \cup \mathcal{N}$ will be returned as a solution case. Together with the solution case, an ABox $\mathcal{R}_i$ will also be returned; the function of $\mathcal{R}_i$ is to record the information which is entailed by $\mathcal{A}$ but removed in $\mathcal{A}'_i$, so that the user can select the best solution according to it.

Algorithm 1. Adaptation($\mathcal{AS}$, k)

Input: an adaptation setting $\mathcal{AS} = (\mathcal{T}, \mathcal{A}, \mathcal{N})$, and a non-negative integer k.
Output: a finite number of pairs $(\mathcal{A}'_1 \cup \mathcal{N}, \mathcal{R}_1)$, ..., $(\mathcal{A}'_n \cup \mathcal{N}, \mathcal{R}_n)$, where $\mathcal{A}'_i \cup \mathcal{N}$
           is a solution case and $\mathcal{R}_i$ records the information been removed.
if $\mathcal{A} \cup \mathcal{N}$ *is consistent w.r.t.* $\mathcal{T}$ **then**
  | return $(\mathcal{A} \cup \mathcal{N}, \emptyset)$;
else
  | $\mathcal{G} :=$ B-MW($\langle \mathcal{T}, \mathcal{A} \rangle, k$);
  | $\mathcal{A}_\mathcal{G} :=$ the ABox representation of $\mathcal{G}$;
  | $S_\mathcal{R} := \{\mathcal{R}_1, ..., \mathcal{R}_n\}$ all the $(\mathcal{A}_\mathcal{G}, \mathcal{N})$-repairs for clashes w.r.t. $\mathcal{T}$;
  | **for** $i \leftarrow 1$ **to** n **do**
  | | $\mathcal{A}_i := \mathcal{A}_\mathcal{G} \setminus \mathcal{R}_i$;
  | | $\mathcal{A}'_i :=$ Rolling($\mathcal{A}_i$, $\mathcal{T}$);
  | return $(\mathcal{A}'_1 \cup \mathcal{N}, \mathcal{R}_1)$, ..., $(\mathcal{A}'_n \cup \mathcal{N}, \mathcal{R}_n)$.

Example 9. Consider the adaptation setting $\mathcal{AS}_1 = (\mathcal{T}_1, \mathcal{A}_1, \mathcal{N}_1)$ described in Example 4. Since $\max\{depth(\mathcal{T}_1), depth(\mathcal{A}_1), depth(\mathcal{N}_1)\} = 1$, we select $k{=}1$ and execute the algorithm Adaptation($\mathcal{AS}_1, 1$).

Firstly, by calling the procedure B-MW($\langle \mathcal{T}_1, \mathcal{A}_1 \rangle, 1$), we get a revision graph $\mathcal{G}_1$ for which the ABox representation is $\mathcal{A}_{\mathcal{G}_1} = \{A(a), C(a), R(a, x_1), A(x_1), C(x_1), R(x_1, x_2), A(x_2), C(x_2)\}$.

Secondly, for the clash $\langle \mathcal{T}_1, \mathcal{A}_{\mathcal{G}_1} \cup \mathcal{N}_1 \rangle \models \top \sqsubseteq \bot$, there are two $(\mathcal{A}_{\mathcal{G}_1}, \mathcal{N}_1)$-justifications $\mathcal{J}_1 = \{A \sqsubseteq \exists R.A, E \sqcap \exists R.A \sqsubseteq \bot, E(a), A(a)\}$ and $\mathcal{J}_2 = \{E \sqcap \exists R.A \sqsubseteq \bot, E(a), A(a), R(a, x_1), A(x_1)\}$. From them we get two $(\mathcal{A}_{\mathcal{G}_1}, \mathcal{N}_1)$-repairs $\mathcal{R}_1 = \{A(a), A(x_1)\}$ and $\mathcal{R}_2 = \{A(a), R(a, x_1)\}$.

Thirdly, from $\mathcal{R}_1$ we get $\mathcal{A}'_1 =$ Rolling($\mathcal{A}_{\mathcal{G}_1} \setminus \mathcal{R}_1, \mathcal{T}_1$) $= \{C(a), \exists R.(C \sqcap \exists R.(A \sqcap C))(a)\}$. From $\mathcal{R}_2$ we get $\mathcal{A}'_2 =$ Rolling($\mathcal{A}_{\mathcal{G}_1} \setminus \mathcal{R}_2, \mathcal{T}_1$) $= \{C(a)\}$.

Finally, the algorithm returns $(\mathcal{A}_1' \cup \mathcal{N}_1, \mathcal{R}_1)$ and $(\mathcal{A}_2' \cup \mathcal{N}_1, \mathcal{R}_2)$, from them the user can select the best solution case with the help of $\mathcal{R}_1$ and $\mathcal{R}_2$. For example, since $\mathcal{R}_2$ contains a role assertion $R(a, x_1)$ which indicates that all the information related to x_1 is lost in $\mathcal{A}_2'$, the first choice for the user is $\mathcal{A}_1' \cup \mathcal{N}_1$. $\square$

Example 10. Consider the adaptation setting $\mathcal{AS} = (\mathcal{T}, \mathcal{A}_{sol}, \mathcal{N}_{pb})$ described in Example 3. Since $\max\{depth(\mathcal{T}), depth(\mathcal{A}_{sol}), depth(\mathcal{N}_{pb})\} = 2$, we select $k = 2$ and execute the algorithm Adaptation($\mathcal{AS}, 2$).

The algorithm will return two results $(\mathcal{A}_1' \cup \mathcal{N}_{pb}, \mathcal{R}_1)$ and $(\mathcal{A}_2' \cup \mathcal{N}_{pb}, \mathcal{R}_2)$, where $\mathcal{A}_1' = \{\exists TreatBy.\,(Anti\text{-}oestrogen \sqcap \exists metabolizedTo.(Compounds \sqcap \exists bindto.OestrogenReceptor))\,(Mary)\}$, $\mathcal{R}_1 = \{Tamoxifen(y)\}$, $\mathcal{A}_2'' = \emptyset$, and $\mathcal{R}_2 = \{TreatBy(Mary, y)\}$.

Since $\mathcal{R}_2$ contains a role assertion $TreatBy(Mary, y)$ which indicates that all the information related to y is lost in $\mathcal{A}_2'$, the first choice for the user is the solution case $\mathcal{A}_1' \cup \mathcal{N}_{pb}$. This solution case states that, for the target problem described by $\mathcal{N}_{pb}$, a solution is to treat $Mary$ by something that is not only anti-oestrogen but also can be metabolized into some compounds which can be bound to oestrogen receptors. $\square$

The following theorems state that our algorithm satisfies **R1-R4**.

Theorem 1. *Let* $(\mathcal{A}_i'', \mathcal{R}_i)$ $(1 \leq i \leq n)$ *be the pairs returned by Adaptation($\mathcal{AS}, k$) for* $\mathcal{AS} = (\mathcal{T}, \mathcal{A}, \mathcal{N})$. *Then the following statements hold for every* $1 \leq i \leq n$: (1) $\langle \mathcal{T}, \mathcal{A}_i'' \rangle \models \mathcal{N}$; (2) $\mathcal{A}_i'' = \mathcal{A} \cup \mathcal{N}$ *if* $\mathcal{A} \cup \mathcal{N}$ *is consistent w.r.t.* $\mathcal{T}$; *and* (3) *if* $\mathcal{N}$ *is consistent w.r.t.* $\mathcal{T}$ *then* $\mathcal{A}_i''$ *is also consistent w.r.t.* $\mathcal{T}$.

Theorem 2. *Given two adaptation settings* $\mathcal{AS}_i = (\mathcal{T}, \mathcal{A}_i, \mathcal{N}_i)$ $(i = 1, 2)$ *and an integer* k, *where* $\langle \mathcal{T}, \mathcal{A}_1 \rangle \equiv \langle \mathcal{T}, \mathcal{A}_2 \rangle$, $\langle \mathcal{T}, \mathcal{N}_1 \rangle \equiv \langle \mathcal{T}, \mathcal{N}_2 \rangle$, $k \geq \max\{depth(\mathcal{T}), depth(\mathcal{A}_1), depth(\mathcal{A}_2), depth(\mathcal{N}_1), depth(\mathcal{N}_2)\}$. *If* $(\mathcal{A}_1'', \mathcal{R}_1)$ *is a pair returned by the algorithm Adaptation($\mathcal{AS}_1, k$), then there must be a pair* $(\mathcal{A}_2'', \mathcal{R}_2)$ *returned by Adaptation($\mathcal{AS}_2, k$) such that* $\langle \mathcal{T}, \mathcal{A}_1'' \rangle \equiv \langle \mathcal{T}, \mathcal{A}_2'' \rangle$ *and* $\mathcal{R}_2 = \sigma(\mathcal{R}_1)$ *for some substitution* σ *of variables.*

Theorem 2 is based on the following fact: let $\mathcal{G}_i = \text{B-MW}(\langle \mathcal{T}, \mathcal{A}_i \rangle, k)$ $(i = 1, 2)$, then $\mathcal{G}_1$ and $\mathcal{G}_2$ are identical up to variable renaming in the case that k is sufficiently large. For Theorem 1 there is no requirement on the value of k.

In our algorithm, the revision graph $\mathcal{G}$ constructed by the procedure B-MW($\langle \mathcal{T}, \mathcal{A} \rangle, k$) is in fact a non-redundant k-depth-bounded model for the KB $\langle \mathcal{T}, \mathcal{A} \rangle$. Therefore, our revision process works on fine-grained representation of models and guarantees the minimal change principle in a fine-grained level. So, our algorithm also satisfies the property specified by **R5**.

The following theorem states that our algorithm is in exponential time.

Theorem 3. *For any adaptation setting* $\mathcal{AS} = (\mathcal{T}, \mathcal{A}, \mathcal{N})$, *assume the role depth of every concept occurring in* $\mathcal{AS}$ *is bounded by some integer* k, *then the algorithm Adaptation($\mathcal{AS}, k$) runs in time exponential with respect to the size of* $\mathcal{AS}$.

6   Discussion and Related Work

The idea of applying KB revision theory to adaptation in CBR was proposed by Lieber [17]. Based on a classical revision operator in propositional logic, a framework for adaptation was presented and it was demonstrated that the adaptation process should satisfy the AGM postulates. This idea was extended by Cojan and Lieber [7] to deal with adaptation based on the DL $\mathcal{ALC}$. Based on an extension of the classical tableau method used for deductive inferences in $\mathcal{ALC}$, an algorithm for adapting cases represented in $\mathcal{ALC}$ was proposed. It was shown that, except for the requirements on syntax-independence and minimality of change (i.e., **R4** and **R5** in our paper), all the other requirements specified by the AGM postulates (i.e., **R1-R3** in our paper) are satisfied by their algorithm.

From the point of view of KB revision in DLs, it is a great challenge to design revision operators or algorithms that satisfy the requirements specified by the AGM postulates. In the literature, there are two kinds of approaches, i.e., MBAs [14] and FBAs [4,16,20], for the instance-level KB revision problem in DLs. As we analyzed in Section 4, they either do not satisfy the requirements specified by **R4** and **R5**, or only work well for DLs of the *DL-Lite* family.

Our method is closer in spirit to the formula-based approaches, but it also inherits some ideas of model-based ones. On the one hand, in our algorithm, the revision graph $\mathcal{G}$ constructed by the procedure B-MW($\langle \mathcal{T}, \mathcal{A} \rangle, k$) can be seen as a non-redundant, k-depth-bounded model for the KB $\langle \mathcal{T}, \mathcal{A} \rangle$, and therefore our revision process essentially works on models. On the other hand, our revision process makes use of $(\mathcal{A}_{\mathcal{G}}, \mathcal{N})$-repairs which inherits some ideas of FMAs based on justifications. As a result, our algorithm not only satisfies the requirements **R4** and **R5**, but also works for the DL $\mathcal{EL}_\perp$.

Given an adaptation setting, our algorithm will return a finite number of pairs $(\mathcal{A}_i'', \mathcal{R}_i)$ $(1 \le i \le n)$, and it is left to the user to select the best solution case according to the sets $\mathcal{R}_i$. We can extend the algorithm to sort all the solution cases by priority. For example, if some $\mathcal{R}_i$ contains a role assertion $R(a, x)$ with x a variable, then the corresponding solution case $\mathcal{A}_i''$ will has a lower priority. Furthermore, we can define a selection function according to the user's selection criteria, and enable our algorithm to return only one best solution case.

7   Conclusion and Future Work

We studied the adaptation problem of CBR in the DL $\mathcal{EL}_\perp$. A formalism for adaptation based on $\mathcal{EL}_\perp$ was presented, and in this formalism the adaptation task was modeled as the instance-level KB revision problem in $\mathcal{EL}_\perp$. We illustrated that existing revision operators and algorithms in DLs did not work for the adaptation setting based on $\mathcal{EL}_\perp$ and then presented a new algorithm. We showed that our algorithm behaves well for $\mathcal{EL}_\perp$ in that it satisfies the requirements proposed in the literature for revision operators.

For future work, we will extend our method to support adaptation based on $\mathcal{EL}^{++}$ [3]. Another work is to implement and optimize our algorithm and test its feasibility in practice. Finally, we will formalize the notion of minimality of change under the framework of adaptation.

Acknowledgments. The authors would like to thank the reviewers for their helpful comments and suggestions. This work was partially supported by the National Natural Science Foundation of China (Nos. 61363030, 61262030), the Natural Science Foundation of Guangxi Province (No.2012GXNSFBA053169) and the Science Foundation of Guangxi Key Laboratory of Trusted Software.

References

1. Alchourrón, C.E., Gärdenfors, P., Makinson, D.: On the logic of theory change: Partial meet contraction and revision functions. J. Symb. Log. 50(2), 510–530 (1985)
2. Baader, F., Calvanese, D., McGuinness, D., Nardi, D., Patel-Schneider, P.F.: The Description Logic Handbook: Theory, Implementation and Applications. Cambridge University Press, Cambridge (2003)
3. Baader, F., Brandt, S., Lutz, C.: Pushing the $\mathcal{EL}$ envelope. In: Proc. of the 19th International Joint Conference on Artificial Intelligence, pp. 364–369. Morgan Kaufmann (2005)
4. Calvanese, D., Kharlamov, E., Nutt, W., Zheleznyakov, D.: Evolution of *DL-lite* knowledge bases. In: Patel-Schneider, P.F., Pan, Y., Hitzler, P., Mika, P., Zhang, L., Pan, J.Z., Horrocks, I., Glimm, B. (eds.) ISWC 2010, Part I. LNCS, vol. 6496, pp. 112–128. Springer, Heidelberg (2010)
5. Chang, L., Sattler, U., Gu, T.L.: Algorithm for adapting cases represented in a tractable description logic. arXiv: 1405.4180 (2014)
6. Chang, L., Sattler, U., Gu, T.L.: An ABox revision algorithm for the description logic $\mathcal{EL}_\bot$. In: Proc. of the 27th International Workshop on Description Logics (2014)
7. Cojan, J., Lieber, J.: An algorithm for adapting cases represented in an expressive description logic. In: Bichindaritz, I., Montani, S. (eds.) ICCBR 2010. LNCS, vol. 6176, pp. 51–65. Springer, Heidelberg (2010)
8. Consortium, T.G.O.: Gene Ontology: Tool for the unification of biology. Nature Genetics 25, 25–29 (2000)
9. d'Aquin, M., Lieber, J., Napoli, A.: Decentralized case-based reasoning for the semantic web. In: Gil, Y., Motta, E., Benjamins, V.R., Musen, M.A. (eds.) ISWC 2005. LNCS, vol. 3729, pp. 142–155. Springer, Heidelberg (2005)
10. Gómez-Albarrán, M., González Calero, P.A., Díaz-Agudo, B., Fernández-Conde, C.J.: Modelling the CBR life cycle using description logics. In: Althoff, K.-D., Bergmann, R., Karl Branting, L. (eds.) ICCBR 1999. LNCS (LNAI), vol. 1650, pp. 147–161. Springer, Heidelberg (1999)
11. Horrocks, I., Patel-Schneider, P.F., Harmelen, F.V.: From SHIQ and RDF to OWL: the making of a web ontology language. J. Web Semantics 1(1), 7–26 (2003)
12. Horrocks, I., Sattler, U.: A tableau decision procedure for $\mathcal{SHOIQ}$. J. Autom. Reasoning 39(3), 249–276 (2007)
13. Katsuno, H., Mendelzon, A.O.: Propositional knowledge base revision and minimal change. Artificial Intelligence 52(3), 263–294 (1991)
14. Kharlamov, E., Zheleznyakov, D., Calvanese, D.: Capturing model-based ontology evolution at the instance level: The case of DL-Lite. J. Comput. Syst. Sci. 79(6), 835–872 (2013)

15. Lehmann, K., Turhan, A.-Y.: A Framework for Semantic-Based Similarity Measures for $\mathcal{ELH}$-Concepts. In: del Cerro, L.F., Herzig, A., Mengin, J. (eds.) JELIA 2012. LNCS, vol. 7519, pp. 307–319. Springer, Heidelberg (2012)
16. Lenzerini, M., Savo, D.F.: On the evolution of the instance level of DL-Lite knowledge bases. In: Proc. of the 24th International Workshop on Description Logics (2011)
17. Lieber, J.: Application of the revision theory to adaptation in case-based reasoning: the conservative adaptation. In: Weber, R.O., Richter, M.M. (eds.) ICCBR 2007. LNCS (LNAI), vol. 4626, pp. 239–253. Springer, Heidelberg (2007)
18. Sánchez-Ruiz, A.A., Ontañón, S., González-Calero, P.A., Plaza, E.: Measuring similarity in description logics using refinement operators. In: Ram, A., Wiratunga, N. (eds.) ICCBR 2011. LNCS, vol. 6880, pp. 289–303. Springer, Heidelberg (2011)
19. Spackman, K.: Managing clinical terminology hierarchies using algorithmic calculation of subsumption: Experience with SNOMED-RT. J. American Medical Informatics Assoc., Fall Symposium Special Issue (2000)
20. Wiener, C.H., Katz, Y., Parsia, B.: Belief base revision for expressive description logics. In: Proc. of the 4th International Workshop on OWL: Experiences and Directions (2006)

Sentiment and Preference Guided Social Recommendation

Yoke Yie Chen[1], Xavier Ferrer[2,3], Nirmalie Wiratunga[1], and Enric Plaza[2]

[1] IDEAS Research Institute, Robert Gordon University, Aberdeen, Scotland
{y.y.chen,n.wiratunga}@rgu.ac.uk
[2] Artificial Intelligence Research Institute (IIIA-CSIC)
Spanish National Research Council (CSIC)
Campus UAB, Bellaterra, Catalonia, Spain
[3] Universitat Autònoma de Barcelona
Bellaterra, Catalonia, Spain
{xferrer,enric}@iiia.csic.es

Abstract. Social recommender systems harness knowledge from social experiences, expertise and interactions. In this paper we focus on two such knowledge sources: sentiment-rich user generated reviews; and preferences from purchase summary statistics. We formalise the integration of these knowledge sources by mixing a novel aspect-based sentiment ranking with a preference ranking. We demonstrate the utility of our proposed formalism by conducting a comparative analysis on data extracted from Amazon.com. In particular we show that the performance of the proposed aspect based sentiment analysis algorithm is superior to existing aspect extraction algorithms and that combining this with preference knowledge leads to better recommendations.

Keywords: social recommender systems, sentiment analysis, aspect extraction, preference graph.

1 Introduction

Recommender systems have traditionally relied on improving their ranked lists by exploiting knowledge about user preferences [26], their information needs [11] or by exploiting similar behavior of other users [1]. The huge success of these systems in the retail sector is also its main driving force towards innovative and improved recommendation algorithms. Representation, similarity and ranking algorithms from the Case-Based Reasoning (CBR) community has naturally made a significant contribution to recommender systems research [18,23]. The dawn of the social web creates many new opportunities for recommendation algorithms and so the emergence of social recommender systems [9,12].

Consider a typical product recommendation scenario on Amazon (see Figure 1 and Figure 2). Here in addition to typical information about an artefact (e.g. Camera image and textual description), there is also information generated or derived from user interactions (e.g. reviews and what users actually buy after

L. Lamontagne and E. Plaza (Eds.): ICCBR 2014, LNCS 8765, pp. 79–94, 2014.
© Springer International Publishing Switzerland 2014

viewing this camera). Specifically we observe that there is explicit knowledge in the form of user opinion and implicit knowledge in the form of user preferences. Here by preference we are referring to preference of users over viewed products. Opinion knowledge is often used to improve recommendations by incorporating the sentiment expressed in opinions to bias the retrieved list [9]. Similarly preference knowledge has also separately been applied to successfully influence recommendations [26]. Our focus in this paper is to harness social content by uniting both these knowledge sources to generate better recommendation rankings.

Our contribution is three-fold: firstly we demonstrate how the choice of sentiment analysis algorithms can impact the quality of recommendations; and secondly show how a page-rank type algorithm can be effortlessly incorporated to derive rankings on the basis of preference relationships; and finally provide a formalism to combine sentiment and preference knowledge. Our results confirm that aspect-based sentiment analysis is far superior to one that is agnostic of aspects. This is because purchase choices are based on comparison of artefacts; which implicitly or explicitly involves comparison of characteristics or aspects of these artefacts. Lastly, we introduce a novel algorithm to infer aspect importance weights guided by knowledge represented in a preference graph. Here the insight is that preference relationships modeled in this graph allow us to identify aspects that are likely to have influenced the users' purchase decision. In particular a users purchase decision hints at the aspects that are likely to have influenced their decision and as such be deemed more important.

The rest of the paper is organized as follows: In Section 2 we present the background research related to this work. Next in Section 3 we describe how preference graphs can be generated by using a case study from Amazon.com. The process of aspect extraction and weight learning for sentiment analysis is presented in Section 4. Finally, our evaluation results are presented in Section 5 followed by conclusions in Section 6.

Fig. 1. Product information (explicit)

Fig. 2. Product reviews (implicit)

2   Related Work

There exist numerous applications of machine learning and CBR in the area of recommendation systems. Content-based filtering approaches exploit the past and current preferences of the user to build new recommendations. Early work in this area has analysed user logs and sessions [5,17]; reused similar user trails [2, 24]; and exploited click-through data [6] to improve recommendations.

Collaborative filtering [15] unlike content-based filtering emphasises the social dimensions of user similarities, preferences and experiences. A good example is Amazon.com, where user ratings and preferences are combined in different ways [16]. In Gupta and Chakraborti [13], graphs are used to model user decisions and combine estimated utilities in the context of preference based recommenders. In Vasudevan and Chakraborti [26], preference models of artefacts are induced based on the trails left by users when critiquing an artefact's aspect. In our work, we do not rely on a preference graph that explicitly defines which aspects are preferred, but instead infer this information from comparing the sentiment-rich content generated by users. Furthermore, unlike other preference based recommender systems which utilise user profile to generate user preferences, we obtained preference knowledge from user interactions. As such we provide a more flexible alternative to preference analysis.

Social recommenders recognise the important role of sentiment analysis of user reviews [9]. Extracting sentiment from natural language constructs is a challenge. Lexicons are often used to ascertain the polarity (positive or negative) and strength of sentiment expressed at word-level (e.g. SentiWordNet [10]). However sophisticated methods are needed to aggregate these scores at the sentence, paragraph and document level to account for negation and other forms of sentiment modifiers [21]. Increasingly aggregation is organised at the aspect level, since the distribution of a user's sentiment is typically mixed and expressed over the aspects of the artefact (e.g. I love the *colour* but not too keen on *size*). Hu and Liu [14] propose an association mining driven approach to identify frequent nouns or noun phrases as aspects. Thereafter sentences are grouped by these aspects and sentiment scores assigned to each aspect group. Whilst there are many other statistical approaches to frequent noun extraction [3,14,19,22]; others argue that identifying semantic relationships in text provides significant improvements in aspect extraction [20]. Here we explore how semantic based extraction can be augmented by frequency counts.

3   Social Recommendation Process

An overview of our proposed process appears in Figure 3. The final outcome is a recommendation of products that are retrieved and ranked on their product rank score, *ProdRank*, with respect to a given query. Central to this ranking is the integration of sentiment scores derived from user reviews with dominant products inferred from the preference graph.

$$ProdRank(p_i) = \alpha * PrefRank(p_i) + (1 - \alpha) * SentiRank(p_i) \qquad (1)$$

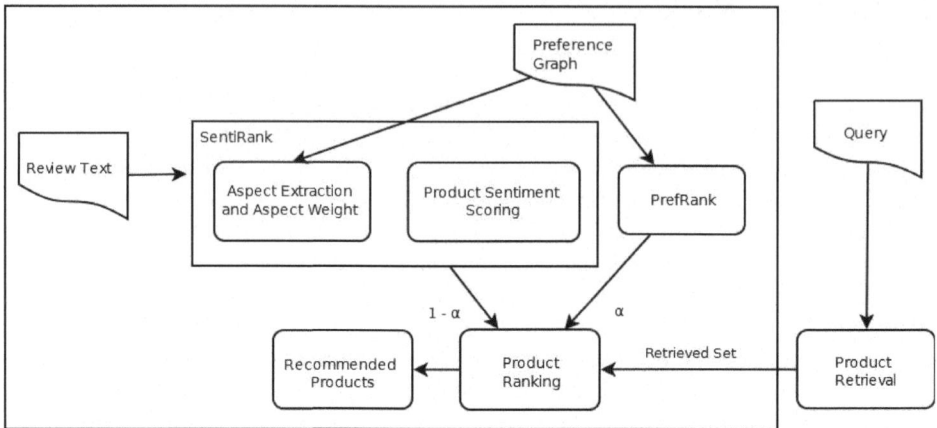

Fig. 3. Overview of social product recommender process

Where *PrefRank* assigns a score to product, p_i, by applying PageRank [4] to a preference graph; and *SentiRank* assigns a score based on aspect level sentiment analysis. The graph is generated from viewed and purchased product pairs; whilst the sentiment scores are derived from product reviews. We also advocate the use of weighted aspect level sentiment analysis and learn these weights by comparing the sentiment difference between node pairs in the graph (see Section 4). Here α is used as a mixture parameter to study the impact of preference-only ($\alpha = 1$), sentiment-only ($\alpha = 0$), or a mixture of both ($1 < \alpha < 0$).

3.1 Preference-Based Product Ranking

A preference relation between a pair of products denotes the preference of one product over the other through the analysis of viewed and purchased product relationships. To illustrate the generation of preference relation from Amazon dataset, consider a snapshot of Amazon product web page in Figure 1. Here *Canon EOS 1100D Digital SLR Camera (CanonSLR)* is the viewed product. At the bottom of this page, we can observed that *Nikon D3100 Digital SLR Camera (NikonSLR)* and *Samsung WB250F Smart Camera (SamsungSmart)* are the products that some of the users purchased after viewing *CanonSLR*. This list of purchased products hints at the preference of users. Using these information, we generate two preference relations in which *NikonSLR* is preferred over *CanonSLR* and *SamsungSmart* is preferred over *CanonSLR*. A preference graph, $G = (\mathcal{P}, \mathcal{E})$, is generated from such product relation (see Figure 4). The set of nodes, $p_i \in \mathcal{P}$, represent products, and the set of directed edges, $\mathcal{E}$, are preference relations, $p_j \succ p_i$, such that a directed edge from product p_i to p_j with $i \neq j$ represents that, for some users, p_j is preferred over product p_i. In some cases where $p_j \succ p_i$ and $p_i \succ p_j$, a bidirectional preference relation can be observed. For any p_i, we use $\mathcal{E}^i$ to denote in-coming and $\mathcal{E}_i$ for outgoing product sets.

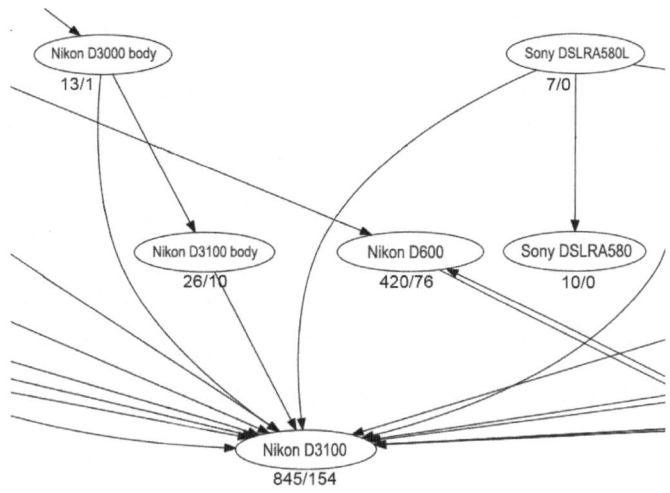

Fig. 4. Preference sub-graph for *Amazon Digital SLR Cameras*

Figure 4 illustrates a preference graph generated from a sample of Amazon data on *Digital SLR Camera*. The number of reviews/questions for a product is shown below each product node. Typically top ranked products have many more incoming edges, while less popular products tend to have more outgoing edges. It is not surprising that product *Nikon D3100* in Figure 4 is listed in Amazon's top 10 *Best Seller* ranking. Based on these observations, the higher the number of incoming edges (quantity) from preferred products (quality), the higher the preference rank, *PrefRank*, is for product, p_i.

$$PrefRank(p_i) = \sum_{p_j \in \mathcal{E}^i} \frac{PrefRank(p_j)}{|\mathcal{E}_j|} \qquad (2)$$

Where $\mathcal{E}^i$ is the set of all viewed products over which p_i is preferred and $\mathcal{E}_j$ is the set of products that are preferred after viewing p_i.

The main assumption here is similar to PageRank where the score captures the importance of pages by the analysis of the quantity and quality of edges to pages [4]. However we also observed that while our assumption is true with most studied products, it is not always the case that a product with higher *PrefRank* will also have a higher rank in Amazon's *Best Seller*. For example, Amazon's third ranked camera has a higher *PrefRank* score compared to the second ranked camera (see Table 1). Similar observations also hold when comparing *PrefRank* scores with user generated overall product ratings. This motivates the need to leverage on further dimensions of knowledge sources such as sentiment from product reviews for product recommendation.

Table 1. Summary of the averages of products organised by Amazon Best Seller rank

Amazon Bestseller Ranking	PREFERENCE				SENTIMENT		
	PrefRank	Incoming Edges	Outgoing edges	#Reviews per Day	*SentiRank* aspects	#Reviews	Product Rating
1	0.1240	26.0	1.0	1.136	0.087	1351.0	4.7
2	0.0037	2.0	2.0	0.770	0.106	242.0	4.7
3	0.0178	10.0	0.0	0.621	0.097	847.0	4.6
4-5	0.0053	1.5	0.5	0.589	0.102	129.0	4.8
6-10	0.0435	5.5	1.25	0.453	0.089	354.75	4.72
11-50	0.0048	2.071	0.892	0.506	0.080	407.35	4.45
51-100	0.0021	0.529	1.647	0.366	0.124	393.64	4.52
N/A	0.0015	0.439	1.469	0.143	0.092	175.34	4.61

4   Aspect Weighted Sentiment-Based Product Ranking

Users author reviews following the purchase of products. These contain user opinion in the form of positive and negative sentiment. Strength of sentiment expresses the intensity with which an opinion is stated with reference to a product [25]. We exploit this information as a means to rank our products, such that products ranked higher denote higher positive sentiment. *SentiRank* of a product, p_i, given a set of related reviews $\mathcal{R}^i$ is computed as follows:

$$SentiRank(p_i) = \frac{\sum_{k=1}^{|\mathcal{R}^i|} SentiScore(r_k)}{|\mathcal{R}^i|} \tag{3}$$

Here *SentiScores* are generated by the SMARTSA system [21] for each $r_k \in \mathcal{R}^i$. The SMARTSA system obtains the sentiment score of sentiment-bearing words from SentiWordNet [10]. The score will be modified to take into consideration of negation terms and lexical valence shifters (e.g. intensifier and diminish terms). The negative and positive strength is expressed as a value in the range [-1:1].

A finer-grained analysis of reviews is achieved by computing sentiment at the aspect level. It allows the sentiment of product, p_i, to be associated with individual aspects $a_j \in \mathcal{A}^i$ where $\mathcal{A}^i \subseteq \mathcal{A}$. The aspects of a product are extracted by using Algorithm 1. Accordingly we have the following rewrite for *SentiRank*.

$$SentiRank(p_i, a_j) = \frac{\sum_{j=1}^{|\mathcal{A}^i|} AspectWeight(a_j) * AspectSentiScore(p_i, a_j)}{|\mathcal{A}^i|} \tag{4}$$

This new formalisation is a weighted summation of sentiment expressed at the aspect level. Once aspects are extracted the sentiment of a product's reviews can be expressed as a distribution over each aspect. Accordingly, the aspect-level sentiment score is:

$$AspectSentiScore(p_i, a_j) = \frac{\sum_{m=1}^{|\mathcal{R}_j^i|} SentiScore(r_m)}{|\mathcal{R}_j^i|} * (1 - Gini) \qquad (5)$$

Where $\mathcal{R}_j^i$ is a set of reviews for product p_i related to aspect a_j and $r_m \in \mathcal{R}_j^i$. We use the Gini coefficient [27] to assign higher sentiment scores to an aspect when there is consensus about the distribution of the sentiment and otherwise is penalised accordingly.

A product purchase choice is a preference made on the basis of one or more aspects. The notion of aspect importance arises when the same set of aspects contribute to similar purchase decisions. Using this same principle, aspects weights are derived by comparing the aspect sentiment score differences between purchased and viewed product pairs in which $(p_x, p_y) \in \{(p_x, p_y)\}_{x,y=1 \wedge x \neq y}^t$

$$AspectWeight(a_j) = \frac{\sum_{x=1}^{|\mathcal{P}|} \sum_{y=1}^{|\mathcal{P}|} \delta(a_j, p_x, p_y)}{|t \in \mathcal{E}|} \qquad (6)$$

where either $p_x \succ p_y$ or $p_y \succ p_x$ or both, and t is the set of product preference pairs containing aspect a_j. The preference difference between any pairs of products is computed as:

$$\delta(a_j, p_x, p_y) = |L_{min}(\mathcal{A}, \mathcal{E})| + \delta'(a_j, p_x, p_y) \qquad (7)$$

$$\delta'(a_j, p_x, p_y) = (AspectSentiScore(a_j, p_x) - AspectSentiScore(a_j, p_y)) \qquad (8)$$

Here $|L_{min}(\mathcal{A}, \mathcal{E})|$ is the lowest preference difference scores obtained over all the aspect and product preference pairs. This is required to avoid negative aspect weights while preserving the importance of the aspects.

4.1 Sentiment Aspect Extraction

Grammatical extraction rules [20] are used to identify a set of candidate aspect phrases from sentences. These rules operate on dependency relations in parsed sentences[1]. Figure 5 lists the rules that we have employed in this work. Here N is a noun, A an adjective, V a verb, h a head term, m a modifier and $\langle h, m \rangle$ is a candidate phrase. Examples of how these rules apply to sample sentences appear in Table 2. Consider the first example sentence which according to Algorithm 1 applies to rule three: $cop(good, is) + nsubj(good, lens) \rightarrow \langle lens, good \rangle$. Next, if a Noun Compound Modifier (nn) exists in the sentence, rules five and six apply; and in this example rule five applies resulting in the following candidate aspects: $(lens, good) + nn(lens, camera) \rightarrow \langle camera\ lens, good \rangle$. In this way given a set of reviews a set of candidate phrases are extracted. For each candidate, non noun (N) words are eliminated. Thereafter frequency of each candidate is calculated

---

[1] Sentences are parsed using the Stanford Dependency parser [7].

according to its N and NN phrase; retaining only those candidates above a frequency cut-off (e.g. greater than 1% of the maximum frequency occurrence of a noun).

$$DP = \text{set of dependency pattern rules}$$
$$\{$$
$$dp_1 : amod(N, A) \rightarrow \langle N, A \rangle,$$
$$dp_2 : acomp(V, A) + nsubj(V, N) \rightarrow \langle N, A \rangle,$$
$$dp_3 : cop(A, V) + nsubj(A, N) \rightarrow \langle N, A \rangle,$$
$$dp_4 : dobj(V, N) + nsubj(V, N') \rightarrow \langle N, V \rangle,$$
$$dp_5 : \langle h, m \rangle + nn(h, N) \rightarrow \langle N + h, m \rangle,$$
$$dp_6 : \langle h, m \rangle + nn(N, h) \rightarrow \langle h + N, m \rangle$$
$$\}$$

Fig. 5. Extraction rules

Algorithm 1. Aspect Selection by Dependency Patterns (FQDPRULES)

1: **INPUT:** S = sentences
2: **for all** s_j in S **do**
3:      $g = grammaticalRelations(s_j)$
4:      $candidateAspects = \{\}$
5:      **for** $dp_i \in DP$ and $1 \leq i \leq 4$ **do**
6:          **if** $g.matches(dp_i) \wedge g.contains(nn)$ **then**
7:              $candidateAspects. \leftarrow g.apply(\{dp_5, dp_6\})$ ▷ Apply rules dp_5, dp_6
8:          **end if**
9:      **end for**
10:     $aspects \leftarrow candidateAspects.select(N, NN)$ ▷ select nouns and compound nouns
11: **end for**
12: $filterByFrequency(aspects)$ ▷ ignore low frequency aspects
13: **return** $aspects$

4.2   Aspect Weight Extraction

Learning of aspect weights relies on preference graph and aspect level sentiment knowledge. Figure 6 and 7 illustrates the notion of preference difference calculations using a trivial three node preference graph. In Figure 6, the relation $p_3(+0.1) \succ p_1(+0.8)$ denotes product p_3 is preferred over p_1 and they have an aspect sentiment score of +0.1 and +0.8 respectively for aspect *lens* (see Equation 5). Corresponding preference difference scores are also shown in Table 3 for two aspects. Here *lens* and *screen* have a normalised aspect weights of 0.03 and 0.97 respectively. Therefore, we suggest that aspect *screen* is more important than aspect *lens*.

Table 2. Definition of grammatical relations

Example	Grammatical relations	Dependencies
The camera lens is good	nsubj (Nominal Subject)	nsubj(good, lens)
	cop (Copula)	cop(good, is)
	nn (Noun Compound Modifier)	nn(lens, camera)
The screen is bright with nice colors	amod (Adjectival Modifier)	amod(colors, nice)
She looks amazing	acomp (Adjectival Complement)	acomp(looks, amazing)
I like the camera lens	dobj (Direct Object)	dobj(like, lens)

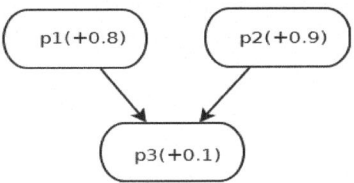 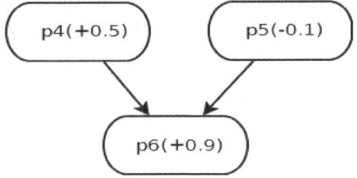

Fig. 6. Sub-graph *lens* aspect   **Fig. 7.** Sub-graph *screen* aspect

Table 3. Aspect preference scores

Aspects	Preference Relations	Preference Difference Scores	Normalised Aspect Weights
lens	$p_3 \succ p_1$	$= 0.1$	$\frac{0.1}{3.1} = 0.03$
	$p_3 \succ p_2$	$= 0.0$	
screen	$p_6 \succ p_4$	$= 1.2$	$\frac{3.0}{3.1} = 0.97$
	$p_6 \succ p_5$	$= 1.8$	

5   Evaluation

In this section we evaluate our proposed integration of sentiment and preference knowledge applied to product recommendation. Since the quality of extracted aspects in sentiment analysis will have a direct impact on the quality of recommendations; we first conduct a pilot study to evaluate the quality of aspects extracted by our algorithm with the state-of-the-art. Thereafter, we evaluate how well the recommendation system works in practice on Amazon.com data using two derived benchmark rankings. Here we are keen to study the complimentary roles of sentiment and preference knowledge and expect to discover a synergistic relation between them.

5.1   Comparative Pilot Study - Aspect Extraction Analysis

We use a public dataset on product reviews containing manually marked-up product aspects [8, 14]. For this study we use phone category products with at least hundred reviews. Precision and recall is used to compare manually labeled aspects with extracted ones using the following alternative extraction algorithms:

- FQITEMS uses shallow NLP to identify single nouns as candidate aspects that are then pruned using a frequency cut-off threshold.
- FQPOS uses Part-of-Speech(POS) extraction patterns that are then pruned using sentiment informed frequency cut-off threshold [9]
- DPRULES uses the dependency extraction rules in Figure 5 [20].
- FQDPRULES same as DPRULES but prunes candidate aspects using a frequency cut-off (See Algorithm 1).

Table 4. Results for aspect extraction pilot study

Approach	Precision	Recall	F-score
FQITEMS	0.60	0.34	0.43
FQPOS	0.71	0.11	0.20
DPRULES	0.28	0.66	0.39
FQDPRULES	0.76	0.25	0.37

Precision of all frequency based extraction approaches are significantly better compared to DPRULES (see Table 4). We also confirm that FQPOS improves over FQITEMS. As expected best results are achieved with FQDPRULES when deep NLP semantics is combined with frequency prunning. Here we observe a 26% and 7% improvement in precision over FQITEMS and FQPOS respectively. Recall trends suggests that FQDPRULES must have many false negatives and so missed extraction opportunities compared to FQITEMS and FQPOS. However a lower precision is more damaging as it is likely to introduce aspect sparsity problems which have detrimental effect on sentiment difference computations. Therefore on the basis of these results we use FQDPRULES to extract product aspects for sentiment analysis in the social recommender experiments in the next sections.

5.2   Amazon Dataset

We crawled 2264 Amazon products during April 2014. From this we use the *Digital SLR Camera* category containing more than 20,000 user generated reviews. Since we are not focusing on the cold-start problem, newer products and those without many user reviews are removed. Here we use 1st January 2008 and less than 15 reviews as the prunning factor for products. Finally, any synonymous products are united leaving us data for 80 products (see Table 5).

The FQDPRULES algorithm extracted 981 unique aspects and on average 128 different aspects for each product. Importantly more than 50% of the products shared at least 70 different aspects, while 30% shared more than 90 aspects on average. The fact that there are many shared aspects is reassuring for product comparison.

Table 5. Amazon Digital SLR Camera dataset

#Products	2,264
#Products (filtered)	80
#Reviews (filtered)	21,034
#Aspects Mean (Std. Dev.)	128.66 (43.84)
#Different aspects	981

5.3   Ranking Strategies

The retrieval set of a query product consists of products that share a similar number of aspects. This retrieval set is ranked using the following sentiment-based recommendation strategies in which the formalisations were presented in Section 4:

- BASE: recommend using sentiment in reviews (see Equation 3);
- ASPECTG: recommend using aspect sentiment analysis (see Equation 5);
- ASPECT: same as ASPECTG but without the *Gini coefficient* weighting in Equation 5; and
- ASPECTG*: same as ASPECTG but with the additional aspect weighting component in Equation 4.

We also present each of the above strategies in combination with preference knowledge (see Equation 2). Finally, the impact of increasing α values is analysed according to Equation 1 to study the relationship between sentiment and preference knowledge.

5.4   Evaluation Metrics

In the absence of a manual qualitative estimate of recommendation or access to user specific purchase trails, we derived approximations from the Amazon data we had crawled. For this purpose, using a *leave-one-out* methodology, the average gain in rank position of recommended products over the left-out query product is computed relative to a benchmark product ranking.

$$\%RankGain = \frac{\displaystyle\sum_{i=1}^{n=3} benchmark(P_q) - benchmark(P_i)}{n * |\mathcal{P} - 1|} \tag{9}$$

where n is the size of the retrieved set and *benchmark* returns the position on the benchmark list. The greater the gain over the query product the better. Suppose the query product is ranked 40th on the benchmark list of 81 unique products $\mathcal{P}$, and the recommended product is ranked 20th on this list, then the recommended product will have a relative benchmark rank improvement of 25%.

We generate two benchmark lists according to the following dimensions:

– *Star-Rating*: Use the Amazon star rating. However, this benchmark is static and tends to be less reliable as it does not necessarily represent the current trends about products. For example, the LCD display on a camera may have been a good reason for a high star rating in the past whilst now almost every camera possesses it.

– *Popular*: Derived from Amazon's reviews, questions and timeline data.

$$Popular(p) = \frac{nReviews + nQuestions}{days\_online} \qquad (10)$$

where *nReviews* and *nQuestions* refer to the number of reviews and questions of a product respectively, and *days_online* is the number of days the product has been on Amazon's website. We found that this formula has some correlation with the top 100 Amazon Best Seller ranking (*Spearman correlation* of -0.4381). Unlike Amazon's benchmark this allows us to experiment with query products that may not necessarily be in the Top 100[2].

5.5 Results

The graphs in Figures 8 and 9 show for increasing numbers of shared aspects the performance of each algorithm in terms of the average %*RankGain* on benchmarks Popular and Star-Rating. In general ASPECT and ASPECTG perform best, with both strategies recommending products with an average rank gain of close to 15% compared to BASE with no aspect extraction. It is worth pointing out that the improvement with applying the *Gini coefficient* is approximately 5% more on average when compared to not using it.

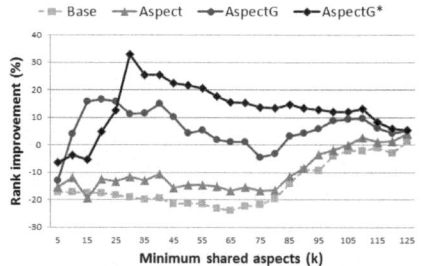

Fig. 8. Benchmark Popular

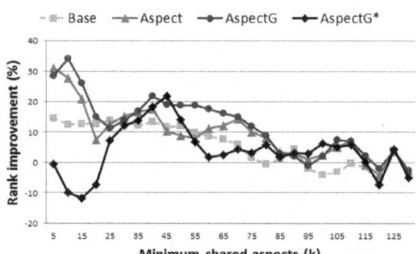

Fig. 9. Benchmark Star-Rating

Results when using aspect weights with ASPECTG* is mixed and seems dependent on the benchmark used. For instance it does very poorly on the static Star-Rating benchmark yet is the winning algorithm on the Popular benchmark. In fact it achieves up to 30% gain (with 30 shared aspects or more with the query product) on the Popular benchmark. One explanation for its poor performance on Star-Rating might be explained by the fact that user preferences

---

[2] http://www.amazon.co.uk/Best-Sellers-Electronics-Digital-SLR-Cameras

Table 6. Preferred aspects for Digital SLR Cameras in Amazon

	Top 5 Popular Aspects			Bottom 5 Popular Aspects	
Aspect	**Weight** (normalised)	**Frequency** (No. of products)	**Aspect**	**Weight** (normalised)	**Frequency** (No. of products)
shutter	0.00229	61	deal	0	67
photography	0.00181	70	lcd	$8.56 \cdot 10^{-5}$	59
point	0.00179	70	control	$2.06 \cdot 10^{-4}$	66
system	0.00176	54	day	$2.92 \cdot 10^{-4}$	70
video	0.00166	68	iso	$3.53 \cdot 10^{-4}$	67

about camera aspects may have changed during the period of 2008-2014. Since aspect weights are learnt on the basis of preference knowledge it may well be that similar information is implicitly influencing the Popular benchmark.

Table 6 shows the top 5 most and less preferred aspects together with their weights and frequency of appearance in products. Note that the sum of aspects weights is 1. Here the *shutter* aspect seems to be very important for Amazon users whilst *deal*, found in 67 different products, is about 30 times less appreciated. It is interesting to note that aspect frequencies are equally distributed, ranging from 59 to 70 in our set formed by 80 products, however the weight distribution provides a finer-grained differentiation.

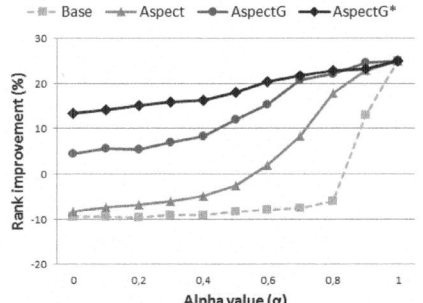

Fig. 10. *ProdRank* on Popular   **Fig. 11.** *ProdRank* on Star-Rating

Next we study the mixing parameter α using 90 shared aspects in Figure 10 and Figure 11; where high values of α mean *PrefRank* having a greater influence on *ProdRank* compared to that of *SentiRank*. In general we observe that increasing values of α lead to increasing gain in benchmark positions. This is most notable on the Popular benchmark, where it achieves a maximum of 24% with all four strategies; whilst showing only a very modest gain of 3% with $\alpha = 0.9$ on the Star-Rating benchmark.

6 Conclusion

Social recommenders have created new opportunities for CBR research based on harnessing and reusing people's online experiences. The volume and variety of social media calls for innovative extraction, representation, retrieval and knowledge integration algorithms.

In this paper we formalised the extraction and integration of two social media sources: sentiment knowledge in product reviews; and preferences from purchase summary statistics. The benefits are demonstrated in a realistic recommendation setting using benchmarks generated from social media. We show that higher precision in aspect extraction was achieved when grammatical rules are combined with support statistics. Importantly there was over 50% shared aspects on average between any product pair, providing for non-sparse aspect-level representations. We confirm that preference knowledge can be conveniently exploited using the PageRank algorithm, and demonstrated the benefits of inferring aspect weights from preference graphs.

In the absence of ground truth data, generation of benchmark data becomes important for recommender research in general. Whilst there are a variety of social data dimensions that can be used to derive benchmark metrics, further work is needed to understand the interactions between these dimensions. Our results show that the combination of sentiment and preference knowledge are promising, but further work is needed to study closely the role of each on different domains and benchmarks. Finally it would also be interesting to explore how aspect importance weights are likely to evolve with context and time.

Acknowledgments. This research has been partially supported by AGAUR Scholarship (2013FI-B 00034) and Project Cognitio TIN2012-38450-C03-03.

References

1. Adomavicius, G., Tuzhilin, A.: Toward the next generation of recommender systems: A survey of the state-of-the-art and possible extensions. IEEE Transactions on Knowledge and Data Engineering 17 (2005)
2. Billsus, D., Pazzani, M.: A hybrid user model for news story classification. Courses and Lectures - Inter. Center for Mechanical Sciences 99 (1999)
3. Blair-Goldensohn, S., Hannan, K., McDonald, R., Neylon, T., Reis, G., Reynar, J.: Building a sentiment summarizer for local service reviews. In: WWW Workshop on NLP in the Information Explosion Era, p. 14 (2008)
4. Brin, S., Page, L.: The anatomy of a large-scale hypertextual web search engine. Computer Networks and ISDN Systems 30(1), 107–117 (1998)
5. Burke, R.: Hybrid recommender systems: Survey and experiments. User Modeling and User-Adapted Interaction, 331–370 (2002)
6. Cao, B., Shen, D., Wang, K., Yang, Q.: Clickthrough log analysis by collaborative ranking. In: AAAI Conf. on Artificial Intelligence (2010)

7. De Marneffe, M., MacCartney, B., Manning, C., et al.: Generating typed dependency parses from phrase structure parses. In: Proc. of Language Resources and Evaluation Conference, pp. 449–454 (2006)
8. Ding, X., Liu, B., Yu, P.S.: A holistic lexicon-based approach to opinion mining. In: Proc. Int. Conf. on Web Search and Data Mining (2008)
9. Dong, R., Schaal, M., O'Mahony, M.P., McCarthy, K., Smyth, B.: Opinionated product recommendation. In: Delany, S.J., Ontañón, S. (eds.) ICCBR 2013. LNCS, vol. 7969, pp. 44–58. Springer, Heidelberg (2013)
10. Esuli, A., Sebastiani, F.: Sentiwordnet: A publicly available lexical resource for opinion mining. In: Proc. Language Resources and Evaluation Conference, pp. 417–422 (2006)
11. Gauch, S., Speretta, M., Chandramouli, A., Micarelli, A.: User profiles for personalized information access. In: Brusilovsky, P., Kobsa, A., Nejdl, W. (eds.) Adaptive Web 2007. LNCS, vol. 4321, pp. 54–89. Springer, Heidelberg (2007)
12. Groh, G., Birnkammerer, S., Köllhofer, V.: Social recommender systems. In: Pazos Arias, J.J., Fernández Vilas, A., Díaz Redondo, R.P. (eds.) Recommender Systems for the Social Web. ISRL, vol. 32, pp. 3–42. Springer, Heidelberg (2012)
13. Gupta, S., Chakraborti, S.: UtilSim: Iteratively helping users discover their preferences. In: Huemer, C., Lops, P. (eds.) EC-Web 2013. LNBIP, vol. 152, pp. 113–124. Springer, Heidelberg (2013)
14. Hu, M., Liu, B.: Mining and summarising customer reviews. In: Proc. of ACM SIGKDD Inter. Conf. on Knowledge Discovery and Data Mining, KDD 2004, pp. 168–177 (2004)
15. Kim, S., Pantel, P., Chklovski, T., Pennacchiotti, M.: Automatically assessing review helpfulness. In: Proc. Conf. on Empirical Methods in Natural Language Processing, pp. 423–430 (2006)
16. Linden, G., Smith, B., York, J.: Amazon.com recommendations: Item-to-item collaborative filtering. IEEE Internet Computing 7(1), 76–80 (2003)
17. McCarthy, K., Salem, Y., Smyth, B.: Experience-based critiquing: Reusing critiquing experiences to improve conversational recommendation. In: Bichindaritz, I., Montani, S. (eds.) ICCBR 2010. LNCS, vol. 6176, pp. 480–494. Springer, Heidelberg (2010)
18. McGinty, L., Smyth, B.: Collaborative case-based reasoning: Applications in personalised route planning. In: Aha, D.W., Watson, I. (eds.) ICCBR 2001. LNCS (LNAI), vol. 2080, pp. 362–376. Springer, Heidelberg (2001)
19. Moghaddam, S., Ester, M.: Opinion digger: An unsupervised opinion miner from unstructured product reviews. In: Proc. Inter. Conf. on Information and Knowledge Management, CIKM 2010 (2010)
20. Moghaddam, S., Ester, M.: On the design of lda models for aspect-based opinion mining. In: Proc. Inter. Conf. on Information and Knowledge Management, CIKM 2012 (2012)
21. Muhammad, A., Wiratunga, N., Lothian, R., Glassey, R.: Contextual sentiment analysis in social media using high-coverage lexicon. In: Research and Development in Intelligent Systems, pp. 79–93 (2013)
22. Popescu, A., Etzioni, O.: Extracting product features and opinions from reviews. In: Natural Language Processing and Text Mining, pp. 9–28. Springer, London (2007)
23. Quijano-Sánchez, L., Bridge, D., Díaz-Agudo, B., Recio-García, J.A.: Case-based aggregation of preferences for group recommenders. In: Agudo, B.D., Watson, I. (eds.) ICCBR 2012. LNCS, vol. 7466, pp. 327–341. Springer, Heidelberg (2012)

24. Singla, A., White, R., Huang, J.: Studying trailfinding algorithms for enhanced web search. In: Proc. ACM SIGIR Conf. on Information Retrieval, pp. 443–450 (2010)

25. Turney, P.: Thumbs up or thumbs down?: semantic orientation applied to unsupervised classification of reviews. In: Proc. Annual Meeting on Association for Computational Linguistics, pp. 417–424 (2002)

26. Vasudevan, S., Chakraborti, S.: Mining user trails in critiquing based recommenders. In: Proc. Inter. Conf. on World Wide Web Companion, pp. 777–780 (2014)

27. Yitzhaki, S.: Relative deprivation and the Gini coefficient. The Quarterly Journal of Economics, 321–324 (1979)

A Hybrid CBR-ANN Approach
to the Appraisal of Internet Domain Names

Sebastian Dieterle and Ralph Bergmann

Business Information Systems II
University of Trier
54286 Trier, Germany
sebastiandieterle@web.de, bergmann@uni-trier.de
www.wi2.uni-trier.de

Abstract. Good domain names have become rare and trading with premium domain names has developed into a profitable business. Domain appraisals are required for many different reasons, e.g., in connection with a loan on a domain name. The aim of this paper is to analyze various methods for estimating prices for domain names. The criteria for this are predictive accuracy, traceability and speed of the appraisal. First, the scientific relevance of the topic is demonstrated based on intensive literature and Internet research. Several approaches based on artificial neural networks (ANNs) and case-based reasoning (CBR) are developed for estimating domain name prices. In addition, hybrid appraisal approaches are introduced that are built up on CBR and which use ANN for improved adaptation and similarity determination. The approaches are evaluated in several configurations using a training set of 4,231 actual domain transactions, which demonstrates their high usefulness.

Keywords: Internet Domain Names, Artificial Neural Networks, Hybrid Application, Appraisal, Similarity, Adaptation.

1   Introduction

Domain names are often seen as the land, and web sites as the buildings, of the virtual world. Due to their uniqueness, premium domain names achieve high market prices. There are many reasons why the fair market value of domain names must be appraised. The manual appraisal of domain names is subjective, time-consuming and expensive [1]. Existing approaches to the application of case-based reasoning (CBR) in the appraisal of assets [6–9] show three fundamental weaknesses: First, the rules for price adjustment often rely on the experience of experts and are therefore not empirically justified. Second, optimized weights are not used in determining similarity, or, third, optimized weights are only applicable to numeric attributes. Knowledge-intensive similarity measures are not used in appraisal methods based on locally weighted regression (LWR). The realization of an appraisal by means of an artificial neural network (ANN) [11, 12] is not traceable for users due to its black box character and is thus not allowed for

L. Lamontagne and E. Plaza (Eds.): ICCBR 2014, LNCS 8765, pp. 95–109, 2014.
© Springer International Publishing Switzerland 2014

an official legal appraisal [13]. In order to address these weaknesses, we propose
hybrid approaches that combine CBR and ANN in different ways. We apply the
traditional CBR approach to retrieve recently sold, similar domain names from
a case base, and adapt the sales price of the cases with respect to the relevant
differences to the query, i.e., the domain name to be appraised [9]. In this process,
the ANN is used for two purposes: a) the weights for determining similarity are
learned by an ANN, and b) the parameters of the adaptation rules for adjusting
the price are also determined by an ANN. In addition, a different combination
of CBR and ANN is proposed in which CBR is applied to pre-select cases to
train an ANN being used for an appraisal based on LWR.

In Section 2, we introduce the related work. The basic approaches to case-
based and neural domain appraisal are described in Section 3. In Section 4, three
hybrid approaches are presented. An empirical evaluation in Section 5 based on
4,231 cases of domain transactions tests various hypotheses by experiments. The
last section concludes with a summary and an outlook on future work.

2   Related Work

An asset can be evaluated in three different ways: based on acquisition costs, on
income, or on market price [2]. The value of an asset (such as real estate) is determined by an appraiser finding recently sold real estate with similar characteristics
in the neighborhood. These prices must be adjusted to increases and deductions,
since no two properties can be compared exactly [3]. The 3Cs appraisal model
from GreatDomains.com was the first to describe factors for domain appraisal.
By means of a matrix, the criteria of *characters* (number of characters), *commerce* (commercial potential) and *.com* (value relevance of the TLD) determine
the value of a domain. Multiple linear regression analyses were generally used.
In the hedonic regression according to Phillips [4], the time factor is taken into
account by pre-processing the data with the Morgan Stanley Internet Index. The
distinguishing feature of Jindra's regression analysis [5] is that more than one
regression model is calculated. Instead, the data are split up into four clusters
and a regression equation is determined for each cluster.

The first use of CBR for the appraisal of real estate was published by Gonzalez & Laureano-Oritz [6]. Compared to regression analysis, this technique more
closely resembles the way real estate appraisers work and is easier for users to
understand and trace. The prices of the most similar previous cases are adjusted
in accordance with heuristic rules and a weighted average value is calculated. The
case-based appraisal of rental prices in the retail trade was studied by O'Roarty
et al. [7]. They transform the rental prices from different years by means of a
rental price index to a standardized level and leave any further price adjustment
to the user. McSherry [8] presents a domain-independent adaptation heuristic
based on the assumption of an additive valuation function and the existence of
certain specific case pairs. The case-based appraisal of domain names by Dieterle
& Bergmann [9] uses knowledge-intensive similarity functions. The adjustment
of the prices is based, among other things, on the Internet Domain Name Index

(IDNX). The case-based appraisal approaches often use heuristic domain knowledge for adjusting prices and determining similarity. Furthermore, price indices consider only the time aspect; standardized regression coefficients are restricted to numeric attributes. In this paper, we shall introduce approaches for the learning of adaptation rules that do not require any specific case pairs and for the learning of weights for the aggregation of arbitrary local similarity measures.

ANN constitute a model which is inspired by the nerve activities of the brain and which allows, among other things, the approximation of linear and nonlinear functions [10]. In the application field of the appraisal of real estate, linear regression is compared to a multi-layer perceptron (MLP) trained with backpropagation. Whereas Rossini [11] came to the conclusion that linear regression produces lower errors, Peterson & Flanagan [12] came to the opposite conclusion on a significantly larger data set. In comparison with CBR, however, both methods are less easy for the user to understand and trace, since the appraisal is based only implicitly on previous sales transactions. Because of the black box nature of ANNs, people are unable to trace the appraisal process and it must not be used as the basis for an official legal appraisal [13].

Hybrid methods make use of the existence of different strengths and weaknesses of the individual methods. On the basis of a linear regression analysis, Rossini [14] calculates additive adjustment rules, to use these by means of the technique of the k nearest neighbors to adjust previous transactions in real estate appraisal. Al-Akhras [15] uses an evolutionary genetic algorithm in order to determine the best topology for an MLP for real estate appraisal, and then trains this MLP with backpropagation. Jalali & Leake propose a hybrid approach to estimate car and house prices [16]. Here, the squared error weighted by an exotic distance function over the k nearest neighbors to a query case is minimized to determine a regression model and thus to appraise a query. One difference between the mentioned hybrid approaches and the approach we present in this paper is the fact that we use a knowledge-intensive similarity measure, so that the determined cases are also similar from a semantic point of view.

3 Case-Based and Neural Network Approaches

3.1 CBR for Domain Appraisal

In our previous work [9], we presented an approach to the case-based appraisal of domain names based on a case base with previous sales transactions. Case attributes are selected such that they allow an appropriate determination of similarity between two domain names. The sales price attribute is the solution attribute stored in each case. The overall CBR approach applied is shown in the top box of Fig. 1. In order to estimate the value of a domain name, it is entered as a query. In the *recall* step, the relevant features of the corresponding domain name are derived and further data are extracted from the Internet, leading to an enriched query (see also the left column of Table 1). In the *retrieve* step, a number of k similar cases are determined from a case base. For this purpose, a knowledge-intensive similarity measure based on the derived features is used.

A weighted average of the local similarity values is used to compute the global similarity. The weights are determined in a heuristic process by experimenting with several variants. In the *reuse* step, the solution attribute (i.e., the sales price) is adapted for each of the k most similar cases. The price index IDNX is used as heuristic expert knowledge for adaptation. In the *remove* step outliers are excluded from further calculations and in the *reckon* step a weighted average of the remaining values is calculated. The weights used for this purpose are derived from the similarity between case and query. This weighted average value represents the estimated value for the query.

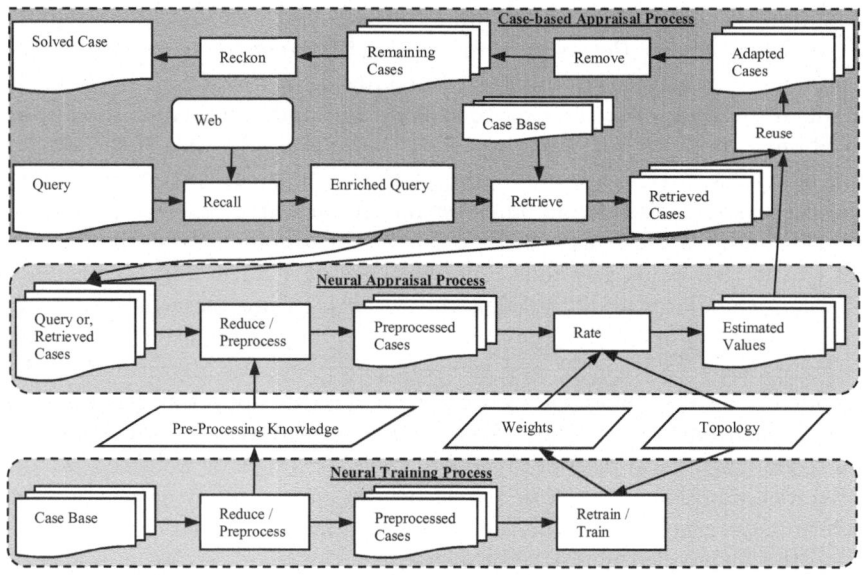

Fig. 1. Hybrid CBR/ANN approach for domain appraisal (this may also serve as an illustration for individual methods)

3.2 ANN for Domain Appraisal

Our overall neural approach to the appraisal of domain names [17] consists of a training process and an appraisal process (see the two bottom boxes in Fig. 1). In the *training process*, a *reduce* step transforms the case base from its original symbolic representation (Section 3.3) into a numeric vector representation suitable for an ANN. This transformation performs a stemming (Section 3.4), the derivation of binary attributes, and a normalization (Section 3.5). Thereby, pre-processing knowledge is generated and stored (e.g., attribute specific normalization parameters or correspondences between binary attributes and stemmed words). The pre-processed training set is used to *retrain* or train a given network structure of an ANN, such as an (adaptive) linear neuron (Adaline) or a MLP with supervised learning, such as resilient propagation (Section 3.6). This method should be able to replicate the prices of the training examples as accurately as possible and at the same time be able to generalize to new problems. The network weights learned during the training are saved. The *appraisal process* describes the procedure for estimating the price of a query domain name. For this purpose, the query is enriched with the data *recalled* from the web (as in the case-based approach) leading to the enriched case representation of the query. In the *reduce* step, this query representation is transformed into the vector format using the same processing steps as in the training phase. Finally, in the *rate* step, the pre-processed query is presented as input to the trained network and the computed network output produces the estimated price for the query domain.

3.3   Case Representation

For the neural and case-based approach, we use two different case/data representations; however, in the hybrid models (Section 4) we combine the two models. The case representation follows a structural CBR approach and contains multi-valued, taxonomic, numeric and textual attributes, in order to enable knowledge-intensive similarity measures [9]. However, the neural data model consists of a numeric vector with a large number of binary and numeric attributes. In Table 1, the two models are compared, showing the case of "winterreise.de" (German for "winter journey").

The attribute "sales price", which describes at what price the domain name was traded, is the only solution attribute and output neuron. Differently, the transaction year is taken into account in the neural data model, with a range of seven binary attributes for each year, in order to take into account fluctuations in price level over time. The length of a domain indicates the number of characters contained in the second-level domain (SLD). This feature is considered to be a key criterion for the domain value, since short domain names can be more easily remembered and typed [4]. The case-based approach contains the multi-valued attribute "categories" (in the Open Directory Project (ODP)), which allows a hierarchical classification of all websites worldwide and the use of taxonomic similarity measures.

The attribute "word components" differs between the two models: Whereas in the case-based model the word components "winter" and "reise" are the values of this multi-valued attribute and allow the use of textual similarity measures, in the neural data model only binary attributes are used for frequent word stems.

Additional attributes describe how many results there are for a term in a search engine, whether the term contains hyphens, special characters or numbers and the age of the domain. Furthermore, consideration is given to how often a term is searched for in Google worldwide and in the region related to the top-level domain (TLD), the average cost per click (CPC), and the number of clicks.

Our approaches can also be applied to other data sets and application domains (e.g. the appraisal of businesses, art, cars, etc.). For this purpose, a different case description (and appropriate local similarity measures) must be defined.

Table 1. Description of the training example "winterreise.de" from 2007

Attribute	Example (CBR)	Example (ANN)
Domain name	winterreise.de	
SLD	winterreise	
TLD	de	
Transaction year /	2007	
2006, 2007, …, 2011, 2012		0; 1; 0; 0; 0; 0; 0
Length	11	11
Single- / 2- / 3-letter domain		0; 0; 0
Number of words	1	1
Number of word components	2	2
Categories (in the ODP)	World: German: Recreation:	
	Travel: Travelogues;	
	World: German: Sports: Winter	
	Sports: Skiing: Journeys	
Words	Winterreise	
Word components /	Winter; Reise	
24, angebo, …, reis, …, www		0; 0; …; 1; …; 0
Search results	158,000	158,000
Contains hyphen	false	0
Number of hyphens	0	0
Contains special characters	false	0
Contains numbers	false	0
Domain age	1999.08333333	1999.08333333
Global monthly searches	5,400	5,400
Local monthly searches	1,600	1,600
Avg. CPC in €	0.81	0.81
Daily clicks	4.52	4.52
Daily cost in €	3.64	3.64
Sales price in €	10,000	10,000

3.4 Reduction Step and Stemming

In the reduction step, the training examples and the query are transformed into a vector form applicable to the ANN. For this purpose, binary attributes are introduced, on the one hand for the transaction year, and on the other hand for frequent word stems. Pre-processing knowledge is saved in regards to which binary attributes exist and occur in at least ten cases. Ten is a usual rule of thumb (see [18]) and represents a compromise between the regression model having a good capability to generalize and taking into account the maximum amount of information possible. First of all, the SLD was decomposed into the morphemes contained within it. For this purpose, the word is cut into two parts ($bookworm, b$ookworm, bo$okworm, etc.) and in each case the number of hits in the search engine is determined. Assuming that it exceeds a predefined threshold, the second most frequent spelling contains the SLD, separated into morphemes. To reduce the morphemes to their stem form (stemming), on one hand the approach in accordance with Caumanns [19] and on the other hand an algorithm defined by a snowball script[1] is used for German words running one after the other. In the first step, the process replaces particular strings in the words to be stemmed, e.g., replacing mutated vowels by the corresponding vowel. In the second step, suffixes are pruned away using a set of rules.

3.5   Normalization by Logarithmizing

Besides the semantic attributes, the numeric attributes must also be transformed into a form applicable to the ANN. This concerns, on the one hand, the logarithmic transformation of the data and, on the other hand, the compression of the data into a specific values range. The log of all the input values and of the output value is calculated and in addition an attribute-specific constant is included. This frequently used pre-processing step smoothes rapidly increasing values ranges.

3.6   Training an ANN with Resilient Propagation

Due to our data model, the ANN has 112 input neurons and one output neuron for the estimated price. The linear neuron (Adaline) and the MLP are considered as a topology. The MLP contains inner layers with the hyperbolic tangent function as an activation function and allows for the approximation of non-linear functions (non-linear regression). The linear neuron has no inner layers and allows a linear regression analysis. The logarithmic pre-processing of the attribute values results in a multiplicative value relationship between input and output values.

A variant of resilient propagation [21], iRPROP+ [20], is used as a supervised learning method to train the neural network. Here, the harmful influence of the partial derivation (i.e., the risk of a too great weight adjustment by a gradient, which is too steep at particular positions) is avoided by taking into account only the algebraic sign of the gradient. For each weight, this iterative method has its own weight-specific update value, which changes during the learning process.

4   Hybrid Approaches to Domain Appraisal

We now describe three variants for the integration of CBR and ANN. A common characteristic of our hybrid approaches is that an extended query description is generated and that similar cases are retrieved and reused in some way. Moreover, a neural network is trained and used for different purposes.

4.1   Hybrid ANN Adaptation

The first hybrid method applies the ANN using the Adaline topology to determine multipliers for the adaptation of the solution attribute, i.e., the sales prices

---

[1] http://snowball.tartarus.org/algorithms/german2/stemmer.html

of the retrieved cases. Figure 1 as a whole – and in particular the links between the case-based and the neural appraisal box – show the links between both methods. This hybrid appraisal process begins when a new query domain q is entered. As in the pure CBR approach, the *recall* and the *retrieve* step are performed and the k most similar cases for the query are determined using a knowledge-intensive similarity measure. The subsequent *reuse* phase is supported by the ANN. As described in Section 3, the neural appraisal process is capable of rating domain names on a stand-alone basis. The query and the k most similar cases are now assessed by means of the neural appraisal process. Thus, the *rate* step generates as output the estimated value for them. $v_n(q)$ is the estimated value for the query and $v_n(c)$ is the estimated value for a retrieved case c. These estimated values are now used in the *reuse* phase of the case-based appraisal process to adjust the price attribute of the k most similar cases according to the query. The adapted sales price of a case c, called c'_p, results from the sales price obtained from a case c_p multiplied by the ratio between the neural estimated value of query q and case c:

$$c'_p := c_p \cdot \frac{v_n(q)}{v_n(c)} \tag{1}$$

Due to the logarithmic pre-processing of the input and output data in the linear neuron, the estimated values $v_n(q)$ and $v_n(c)$ can be split into a basic value (bias) and local multipliers for every attribute (cf. [17]). This applies to the quotient of the above equation, leading to the following adaptation formula:

$$c'_p := c_p \cdot \prod_{a \in A} \underbrace{\frac{(i_a(q_a))^{w_a}}{(i_a(c_a))^{w_a}}}_{local\ multiplier} \tag{2}$$

The actual sales price of a case is thus adapted by a number of local multipliers, one for each attribute $a \in A$. A local multiplier results from the quotient of an attribute-specific value for query and case. Here, $i_a(.)$ is the normalization function applied to the case/query value q_a/c_a of attribute a. w_a is the weight of attribute a that results from the training of the Adaline. A valuation of the two domain names "asienurlaub.de" (query) and "ayuvedareisen.de" (case) is illustrated in Fig. 2. For the case and the query, the second and third column, respectively, show the multipliers per attribute. The last line shows the resulting price estimated by the output neuron.

The following steps proceed according to the CBR appraisal approach, i.e., outliers are *removed* and the weighted average value of the remaining similar cases is determined (*reckon*). Fig. 3 shows results of the hybrid appraisal of the query "asienurlaub.de". The eleven most similar cases are determined, the relevant differences between case and query are adjusted by means of multipliers, and a weighted average price is calculated.

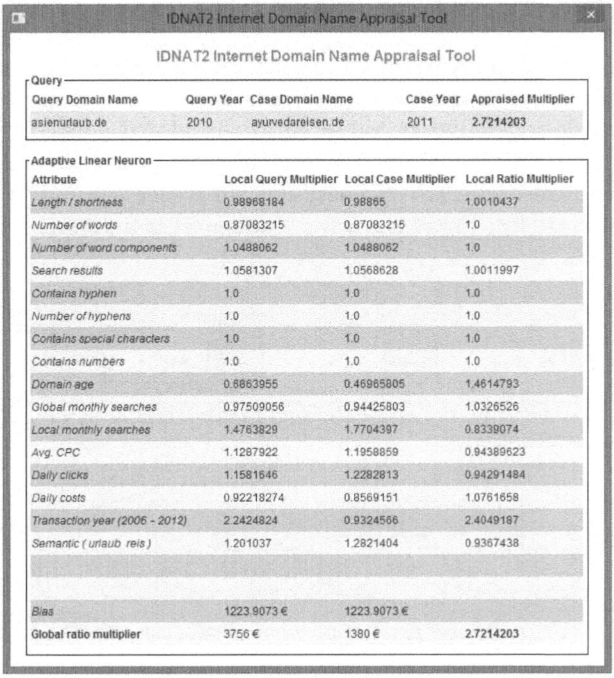
IDNAT2 Internet Domain Name Appraisal Tool
IDNAT2 Internet Domain Name Appraisal Tool
Query
Query Domain Name Query Year Case Domain Name Case Year Appraised Multiplier
asienurlaub.de 2010 ayurvedareisen.de 2011 2.7214203
Adaptive Linear Neuron
Attribute Local Query Multiplier Local Case Multiplier Local Ratio Multiplier
Length / shortness 0.98968184 0.98865 1.0010437
Number of words 0.87083215 0.87083215 1.0
Number of word components 1.0488062 1.0488062 1.0
Search results 1.0581307 1.0568628 1.0011997
Contains hyphen 1.0 1.0 1.0
Number of hyphens 1.0 1.0 1.0
Contains special characters 1.0 1.0 1.0
Contains numbers 1.0 1.0 1.0
Domain age 0.6863955 0.46965805 1.4614793
Global monthly searches 0.97509056 0.94425803 1.0326526
Local monthly searches 1.4763829 1.7704397 0.8339074
Avg. CPC 1.1287922 1.1958859 0.94389623
Daily clicks 1.1581646 1.2282813 0.94291484
Daily costs 0.92218274 0.8569151 1.0761658
Transaction year (2006 - 2012) 2.2424824 0.9324566 2.4049187
Semantic (urlaub reis) 1.201037 1.2821404 0.9367438
Bias 1223.9073 € 1223.9073 €
Global ratio multiplier 3756 € 1380 € 2.7214203

Fig. 2. Determination of multipliers for a query and a similar case

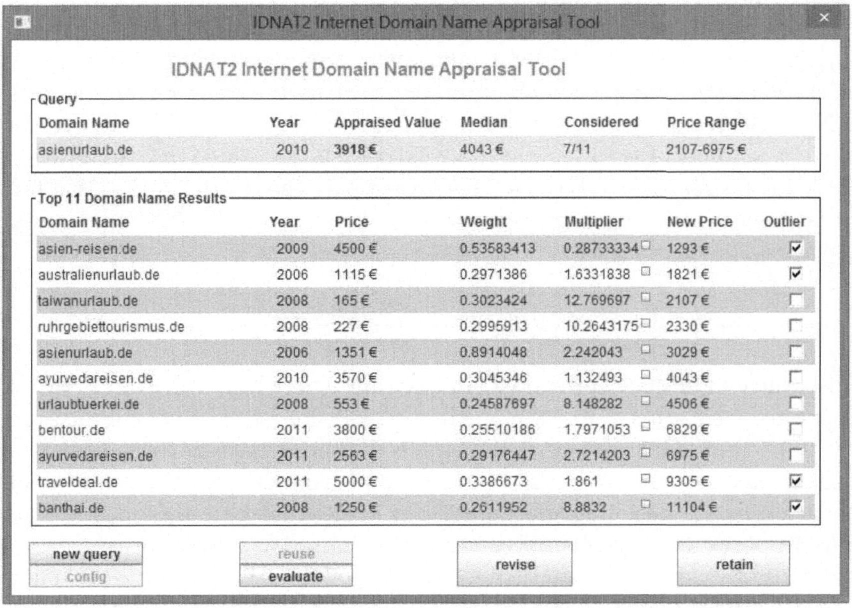
IDNAT2 Internet Domain Name Appraisal Tool
IDNAT2 Internet Domain Name Appraisal Tool
Query
Domain Name Year Appraised Value Median Considered Price Range
asienurlaub.de 2010 3918 € 4043 € 7/11 2107-6975 €
Top 11 Domain Name Results
Domain Name Year Price Weight Multiplier New Price Outlier
asien-reisen.de 2009 4500 € 0.53583413 0.28733334 1293 €
australienurlaub.de 2006 1115 € 0.2971386 1.6331838 1821 €
taiwanurlaub.de 2008 165 € 0.3023424 12.769697 2107 €
ruhrgebiettourismus.de 2008 227 € 0.2995913 10.2643175 2330 €
asienurlaub.de 2006 1351 € 0.8914048 2.242043 3029 €
ayurvedareisen.de 2010 3570 € 0.3045346 1.132493 4043 €
urlaubtuerkei.de 2008 553 € 0.24587697 8.148282 4506 €
bentour.de 2011 3800 € 0.25510186 1.7971053 6829 €
ayurvedareisen.de 2011 2563 € 0.29176447 2.7214203 6975 €
traveldeal.de 2011 5000 € 0.3386673 1.861 9305 €
banthai.de 2008 1250 € 0.2611952 8.8832 11104 €
new query
config
reuse
evaluate
revise
retain

Fig. 3. GUI of the hybrid price adjustment of domain names

4.2   Hybrid ANN Similarity

The purpose of the second hybrid approach is to improve the assessment of similarity by the optimization of the weights used in the weighted average aggregation of the local similarity values into the global similarity. Please note that we use various knowledge-intensive local similarity measures, such as numeric, taxonomic, and textual similarity measures (see 3.1 and [9]). The hybrid approach is illustrated in Fig. 4, showing the neural training and the case-based appraisal process.

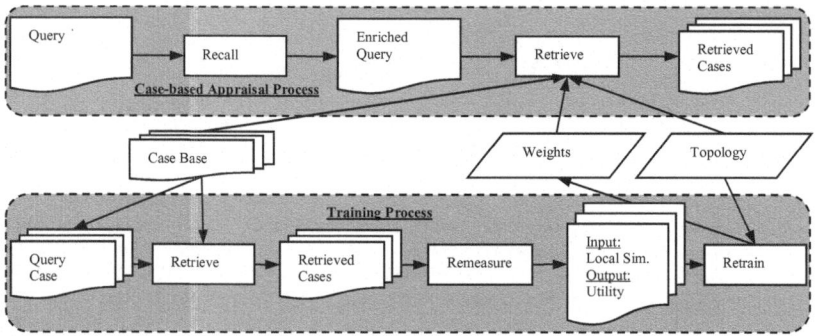

Fig. 4. Optimization of weights by a neural network

The training process makes use of the existing case base. Iteratively, each case in the case base is used as a query q and the k most similar cases to it are determined (*retrieve* step) by means of the non-optimized (initial) similarity measure. The retrieved query-case pairs (q, c) are used to derive training data for the ANN (which is different to the ANN of the first hybrid approach). To ensure a sufficient amount of training data, a relatively large value for k is chosen (e.g., $k = 50$). In the *remeasure* step, for each case pair the *utility* of the case for the query is determined, i.e., how well the sales price c_p of the case c predicts the sales price q_p of the query q.

$$utility(q, c) := \frac{min(q_p, c_p)}{max(q_p, c_p)} \tag{3}$$

This utility value measures how similar (and therefore how useful) two solutions are to each other, i.e., it is a kind of solution similarity. For instance, a utility of 0.5 results if the price of the case is double or half the price of the query. The subsequent *retrain* step uses the linear neuron as a topology and resilient propagation as the learning method. Each query-case pair provides a training sample. In particular, the vector of the local similarity values between case and query attribute values represent the net input. The utility value represents the desired net output.

Due to the optimization process which the ANN performs during learning, the attribute weights are adjusted as follows: given a query-case pair with a high utility value, the attributes with a high local similarity will get a high weight. Conversely, attributes with a low similarity will receive a low weight. Thus, the *utility* value and the weighted sum of the local similarity measures should approximate one another. The network thus minimizes the difference between these two values across all query-case pairs. The purpose of this is, on the one hand, to weight cases that have achieved similar sales prices in such a way that they achieve the highest possible similarity to one another and, on the other hand, to weight cases that show different sales prices in such a way that they achieve the lowest possible similarity to one another. This should ensure that the most similar cases are also the cases which have, as far as possible, achieved the most similar prices.

Similarly, an MLP trained with utility values can also be used to transform the local similarity measures into a global similarity measure by forward propagation. However, the resulting similarity assessment is less transparent, as the forward propagation cannot be traced.

4.3   Hybrid LWR Based on CBR

The last hybrid approach uses case-based retrieval, i.e., a pre-selection by means of CBR, as a pre-processing step for an ANN. This results in a form of LWR, in which cases which lie further away from the query have no influence on the determined regression model. The approach consists of three phases, as illustrated in Fig. 5: a case-based retrieval process, a neural appraisal process, and a neural training process.

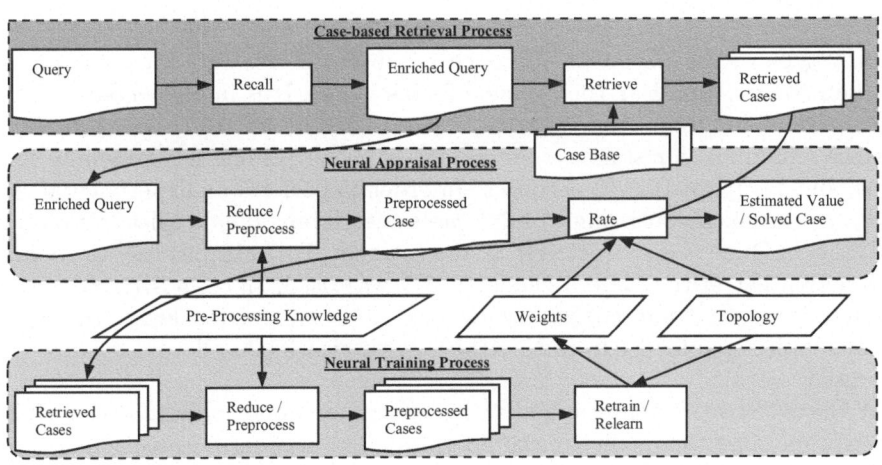

Fig. 5. Case-based pre-selection for a neural network

In the case-based *retrieval* process, the k most similar cases are determined for the query by means of a knowledge-intensive similarity measure. For this purpose, a relatively large value for k is selected (e.g., 20 % of the case base), in order to be able to determine a regression model capable of generalization.

The reuse step is now implemented in the form of the neural appraisal and training process. In the *reduce* step, the query on the one hand and the k most similar cases on the other hand are transformed from the case-based format into the numeric vector format, as described before. In the *retrain* phase the ANN is trained for each query with the preselected training set, i.e., the determined k most similar cases. This repeated training for each individual query is a major difference to the two previous hybrid approaches, in which the training is executed only once. This has the consequence that the linear neuron is especially optimized for the local (and semantic) neighborhood of the query. The training processes minimizes the following error function:

$$E := \sum_{c \in Ret(q)} (f(c) - c_p)^2 \tag{4}$$

Thereby, $Ret(q)$ represents the k most similar cases retrieved for the query q, c_p is the actual sales price of case c, and $f(c)$ is the value predicted by the ANN for c. The LWR only requires a simple topology, and often even a linear model is sufficient. In the *rate* step which follows, the pre-processed query is presented to the retrained ANN as input and thus an estimated value for the query is calculated by forward propagation. A fundamental difference to the ordinary LWR is that, in this approach, semantically similar cases are retrieved by a knowledge-intensive similarity measure.

5   Empirical Evaluation

A prototype of the proposed approach has been implemented, called Internet Domain Name Appraisal Tool Version 2 (IDNAT2). It is implemented in JAVA using various program libraries, such as jCOLIBRI2 for CBR, Encog3 for ANN, Apache Commons for statistical functions, Apache Lucene for stemming functions, and Jsoup for HTML decoding. In order to perform an experimental evaluation, a case base consisting of 4,231 cases describing domain sales transactions with the TLD .de was extracted from the Internet. For this purpose, the domain transaction list of the United-Domains AG was used, since it is easily accessible and relatively comprehensive with over 1,000 .de entries. Furthermore, the world's largest public list from Namebio - with over 3,000 relevant entries - was also used.

We have considered three quantitative criteria and one qualitative criterion in order to measure the quality of the appraisal. The standard evaluation criterion of the predictive accuracy of a regression analysis is the squared correlation coefficient R^2 (in this study, between the predicted value calculated by IDNAT2 and the actual sales price). However, there are some very highly priced values among the domain prices, which is why we use logarithmic values ($log(v+10)$) to keep the positive or negative influence of single values on the R^2 small. Moreover, we have measured the time required to solve an appraisal query and the time required to train the ANN. As a qualitative criterion, we have considered how traceable - and thus how reliable - the solution is. As a validation design for all the experiments, an n-fold cross validation was used with four test sets performed on an ASUS desktop PC (CM6870 Series).

In the context of the evaluation, the following three hypotheses were tested:
H1: The hybrid ANN adaptation achieves higher predictive accuracy than CBR without adaptation.
H2: The hybrid ANN similarity (with optimized weights) achieves higher predictive accuracy than CBR with heuristic, manually assigned weights.
H3: The hybrid LWR based on CBR achieves higher predictive accuracy than a pure neural network (Adaline or MLP).

Table 2. Comparison of the appraisal approaches on the test set

Approach	R^2	Query time / training time	Trace- ability
Pure CBR no adaptation	0.315	5.4 s / -	very good
Hybrid ANN adaptation	0.492	5.6 s / 5 s	excellent
Hybrid ANN similarity	0.453	5.4 s / 4.7 h	good
Hybrid ANN adaptation and similarity	0.516	5.6 s / 4.7 h	very good
Hybrid LWR based on CBR ($k = 10$)	0.062	5.7 s / -	below avg.
Hybrid LWR based on CBR ($k = 100$)	0.334	5.8 s / -	below avg.
Hybrid LWR based on CBR ($k = 800$)	0.532	7.0 s / -	below avg.
Pure Adaline	0.519	0.0 s / 5 s	avg.
Pure MLP	0.506	0.0 s / 10 s	none

The results of the evaluation in Table 2 confirm the hypothesis **H1**, which implies that the pure case-based approach without adaptation has lower predictive accuracy. The R^2 is considerably higher with the ANN adaptation process. The process with adaptation is only slightly slower regarding the calculation time for a query. An advantage relating to traceability is that the k most similar prices and their weights are displayed as an explanation. The reliability is increased because the relevant differences between case and query are compensated in a manner which is transparent to the user.

Additionally, the hybrid ANN similarity approach outperforms the pure CBR approach with heuristically assigned weights in terms of predictive accuracy, and thus confirms hypothesis **H2**. This advantage, however, works to the disadvantage of the other two criteria. The query time is unchanged but the ANN similarity approach has a very long training time. Moreover, the traceability of the results decreases, since in the experiments it turned out that the attributes reflecting the content of the domain received a smaller weight through the training, and thus the cases found appear less similar to the user.

Finally, the hybrid LWR approach is compared with a pure Adaline approach (10,000 iterations) and a pure MLP approach (two inner layers, each of 100 neurons, 100 iterations). The results in Table 2 confirm the hypothesis **H3** since, with 800 cases, a slightly higher predictive accuracy can be achieved with the hybrid LWR. However, the hybrid LWR has the highest query time of all the approaches presented, since the retrieval of a large number of similar cases is time-consuming and ANN training occurs for every query. Since the ANN is trained for every query, two domain names with an equal attribute value may receive a different local multiplier. In addition, the basic value (bias) is no longer the same for every domain. Hence, the hybrid LWR appraisal is very difficult to trace.

6   Conclusion and Future Work

In this paper, three hybrid approaches were introduced for the first time to domain appraisal. It has been shown that the predictive accuracy of a case-based system can be clearly increased by neural price adjustment and/or by the learning of weights. If a neural network is trained with cases in the local neighborhood of the query, then the highest predictive accuracy is achieved. For the user to have confidence in the appraisal, it is important that the method used to determine the estimated price on the basis of semantically similar cases should be traceable. Therefore, the combination of knowledge-intensive similarity measures with the adaptive character of neural networks is a key feature of the approaches presented in this paper. Depending on the application scenario (such as the mass appraisal of large domain portfolios), it is necessary to select the approach in which the criteria of traceability, predictive accuracy, and speed stand in the best relationship one to another. Genetic algorithms [15] could be used in future studies to optimize local similarity functions and local multipliers for adaptation. Parallelization with Hadoop or Amazon EC2, and improved index techniques (such as the Mac/Fac model [22] or cluster-based retrieval [23]), could in future extend the sequential retrieval. This seems particularly promising when the case base grows or when a mass appraisal of large domain portfolios needs to be performed. The applicability for car or real estate data could also be assessed.

References

1. Salvador, P., Nogueira, A.: Analysis of the Internet Domain Names Re-registration Market. Procedia Computer Science 3, 325–335 (2011)
2. Kahn, S.A., Case, F.E.: Real Estate Appraisal and Investment. The Ronald Press Company, New York (1977)
3. Pagourtzi, E., Assimakopoulos, V., Hatzichristos, T., French, N.: Real Estate Appraisal: a Review of Valuation Methods. Journal of Property Investment & Finance 21(4), 383–401 (2003)
4. Phillips, G.M.: A Hedonic Regression Model for Internet Domain Name Valuation. Business Valuation Review 22(2), 90–98 (2003)
5. Jindra, M.: The Market for Internet Domain Names. In: 16th European Regional Conference, September 4-6, Porto, Portugal (2005)
6. Gonzalez, A.J., Laureano-Ortiz, R.: A Case-Based Reasoning Approach to Real Estate Property Appraisal. Expert Systems with Applications 4(2), 229–246 (1992)

7. O'Roarty, B., McGreal, S., Adair, A., Patterson, D.: Case-Based Reasoning and Retail Rent Determination. Journal of Property Research 14(4), 309–328 (1997)
8. McSherry, D.: An Adaptation Heuristic for Case-Based Estimation. In: Smyth, B., Cunningham, P. (eds.) EWCBR 1998. LNCS (LNAI), vol. 1488, pp. 184–195. Springer, Heidelberg (1998)
9. Dieterle, S., Bergmann, R.: Case-Based Appraisal of Internet Domains. In: Díaz Agudo, B., Watson, I. (eds.) ICCBR 2012. LNCS, vol. 7466, pp. 47–61. Springer, Heidelberg (2012)
10. Samarasinghe, S.: Neural Networks for Applied Sciences and Engineering: from Fundamentals to Complex Pattern Recognition. Auerbach Publications, Boca Raton (2007)
11. Rossini, P.: Application of Artificial Neural Networks to the Valuation of Residential Property. In: Third Annual Pacific-Rim Real Estate Society Conference, Palmerston North, New Zealand (1997)
12. Peterson, S., Flanagan, A.B.: Neural Network Hedonic Pricing Models in Mass Real Estate Appraisal. Journal of Real Estate Research 31(2), 147–164 (2009)
13. McCluskey, W., Anand, S.: The Application of Intelligent Hybrid Techniques for the Mass Appraisal of Residential Properties. Journal of Property Investment & Finance 17(3), 218–239 (1999)
14. Rossini, P.: Using Expert Systems and Artificial Intelligence for Real Estate Forecasting. In: Sixth Annual Pacific-Rim Real Estate Society Conference, Sydney, Australia, pp. 24–27 (2000)
15. Al-Akhras, M.: An Evolutionary-Optimised Artificial Neural Network Approach for Automatic Appraisal of Jordanian Lands and Real Properties. In: NN 2010/EC 2010/FS 2010 Proceedings of the 11th WSEAS International Conference on Neural Networks and 11th WSEAS International Conference on Evolutionary Computing and 11th WSEAS International Conference on Fuzzy Systems, pp. 203–208 (2010)
16. Jalali, V., Leake, D.: An Ensemble Approach to Instance-Based Regression Using Stretched Neighborhoods. In: The Twenty-Sixth International FLAIRS Conference, pp. 381–386 (2013)
17. Dieterle, S.: Neuronale und fallbasierte Bewertung von Internet-Domainnamen. Master's thesis, University of Trier (2013)
18. Warner, R.M.: Applied Statistics: From Bivariate Through Multivariate Techniques. SAGE Publications (2013)
19. Caumanns, J.: A Fast and Simple Stemming Algorithm for German Words. Technical report, Department of Mathematics and Computer Science, Free University of Berlin (1999)
20. Igel, C., Hüsken, M.: Improving the Rprop Learning Algorithm. In: Proceedings of the Second International Symposium on Neural Computation, NC 2000, pp. 115–121. ICSC Academic Press (2000)
21. Riedmiller, M., Braun, H.: RPROP - A Fast Adaptive Learning Algorithm. In: Proceedings of ISCIS VII. University of Karlsruhe (1992)
22. Gentner, D., Forbus, K.D.: MAC/FAC: A Model of Similarity-based Retrieval. In: Proceedings of the Thirteenth Annual Conference of the Cognitive Science Society, pp. 504–509. Lawrence Erlbaum (1991)
23. Can, F., Altingövde, I.S., Demir, E.: Efficiency and Effectiveness of Query Processing in Cluster-Based Retrieval. Information Systems 29(8), 697–717 (2004)

Further Experiments
in Opinionated Product Recommendation

Ruihai Dong[1], Michael P. O'Mahony[2], and Barry Smyth[2]

[1]CLARITY Centre for Sensor Web Technologies
[2]Insight Centre for Data Analytics
School of Computer Science and Informatics
University College Dublin, Ireland

Abstract. In this paper we build on recent work on case-based product recommendation focused on generating rich product descriptions for use in a recommendation context by mining user-generated reviews. This is in contrast to conventional case-based approaches which tend to rely on case descriptions that are based on available meta-data or catalog descriptions. By mining user-generated reviews we can produce product descriptions that reflect the opinions of real users and combine notions of similarity and opinion polarity (sentiment) during the recommendation process. In this paper we compare different variations on our review-mining approach, one based purely on features found in reviews, one seeded by features that are available from meta-data, and one hybrid approach that combines both approaches. We evaluate these approaches across a variety of datasets form the travel domain.

Keywords: recommender systems, opinion mining, sentiment analysis.

1   Introduction

Case-based recommender systems [1,2] generate recommendations based on the similarity between some target query and a case base of product cases, preferring those cases that best match the target query. Usually, these cases are based on available features or meta-data, the type of features that might be found in a product catalog. This approach has served us well but its connection to the *experience-based* origins of case-based reasoning (CBR) are at best tenuous. After all, the type of product descriptions that are found in catalogs hardly capture the experiences of users of the products. Instead, they usually describe the physical properties of a product or its technical features.

Recently, we have proposed harnessing user-generated product reviews as a source of product information for use in novel approach to case-based recommendation, one that does rely on the experiences of users, as expressed through the opinions of real users in the reviews that they write [3,4]. As a result products can be recommended based on a combination of feature similarity and opinion polarity (or sentiment). So, for example, a traveller looking for accommodation options with a business centre can receive recommendations for hotels, not just

L. Lamontagne and E. Plaza (Eds.): ICCBR 2014, LNCS 8765, pp. 110–124, 2014.
© Springer International Publishing Switzerland 2014

with business centres or related services (similarity), but for hotels that have excellent business centres (sentiment).

In this paper, we consider a question that was commonly asked in relation to the work of [4]; namely, instead of relying purely on user-generated reviews as a source of description information, why not incorporate existing meta-data when it is available? Consequently we describe two variations on our opinionated recommendation approach. First, instead of sourcing features exclusive from user reviews, we use existing meta-data features, where available, as seed features, and look to the user reviews as a source of frequency and sentiment information. Second, we implement a hybrid approach that uses meta-data features in addition to those that can be mined from user reviews. Further, we compare these two approaches to our existing approach which relies exclusively on features mined from user reviews. This comparison is based on large-scale TripAdvisor hotel information across multiple cities, and uses TripAdvisor's existing *more-like-this* style recommender as a baseline against which to judge the quality of recommendations produced by the above approaches.

2   Related Work

Today consumer reviews have become an important part of our shopping experience. Lee et al. report that 84 percent of Americans are influenced by online reviews when they are making purchase decisions [5]; see also [6,7]. Many companies have now recognised that consumer reviews represent a new and important communication channel with their consumers, and they have begun monitoring online consumer reviews as a crucial source of product feedback [8]. Further, companies can predict their performance or sales according to this online feedback; for example, Duan et al. used Yahoo movie reviews and box office returns to examine the persuasive and awareness effects of online user reviews on the daily box office performance [8].

Consumer reviews have also been recognised as an important source of recommendation knowledge [9]. Reviews have been used, for example, to infer user preferences to elicit additional user preference data for use in collaborative filtering algorithms. In this regard, Zhang et al. propose the SElf-Supervised, Lexicon-based and Corpus-based (SELC) model to generate virtual ratings from consumer reviews and integrated them into a collaborative filtering recommender system [10]. The results show that such virtual ratings have a significant impact on recommendation accuracy; see also [11]. Further, reviews can be leveraged for user profiling purposes. For example, Musat et al. propose a technique to build profiles based on the review texts previously authored by users, and to use these profiles to identify other reviews that users are likely to be interested in [12]. Further, Esparza et al. aggregated the real-time short textual reviews available on the Bilppr service to build term-based user profiles for use in a content-based recommender system [13].

For the most part, classical case-base recommender systems are limited by their use of static case representations, based on technical meta-data or catalogue features. Recently, researchers have explored new approaches to building product case descriptions by leveraging opinion data from consumer reviews. For example, Huang et al. propose a method to represent products by a set of feature-sentiment value pairs extracted from consumer reviews, and recommendations are then re-ranked by the sentiment score of distinguishing features [14]. In this work, the cases that we produce from reviews are experiential in nature because they are intimately derived from the opinions of real users. Moreover, the availability of sentiment information accommodates a novel approach to case retrieval, one that combines feature similarity and opinion sentiment, than might normally be found in case based recommenders.

As mentioned previously the main contribution of this paper is to extend the work of [4] and [3] by considering the role of opinions and more classical metadata type technical features. To this end we describe one way to combine both types of features within a single recommendation framework. A second contribution is to demonstrate the applicability of this general approach to review mining and recommendation on an entirely new dataset in a very different product domain, as we move away from consumer electronics to the travel domain and hotel reviews.

3   Mining Experiential Product Cases

A summary of our overall approach is presented in Figure 1, including how we mine experiential cases and how we generate recommendations. There are 4 basic steps: (1) identifying useful product features; (2) associating these features with sentiment information based on the content of user-generated reviews; (3) aggregating features and sentiment to produce experiential cases; and (4) the retrieval and ranking of cases for recommendation given a target query.

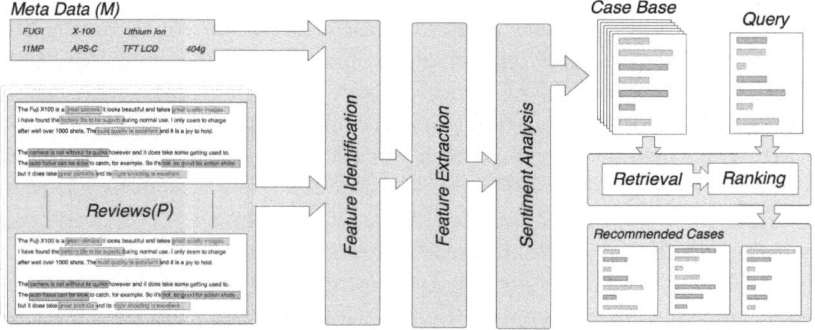

Fig. 1. An overview of the experiential product recommendation architecture

3.1 Identifying Review Features

In this work we consider two ways to identify product features. First we apply the technique described in [15] to automatically extract features from reviews. Second, we look to product meta-data as an external source of features (i.e. external to product reviews) which can then be identified within reviews for the purpose of frequency and sentiment calculation.

Mining Product Features from Review Text. Briefly, the approach described in [15] considers *bi-gram* features and *single-noun* features and uses a combination of shallow NLP and statistical methods to mine them [16,17]. For example, bi-grams in reviews which conform to one of two basic part-of-speech co-location patterns are considered — an adjective followed by a noun (AN) or a noun followed by a noun (NN) — excluding bi-grams whose adjective is a sentiment word (e.g. *excellent, terrible* etc.) in the sentiment lexicon [18]. Separately, single-noun features are validated by eliminating nouns that are rarely associated with sentiment words in reviews as per [18], since such nouns are unlikely to refer to product features; we will refer to features that are identified in this way as *review features* or *RF*.

Using Meta-Data as Product Features. One of the limitations of the above approach is that it can generate some unusual features that are unlikely to matter in any meaningful way during product recommendation; sometimes reviews wander off topic, for example, or address rarely relevant, or downright incorrect, aspects of a product. If this was to occur frequently, then recommendation effectiveness could be compromised. Thus, in this paper we also consider available meta-data as an original source of features that matter. For example, in the case of the TripAdvisor data that we use in this work, the hotels themselves are accompanied by meta-data in the form of an edited set of *amenities* (for example, spa, swimming pool, business centre, etc.) that are available at the hotel. These amenities can serve as product features in their own right and are used as such in this alternative approach. We will refer to features that are identified in this manner as *amenity features* or *AF*.

3.2 Evaluating Feature Sentiment

For each feature (whether RF or AF in origin) we evaluate its sentiment based on the sentence containing the feature within a given review. We use a modified version of the *opinion pattern mining* technique proposed by Moghaddam and Ester [19] for extracting opinions from unstructured product reviews. Once again we use Hu and Liu's sentiment lexicon [18] as the basis for this analysis. For a given feature F_i and corresponding review sentence S_j from review R_k, we determine whether there are any sentiment words in S_j. If there are not then this feature is marked as *neutral* from a sentiment perspective. If there are sentiment words then we identify the word w_{min} which has the minimum word-distance to F_i.

Next we determine the part-of-speech (POS) tags for w_{min}, F_i and any words that occur between w_{min} and F_i. The POS sequence corresponds to an *opinion pattern*. For example, in the case of the bi-gram feature *noise reduction* and the review sentence, *"...this camera has great noise reduction..."* then w_{min} is the word *"great"* which corresponds to an opinion pattern of *JJ-FEATURE* as per Moghaddam and Ester [19]. After a complete pass of all features over all reviews, we can compute the frequency of all opinion patterns that have been recorded. A pattern is deemed to be valid (from the perspective of our ability to assign sentiment) if it occurs more than the average number of times. For valid patterns we assign sentiment to F_i based on the sentiment of w_{min} and subject to whether S_j contains any negation terms within a 4-word-distance of w_{min}. If there are no such negation terms then the sentiment assigned to F_i in S_j is that of the sentiment word in the sentiment lexicon; otherwise this sentiment is reversed. If an opinion pattern is deemed not to be valid (based on its frequency), then we assign a *neutral* sentiment to each of its occurrences within the review set.

3.3   Generating Experiential Cases

For each product P we have a set of features $F(P) = \{F_1, ..., F_m\}$ that have been either identified from the meta-data associated with P or that have been discussed in the various reviews of P, $Reviews(P)$. And for each feature F_i we can compute various properties including the fraction of reviews it appears in (its *popularity*, see Equation 1) and the degree to which reviews mention it in a positive, neutral, or negative light (its *sentiment*, see Equation 2, where $Pos(F_i, P)$, $Neg(F_i, P)$, and $Neut(F_i, P)$ denote the number of times that feature F_i has positive, negative and neutral sentiment in the reviews for product P, respectively). Thus, each product can be represented as a *product case*, $Case(P)$, which aggregates product features, popularity and sentiment data as in Equation 3.

$$Pop(F_i, P) = \frac{|\{R_k \in Reviews(P) : F_i \in R_k\}|}{|Reviews(P)|} \tag{1}$$

$$Sent(F_i, P) = \frac{Pos(F_i, P) - Neg(F_i, P)}{Pos(F_i, P) + Neg(F_i, P) + Neut(F_i, P)} \tag{2}$$

$$Case(P) = \{[F_i, Sent(F_i, P), Pop(F_i, P)] : F_i \in F(P)\} \tag{3}$$

4   Recommending Products

Unlike traditional content-based recommenders — which tend to rely exclusively on similarity in order to rank products with respect to some user profile or query — the above approach accommodates the use of feature sentiment, as well as feature similarity, during recommendation; see [3,4]. Briefly, a candidate recommendation product C can be scored against a query product Q according to a weighted combination of similarity and sentiment as per Equation 4. $Sim(Q, C)$ is a traditional similarity metric such as cosine similarity, producing a value between 0 and 1, while $Sent(Q, C)$ is a sentiment metric producing a value between -1 (negative sentiment) and +1 (positive sentiment).

$$Score(Q, C) = (1 - w) \times Sim(Q, C) + w \times \left(\frac{Sent(Q, C) + 1}{2} \right) \qquad (4)$$

4.1   Similarity Assessment

For the purpose of similarity assessment we use a standard cosine similarity metric based on feature popularity scores as per Equation 5; This is inline with standard approaches to content-based similarity; see, for example [20].

$$Sim(Q, C) = \frac{\displaystyle\sum_{F_i \epsilon F(Q) \cup F(C)} Pop(F_i, Q) \times Pop(F_i, C)}{\sqrt{\displaystyle\sum_{F_i \epsilon F(Q)} Pop(F_i, Q)^2} \times \sqrt{\displaystyle\sum_{F_i \epsilon F(C)} Pop(F_i, C)^2}} \qquad (5)$$

4.2   Sentiment Assessment

As mentioned earlier, sentiment information is unusual in a recommendation context but its availability offers a second way to compare products, based on a feature-by-feature sentiment comparison as per Equation 6. We can say that F_i is *better* in C than Q if F_i in C has a higher sentiment score than it does in Q.

$$better(F_i, Q, C) = \frac{Sent(F_i, C) - Sent(F_i, Q)}{2} \qquad (6)$$

We can then calculate an overall better score at the product level by aggregating the individual better scores for the product features. We can do this in one of two ways as follows.

The first approach, which we shall refer to as $B1$, calculates an average better score across the shared features of Q and C as per Equation 7. A potential shortcoming of this approach is that it remains silent about those features which are not common to Q and C, the so-called *residual* features.

$$B1(Q, C) = \frac{\sum_{F_i \in F(Q) \cap F(C)} better(F_i, Q, C)}{|F(Q) \cap F(C)|} \qquad (7)$$

The second approach, which we shall refer to as $B2$, computes the average better scores across the *union* of features of Q and C, assigning non-shared features a neutral sentiment score of 0; see Equation 8. Unlike $B1$, this second approach does give due consideration to the *residual* features in the the query and candidate cases. Whether or not these residual features play a significant role remains to be seen and we will return to this question as part of the evaluation later in this paper.

$$B2(Q, C) = \frac{\sum_{F_i \in F(Q) \cup F(C)} better(F_i, Q, C)}{|F(Q) \cup F(C)|} \qquad (8)$$

Note that in Equation 4, $Sent(Q, C)$ is set to either $B1$ or $B2$ depending on the particular recommender system variation under evaluation.

5   Evaluation

Previous work has evaluated some elements of the above approaches. In particular, the work presented in [4] focused on evaluating the RF variation on a set of tens of thousands of Amazon product reviews for a few hundred consumer electronics products. In this paper, we extend this work in two important ways. First, we expand the evaluation considerably to cover a large set of TripAdvisor hotel reviews, covering more than a hundred thousand reviews across thousands of hotels in 6 international cities. The importance of this is not just to evaluate a larger set of reviews and products, but also to look at reviews that have written for very different purposes (travel versus consumer electronics). The second way that we add to previous work is to consider the new AF variation described above as an alternative way to source product features (from meta-data); indeed, we also consider a hybrid RF-AF approach as a third algorithmic variation.

5.1   Datasets

The data for this experiment was sourced from TripAdvisor during September 2013. We focused on 6 different cities across Europe, Asia, and the US. We extracted 148,704 reviews across 1,701 hotels. This data is summarised in Table 1, where we show the total number of reviews per city (*#Reviews*), the number of hotels per city (*#Hotels*), as well as including statistics (mean and standard deviation) on the number of amenities per hotel (A), the number of amenity features extracted from reviews per hotel (AF), and the number of review features extracted from the reviews per hotel (without seeding with amenities)(RF). We can immediately see that using the AF technique to identify features produces much smaller feature-sets for cases than using the RF approach, owing to the limited amount of amenity meta-data availability for each hotel.

Table 1. Dataset statistics

City	#Reviews	#Hotels	$\mu(\sigma)_A$	$\mu(\sigma)_{AF}$	$\mu(\sigma)_{RF}$
Dublin	13,019	138	5.7 (2.6)	4.1 (1.0)	30.2 (4.6)
New York	31,881	337	6.1 (2.5)	4.0 (1.4)	32.9 (4.8)
Singapore	14,576	186	5.7 (3.4)	3.7 (1.5)	28.8 (6.2)
London	62,632	717	4.5 (2.7)	3.9 (1.2)	31.8 (5.5)
Chicago	11,091	125	7.6 (2.2)	4.4 (1.3)	28.6 (5.0)
Hong Kong	15,505	198	6.2 (3.0)	4.1 (1.6)	33.8 (6.1)

5.2   Methodology

We adopt a standard *leave-one-out* approach to recommendation. For each city dataset, we treat each hotel in turn as a query Q and generate a set of top-5 recommendations according to Equation 4 using different values of w (0 to 1 in increments of 0.1) in order to test the impact of different combinations of similarity and sentiment. We do this for hotel cases that are based on amenity features and review features to produce a set of recommendations that derive from amenity features AF and a set that derive from review features RF. We also implement a hybrid approach (denoted AF-RF) that combines AF and RF by simply combining the features identified by AF and RF into a single case structure. Finally we also implement the $B1$ and $B2$ variations when it comes to computing $Sent(Q, C)$ in Equation 4. This provides a total of 6 difference algorithmic variants for generating recommendations.

To evaluate the resulting recommendation lists, we compare our recommendations to those produced natively by TripAdvisor (TA) using two comparison metrics. First, we calculate the average *query similarity* between each set of recommendations (AF, RF, AF-RF and TA) and Q. To do this we use a Jaccard similarity metric based on an expanded set of hotel features that is made up of the hotel amenities plus hotel cost, star rating, and size (number of rooms). Query similarity indicates how similar recommendations are to the query case and, in particular, whether there is any difference in similarity between those recommendations generated by our approaches and those produced by TA.

The second comparison metric is the average *ratings benefit*. This compares two sets of recommendations based on their overall TripAdvisor user ratings (see Equation 9). We calculate a ratings benefit for each of our 6 recommendation lists (denoted by R in Equation 9) compared to the recommendations produced by TA; a ratings benefit of 0.1 means that our recommendation list enjoys an average rating score that is 10% higher that those produced by the default TripAdvisor approach (TA).

$$Ratings\ Benefit(R, TA) = \frac{\overline{Rating(R)} - \overline{Rating(TA)}}{\overline{Rating(TA)}} \qquad (9)$$

5.3   Experience Case Mining

To begin with it is worth gaining an understanding of the extent to which the AF and RF approaches are able to generate rich experiential case descriptions, in terms of the number of features that can be extracted on a product-by-product basis. To this end Figure 2 presents features histograms showing the number of cases with different numbers of amenity features (AF) and review features (RF) as extracted from reviews, and the number of amenities (A) available for each hotel as sourced from TripAdvisor.

As expected there is a significant different between the number of amenity features and the number of review features extracted. Clearly, cases that are based on review features enjoy much richer descriptions than those that rely

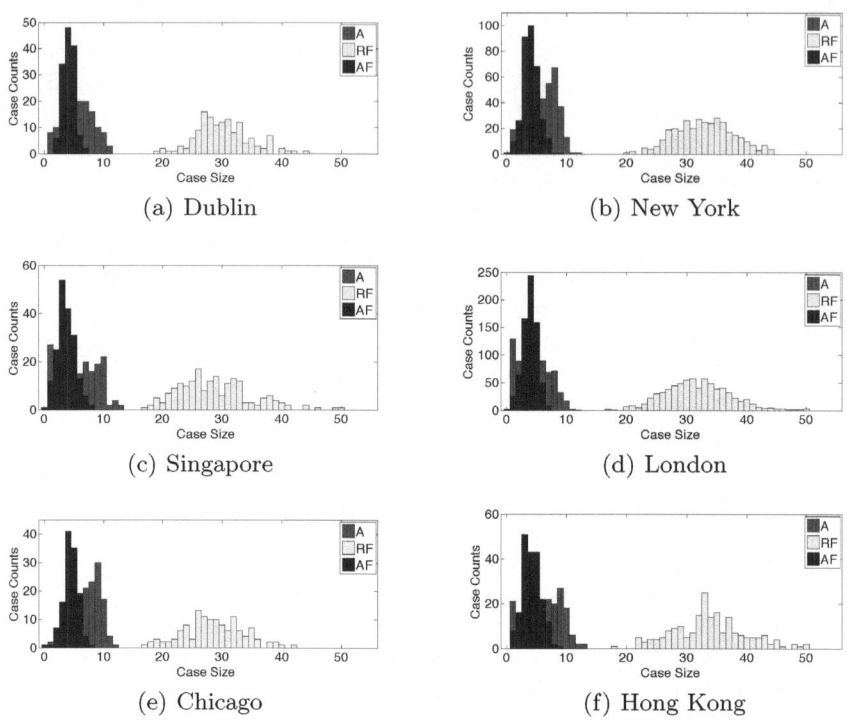

Fig. 2. Hotel case size histograms

only on amenity features. Moreover, it can be seen that, on average, only 4 of the 6 amenity features associated with hotels are extracted from reviews using the AF approach, which further highlights the limitations of this approach from a case representation perspective.

5.4   Recommendation Results

The richness of cases aside, the true test of these approaches of course relates to their ability to generate recommendations that are likely to be of interest to end-users. With this in mind, and as mentioned above, we evaluate the quality of recommendation lists based on their average query similarity and their average ratings benefit. Figures 3 and 4 show the results when the $B1$ and $B2$ metrics are used to score recommendation candidates, respectively (see Section 4). Six graphs are shown in each figure, one for each of the cities considered. Each individual graph shows plots for the 3 different algorithmic techniques (AF, RF, and AF-RF), and each algorithmic technique is associated with two plots: a plot of average query similarity (dashed lines) and a plot of average ratings benefit (solid lines) against w.

Each graph also shows the average query similarity for the TA default TripAdvisor recommendations (the black horizontal solid line), and the region between

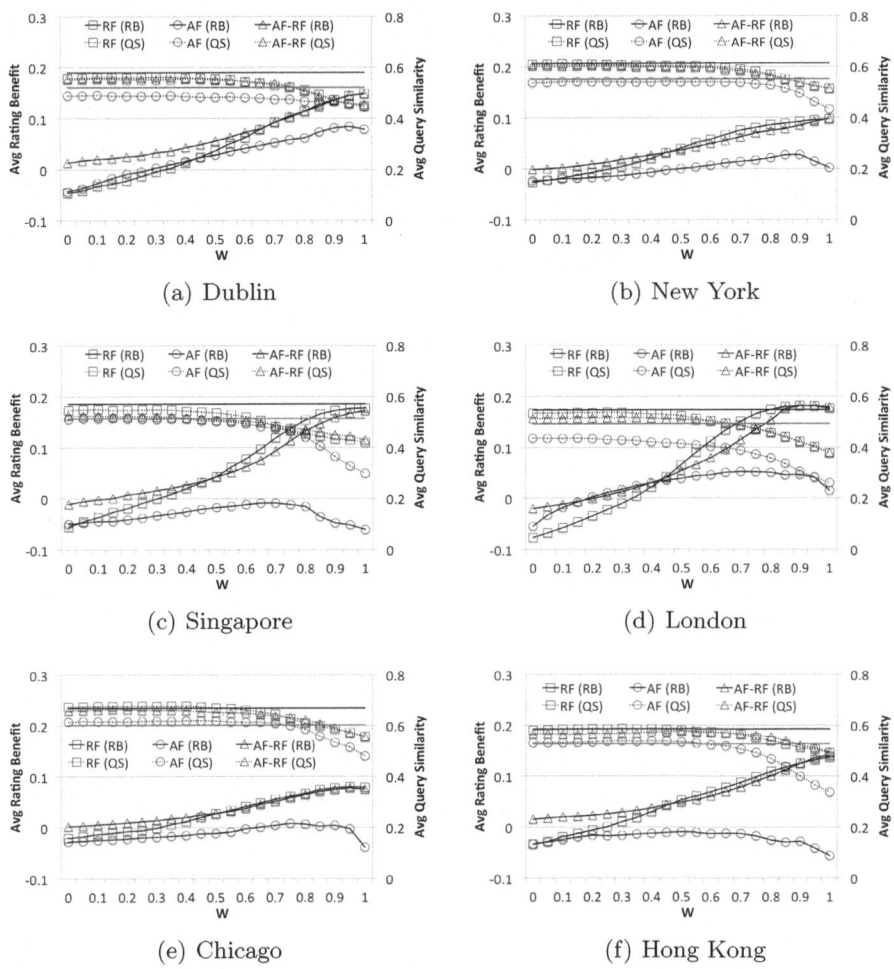

Fig. 3. Ratings benefit (RB) and query similarity (QS) using the $B1$ sentiment metric

the black and red lines corresponds to the region of 90% similarity; that is, query similarity scores that fall within this region are 90% as similar to the target query as the default recommendations produced by TA. The intuition here is that query similarity scores which fall below this region run the risk of compromising too much query similarity to be useful as *more-like-this* recommendations.

5.5   Ratings Benefit vs. w

There are a number of general observations that can be made about these results. First, as w increases we can see that there is a steady increase in the average ratings benefits and this is consistent across all of the algorithmic and dataset variations. In other words, as we increase the influence of sentiment in

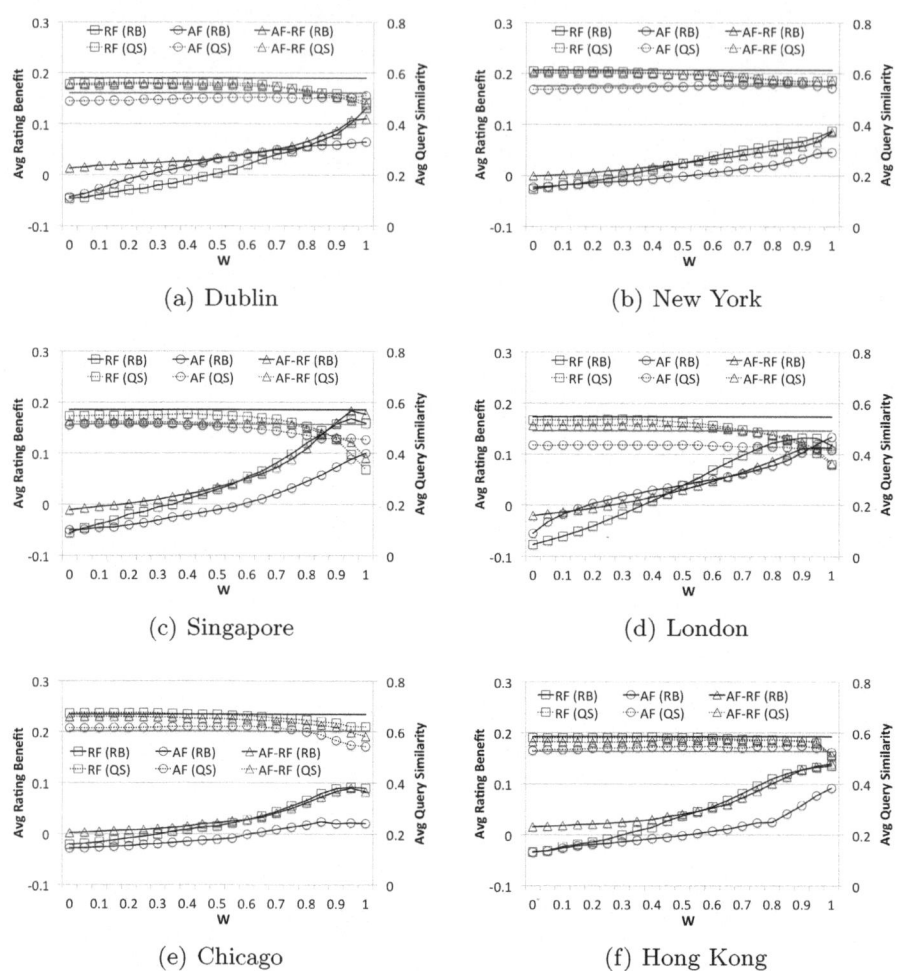

Fig. 4. Ratings benefit (RB) and query similarity (QS) using the $B2$ sentiment metric

the scoring function (Equation 4), we tend to produce recommendations that offer better overall ratings than those produced by TA; thus combining similarity and sentiment in recommendation delivers a positive effect overall.

Generally speaking this effect is less pronounced for the AF only variations, especially for values of w above 0.5. For example, in Figure 3(d), for London hotels (and using $B1$ for sentiment analysis), we can see that the ratings benefit for AF grows from -0.05 ($w = 0$) to a maximum of 0.05 ($w = 0.7$), whilst the ratings benefit grows from -0.07 ($w = 0$) to 0.18 (at $w = 0.9$) for RF. This suggests that the review features are playing a more significant role in influencing recommendation quality (in terms of ratings benefit) than the amenity features. This is not surprising given the difference in the numbers of amenity and review features extracted from reviews; on average, 4 features were extracted per hotel using the AF approach, compared to 30 features using RF (see Table 1).

5.6   Query Similarity vs. w

We can also see that as w increases there is a gradual drop in query similarity. In other words, as we increase the influence of sentiment (and therefore decrease the influence of similarity) in the scoring function (Equation 4), we tend to produce recommendation lists that are increasingly less similar to the target query. On the one hand, this is a way to introduce more diversity [21] into the recommendation process with the added benefit, as above, that the resulting recommendations tend to enjoy a higher ratings benefit compared to the default TripAdvisor recommendations (TA). But on the other hand, there is the risk that too great a query similarity drop may lead to products that are no longer deemed to be relevant by the end-user. For this reason, we have (somewhat arbitrarily) chosen to prefer query similarities that remain within 90% of those produced by TA.

Once again there is a marked difference between the AF approach and those approaches that include review features (RF and AF-RF). The former tends to produce recommendation lists with lower query similarity than either of RF or AF-RF, an effect that is consistent across all 6 cities and regardless of whether $B1$ or $B2$ is used in recommendation. For example, consider Figure 4(d) for London hotels (and using $B2$ for sentiment analysis). In this case, we can see that the average query similarity for AF starts at about 0.44 (at $w = 0$) and drops to about 0.41 (at $w = 1$), compared to a TA query similarity of about 0.55. In contrast, the RF and AF-RF techniques deliver query similarities in the range 0.36 to 0.54, often within the 90% query similarity range.

5.7   Shared vs. Residual Features

In this study we have also tested two variations on how to calculate the sentiment differences between cases: $B1$ focused just on those features common to both cases whereas $B2$ considered all features of the cases. In general, the graphs in Figure 3 and 4 make it difficult to discern any major difference between these two options across the AF, RF, or AF-RF approaches. Any differences that are found probably reflect the relative importance of shared and residual features among the different city datasets. For example, in the London dataset, $B1$ seems to produce marginally better ratings benefits at least for RF and AF-RF, whereas the reverse is true for Chicago. It is therefore difficult to draw any significant conclusions at this stage, although in what follows we will argue for a slight advantage for the $B2$ approach.

5.8   A Fixed-Point Comparison

To aid in the evaluation of the different recommendation approaches across the various datasets it is useful to compare the ratings benefits by establishing a fixed point of query similarity. We have highlighted above how favourable ratings benefits tend to come at a query similarity cost, and we have suggested that we might reasonably be wary when query similarity drops below 90% of the level

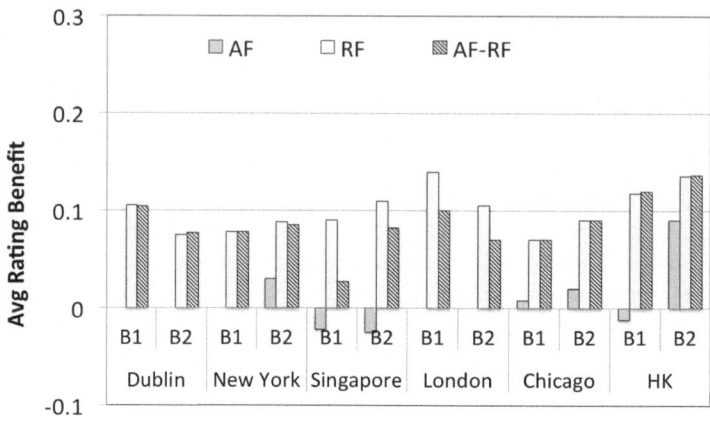

Fig. 5. Summary ratings benefits at the 90% query similarity level

found for the default TA recommendations. With this in mind, we can usefully compare the various recommendation approaches by noting the average ratings benefit available at the value of w for which the query similarity of a given approach falls below the 90% default (TA) query similarity level. For example, in Figure 3(c), for Singapore hotels and the $B1$ sentiment analysis technique, we can see that the query similarity for the RF approach falls below the 90% threshold at about $w = 0.625$ and this corresponds to a ratings benefit of 0.09.

Performing this analysis for each of the 6 recommendation approaches across the different city datasets gives the ratings benefits represented by the bar chart in Figure 5. This helps to clarify the relative differences between the various techniques. For example, the RF technique delivers an average ratings benefit of approximately 0.1 (and as high as 0.14 in the case of London and Hong Kong).

One of the key questions for this work was the utility of meta-data as a source of review features, which corresponds to the AF approach. In comparison to the above, the AF approaches offer an average ratings benefit of only 0.01, with a maximum benefit of 0.09 (Hong Kong), and sometimes leading to a lower ratings benefit that is available from the default recommendations (TA), as is the case with Singapore. In fact, for London, Dublin, and New York, the AF approach often delivers query similarities that are consistently below the 90% threshold and so do not register any ratings benefit in these cases. Clearly the amenity features used by AF are not providing any significant benefit, and certainly nothing close to that offered by RF, likely because of the relative lack of amenity features compared to review features.

Indeed combining amenity and review features, as in the hybrid AF-RF approach, does not generally offer any real advantage over RF alone. The average ratings benefit for AF-RF is 0.087, better than AF but not as good as RF on its own. At best AF-RF provides a ratings benefit that is comparable to that provided by RF (as is the case for Chicago, Dublin, New York, and Hong Kong), but in some cases (Singapore and London) it performs worse than RF.

6 Conclusions

In this paper we have extended an approach to producing product cases from user-generated reviews for the purpose of recommendation. In particular, we have evaluated a number of different approaches to review mining (both with and without meta-data) and have described the results of a large-scale evaluation on TripAdvisor hotel reviews across 6 different cities. The results demonstrate the potential for this strategy as a viable approach for product recommendation, and they compare closely to similar results observed in other domains for related techniques [3,4]. Our analysis of the relative merits of leveraging meta-data (hotel amenities in the case of the TripAdvisor datasets) concludes that these features offer little or no real benefit when it comes to recommendation, at least in comparison to a pure review-mining approach.

Acknowledgments. This work is supported by Science Foundation Ireland through the CLARITY Centre for Sensor Web Technologies under grant number 07/CE/I1147 and through the Insight Centre for Data Analytics under grant number SFI/12/RC/2289.

References

1. Bridge, D., Göker, M.H., McGinty, L., Smyth, B.: Case-based recommender systems. Knowl. Eng. Rev. 20, 315–320 (2005)
2. Smyth, B.: Case-based recommendation. In: Brusilovsky, P., Kobsa, A., Nejdl, W. (eds.) Adaptive Web 2007. LNCS, vol. 4321, pp. 342–376. Springer, Heidelberg (2007)
3. Dong, R., Schaal, M., O'Mahony, M.P., McCarthy, K., Smyth, B.: Opinionated product recommendation. In: Delany, S.J., Ontañón, S. (eds.) ICCBR 2013. LNCS, vol. 7969, pp. 44–58. Springer, Heidelberg (2013)
4. Dong, R., O'Mahony, M.P., Schaal, M., McCarthy, K., Smyth, B.: Sentimental product recommendation. In: Proceedings of the 7th ACM Conference on Recommender Systems, RecSys 2013, pp. 411–414. ACM, New York (2013)
5. Lee, J., Park, D.-H., Han, I.: The different effects of online consumer reviews on consumers' purchase intentions depending on trust in online shopping mall: An advertising perspective. Internet Research 21(2), 187–206 (2011)
6. Zhu, F., Zhang, X.M.: Impact of online consumer reviews on sales: The moderating role of product and consumer characteristics. Journal of Marketing 74(2), 133–148 (2010)
7. Dhar, V., Chang, E.A.: Does chatter matter? the impact of user-generated content on music sales. Journal of Interactive Marketing 23(4), 300–307 (2009)
8. Dwyer, P.: Measuring the value of electronic word of mouth and its impact in consumer communities. Journal of Interactive Marketing 21(2), 63–79 (2007)

9. Aciar, S., Zhang, D., Simoff, S., Debenham, J.: Informed recommender: Basing recommendations on consumer product reviews. IEEE Intelligent Systems 22(3), 39–47 (2007)
10. Zhang, W., Ding, G., Chen, L., Li, C., Zhang, C.: Generating virtual ratings from Chinese reviews to augment online recommendations. ACM Trans. Intell. Syst. Technol. 4, 9:1–9:17 (2013)
11. Ganu, G., Kakodkar, Y., Marian, A.: Improving the quality of predictions using textual information in online user reviews. Information Systems 38(1), 1–15 (2013)
12. Musat, C.-C., Liang, Y., Faltings, B.: Recommendation using textual opinions. In: Proceedings of the 23rd International Joint Conference on Artificial Intelligence, pp. 2684–2690. AAAI Press, Menlo Park (2013)
13. Esparza, S., O'Mahony, M., Smyth, B.: Effective product recommendation using the real-time web. In: Bramer, M., Petridis, M., Hopgood, A. (eds.) Research and Development in Intelligent Systems XXVII, pp. 5–18. Springer, London (2011)
14. Huang, J., Etzioni, O., Zettlemoyer, L., Clark, K., Lee, C.: Revminer: An extractive interface for navigating reviews on a smartphone. In: Proceedings of the 25th Annual ACM Symposium on User Interface Software and Technology, UIST 2012, pp. 3–12. ACM, New York (2012)
15. Dong, R., Schaal, M., O'Mahony, M.P., Smyth, B.: Topic extraction from online reviews for classification and recommendation. In: Proceedings of the 23rd International Joint Conference on Artificial Intelligence, IJCAI 2013. AAAI Press, Menlo Park (2013)
16. Hu, M., Liu, B.: Mining and summarizing customer reviews. In: Proceedings of the Tenth ACM SIGKDD International Conference on Knowledge Discovery and Data Mining, KDD 2004, pp. 168–177. ACM, New York (2004)
17. Justeson, J., Katz, S.: Technical terminology: some linguistic properties and an algorithm for identification in text. Natural Language Engineering, 9–27 (1995)
18. Hu, M., Liu, B.: Mining opinion features in customer reviews. In: Proceedings of the 19th National Conference on Artifical Intelligence, AAAI 2004, pp. 755–760. AAAI Press (2004)
19. Moghaddam, S., Ester, M.: Opinion digger: an unsupervised opinion miner from unstructured product reviews. In: Proceedings of the 19th ACM International Conference on Information and Knowledge Management, CIKM 2010, pp. 1825–1828. ACM, New York (2010)
20. Pazzani, M., Billsus, D.: Content-based recommendation systems. In: Brusilovsky, P., Kobsa, A., Nejdl, W. (eds.) Adaptive Web 2007. LNCS, vol. 4321, pp. 325–341. Springer, Heidelberg (2007)
21. Smyth, B., McClave, P.: Similarity vs. diversity. In: Aha, D.W., Watson, I. (eds.) ICCBR 2001. LNCS (LNAI), vol. 2080, pp. 347–361. Springer, Heidelberg (2001)

How Much Do You Trust Me?
Learning a Case-Based Model of Inverse Trust

Michael W. Floyd[1], Michael Drinkwater[1], and David W. Aha[2]

[1] Knexus Research Corporation, Springfield, VA, USA
[2] Navy Center for Applied Research in Artificial Intelligence,
Naval Research Laboratory (Code 5514), Washington, DC, USA
{first.last}@knexusresearch.com, david.aha@nrl.navy.mil

Abstract. Robots can be important additions to human teams if they improve team performance by providing new skills or improving existing skills. However, to get the full benefits of a robot the team must trust and use it appropriately. We present an agent algorithm that allows a robot to estimate its trustworthiness and adapt its behavior in an attempt to increase trust. It uses case-based reasoning to store previous behavior adaptations and uses this information to perform future adaptations. We compare case-based behavior adaptation to behavior adaptation that does not learn and show it significantly reduces the number of behaviors that need to be evaluated before a trustworthy behavior is found. Our evaluation is in a simulated robotics environment and involves a movement scenario and a patrolling/threat detection scenario.

Keywords: trust, behavior adaptation, human-robot interaction.

1   Introduction

Robots can be important members of human teams if they provide capabilities that are critical for accomplishing team goals and complement those of their human teammates. These could include improved sensory capabilities, communication capabilities, or an ability to operate in environments humans can not (e.g., rough terrain or dangerous situations). Including these robots might be necessary for the team to meet its objectives and reduce human risk. However, to make full use of these robots the human teammates will need to trust them.

This is especially important for robots that operate autonomously or semi-autonomously. In these situations, their human operator(s) would likely issue commands or delegate tasks to the robot to reduce their workload or more efficiently achieve team goals. A lack of trust in the robot could result in the humans under-utilizing it, unnecessarily monitoring the robot's actions, or possibly not using it at all [1].

A robot could be designed so that it operates in a sufficiently trustworthy manner. However, this may be impractical because the measure of trust might be task-dependent, user-dependent, or change over time [2]. For example, if a robot receives a command from an operator to navigate between two locations

L. Lamontagne and E. Plaza (Eds.): ICCBR 2014, LNCS 8765, pp. 125–139, 2014.
© Springer International Publishing Switzerland 2014

in a city, one operator might prefer the task be performed as quickly as possible whereas another might prefer the task be performed as safely as possible (e.g., not driving down a road with heavy automobile traffic or large potholes). Each operator has distinct preferences that influence how they will trust the robot's behavior, and these preferences may conflict. Even if these user preferences were known in advance, a change in context could also influence what behaviors are trustworthy. An operator who generally prefers a task to be performed quickly would likely change that preference if the robot was transporting hazardous material, whereas an operator who prefers safety would likely change their preferences in an emergency situation. Similarly, it may be infeasible to elicit a complete knowledge base of rules defining trustworthy behavior if the experts do not know the explicit rules or there are so many rules it is impractical to extract them all.

The ability of a robot to behave in a trustworthy manner regardless of the operator, task, or context requires that it can evaluate its trustworthiness and adapt its behavior accordingly. The robot may not always get explicit feedback about its trustworthiness but will instead need to estimate its trustworthiness based on its interactions with its operator. Such an estimate, which we refer to as an *inverse trust estimate*, differs from traditional computational trust metrics in that it measures how much trust another agent has in the robot rather than how much trust the robot has in another agent. In this paper we examine how a robot can estimate the trust an operator has in it, adapt its behavior to become more trustworthy, and learn from previous adaptations so it can perform trustworthy behaviors more quickly. We use case-based reasoning (CBR) to allow the robot to learn from previous behavior adaptations. The robot stores previous behavior adaptation information as cases in its case base and uses those cases to perform future behavior adaptations.

In the remainder of this paper we describe our behavior adaptation approach and evaluate it in a simulated robotics domain. We describe the robot's behavior and the aspects that it can modify in Section 2. Section 3 presents the inverse trust metric and Section 4 describes how it can be used to guide the robot's behavior. In Section 5, we evaluate our case-based behavior adaptation strategy in a simulated robotics domain and report evidence that it can efficiently adapt the robot's behavior to the operator's preferences. Related work is examined in Section 6 followed by a discussion of future work and concluding remarks in Section 7.

2   Agent Behavior

We assume the robot can control and modify aspects of its behavior. These modifiable components could include changing a module (e.g., switching between two path planning algorithms), its parameter values, or its data (e.g., using a different map of the environment). By modifying these components the robot can immediately change its behavior.

We define each modifiable behavior component i to have a range of selectable values $\mathcal{C}_i$. If the robot has m modifiable components, its current behavior B

will be a tuple containing the currently selected value c_i for each modifiable component ($c_i \in \mathcal{C}_i$):

$$B = \langle c_1, c_2, \ldots, c_m \rangle$$

By changing one or more of its behavior components, the robot switches from using its current behavior B to a new behavior B'. While operating in the environment, the robot might change its behavior several times, resulting in a sequence of behaviors $\langle B_1, B_2, \ldots, B_n \rangle$. Since the goal of the robot is to perform trustworthy behavior, behavior changes will occur because a current behavior B was found to be untrustworthy and it is attempting to perform a more trustworthy behavior.

3   Inverse Trust Estimate

Traditional trust metrics are used to estimate the trust an agent should have in other agents [3]. The agent can use prior interactions with those agents or feedback from others to determine their trustworthiness. The information this agent uses is likely internal to it and not directly observable by a third party. In a robotics context, the robot will not be able to observe the information a human operator uses to assess their trust in it. Instead, the robot will need to acquire this internal information to estimate operator trust.

One option would be to directly ask the operator, either as it is interacting with the robot [4] or after the task has been completed [5,6], about how trustworthy the robot was behaving. However, this might not be practical in situations that are time-sensitive or where there would be a significant delay between when the robot wishes to evaluate its trustworthiness and the next opportunity to ask the operator (e.g., during a multi-day search and rescue mission). An alternative that does not require direct operator feedback is for the robot to *infer* the trust the operator has in it.

Factors that influence human-robot trust can be grouped into three main categories [1]: robot-related factors (e.g., performance, physical attributes), human-related factors (e.g., engagement, workload, self-confidence), and environmental factors (e.g., group composition, culture, task type). Although these factors have all been shown to influence human-robot trust, the strongest indicator of trust is robot performance [7,8]. Kaniarasu et al. [9] have used an inverse trust metric that estimates robot performance based on the number of times the operator warns the robot about its behavior and the number of times the operator takes manual control of the robot. They found this metric aligns closely with the results of trust surveys performed by the operators. However, this metric does not take into account factors of the robot's behavior that increase trust.

The inverse trust metric we use is based on the number of times the robot completes an assigned task, fails to complete a task, or is interrupted while performing a task. An interruption occurs when the operator instructs the robot to stop its current autonomous behavior. Our robot infers that any interruptions are a result of the operator being unsatisfied with the robot's performance.

Similarly, our robot assumes the operator will be unsatisfied with any failures and satisfied with any completed tasks. Interrupts could also be a result of a change in the operator's goals, or failures could be a result of unachievable tasks, but the robot works under the assumption that those situations occur rarely.

Our control strategy estimates whether trust is increasing, decreasing, or remaining constant while the current behavior B' is being used by the robot. We estimate this value as follows:

$$Trust_{B'} = \sum_{i=1}^{n} w_i \times cmd_i,$$

where there were n commands issued to the robot while it was using its current behavioral configuration. If the ith command ($1 \leq i \leq n$) was interrupted or failed it will decrease the trust value and if it was completed successfully it will increase the trust value ($cmd_i \in \{-1, 1\}$). The ith command will also receive a weight ($w_i = [0, 1]$) related to the command (e.g., a command that was interrupted because the robot performed a behavior slowly would likely be weighted less than an interruption because the robot injured a human).

4 Trust-guided Behavior Adaptation Using CBR

The robot uses the inverse trust estimate to infer if its current behavior is trustworthy, is not trustworthy, or it does not yet know. We use two threshold values to identify trustworthy and untrustworthy behavior: the trustworthy threshold (τ_T) and the untrustworthy threshold (τ_{UT}). Our robot uses the following tests:

- If the trust value reaches the trustworthy threshold ($Trust_{B'} \geq \tau_T$), the robot will conclude it has found a sufficiently trustworthy behavior (although it may continue evaluating trust in case any changes occur).
- If the trust value falls to or below the untrustworthy threshold ($Trust_{B'} \leq \tau_{UT}$), the robot will modify its behavior in an attempt to be more trustworthy.
- If the trust value is between the two thresholds ($\tau_{UT} < Trust_{B'} < \tau_T$), the robot will continue to evaluate the operator's trust.

In the situations where the trustworthy threshold has been reached or neither threshold has been reached, the robot will continue to use its current behavior. However, when the untrustworthy threshold has been reached the robot will modify its behavior in an attempt to behave in a more trustworthy manner.

When a behavior B is found by the robot to be untrustworthy it is stored as an evaluated pair E that also contains the time t it took the behavior to be labeled as untrustworthy:

$$E = \langle B, t \rangle$$

The time it took for a behavior to reach the untrustworthy threshold is used to compare behaviors that have been found to be untrustworthy. A behavior B'

that reaches the untrustworthy threshold more quickly than another behavior B'' ($t' < t''$) is assumed to be less trustworthy than the other. This is based on the assumption that if a behavior took longer to reach the untrustworthy threshold then it was likely performing some trustworthy actions or was not performing untrustworthy actions as quickly.

As the robot evaluates behaviors, it stores a set $\mathcal{E}_{past}$ of previously evaluated behaviors ($\mathcal{E}_{past} = \{E_1, E_2, \ldots, E_n\}$). It continues to add to this set until it locates a trustworthy behavior B_{final} (when the trustworthy threshold is reached), if a trustworthy behavior exists. The sets of evaluated behaviors can be thought of as the search path that resulted in the final solution (the trustworthy behavior). The search path information is potentially useful because if the robot can determine it is on a similar search path that it has previously encountered (similar behaviors being labeled untrustworthy in a similar amount of time) then the robot can identify what final behavior it should attempt.

To allow for the reuse of past behavior adaptation information we use case-based reasoning. Each *case* C is composed of a problem and a solution. In our context, the *problem* is the previously evaluated behaviors and the *solution* is the final trustworthy behavior:

$$C = \langle \mathcal{E}_{past}, B_{final} \rangle$$

These cases are stored in a *case base* and represent the robot's knowledge about previous behavior adaptation.

When the robot modifies its behavior it selects new values for one or more of the modifiable components. The new behavior B_{new} is selected as a function of all behaviors that have been previously evaluated for this operator and its case base CB:

$$B_{new} = selectBehavior(\mathcal{E}_{past}, CB)$$

The *selectBehavior* function (Algorithm 1) attempts to use previous adaptation experience to guide the current adaptation. The algorithm iterates through each case in the case base (line 2) and checks to see if that case's final behavior has already been evaluated (line 3). If so, the robot has already found the behavior to be untrustworthy and does not try to use it again. Algorithm 1 then compares the sets of evaluated behaviors of the remaining cases ($C_i.\mathcal{E}_{past}$) to the robot's current set of evaluated behaviors ($\mathcal{E}_{past}$) using a similarity metric (line 4). The most similar case's final behavior is returned and will be used by the robot (line 10). If no such behaviors are found (the final behaviors of all cases have been examined or the case base is empty), the *modifyBehavior* function is used to select the next behavior to perform (line 9). It selects an evaluated behavior E_{max} that took the longest to reach the untrustworthy threshold ($\forall E_i \in \mathcal{E}_{past}(E_{max}.t \geq E_i.t)$) and performs a random walk (without repetition) to find a behavior B_{new} that required the minimum number of changes from $E_{max}.B$ and has not already been evaluated ($\forall E_i \in \mathcal{E}_{past}(B_{new} \neq E_i.B)$). If all possible behaviors have been evaluated and found to be untrustworthy the robot will stop adapting its behavior and use the behavior from E_{max}.

Algorithm 1. Selecting a New Behavior

Function: $selectBehavior(\mathcal{E}_{past}, CB)$ **returns** B_{new};

---

1 $bestSim \leftarrow 0;\ B_{best} \leftarrow \varnothing;$
2 **foreach** $C_i \in CB$ **do**
3   **if** $C_i.B_{final} \notin \mathcal{E}_{past}$ **then**
4          $sim_i \leftarrow sim(\mathcal{E}_{past}, C_i.\mathcal{E}_{past});$
5   **if** $sim_i > bestSim$ **then**
6              $bestSim \leftarrow sim_i;$
7              $B_{best} \leftarrow C_i.B_{final};$

8 **if** $B_{best} = \varnothing$ **then**
9      $B_{best} \leftarrow modifyBehavior(\mathcal{E}_{past});$
10 **return** $B_{best};$

The similarity between two sets of evaluated behaviors (Algorithm 2) is complicated by the fact that the sets may vary in size. The size of the sets depends on the number of previous behaviors that were evaluated by the robot in each set and there is no guarantee that the sets contain identical behaviors. To account for this, the similarity function looks at the overlap between the two sets and ignores behaviors that have been examined in only one of the sets. Each evaluated behavior in the first set is matched to an evaluated behavior E_{max} in the second set that contains the most similar behavior (line 3, $sim(B_1, B_2) = \frac{1}{m} \sum_{i=1}^{m} sim(B_1.c_i, B_2.c_i)$, where the similarity function will depend on the specific type of behavior component). If those behaviors are similar enough, based on a threshold λ (line 4), then the similarity of the time components of these evaluated behaviors are included in the similarity calculation (line 5). This ensures that only matches between evaluated behaviors that are highly similar (i.e., similar behaviors exist in both sets) are included in the similarity calculation (line 9). The similarity metric only includes comparisons between time components because the goal is to find when similar behaviors were found to be untrustworthy in a similar amount of time.

5   Evaluation

In this section, we describe an evaluation for our claim that our case-based reasoning approach can adapt, identify, and perform trustworthy behaviors more quickly than a random walk approach. We conducted this study in a simulated environment with a simulated robot and operator. We examined two robotics scenarios: movement and patrolling for threats.

5.1   eBotWorks Simulator

Our evaluation uses the eBotworks simulation environment [10]. eBotworks is a multi-agent simulation engine and testbed that allows for multimodal command

Algorithm 2. Similarity between sets of evaluated behaviors

Function: $sim(\mathcal{E}_1, \mathcal{E}_2)$ ***returns*** sim;

---

1 $totalSim \leftarrow 0$; $num \leftarrow 0$;
2 **foreach** $E_i \in \mathcal{E}_1$ **do**
3      $E_{max} \leftarrow \underset{E_j \in \mathcal{E}_2}{\arg\max}\,(sim(E_i.B, E_j.B))$;
4   **if** $sim(E_i.B, E_{max}.B) > \lambda$ **then**
5          $totalSim \leftarrow totalSim + sim(E_i.t, E_{max}.t)$;
6          $num \leftarrow num + 1$;
7 **if** $num = 0$ **then**
8   **return** 0;
9 **return** $\frac{totalSim}{num}$;

and control of unmanned systems. It allows for autonomous agents to control simulated robotic vehicles while interacting with human operators, and for the autonomous behavior to be observed and evaluated. We chose to use eBotworks based on its flexibility in autonomous behavior modeling, the ability for agents to process natural language commands, and built-in experimentation and data collection capabilities.

The robot operates in a simulated urban environment containing landmarks (e.g., roads) and objects (e.g., houses, humans, traffic cones, vehicles, road barriers). The robot is a wheeled unmanned ground vehicle (UGV) and uses eBotwork's built-in natural language processing (for interpreting user commands), locomotion, and path-planning modules. The actions performed by a robot in eBotworks are non-deterministic (e.g., the robot cannot anticipate its exact position after moving).

5.2   Experimental Conditions

We use simulated operators in our study to issue commands to the robot. In each experiment, one of these operators interacts with the robot for 500 *trials*. The simulated operators differ in their preferences, which will influence how they evaluate the robot's performance (when an operator allows the robot to complete a task and when it interrupts). At the start of each trial, the robot randomly selects (with a uniform distribution) initial values for each of its modifiable behavior components. Throughout the trial, a series of experimental *runs* will occur. Each run involves the simulated operator issuing a command to the robot and monitoring the robot as it performs the assigned task. During a run, the robot might complete the task, fail to complete the task, or be interrupted by the operator. At the end of a run the environment will be reset so a new run can begin. The results of these runs will be used by the robot to estimate the operator's trust in it and to adapt its behavior if necessary. A trial concludes when the robot successfully identifies a trustworthy behavior or it has evaluated all possible behaviors.

For the case-based behavior adaptation, at the start of each experiment the robot will have an empty case base. At the end of any trial where the robot has found a trustworthy behavior and has performed at least one random walk adaptation (i.e., the agent could not find a solution by only using information in the case base), a case will be added to the case base and can be used by the robot in subsequent trials. The added case represents the trustworthy behavior found by the robot and the set of untrustworthy behaviors that were evaluated before the trustworthy behavior was found.

We set the robot's trustworthy threshold $\tau_T = 5.0$ and its untrustworthy threshold $\tau_{UT} = -5.0$. These threshold values were chosen to allow some fluctuation between increasing and decreasing trust while still identifying trustworthy and untrustworthy behaviors quickly. To calculate the similarity between sets of evaluated behaviors we set the similarity threshold to be $\lambda = 0.95$ (behaviors must be 95% similar to be matched). This threshold was used so that only highly similar behaviors will be matched together.

5.3   Scenarios

The scenarios we evaluate, movement and patrolling for threats, were selected to demonstrate the ability of our behavior adaptation technique when performing increasingly complex tasks. While the movement scenario is fairly simple, the patrolling scenario involves a more complex behavior with more modifiable behavior components.

Movement Scenario: The initial task the robot is required to perform involves moving between two locations in the environment. The simulated operators used in this scenario assess their trust in the robot using three performance metrics:

- **Task Duration:** The simulated operator has an expectation about the amount of time that the task will take to complete ($t_{complete}$). If the robot does not complete the task within that time, the operator may, with probability p_α, interrupt the robot and issue another command.
- **Task Completion:** If the operator determines that the robot has failed to complete the task (e.g., the robot is stuck), it will interrupt.
- **Safety:** The operator may interrupt the robot, with probability p_γ, if the robot collides with any obstacles along the route.

We use three simulated operators:

- **Speed-Focused Operator:** This operator prefers the robot to move to the destination quickly regardless of whether it hits any obstacles ($t_{complete} = 15$ seconds, $p_\alpha = 95\%$, $p_\gamma = 5\%$).
- **Safety-Focused Operator:** This operator prefers the robot to avoid obstacles regardless of how long it takes to reach the destination ($t_{complete} = 15$ seconds, $p_\alpha = 5\%$, $p_\gamma = 95\%$).

- **Balanced Operator:** This operator prefers a balanced mixture of speed and safety ($t_{complete} = 15$ seconds, $p_\alpha = 95\%$, $p_\gamma = 95\%$).

The robot has two modifiable behavior components: *speed* (meters per second) and *obstacle padding* (meters). Speed relates to how fast the robot can move and obstacle padding relates to the distance the robot will attempt to maintain from obstacles during movement. The set of possible values for each modifiable component (C_{speed} and $C_{padding}$) are determined from minimum and maximum values (based on the robot's capabilities) with fixed increments.

$$C_{speed} = \{0.5, 1.0, \ldots, 10.0\}$$
$$C_{padding} = \{0.1, 0.2, 0.3, \ldots, 2.0\}$$

Patrolling Scenario: The second task the robot is required to perform involves patrolling between two locations in the environment. At the start of each run, 6 suspicious objects representing potential threats are randomly placed in the environment. Of those 6 suspicious objects, between 0 and 3 (inclusive) denote hazardous explosive devices (selected randomly using a uniform distribution). As the robot moves between the start location and the destination it will scan for suspicious objects nearby. When it identifies a suspicious object it will pause its patrolling behavior, move toward the suspicious object, scan it with its explosives detector, label the object as an explosive or harmless, and then continue its patrolling behavior. The accuracy of the explosives detector the robot uses is a function of how long the robot spends scanning the object (longer scan times result in improved accuracy) and its proximity to the object (smaller scan distances increase the accuracy). The scan time (seconds) and scan distance (meters) are two modifiable components of the robot's behavior whose set of possible values are:

$$C_{scantime} = \{0.5, 1.0, \ldots, 5.0\}$$
$$C_{scandistance} = \{0.25, 0.5, \ldots, 1.0\}$$

The simulated operators in this scenario base their decision to interrupt the robot on its ability to successfully identify suspicious objects and label them correctly (in addition to the task duration, task completion, and safety factors discussed in the movement scenario). An operator will interrupt the robot if it does not scan one or more of the suspicious objects or incorrectly labels a harmless object as an explosive. In the event that the robot incorrectly labels an explosive device as harmless, the explosive will eventually detonate and the robot will fail its task. When determining its trustworthiness, the robot will give higher weights to failures due to missing explosive devices (they will be weighted 3 times higher than other failures or interruptions).

In this scenario we use two simulated operators:

- **Speed-Focused Operator:** The operator prefers that the robot performs the patrol task within a fixed time limit ($t_{complete} = 120$ seconds, $p_\alpha = 95\%$, $p_\gamma = 5\%$).

- **Detection-Focused Operator:** The operator prefers the task be performed correctly regardless of time ($t_{complete} = 120$ seconds, $p_\alpha = 5\%$, $p_\gamma = 5\%$).

5.4   Results

We found that both the case-based behavior adaptation and the random walk behavior adaptation strategies resulted in similar trustworthy behaviors for each simulated operator. In the movement scenario, for the speed-focused operator the trustworthy behaviors had higher speeds regardless of padding ($3.5 \leq speed \leq 10.0$, $0.1 \leq padding \leq 1.9$). The safety-focused operator had higher padding regardless of speed ($0.5 \leq speed \leq 10.0$, $0.4 \leq padding \leq 1.9$). Finally, the balanced operator had higher speed and higher padding ($3.5 \leq speed \leq 10.0$, $0.4 \leq padding \leq 1.9$). These results are consistent with our previous findings [11] that trust-guided behavior adaptation using random walk converges to behaviors that appear to be trustworthy for each type of operator.

In the patrolling scenario, which we have not studied previously, the differences between the trustworthy behaviors for the two operators are not only in the ranges of the values for the modifiable components but also their relations to each other. Similar to what was seen in the movement scenario, since the speed-focused patrol operator has a time preference the robot only converges to higher speed values whereas the detection-focused operator has no such restriction (the speed-focused operator never converges to a speed below 2.0). The speed-focused patrol operator never has both a low speed and a high scan time. This is because these modifiable components are interdependent. If the robot spends more time scanning, it will need to move through the environment at a higher speed. Similarly, both operators converge to scan time and scan distance values that reveal a dependence. The robot only selects a poor value for one of the modifiable components (low scan time or high scan distance) if it selects a very good value for the other component (high scan time or low scan distance). This shows that behavior adaptation can select trustworthy values when the modifiable components are mostly independent or when there is a strong dependence between multiple behavior components.

Both the case-based reasoning and random walk adaptation approaches converged to similar trustworthy behaviors. The only noticeable difference is that final behaviors stored in cases are found to be trustworthy in more trials. This is what we would expect from the case-based approach since these cases are retrieved and their final behaviors are reused. The primary difference between the two behavior adaption approaches was related to the number of behaviors that needed to be evaluated before a trustworthy behavior was found. Table 1 shows the mean number of evaluated behaviors (and 95% confidence interval) when interacting with each operator type (over 500 trials for each operator). The table also lists the number of cases acquired during the case-based behavior adaptation experiments (each experiment started with an empty case base). In addition to being controlled by only a single operator, we also examined a condition in which, for each scenario, the operator is selected at random with equal probability. This represents a more realistic scenario where the robot will be required to interact with a variety of operators without any knowledge about which operator will control it.

Table 1. Mean number of behaviors evaluated before finding a trustworthy behavior

Scenario	Operator	Random Walk	Case-based	Cases Acquired
Movement	*Speed-focused*	20.3 (±3.4)	1.6 (±0.2)	24
Movement	*Safety-focused*	2.8 (±0.3)	1.3 (±0.1)	18
Movement	*Balanced*	27.0 (±3.8)	1.8 (±0.2)	33
Movement	*Random*	14.6 (±2.9)	1.6 (±0.1)	33
Patrol	*Speed-focused*	344.5 (±31.5)	9.9 (±3.9)	25
Patrol	*Detection-focused*	199.9 (±23.3)	5.5 (±2.2)	22
Patrol	*Random*	269.0 (±27.1)	9.3 (±3.2)	25

The case-based approach required significantly fewer behaviors to be evaluated in all seven experiments (using a paired t-test with $p < 0.01$). This is because the case-based approach could learn from previous adaptations and use that information to quickly find trustworthy behaviors. At the beginning of a trial, when the robot's case base is empty, the case-based approach must perform adaptation that is similar to the random walk approach. As the case base size grows, the number of times random walk adaptation is required decreases until the agent generally performs only one case-based behavior adaptation before finding a trustworthy behavior. Even when the case base contains cases from all two (in the patrol scenario) or three (in the movement scenario) simulated operators, the case-based approach can quickly differentiate between the users and select a trustworthy behavior for the current operator. The number of adaptations required for the safety-focused and detection-focused operators were lower than for the other operators in their scenarios because a higher percentage of behaviors are considered trustworthy for those operators.

5.5   Discussion

The primary limitation of the case-based approach is that it relies on the random walk search when it does not have any suitable cases to use. Although the mean number of behaviors evaluated by the case-based approach is low, the situations where random walk is used require an above-average number of behaviors to be evaluated (closer to the mean number of behaviors evaluated when only random walk is used). For example, if we consider only the final 250 trials for each of the patrol scenario operators the mean number of behaviors evaluated is lower than the overall mean (4.2 for the speed-focused, 2.8 for the detection-focused, and 3.3 for the random). This is because the robot performs the more expensive random walk adaptations in the early trials and generates cases that are used in subsequent trials.

Two primary solutions exist to reduce the number of behaviors examined: improved search and seeding of the case base. We used random walk search because it requires no explicit knowledge about the domain or the task. However, a more intelligent search that could identify relations between interruptions and modifiable components (e.g., an interruption when the robot is close to objects requires a change to the padding value) would likely improve adaptation time. Since a higher number of behaviors need to be evaluated when new cases are created, if a set of initial cases were provided to the robot it would be able to decrease the number of random walk adaptations (or adaptations requiring a different search technique) it would need to perform. These two solutions introduce their own potential limitations. A more informed search requires introducing domain knowledge, which may not be easy to obtain, and seeding the case base requires an expert to manually author cases (or another method for case acquisition). The specific requirements of the application domain will influence whether faster behavior adaptation or lower domain knowledge requirements are more important.

6   Related Work

In addition to Kaniarasu et al. [9], Saleh et al. [12] have also proposed a measure of inverse trust and use a set of expert-authored rules to measure trust. Unlike our own work, while these approaches measure trust, they do not use this information to adapt behavior. The limitation of using these trust metrics to guide our behavior adaptation technique is that one of the metrics only measures decreases in trust [9] and the other requires expert-authored rules [12].

The topic of trust models in CBR is generally examined in the context of recommendation systems [13] or agent collaboration [14]. Similarly, the idea of case provenance [15] is related to trust in that it involves considering the source of a case and if that source is a reliable source of information. These investigations consider traditional trust, where an agent determines its trust in another agent, rather than inverse trust, which is our focus.

Case-based reasoning has been used for a variety of robotics applications and often facilitates action selection [16] or behavior selection [17]. Existing work on CBR and robotics differs from our own in that most systems attempt to optimize the robot's performance without considering that sub-optimal performance may be necessary to gain a human teammate's trust. In an assistive robotics task, a robotic wheelchair uses CBR to learn to drive in a similar manner to its operator [18]. This work differs from our own in that it requires the operator to demonstrate the behavior over several trials, like a teacher, and the robot learns from those observations. The robot in our system received information that is not annotated so it can not benefit from direct feedback or labelling by the operator.

Shapiro and Shachter [19] discuss the need for an agent to act in the best interests of a user even if that requires sub-optimal performance. Their work is on identifying factors that influence the user's utility function and updating the agent's reward function accordingly. This is similar to our own work in that behavior is modified to align with a user's preference, but our robot is not given an explicit model of the user's reasoning process.

Conversational recommender systems [20] iteratively improve recommendations to a user by tailoring the recommendations to the user's preferences. As more information is obtained through dialogs with a user, these systems refine their model of that user. Similarly, learning interface agents observe a user performing a task (e.g., sorting e-mail [21] or schedule management [22]) and learn the user's preferences. Both conversational recommender systems and learning interface agents are designed to learn preferences for a single task whereas our behavior adaptation requires no prior knowledge about what tasks will be performed.

Our work also relates to other areas of learning during human-robot interactions. When a robot learns from a human, it is often beneficial for the robot to understand the environment from the perspective of that human. Breazeal et al. [23] examined how a robot can learn from a cooperative human teacher by mapping its sensory inputs to how it estimates the human is viewing the environment. This allows the robot to learn from the viewpoint of the teacher and possibly discover information it would not have noticed from its own viewpoint. This is similar to preference-based planning systems that learn a user's preferences for plan generation [24]. Like our own work, these systems involve inferring information about the reasoning of a human. However, they differ in that they involve observing a teacher demonstrate a specific task and learning from those demonstrations.

7   Conclusions

In this paper we presented an inverse trust measure that allows a robot to estimate an operator's trust and adapt its behavior to increase trust. As the robot performs trust-guided adaptation, it learns using case-based reasoning. Each time it successfully finds a trustworthy behavior, it can record a case that contains the trustworthy behavior as well as the sequence of untrustworthy behaviors that it evaluated.

We evaluated our trust-guided behavior adaptation algorithm in a simulated robotics environment by comparing it to a behavior adaptation algorithm that does not learn from previous adaptations. Two scenarios were examined: movement and patrolling for threats. Both approaches converge to trustworthy behaviors for each type of operator but the case-based algorithm requires significantly fewer behaviors to be evaluated before a trustworthy behavior is found. This is advantageous because the chances that the operator will stop using the robot increase the longer the robot is behaving in an untrustworthy manner.

Although we have shown the benefits of trust-guided behavior adaptation, several areas of future work exist. In longer scenarios it may be important to not only consider undertrust, as we have done in this work, but also overtrust. In situations of overtrust, the operator may trust the robot too much and allow the robot to behave autonomously even when it is performing poorly. We also plan to include additional trust factors in the inverse trust estimate and add mechanisms that promote transparency between the robot and operator. More generally, adding an ability for the robot to reason about its own goals and the goals of the operator would allow the robot to verify it is trying to achieve the same goals as the operator and identify any unexpected goal changes (e.g., such as when a threat occurs). Examining more complex interactions between the operator and the robot, like providing preferences or explaining the reasons for interruptions, would allow the robot to build a more elaborate operator model and potentially use different learning strategies than those presented here.

Acknowledgments. Thanks to the Naval Research Laboratory and the Office of Naval Research for supporting this research.

References

1. Oleson, K.E., Billings, D.R., Kocsis, V., Chen, J.Y., Hancock, P.A.: Antecedents of trust in human-robot collaborations. In: 1st International Multi-Disciplinary Conference on Cognitive Methods in Situation Awareness and Decision Support, pp. 175–178 (2011)
2. Desai, M., Kaniarasu, P., Medvedev, M., Steinfeld, A., Yanco, H.: Impact of robot failures and feedback on real-time trust. In: 8th International Conference on Human-robot Interaction, pp. 251–258 (2013)
3. Sabater, J., Sierra, C.: Review on computational trust and reputation models. Artificial Intelligence Review 24(1), 33–60 (2005)
4. Kaniarasu, P., Steinfeld, A., Desai, M., Yanco, H.A.: Robot confidence and trust alignment. In: 8th International Conference on Human-Robot Interaction, pp. 155–156 (2013)
5. Jian, J.Y., Bisantz, A.M., Drury, C.G.: Foundations for an empirically determined scale of trust in automated systems. International Journal of Cognitive Ergonomics 4(1), 53–71 (2000)
6. Muir, B.M.: Trust between humans and machines, and the design of decision aids. International Journal of Man-Machine Studies 27(5-6), 527–539 (1987)
7. Hancock, P.A., Billings, D.R., Schaefer, K.E., Chen, J.Y., De Visser, E.J., Parasuraman, R.: A meta-analysis of factors affecting trust in human-robot interaction. Human Factors: The Journal of the Human Factors and Ergonomics Society 53(5), 517–527 (2011)
8. Carlson, M.S., Desai, M., Drury, J.L., Kwak, H., Yanco, H.A.: Identifying factors that influence trust in automated cars and medical diagnosis systems. In: AAAI Symposium on the Intersection of Robust Intelligence and Trust in Autonomous Systems, pp. 20–27 (2014)
9. Kaniarasu, P., Steinfeld, A., Desai, M., Yanco, H.A.: Potential measures for detecting trust changes. In: 7th International Conference on Human-Robot Interaction, pp. 241–242 (2012)
10. Knexus Research Corporation: eBotworks (2013), http://www.knexusresearch.com/products/ebotworks.php (Online; accessed April 9, 2014)
11. Floyd, M.W., Drinkwater, M., Aha, D.W.: Adapting autonomous behavior using an inverse trust estimation. In: Murgante, B., et al. (eds.) ICCSA 2014, Part I. LNCS, vol. 8579, pp. 728–742. Springer, Heidelberg (2014)
12. Saleh, J.A., Karray, F., Morckos, M.: Modelling of robot attention demand in human-robot interaction using finite fuzzy state automata. In: International Conference on Fuzzy Systems, pp. 1–8 (2012)

13. Tavakolifard, M., Herrmann, P., Öztürk, P.: Analogical trust reasoning. In: Ferrari, E., Li, N., Bertino, E., Karabulut, Y. (eds.) IFIPTM 2009. IFIP AICT, vol. 300, pp. 149–163. Springer, Heidelberg (2009)
14. Briggs, P., Smyth, B.: Provenance, trust, and sharing in peer-to-peer case-based web search. In: Althoff, K.-D., Bergmann, R., Minor, M., Hanft, A. (eds.) ECCBR 2008. LNCS (LNAI), vol. 5239, pp. 89–103. Springer, Heidelberg (2008)
15. Leake, D., Whitehead, M.: Case provenance: The value of remembering case sources. In: Weber, R.O., Richter, M.M. (eds.) ICCBR 2007. LNCS (LNAI), vol. 4626, pp. 194–208. Springer, Heidelberg (2007)
16. Ros, R., Veloso, M.M., de Màntaras, R.L., Sierra, C., Arcos, J.-L.: Retrieving and reusing game plays for robot soccer. In: Roth-Berghofer, T.R., Göker, M.H., Güvenir, H.A. (eds.) ECCBR 2006. LNCS (LNAI), vol. 4106, pp. 47–61. Springer, Heidelberg (2006)
17. Likhachev, M., Arkin, R.C.: Spatio-temporal case-based reasoning for behavioral selection. In: International Conference on Robotics and Automation, pp. 1627–1634 (2001)
18. Urdiales, C., Peula, J.M., Fernández-Carmona, M., Sandoval, F.: Learning-based adaptation for personalized mobility assistance. In: Delany, S.J., Ontañón, S. (eds.) ICCBR 2013. LNCS, vol. 7969, pp. 329–342. Springer, Heidelberg (2013)
19. Shapiro, D., Shachter, R.: User-agent value alignment. In: Stanford Spring Symposium - Workshop on Safe Learning Agents (2002)
20. McGinty, L., Smyth, B.: On the role of diversity in conversational recommender systems. In: Ashley, K.D., Bridge, D.G. (eds.) ICCBR 2003. LNCS, vol. 2689, pp. 276–290. Springer, Heidelberg (2003)
21. Maes, P., Kozierok, R.: Learning interface agents. In: 11th National Conference on Artificial Intelligence, pp. 459–465 (1993)
22. Horvitz, E.: Principles of mixed-initiative user interfaces. In: 18th Conference on Human Factors in Computing Systems, pp. 159–166 (1999)
23. Breazeal, C., Gray, J., Berlin, M.: An embodied cognition approach to mindreading skills for socially intelligent robots. International Journal of Robotic Research 28(5) (2009)
24. Li, N., Kambhampati, S., Yoon, S.W.: Learning probabilistic hierarchical task networks to capture user preferences. In: 21st International Joint Conference on Artificial Intelligence, pp. 1754–1759 (2009)

Tuuurbine: A Generic CBR Engine over RDFS*

Emmanuelle Gaillard, Laura Infante-Blanco,
Jean Lieber, and Emmanuel Nauer

Université de Lorraine, LORIA — 54506 Vandœuvre-lès-Nancy, France
CNRS — 54506 Vandœuvre-lès-Nancy, France
Inria — 54602 Villers-lès-Nancy, France
{emmanuelle.gaillard,laura.infanteblanco,
jean.lieber,emmanuel.nauer}@loria.fr

Abstract. This paper presents TUUURBINE, a case-based reasoning (CBR) system for the Semantic Web. TUUURBINE is built as a generic CBR system able to reason on knowledge stored in RDF format; it uses Semantic Web technologies like RDF/RDFS, RDF stores, SPARQL, and optionally Semantic Wikis. TUUURBINE implements a generic case-based inference mechanism in which adaptation consists in retrieving similar cases and in replacing some features of these cases in order to obtain one or more solutions for a given query. The search for similar cases is based on a generalization/specialization method performed by means of generalization costs and adaptation rules. The whole knowledge (cases, domain knowledge, costs, adaptation rules) is stored in an RDF store.

Keywords: generic CBR engine, Semantic Web, RDFS, triple store.

1 Introduction and Motivations

This paper presents TUUURBINE (http://tuuurbine.loria.fr/), a new generic case-based reasoning system (CBR), the reasoning procedure of which is based on a domain ontology. TUUURBINE was created on the basis of the experience acquired during five years with the TAAABLE system [1].

This research work is motivated by the will to develop a generic CBR system. TUUURBINE implements a generic case-based inference. Searching for similar cases is based on a generalization/specialization process performed by means of generalization costs and adaptation rules. The generic adaptation approach is inspired from the TAAABLE's adaptation.

The second major motivation is the development of a system able to exploit the huge and growing amount of knowledge available on the Web, especially

---

* The development of TUUURBINE was supported by an Inria ADT funding from October 2011 to October 2013. The authors would like to thank the reviewers who have helped improving the quality of this paper: it was not possible to take into account all their remarks in the paper, but these remaining remarks point out interesting issues that the authors plan to address as future work.

L. Lamontagne and E. Plaza (Eds.): ICCBR 2014, LNCS 8765, pp. 140–154, 2014.
© Springer International Publishing Switzerland 2014

the knowledge coming from The Linked Data or contained in Semantic Wikis. This is why TUUURBINE is built according to Semantic Web standards (RDF, SPARQL, RDFS), to facilitate the interoperability with Semantic Web knowledge. With TUUURBINE, the knowledge (cases, domain knowledge, costs, adaptation rules) is encoded in a triple store, which may be installed in the same machine as the reasoner or in a remote one, and could be additionally be interfaced with a Semantic Wiki, in order to benefit from collaborative web edition and knowledge management involved in the reasoning process [2].

Storing the knowledge in a triple store provides an important advantage: the knowledge is not managed by the CBR system anymore, but by an external tool that is efficient and based on standards. A clear separation is made between the reasoning inference engine sub-system and the knowledge handling and storage sub-system of the CBR scheme: the knowledge used by the CBR system is not loaded in cache once and for all, but on demand, only the necessary knowledge is retrieved to solve a new problem, using SPARQL language. This approach also ensures the exploitation of up-to-date knowledge to best solve a problem in environments where the knowledge evolves continuously. The choice of such an architecture is the result of the experience acquired with TAAABLE. Indeed, in the first version of TAAABLE, the knowledge was completely loaded in memory from several files encoded in various languages [3]. This caused the problem of impersistence of updated knowledge cache, which appears frequently when tuning the system. The second version of TAAABLE improved the knowledge evolution using a Semantic Wiki, called WIKITAAABLE [2] enabling collaborative edition of the knowledge, but still the TAAABLE CBR system required, at that time, loading a new dump of knowledge in memory, after every single knowledge modification, in order to use up-to-date knowledge. The last version of TAAABLE, which uses now the TUUURBINE CBR engine, benefits from a complete synchronization with the knowledge base thanks to a dynamic access to the triple store.

The paper is organized as follows. Section 2 describes briefly the Semantic Web technologies used in TUUURBINE. Section 3 describes the generic CBR approach of TUUURBINE and details the knowledge the approach is based on and the knowledge representation choices that were made. Section 4 details the architecture of the system. Section 5 presents two use-cases. Section 6 discusses this work and relates it to other works. Section 7 concludes the paper.

2   Semantic Web Technologies Used in Tuuurbine

RDF, RDFS and the Triple Stores. RDF[1], SPARQL[2] and RDFS[3] are three standards recommendations of the W3C (World Wide Web Consortium) for the Semantic Web.

RDF (Resource Description Framework) is a format to encode resources over the Web, existing in various syntaxes. A *resource* is any kind of entity that has

---

[1] http://www.w3.org/RDF
[2] http://www.w3.org/2009/sparql
[3] http://www.w3.org/TR/rdf-schema

been reified (i.e., associated with an identifier). A *literal* is a constant datatype (e.g., an integer, a float, a string, etc.). A *property* relates a resource to another resource or a literal. An RDF base is a set of *triples*, a triple being an expression of the form $\langle s \ p \ o \rangle$. s, p and o are respectively called the *subject* (a resource), the *predicate* (a property) and the *object* (a resource or a literal) of a triple. For example, if romeo and juliet are two resources to be understood as the main characters of [4], if loves and ofTheFamily are the properties meaning "loves" and "is a member of the family", and if age relates a resource to a literal of integer datatype indicating the age of the resource, then the RDF base $\mathcal{B} = \{\langle$romeo ofTheFamily montague$\rangle, \langle$romeo loves juliet$\rangle, \langle$juliet age 13$\rangle\}$ means that Romeo, a member of the family Montague, loves Juliet who is 13.

SPARQL (recursive acronym for SPARQL Protocol And RDF Query Language) is at the same time a protocol and a query language for RDF. It can be likened to SQL: SPARQL is to RDF as SQL is to relational databases. For instance, the following SPARQL query to the base $\mathcal{B}$ above returns the resources x such that $\langle x$ loves juliet$\rangle \in \mathcal{B}$ and $\langle x$ ofTheFamily montague$\rangle \in \mathcal{B}$:

```
SELECT ?x
WHERE { ?x loves juliet .          // "." stands for "and"
        ?x ofTheFamily montague}
```

RDFS (RDF Schema) can be considered as a knowledge representation formalism, the syntax of which is RDF (a formula of RDFS is an RDF triple) and the semantics of which is associated with a set of resources having a predefined semantics, called the RDFS *vocabulary*. In this paper, only the properties rdf:type and rdfs:subClassOf of the RDFS vocabulary are considered and they are abbreviated by type and subc. type can be understood as "is an element of" ($\in$) and subc as "is a subset of" ($\subseteq$). The following inference rules are used for RDFS entailment reduced to the vocabulary $\{$type, subc$\}$:

$$\frac{\langle a \ \text{type} \ C \rangle \quad \langle C \ \text{subc} \ D \rangle}{\langle a \ \text{type} \ D \rangle} \qquad \frac{\langle C \ \text{subc} \ D \rangle \quad \langle D \ \text{subc} \ E \rangle}{\langle C \ \text{subc} \ E \rangle}$$

subc enables to define hierarchies of classes based on the "is more specific than" relation. Moreover, if C denotes an RDFS class, then $\langle C \ \text{subc} \ C \rangle$ is an RDFS axiom. If $\mathcal{B}$ is an RDFS base and τ is a triple, $\mathcal{B} \vdash \tau$ means that τ is a logical consequence of $\mathcal{B}$. Thus, $\mathcal{B} \nvdash \tau$ means that τ cannot be entailed from $\mathcal{B}$.

For example, the hierarchy

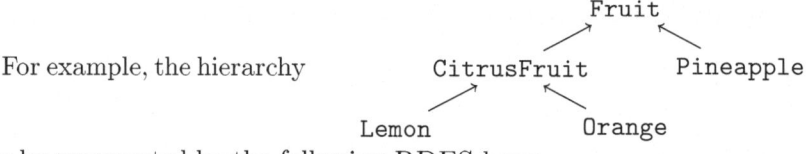

can be represented by the following RDFS base:

$$\mathcal{H} = \{\langle\text{CitrusFruit subc Fruit}\rangle, \quad \langle\text{Pineapple subc Fruit}\rangle,$$
$$\langle\text{Lemon subc CitrusFruit}\rangle, \quad \langle\text{Orange subc CitrusFruit}\rangle\}$$

which entails, e.g., the triple $\langle$Orange subc Fruit$\rangle$, i.e. every orange is a fruit.

Let SQ be a SPARQL query and $\mathcal{B}$ be a RDFS base. Then $\text{Result}_{\vdash}(\text{SQ}, \mathcal{B})$ denotes (in this paper) the result of the SPARQL query on $\mathcal{B}$ using the entailment. For example,

$$\text{Result}_{\vdash}\left(\boxed{\begin{array}{l} \text{SELECT ?x} \\ \text{WHERE } \{ \text{ ?x subc CitrusFruit} \} \end{array}}, \mathcal{H}\right) = \left\{\begin{array}{l} \text{Lemon, Orange,} \\ \text{CitrusFruit} \end{array}\right\}.$$

A *triple store* (also called RDF *store*) engine is a database management system for RDF. Using an RDF store enables in particular to avoid loading the whole RDF knowledge base in cache. Some triple store engines support RDFS entailment. However, for performance reasons, in the current version of TUUURBINE, the triple store chosen by default is 4Store[4] which does not draw entailments, these being implemented by TUUURBINE.

Semantic Wikis. A *semantic wiki* is a wiki the contents of which are not only documents and links between documents (as in classical Wikis) but also machine-processable data. SMW is a Semantic Wiki engine extending MediaWiki.[5] The semantic data in a semantic wiki using SMW are managed via a triple store. From a knowledge engineering viewpoint, a semantic wiki engine can be considered as a cooperative knowledge management tool: the knowledge can be edited thanks to this tool, associated with (non formalized) pieces of knowledge in plain text. To exploit the data of a Semantic Wiki, one can create a dump to an RDF file or query directly the triple store, using a SPARQL endpoint.

3   Tuuurbine Reasoning Principles

This section presents the principles upon which the TUUURBINE engine has been implemented. First, the running example is introduced. This example is then developed in the subsequent sections: representation of pieces of knowledge (cases, domain knowledge, similarity and adaptation knowledge), the representation of TUUURBINE queries, and the retrieval and adaptation procedures.

3.1   Introduction of the Running Example

Let us consider an application of CBR where a case is a recipe (and a case base represents a recipe book). Such application has been developed for the Computer Cooking Contest at ICCBR for the past years.

In this application, let us consider the following query:

$$\text{Q} = \boxed{\text{a cocktail recipe with mint, gin, orange juice but no wine.}} \tag{1}$$

If at least one recipe matches exactly Q, the application returns it. Otherwise, a recipe matching approximately Q is searched (retrieval step) and then is modified in order to answer Q (adaptation step).

---

[4] http://4store.org
[5] http://semantic-mediawiki.org/

Let us assume that no recipe matching exactly Q can be found but that the following source case similar to Q is retrieved:

Source = | **Recipe** "Mexican cocktail in my way"
Dish type: cocktail
Ingredients: 40 cl tequila, 1 l guava juice, 1 l pineapple juice, 30 cl apple juice, 1 pkt vanilla sugar, 3 mint leaves
Preparation: Combine guava juice and pineapple. Add tequila and apple juice. [...] |

Source matches approximately Q since:

- There is an exact match on the dish type (cocktail), the ingredient mint and the absence of the ingredient wine.
- The ingredients Tequila (Source) and Gin (Q) are subclasses of Liquor, and the ingredients GuavaJuice, PineappleJuice, AppleJuice (Source) and OrangeJuice (Q) are subclasses of FruitJuice.

Finally, adaptation modifies Source so that the modified recipe matches exactly Q. The adaptation is usually based on the approximate matching between Source and Q. In this example, it consists in applying the following modification:

$$\boxed{\text{Replace tequila with gin and guava juice, pineapple juice, or apple juice with orange juice.}} \tag{2}$$

Furthermore, there could be available rules which can be applied to adapt the case. In the example, let us consider the following adaptation rule:

$$\text{AR}_1 = \boxed{\begin{array}{l}\text{In the context of a cocktail dish without anise,}\\ \text{guava juice and vanilla sugar can be substituted with}\\ \text{orange juice and sugar cane syrup.}\end{array}}$$

This adaptation rule enables the retrieval of another (hopefully, better) adaptation of Source, given by the following modification:

$$\boxed{\text{Replace tequila with gin, and guava juice and vanilla sugar with orange juice and sugar cane syrup.}} \tag{3}$$

3.2   Knowledge Containers of a Tuuurbine Application

A CBR knowledge base can be split into four knowledge containers [5]. This knowledge is represented by an RDFS base KB and managed by a triple store.

Representation of Cases. In the running example, the reasoning takes into account only the dish type and the ingredients of the recipe (neither its title nor its preparation). Moreover, the ingredient quantities are not taken into account.

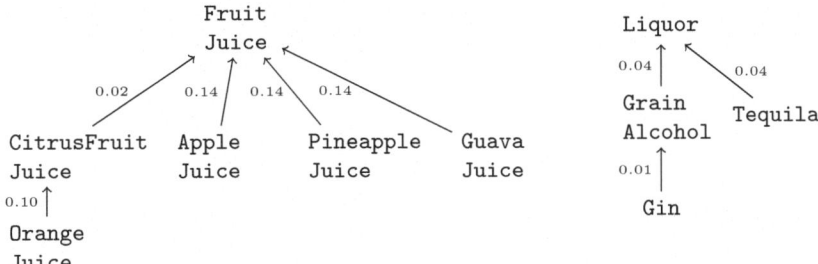

Fig. 1. The hierarchy forming the domain knowledge used in the running example with the generalization costs as retrieval knowledge

Therefore, this recipe can be represented by the following RDFS base:

$$\text{Source} = \begin{cases} \langle s \text{ type SourceCase} \rangle, \langle s \text{ dishType CocktailDish} \rangle, \\ \langle s \text{ ingredient Tequila} \rangle, \langle s \text{ ingredient GuavaJuice} \rangle, \\ \langle s \text{ ingredient PineappleJuice} \rangle, \langle s \text{ ingredient AppleJuice} \rangle, \\ \langle s \text{ ingredient VanillaSugar} \rangle, \langle s \text{ ingredient Mint} \rangle \end{cases}$$

More generally, a case of the case base is identified by a resource s and is defined by an RDFS base `Source` $\subseteq$ `KB` containing the triples of the form $\langle s \text{ prop val} \rangle$: this is a simple feature-value representation (the attribute being `prop` and the value being `val`). The triple $\langle s \text{ type SourceCase} \rangle$ is mandatory to indicate that this case belongs to `CaseBase`, the case base: it enables the obtention of the whole case base via the following SPARQL query:

```
SELECT ?s
WHERE { ?s type SourceCase}
```

Representation of Domain Knowledge. The domain knowledge is represented by an RDFS base `DK` (`DK` $\subseteq$ `KB`) consisting in a set of triples in the form $\langle C \text{ subc } D \rangle$. Fig. 1 represents the domain knowledge for the running example by a hierarchy the edges $C \xrightarrow{x} D$ of which represent the triples $\langle C \text{ subc } D \rangle$ (the meaning of x is explained hereafter).

Representation of Similarity (Retrieval Knowledge). The retrieval knowledge is encoded by a cost function, associating a triple $\langle C \text{ subc } D \rangle \in \text{DK}$ to a positive real number $\text{cost}(\langle C \text{ subc } D \rangle)$ for a given property. If $C \xrightarrow{x} D$ is an edge of the Fig. 1 hierarchy then $\text{cost}(\langle C \text{ subc } D \rangle) = x$. This cost can be understood intuitively as the measure of "the generalization effort" from C to D. The cost function is assumed to be additive:

$$\text{cost}(\langle C \text{ subc } E \rangle) = \text{cost}(\langle C \text{ subc } D \rangle) + \text{cost}(\langle D \text{ subc } E \rangle)$$

Therefore $\text{cost}(\langle \text{OrangeJuice subc FruitJuice} \rangle) = 0.10 + 0.02 = 0.12$ can be deduced from Fig. 1 and, for any RDFS class C, $\text{cost}(\langle C \text{ subc } C \rangle) = 0$.

Since it is a tedious work to set manually all these costs, some default values are computed by TUUURBINE according to the following formula:

$$\text{cost}(\langle C \ \ \texttt{subc} \ \ D\rangle) = K \cdot \frac{\#\texttt{CasesWith}(D) - \#\texttt{CasesWith}(C)}{|\texttt{CaseBase}|}$$

$$\text{where } \#\texttt{CasesWith}(X) = \left| \texttt{Result}_{\vdash} \left(\boxed{\begin{array}{l} \texttt{SELECT ?x} \\ \texttt{WHERE \{ ?x type SourceCase . } \\ \texttt{?x ?p } X\texttt{\}}\end{array}}, \texttt{KB} \right) \right|,$$

K is a coefficient cost factor depending of ?p

and $|A|$ is the cardinal of the set A

A similar formula is used by the TAAABLE system [1].

Representation of Adaptation Knowledge. The adaptation of a case `Source` to answer a query `Q` is performed using domain knowledge, retrieval knowledge and, when available, adaptation knowledge in the form of a finite set `AK` of adaptation rules. Syntactically, an adaptation rule is defined as follows:

$$\text{AR} = \boxed{\begin{array}{l} \textbf{p-context } p_1, p_2, \ldots \\ \textbf{n-context } n_1, n_2, \ldots \\ \textbf{replace } r_1, r_2, \ldots \\ \textbf{with } w_1, w_2, \ldots \end{array}} \quad \begin{array}{l} \textit{// positive context} \\ \textit{// negative context} \\ \textit{// left part} \\ \textit{// right part} \end{array}$$

where p_1, p_2, ..., n_1, n_2, ..., r_1, r_2, ..., w_1, w_2, ... are expressions in the form $\texttt{prop}\!:\!C$, prop being an RDFS property and C being an RDFS class. For example, the adaptation rule introduced in the running example (§3.1) can be formalized by

$$\text{AR}_1 = \boxed{\begin{array}{l} \textbf{p-context } \texttt{dishType:CocktailDish} \\ \textbf{n-context } \texttt{ingredient:Anise} \\ \textbf{replace } \texttt{ingredient:GuavaJuice, ingredient:VanillaSugar} \\ \textbf{with } \texttt{ingredient:OrangeJuice, ingredient:SugarCaneSyrup} \end{array}}$$

Let `Source` be a case identified by s. `AR` is applicable on `Source` if:

- For each term $t = \texttt{prop}\!:\!C \in \{p_1, p_2, \ldots, r_1, r_2, \ldots\}$, KB $\vdash \langle s \ \texttt{prop} \ C\rangle$ (recall that `Source` $\subseteq$ KB);
- For each term $t = \texttt{prop}\!:\!C \in \{n_1, n_2, \ldots\}$, KB $\nvdash \langle s \ \texttt{prop} \ C\rangle$.[6]

If `AR` is applicable on `Source` then the application of `AR` on `Source` gives a case `AR(Source)` obtained by:

---

[6] This amounts to a closed world assumption (CWA) defined by the inference rule $\dfrac{\text{KB} \nvdash \langle s \ \texttt{prop} \ C\rangle}{\neg \langle s \ \texttt{prop} \ C\rangle}$. CWA is justified by the fact that there is no negation in RDFS. From a CBR viewpoint, this means that a case represents a specific situation (it is not a generalized case [6]): every fact τ that is expressible in the KB vocabulary that is not entailed by `Source` is assumed not to hold for `Source`.

- Deleting all the triples of Source matching r_1, r_2, ...;
- Adding all the triples $\langle s$ prop $C \rangle$ for each prop$:C \in \{w_1, w_2, ...\}$;
- Substituting all the occurrences of s by s' (a new case identifier).

For example, the rule AR_1 is applicable on the recipe Source of the running example and $AR_1(\text{Source}) = \text{AS}$ (adapted Source) with:

$$AS = \begin{cases} \langle s' \text{ type SourceCase} \rangle, \langle s' \text{ dishType CocktailDish} \rangle, \\ \langle s' \text{ ingredient Tequila} \rangle, \langle s' \text{ ingredient PineappleJuice} \rangle, \\ \langle s' \text{ ingredient AppleJuice} \rangle, \langle s' \text{ ingredient Mint} \rangle, \\ \langle s' \text{ ingredient OrangeJuice} \rangle, \langle s' \text{ ingredient SugarCaneSyrup} \rangle \end{cases}$$

In other words, the application of AR_1 consists in applying the following substitution to the ingredients:

$$\Sigma = \text{GuavaJuice} \wedge \text{VanillaSugar} \rightsquigarrow \text{OrangeJuice} \wedge \text{SugarCaneSyrup}$$

Any adaptation rule $AR \in AK$ has an associated value $cost(AR) > 0$, used during the adaptation process (see Section 3.5).

Finally, TUUURBINE proposes another kind of adaptation rules called "specific adaptation rules" (SAR) with $cost(SAR) = 0$. Such a rule is associated to a source case which constitutes its context. For the recipe application, this rule can be seen as a way of encoding variants of the recipe. For example, the piece of information "Basil can be used instead of mint in the recipe of the running example" could be represented by the following specific adaptation rule:

$$SAR_1 = \boxed{\begin{array}{l} \textbf{p-context} \text{ "Mexican cocktail in my way" recipe} \\ \textbf{replace} \text{ ingredient:Mint} \\ \textbf{with} \text{ ingredient:Basil} \end{array}}$$

3.3   Representation of Tuuurbine Queries

Syntactically, a TUUURBINE query is a conjunction of expressions in the form sign prop:val where sign $\in \{\epsilon, +, !, -\}$, prop is an RDF property and val is either a resource representing a class or a literal. For example, the following query is a TUUURBINE translation of the query (1):[7]

$$Q = +\text{dishType:CocktailDish} \wedge \text{ingredient:Mint}$$
$$\wedge \text{ingredient:Gin} \wedge \text{ingredient:OrangeJuice} \wedge !\text{ingredient:Wine} \tag{4}$$

The signs ϵ and $+$ are "positive signs": they prefix features that the requested case must have. $+$ indicates that this feature must also occur in the source case whereas ϵ indicates that the source case may not have this feature, thus the adaptation phase has to make it appear in the final case.

---

[7] ϵ denotes the empty word. Thus, ϵprop:val is simply written prop:val. The $+$ in front of dishType:CocktailDish means that the part of the case base searched corresponds to cocktail recipes.

The signs ! and − are "negative signs": they prefix features that the requested case must not have. − indicates that this feature must not occur in the source case whereas ! indicates that the source case may have this feature, and that the adaptation phase has to remove it.

3.4   Case Retrieval in Tuuurbine

Let Q be a TUUURBINE query. The goal of retrieval is to find the cases Source ∈ CaseBase that best match Q. If no source case exactly match Q, then the query is relaxed and an approximate matching is searched.

Exact matching search. Q can be written $Q = \bigwedge_i \text{sign}_i \text{prop}_i : \text{val}_i$. For each

i, let us consider the SPARQL query $SQ_i =$

```
SELECT ?s
WHERE { ?s type SourceCase .
        ?s prop_i  ?x .
        ?x subc  val_i}
```

For example, the element of query +dishType : CocktailDish gives a SPARQL query that can be read "get the source cases with a cocktail dish type" (i.e., the ?s such that ?s is a instance of SourceCase and such that ?s has a dish type ?x which is a subclass of CocktailDish).

The exact matching of the query Q can be done by executing the SPARQL queries SQ_i and combining them as follows (EMS stands for "Exact Matching Search"):

$$EMS(Q) = \bigcap_{i,\text{sign}_i \in \{\epsilon,+\}} \text{Result}_\vdash(SQ_i) \setminus \bigcup_{i,\text{sign}_i \in \{!,-\}} \text{Result}_\vdash(SQ_i)$$

In other words, the result of the exact matching search are the source cases that match the SPARQL queries SQ_i such that sign_i is a positive sign and that does not match the SPARQL queries SQ_i such that sign_i is a negative sign.[8]

Approximate Search. The principle of this search is to find a generalization function Γ with minimal cost such that the execution of the query Q modified by Γ returns at least one source case: $EMS(\Gamma(Q)) \neq \emptyset$.

Let $Q = \bigwedge_i \text{sign}_i \text{prop}_i : \text{val}_i$ be a query. A one step-generalization $\gamma(Q)$ of Q consists in generalizing a term $\text{sign}_i \text{prop}_i : \text{val}_i$ such that $\text{sign}_i \notin \{+, -\}$:

- If $\text{sign}_i = \epsilon$ and val_i is an RDFS class, then the generalizations of this term are the terms $\text{sign}_i \text{prop}_i : \text{val}$ such that $\langle \text{val}_i \text{ subc val} \rangle \in DK$. This one step generalization is written $\gamma = \text{prop}_i : \text{val}_i \rightsquigarrow \text{prop}_i : \text{val}$ or, simply, $\gamma = \text{val}_i \rightsquigarrow \text{val}$. The cost of such a generalization is $\text{cost}(\langle \text{val}_i \text{ subc val} \rangle)$.
- If $\text{sign}_i = !$ or if $\text{sign}_i = \epsilon$ and val_i is a literal, then val_i is directly generalized to the ontology top $\top$. This one step generalization is written $\gamma = \text{val}_i \rightsquigarrow \top$. The cost of such a generalization is 0.

---

[8] In practice, EMS(Q) could be computed thanks to the execution of fewer SPARQL queries thus giving the same result with a lower computational cost.

A generalization function Γ is a composition of one-step generalizations γ_1, γ_2, $\ldots\gamma_n$: $\Gamma = \gamma_n \circ \ldots \circ \gamma_2 \circ \gamma_1$. Its cost is the sum of the costs of γ_i. For example, the generalization function Γ can be applied on Q defined by equation (4):

$$\Gamma = \texttt{Gin} \rightsquigarrow \texttt{Liquor} \circ \texttt{OrangeJuice} \rightsquigarrow \texttt{FruitJuice}$$
$$\text{and } \Gamma(\texttt{Q}) = +\texttt{dishType:CocktailDish} \wedge \texttt{ingredient:Mint}$$
$$\wedge \texttt{ingredient:Liquor} \wedge \texttt{ingredient:FruitJuice}$$
$$\wedge \texttt{!ingredient:Wine} \tag{5}$$
$$\text{cost}(\Gamma) = \text{cost}(\langle\texttt{Gin subc Liquor}\rangle)$$
$$+ \text{cost}(\langle\texttt{OrangeJuice subc FruitJuice}\rangle)$$

The execution of the query $\Gamma(\texttt{Q})$ returns the recipe of the example: Source $\in$ EMS($\Gamma(\texttt{Q})$) so, provided that cost(Γ) is the minimum of the costs of the generalization functions Λ such that EMS($\Lambda(\texttt{Q})$) $\neq \emptyset$, Source is a retrieved case. Technically, Γ is searched by increasing cost in a generalization function space.

Searching for Less Similar Cases. It may occur that a user of a Tuuurbine application wants to find other cases than the ones returned in a first launch. Tuuurbine offers the possibility to do so: it simply consists in resuming the search after Γ has been found. This way, a second generalization function Γ' can be found. For example:

$$\Gamma' = \texttt{Gin} \rightsquigarrow \texttt{Alcohol} \circ \texttt{Mint} \rightsquigarrow \texttt{Herb}$$

3.5 Case Adaptation in Tuuurbine

There are two adaptation processes in Tuuurbine.

The first adaptation process consists in obtaining the matching between Source and Q that has permitted the retrieval of Source, this matching being composed of the matching between Source and $\Gamma(\texttt{Q})$ (given by the fact that Source $\in$ EMS($\Gamma(\texttt{Q})$)) and the matching between $\Gamma(\texttt{Q})$ and Q (given by the generalization function Γ). This first kind of adaptation, for the running example, works as follows. $\Gamma(\texttt{Q})$ defined by equation (5) matches exactly Source (wrt to DK). In fact, there are several matchings between Source and $\Gamma(\texttt{Q})$:

Tequila matches Liquor

each $ing \in \{\texttt{GuavaJuice}, \texttt{PineappleJuice}, \texttt{AppleJuice}\}$ matches FruitJuice

By composing this matching between Source and $\Gamma(\texttt{Q})$ and the matching between $\Gamma(\texttt{Q})$ and Q given by Γ, it comes the following adaptation:

In the "Mexican cocktail in my way" recipe, substitute Tequila by Gin

and substitute $\begin{vmatrix} \texttt{GuavaJuice and/or} \\ \texttt{PineappleJuice and/or} \\ \texttt{AppleJuice} \end{vmatrix}$ by OrangeJuice.

which represents the expected adaptation (2).

The second kind of adaptation uses the adaptation rules $AR \in AK$ *and* the query generalization procedure used for case retrieval. More precisely it searches a pair (Σ, Γ) such that:

- $\Sigma(\texttt{Source})$ matches exactly $\Gamma(\texttt{Q})$ (wrt DK);
- Σ is a composition of adaptation rules that is applicable on Source ($\Sigma = AR_p \circ \ldots \circ AR_2 \circ AR_1$ such that AR_1 is applicable on Source, AR_2 is applicable on $AR_1(\texttt{Source})$, etc.);
- Γ is a generalization function;
- $\text{cost}(\Sigma)+\text{cost}(\Gamma)$ is minimal (where $\text{cost}(\Sigma) = \text{cost}(AR_p)+\ldots+\text{cost}(AR_2)+\text{cost}(AR_1)$).

Technically, this kind of adaptation relies on a best-first search: the states are pairs (Σ, Γ); a final state is such that $\Sigma(\texttt{Source})$ matches exactly $\Gamma(\texttt{Q})$; the state space is searched by increasing $\text{cost}(\Sigma) + \text{cost}(\Gamma)$. It must be noticed that:

- $\text{cost}(\Sigma) + \text{cost}(\Gamma) \leq \text{cost}(\Gamma_{\text{retrieval}})$ where $\Gamma_{\text{retrieval}}$ is the generalization function generated during retrieval (and used by the first kind of adaptation);
- If $AK = \emptyset$ then Σ is the identity function and $\Gamma = \Gamma_{\text{retrieval}}$.

For the example, if $AK = \{AR_1\}$ and $\text{cost}(AR_1) + \text{cost}(\langle \texttt{Gin subc Liquor} \rangle) < \text{cost}(\Gamma_{\text{retrieval}})$ then the adaptation coincides with (3).

4   Implementation

The general architecture of Tuuurbine is presented in Fig. 2. Its CBR engine is composed of three main modules: the RDF store manager, the case manager and the reasoning manager. The case manager implements EMS (the exact matching search).The case retrieval triggers the RDF store manager, in charge of generating and executing a set of SPARQL queries, using a SPARQL endpoint. The results are returned to the case manager which combines them to build the cases.

When no case satisfies the EMS, the reasoning manager is triggered for an approximative search. A first step consists in loading the generalization costs from the triple store (only the costs that are potentially useful). This is done once, at the beginning of the approximative search process. The generalisation process iterates until the EMS returns at least one case for a generalized query. These cases are then adapted using possibly adaptation knowledge and generalization knowledge as explained in §3.5.

Tuuurbine is implemented as a web service. This allows to use HTTP requests to query the CBR engine, but a generic user interface is also provided (an instantiation of this GUI is visible in Fig. 3). Additionally, a Semantic Wiki may be used to manage the whole knowledge in a more convenient way, as it is done for example in WikiTaaable, a semantic wiki in the domain of cooking.

Tuuurbine required about 6000 lines of Java code. The user interface is developed in PHP, javascript, HTML/CSS. The exchange between the interface

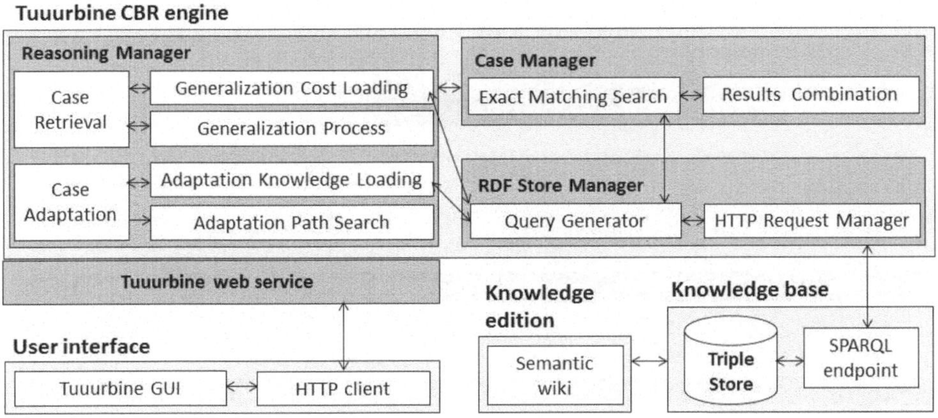

Tuuurbine CBR engine

Fig. 2. Architecture of TUUURBINE

and the CBR engine, as well as the configuration files are based on JSON encoding language. TUUURBINE is distributed under an Affero GPL Licence and is available from http://tuuurbine.loria.fr/.

The retrieval time for the query (4) is 3156 ms on a server Intel Xeon E5520 64 bits with 8×2.27 GHz processors, and a 32 GB RAM, running on Linux Ubuntu 12.4. 85% of this time is for disk access and data communication. It required the execution of 117 SPARQL queries. This example is realistic: the RDF base size is 277 megabytes, it contains more than 10^6 triples, 1641 of them being triples of the form $\langle s$ type Recipe$\rangle$ (hence 1641 source cases) and 2316 triples of the form $\langle C$ subc $D \rangle$ constitute the domain knowledge. A systematic performance study of TUUURBINE remains to be done as future work.

5   Tuuurbine in Action

The first use case is the instantiation of TUUURBINE in the cooking domain, as the new version of TAAABLE, a CBR system which retrieves and creates cooking recipes by adaptation that has been developed to participate in the Computer Cooking Contest[9] since 2008 [3]. Fig. 3 presents the TUUURBINE interface running query 4 and is reachable at http://tuuurbine.loria.fr/taaable/. The TAAABLE knowledge base (http://wikitaaable.loria.fr/) is composed of the four classical knowledge containers: (1) the domain knowledge as an ontology of the cooking domain which includes several hierarchies (about food, dish types, etc.), (2) the case base, which are recipes described by their titles, the dish type they produces, the ingredients which are required, the preparation steps, etc., (3) the adaptation knowledge takes the form of adaptation rules as introduced in this paper, and (4) the retrieval knowledge which is stored as cost values on subclass-of relations and adaptation rules.

---

[9] http://computercookingcontest.net

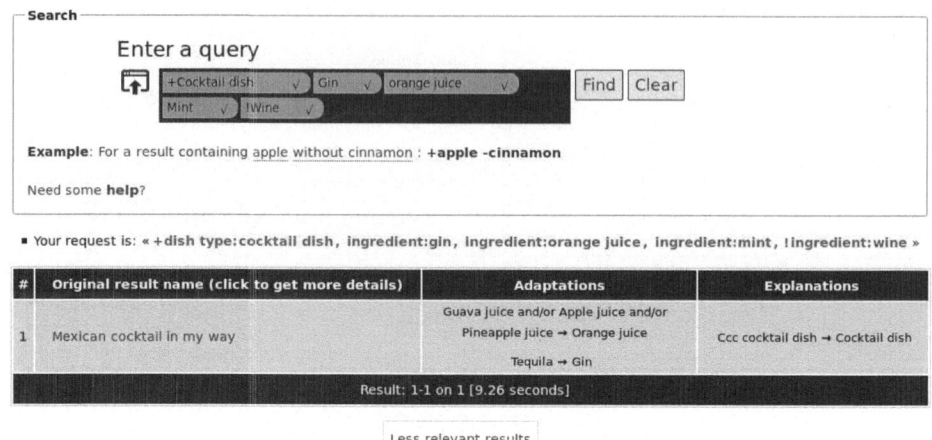

Fig. 3. Generic TUUURBINE interface applied to the cooking domain

To show that TUUURBINE is independent of the application domain, a second instantiation of TUUURBINE has been implemented in the domain of music. This application use case is about searching for a piece of music you may play using a given set of instruments. The cases are pieces of music described by its composer(s), its music genre(s), origin(s) and year of creation, and the instruments required for playing them. The adaptation consists in replacing some instruments by other ones. The music ontology and the TUUURBINE interface for the music CBR system can be respectively reached at http://muuusic.loria.fr/ and http://tuuurbine.loria.fr/muuusic/.

6   Discussion and Related Work

Several tools address CBR in a generic way. myCBR and jCOLIBRI are probably the most famous of them. jCOLIBRI [7] is an object-oriented framework for developing CBR applications. This framework includes connectors with several kinds of data sources (database, XML, etc.) for loading the cases in cache, on which the retrieval, reuse, revise and retain tasks can be performed. Specialized modules take into account various types of CBR applications, like CBR on textual cases, data or knowledge-intensive CBR. With jCOLIBRI, the knowledge intensive CBR approach consists in retrieving cases according to an ontology based similarity measure and in replacing some case features (instance of an ontology concept) with the closest ones in the ontology (based on similarity minimization). TUUURBINE also follows this substitution approach but the way the retrieval and the adaptation of cases are made is very different. Indeed, using the structure of the ontology only in order to compute a numerical measure has a major limitation, due to the lack of semantics of the measure, which can, moreover, be computed with various similarity functions (cosine, fdeep, etc.) which take into account various criteria (depth of the two concepts c_1 and c_2 which are compared, depth and relations of c_1 and c_2 with their least common subsumer, etc.). Such similarity functions do not take into account the distribution of the instances in the set of cases, in comparison with our cost function which represents a generalization effort. Moreover, the cost function is independent of the level of structuration of the ontology. So, having many intermediate concepts (like for example CitrusFruitJuice between OrangeJuice and FruitJuice) do not impact the case retrieval. This point is crucial for improving case retrieval with a better structured ontology [8].

myCBR [9] is a Java open source similarity-based tool. Cases, described by attributes with different weights, may be created thanks to a graphical user interface or through a file import (e.g., CSV file). A "Linked Open Data Connector" provides an access to open data sources for building taxonomies. Many similarity functions are provided and may be combined to rank cases according to their similarity with the query (during the retrieval step). The adaptation consists in replacing some case features with others, using adaptation rules retrieved from the myCBR knowledge models (e.g. similarity tables or taxonomies). So, the myCBR retrieving/adapting procedure is rather similar to the jCOLIBRI one.

Like myCBR and jCOLIBRI, Tuuurbine is a generic CBR engine. The originality of Tuuurbine is the implementation of a CBR based on RDFS and the exploitation of the semantics of the RDF and RDFS models. The whole knowledge (cases, domain knowledge, costs, adaptation rules) is described using RDF and RDFS and is managed outside Tuuurbine using a triple store and SPARQL, two standard tools of the Semantic Web. The edition of the knowledge is also facilitated with the use of a Semantic Wiki, another Semantic Web tool.

7   Conclusion

This paper has presented Tuuurbine, a generic CBR engine based on RDFS and using Semantic Web technologies. The main principles for representation of the four knowledge containers as well as the reasoning processes have been detailed. Information about technical use and configuration of Tuuurbine is available at http://tuuurbine.loria.fr/. Tuuurbine is used in the new version of the Taaable system, a CBR system adapting cooking recipes, which will participate in the 2014's Computer Cooking Contest.

The current version of the Tuuurbine case-based inference engine using query generalizations is based on the subc property (generalization of a class by a super-class). Other constructs of RDFS could be used as well in the Tuuurbine reasoning process: the subproperty relation (subp), domains and ranges of properties, literals (e.g., numerical values). This constitutes a future direction of research. It is planned to address first the use of subp. For instance, if ⟨mainIngredient subp ingredient⟩ ∈ DK (if x is a main ingredient of y then x is an ingredient of y), the query Q = mainIngredient:TomatoJuice can be generalized into $\Gamma(Q) =$ ingredient:TomatoJuice.

Another direction of work related to RDFS expressiveness consists in managing "deeper" cases: in the current implementation, only values directly related to s are taken into account, but such values could be related to other values participating to the case representation. Following this direction would make TUUURBINE evolve from an attribute-value representation to something similar to an object-based representation (like the ones of jCOLIBRI and myCBR).

A third direction of work is the integration of TUUURBINE with other generic CBR systems such as jCOLIBRI and myCBR (cf. Section 6) or Revisor [10].

References

1. Cordier, A., et al.: Taaable: a Case-Based System for personalized Cooking. In: Montani, S., Jain, L.C. (eds.) Successful Case-based Reasoning Applications-2. SCI, vol. 494, pp. 121–162. Springer, Heidelberg (2014)
2. Cordier, A., Lieber, J., Molli, P., Nauer, E., Skaf-Molli, H., Toussaint, Y.: WIK-ITAAABLE: A semantic wiki as a blackboard for a textual case-based reasoning system. In: SemWiki 2009 - 4th Semantic Wiki Workshop at the 6th European Semantic Web Conference - ESWC 2009, Heraklion, Grèce (May 2009)
3. Badra, F., Bendaoud, R., Bentebibel, R., Champin, P.-A., Cojan, J., Cordier, A., Després, S., Jean-Daubias, S., Lieber, J., Meilender, T., Mille, A., Nauer, E., Napoli, A., Toussaint, Y.: TAAABLE: Text Mining, Ontology Engineering, and Hierarchical Classification for Textual Case-Based Cooking. In: Schaaf, M. (ed.) 9th European Conference on Case-Based Reasoning - ECCBR 2008, Workshop Proceedings, Trier, Allemagne, pp. 219–228 (2008)
4. Shakespeare, W.: Romeo and Juliet (1597)
5. Richter, M.M.: Introduction. In: Lenz, M., Bartsch-Spörl, B., Burkhard, H.-D., Wess, S. (eds.) Case-Based Reasoning Technology. LNCS (LNAI), vol. 1400, pp. 1–15. Springer, Heidelberg (1998)
6. Maximini, K., Maximini, R., Bergmann, R.: An investigation of generalized cases. In: Ashley, K.D., Bridge, D.G. (eds.) ICCBR 2003. LNCS (LNAI), vol. 2689, pp. 261–275. Springer, Heidelberg (2003)
7. Recio-García, J.A., González-Calero, P.A., Díaz-Agudo, B.: jcolibri2: A framework for building case-based reasoning systems. Science of Computer Programming (79), 126–145 (2014)
8. Dufour-Lussier, V., Lieber, J., Nauer, E., Toussaint, Y.: Improving case retrieval by enrichment of the domain ontology. In: Ram, A., Wiratunga, N. (eds.) ICCBR 2011. LNCS, vol. 6880, pp. 62–76. Springer, Heidelberg (2011)
9. Bach, K., Althoff, K.-D.: Developing case-based reasoning applications using my-CBR 3. In: Díaz Agudo, B., Watson, I. (eds.) ICCBR 2012. LNCS (LNAI), vol. 7466, pp. 17–31. Springer, Heidelberg (2012)
10. Cojan, J., Lieber, J.: Applying belief revision to case-based reasoning. In: Prade, H., Richard, G. (eds.) Computational Approaches to Analogical Reasoning: Current Trends. SCI, vol. 548, pp. 133–161. Springer, Heidelberg (2014)

How Case-Based Reasoning on e-Community Knowledge Can Be Improved Thanks to Knowledge Reliability*

Emmanuelle Gaillard[1], Jean Lieber[1], Emmanuel Nauer[1], and Amélie Cordier[2]

[1] Université de Lorraine, LORIA — 54506 Vandœuvre-lès-Nancy, France
CNRS — 54506 Vandœuvre-lès-Nancy, France
Inria — 54602 Villers-lès-Nancy, France
firstname.lastname@loria.fr
[2] Université de Lyon, CNRS, France
Université Lyon 1, LIRIS, UMR5205, F-69622, France
firstname.lastname@liris.cnrs.fr

Abstract. This paper shows that performing case-based reasoning (CBR) on knowledge coming from an e-community is improved by taking into account knowledge reliability. MKM (meta-knowledge model) is a model for managing reliability of the knowledge units that are used in the reasoning process. For this, MKM uses meta-knowledge such as belief, trust and reputation, about knowledge units and users. MKM is used both to select relevant knowledge to conduct the reasoning process, and to rank results provided by the CBR engine according to the knowledge reliability. An experiment in which users perform a blind evaluation of results provided by two systems (with and without taking into account reliability, i.e. with and without MKM) shows that users are more satisfied with results provided by the system implementing MKM.

Keywords: case-based reasoning, meta-knowledge, evaluation, reliability, filtering, ranking, feedback.

1 Introduction

This paper shows experimentally on a use case that taking into account knowledge reliability in case-based reasoning (CBR) improves user satisfaction about the results provided by the system. For managing knowledge reliability, we use a meta-knowledge model, called MKM, that was introduced in a previous work [1].

By analogy with past experiences, or cases, CBR solves new problems [2]. The reasoning process uses knowledge among which cases, domain knowledge,

---

* This work is supported by French National Agency for Research (ANR), program Contint 2011 through the Kolflow project. More information about Kolflow is available on the project website: http://kolflow.univ-nantes.fr/. The authors wish also to thank the persons who have participated to the evaluation, and especially students of the DUT Informatique of Université Lyon 1 and students of the Licence Informatique of Université de Lorraine.

L. Lamontagne and E. Plaza (Eds.): ICCBR 2014, LNCS 8765, pp. 155–169, 2014.
© Springer International Publishing Switzerland 2014

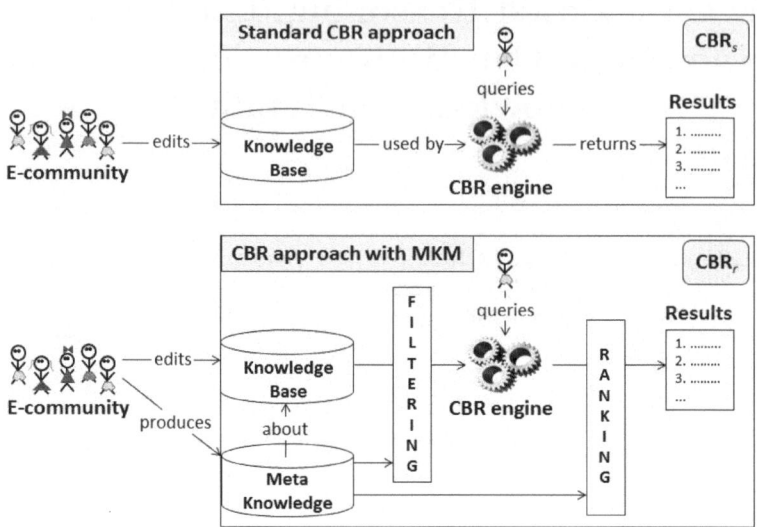

Fig. 1. Extending a standard CBR approach with MKM

similarity knowledge and adaptation knowledge [3]. Acquiring enough knowledge to perform quality reasoning is cumbersome and tedious. This is the reason why the Web, and more specifically, e-communities, is more and more explored to build knowledge bases. An e-community is a group of people communicating over the Internet to share common ideas, goals, interests, hobbies, etc. Mining an e-community is an efficient way to acquire knowledge on a specific domain (such as video games, programming languages, or cooking). However, due to several factors (e.g. the expertise level of users, their points of view, etc.), the quality of this knowledge is questionable. For example, some people will consider that a tomato is a fruit, but this assertion is not relevant in the cooking domain because adapting a fruit salad recipe by replacing a fruit by a tomato is not really a good (a tasty) idea. Therefore, if we want this knowledge to be usable by a reasoner, while ensuring the quality of the results, we need to estimate the reliability of the knowledge originating from the e-communities. For this, we use MKM, which was designed to capture the reliability of each individual piece of knowledge, denoted KU (knowledge unit) in the following. MKM manages a new knowledge container: the meta-knowledge container, in which each KU is associated to its reliability computed from three other types of meta-knowledge: belief, trust and reputation.

To demonstrate that using MKM on a CBR system using knowledge from the Web improves user satisfaction, two CBR systems are compared. These two CBR systems use the same inference engine and the same knowledge base, built by an e-community. The first system, denoted by CBR$_s$ (s for standard), is the reference system. It implements a standard CBR approach, in which all KUs have the same reliability. The second system extends the reference system with

MKM, where reliability of each KU depends on the community opinion. This second system is denoted by CBR_r (r for reliability).

Fig. 1 shows the common points and differences between CBR_s and CBR_r. The common points are that (1) the two systems are both triggered by queries, (2) the same CBR engine is used to perform reasoning, and (3) the reasoning is based on the same knowledge base coming from an e-community. The main difference between CBR_s and CBR_r is that CBR_r uses an additional container for meta-knowledge. This container is used to filter the KUs exploited by the CBR engine and to rank final CBR results according to the KUs involved in each result.

This paper is organized as follows. Section 2 introduces the use case. Section 3 presents MKM principles. Section 4 presents our evaluation methodology and section 5 presents and discusses the results. Section 6 concludes the paper.

2   The Use-Case: TAAABLE

TAAABLE is a CBR system in the cooking domain which retrieves and creates recipes by adaptation [4]. In this paper, we consider an instance of TAAABLE which uses KUs stored in ATAAABLE, a semantic wiki (in French).

The Domain Knowledge. The domain knowledge (DK) is an ontology composed of a set of atomic classes of several hierarchies (food, dish type, localization, ...). Classes are organized according to subsumption relations. Given two concepts A and B of this ontology, A subsumes B, denoted by A $\sqsupseteq$ B, if the set of instances of B is included in the set of instances of A. For instance, FruitJuice $\sqsupseteq$ OrangeJuice, means that all the instances of orange juice are instances of fruit juice.

Case Base. The case base consists on a set of recipes. Each recipe R of the case base is represented by its index denoted by $idx(R)$ which is a conjunction of classes of the domain ontology. For example, $idx(R) =$ CocktailDish$\wedge$Tequila$\wedge$ PineappleJuice $\wedge$ AppleJuice $\wedge$ Mint is the index of a cocktail recipe which ingredients are tequila, pineapple juice, apple juice and mint.

Query. In TAAABLE, a query Q is also a conjunction of classes [1]. For example, $Q =$ CocktailDish$\wedge$Gin$\wedge$OrangeJuice means "I want a cocktail with gin and orange juice."

Case Retrieval. The retrieval process consists in searching cases that best match the query. If an exact matching exists, the corresponding cases are returned. Otherwise, the query is relaxed using a generalization function Γ composed of one-step generalizations, which transforms Q (with a minimal cost) until at least one recipe of the case base matches $\Gamma(Q)$.

---

[1] Actually, the query language is more complex, but this has no importance in this paper.

A one step-generalization is denoted by $\gamma = A \rightsquigarrow B$, where A and B are classes and $B \sqsupseteq A$ belongs to the domain ontology. Each one-step generalization is associated to a cost denoted by $cost(A \rightsquigarrow B)$. The generalization Γ of Q is a composition of one-step generalizations $\gamma_1, \gamma_2, \ldots \gamma_n$: $\Gamma = \gamma_n \circ \ldots \circ \gamma_2 \circ \gamma_1$, with $cost(\Gamma) = \sum_{i=1}^{n} cost(\gamma_i)$. How this cost is computed is detailed in [4].

In the space of generalization functions Γ, the least costly such that at least one case matches exactly $\Gamma(Q)$ is searched. With the example introduced above, TAAABLE produces $\Gamma = \texttt{OrangeJuice} \rightsquigarrow \texttt{FruitJuice} \circ \texttt{Gin} \rightsquigarrow \texttt{Liquor}$ and so $\Gamma(Q) = \texttt{CocktailDish} \wedge \texttt{FruitJuice} \wedge \texttt{Liquor}$, which matches $idx(R)$.

TAAABLE returns all the adapted cases that match the generalized query. Less similar cases may be retrieved too by resuming the generalization process which will search for the *next* least costly generalization.

Adaptation. TAAABLE implements two adaptation processes. The first one consists in a specialization of the generalized query produced by the retrieval step. According to $\Gamma(Q)$, to R, and to DK, AppleJuice is replaced with OrangeJuice in R because FruitJuice of $\Gamma(Q)$ subsumes both AppleJuice and OrangeJuice. In the same way, Tequila is replaced with Gin in R because Liquor of $\Gamma(Q)$ subsumes both Tequila and Gin.

The second adaptation process consists in using adaptation rules where some ingredients are replaced with others in a given context [5]. For example, in cocktail recipes, replacing Gin and AppleJuice with Tequila and OrangeJuice, is an example of adaptation rule.

Knowledge Unit. Each subsumption relation, each recipe index, each adaptation rule is a KU potentially used by the ATAAABLE reasoning process.

3   Meta-Knowledge Model

MKM is based on [6] and was presented in a previous work [1]. The integration of MKM in a CBR system enables it to take into account reliability of each KU during the reasoning process. When a user queries the system, only the most reliable KUs are used for reasoning.

3.1   State of the Art

Using reliable knowledge elements in a knowledge-based system allows to infer knowledge with an acceptable level of trustworthiness. Knowledge reliability is influenced by several factors, sometimes interrelated, as discussed in [6], where a generic model for representing knowledge generated by online communities is proposed. In an effort to provide a basis for exploiting partially reliable knowledge in a reasoning process, this work identifies a set of dimensions for knowledge reliability like origin, belief, quality, and trust. Several models have been proposed, both for conceptualizing and evaluating trust. The most common systems exploiting trust are based on reputation (e.g. [7]) In the human computer interaction domain, Golbeck [8] asserts that "A trusts B if A commits to an action based on the belief that B's future actions will lead to a good outcome." She used this definition in her recommending system for movies, where users can rate both movies and other users. For a given user, movie recommender scores are computed by taking into account the community opinion: scores rely on movies ratings weighted by user reputation. More recently, [9] has presented a framework for the prediction of trust and distrust.

Meta-knowledge is already used in CBR, mainly for case base maintenance and recommendation. [10] presents a CBR system recommending movies to a group according to movies watched, in the past, by other groups. The similarity between two groups depends on the similarity between their members, one by one, and takes into account the degree of trust in addition to other criteria (age, gender, etc.). In [11], authors integrate trust in addition to provenance in a CBR approach to propose a model of collaborative web search. During a user search, web pages are filtered and ranked using their relevance to the query and the reputation of users having already selected the pages. In this work, a case is a pair (query/web page) and the provenance of the case is an indicator of the quality of the result.

In CBR systems, meta-knowledge is also used for case base maintenance. To maintain the CBR performance by stabilizing the case base size, [12] proposes to model case competencies according to their coverage set and their reachability set. [13] proposes to integrate the provenance of a case, to guide the case base maintenance and to increase the confidence of future results. For example, a repair is propagated through generated cases from the initial case and the quality of a case is measured by the length of the adaptation path. However, the quality of initial cases is set to a same maximum value and does not depend on external factors nor on additional meta-knowledge like, e.g., the provenance of initial cases.

More details about the state of art about meta-knowledge in literature, in particular provenance, quality and belief, and meta-knowledge used in CBR systems are presented in [1].

3.2   Meta-Knowledge Model Principles

Fig. 2 shows the links between the different types of meta-knowledge used in MKM. Users of the e-community may evaluate KUs by giving rates, which will produce two types of meta-knowledge (Fig. 2, white background):

– The belief score, when a user u rates a KU ku and which represents the belief u has in ku.
– The *a priori* trust score, when a user u evaluates another user v and which represents the trust u has towards v, independently of any contributions inside the community.

Theses evaluations are the foundations of the system and allow to compute the intermediate meta-knowledge (Fig. 2, light grey background):

– A quality score of a KU ku, which represents the global quality of ku for the e-community, is inferred from all the belief scores about ku.

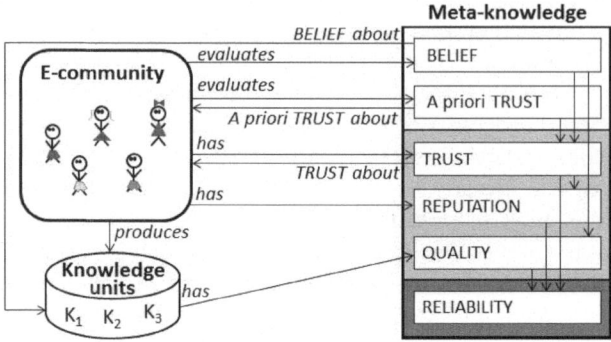

Fig. 2. Dependencies between users, knowledge units and meta-knowledge

– A trust score from a user u towards a user v, which represents how u trusts v, is inferred from the *a priori* trust score that u has assigned to v and from the belief scores that u has assigned to the KUs produced by v.
– A reputation score of a user u, which represents the reputation of u in the e-community, is inferred from all the trust scores about u.

And finally, reliability (Fig. 2, dark gray background) is the meta-knowledge that will be used by the CBR system to filter knowledge and to rank results. Let `reliability`(ku) be the function which returns the reliability score of a KU ku for the e-community, computed from quality, trust, and reputation scores.

3.3 Plugging the Meta-Knowledge on a CBR System

As proposed in [1], MKM may be used to extend an existing CBR system by adding a filtering process and a ranking process.

Filtering is used to select the most *reliable* set of knowledge according to the query. All the KUs with a reliability score higher than a given threshold are selected to be used by the CBR engine (the KUs with a reliability score lower than the threshold are not used by the CBR engine as if there were removed from the knowledge base). In TAAABLE, a threshold of 0.3 gave good results during the experiments. Thus, the threshold is fixed to 0.3 for this work, but a precise method to fix the threshold remains to be studied.

Ranking computation is used to order the set of results according to the meta-knowledge associated to the KUs involved in the computation of the results. For each result R, an *inferred reliability*, denoted by `inferred_reliability`(R) is computed. The general function proposed in the initial model [1] to compute the inferred reliability has been instantiated by a probabilistic approach. The *inferred reliability* of a result can be seen intuitively as the probability that the result will be satisfactory (e.g., for a cooking application, the probability that this is the recipe of a tasty dish); this probability depends on the probability that the retrieved case is satisfactory, that each KU used in the adaptation is

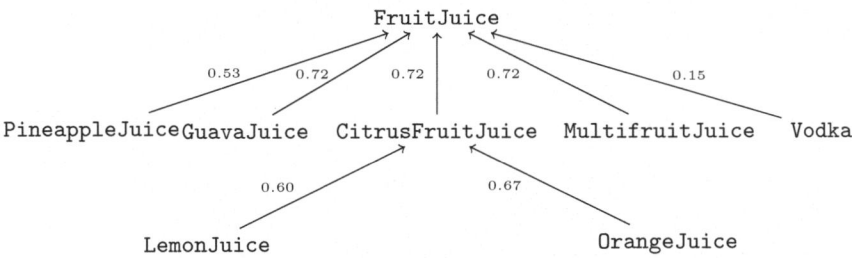

Fig. 3. Part of the food hierarchy with reliability scores of subsumption relations

satisfactory, and we assume that these probabilities are independent one from another. Each probability is equivalent to the reliability score of the KU.

The *inferred reliability* of a result is computed as follows:

- If the result matches exactly the query, the *inferred reliability* is the reliability score of the case.
- If the adaptation process uses an adaptation rule for producing the result R, the *inferred reliability* of R is the product of the case reliability by the adaptation rule reliability. For example, if the retrieve case is C (a cocktail containing apple juice), with `reliability(C)` = 0.76, and if the adaptation rule AR, which proposes to replace apple juice with orange juice in cocktails, has been used, with `reliability(AR)` = 0.8, then `inferred_reliability(R)` = `reliability(C)` × `reliability(AR)` = 0.76 × 0.8 = 0.608.
- If the adaptation process uses a generalization-specialization path to compute the adaptation, the *inferred reliability* is the product of the case reliability by the product of the reliability of all the KUs involved during the generalization-specialization, that is to say, the product of the reliability of each subsumption relation involved. Fig. 3 presents a part of the food hierarchy with the reliability score of the subsumption relations. For example, according to Fig. 3, the reliability of the adaptation A = PineappleJuice ⤳ GuavaJuice may be computed by taking into account that PineappleJuice has been generalized in FruitJuice and that FruitJuice has been specialized in GuavaJuice: `reliability(PineappleJuice ⤳ GuavaJuice)` = 0.53×0.72 ≃ 0.38. Let R be the result of case C adapted using the substitution A, the inferred reliability of R is computed as the product of the reliabilities of C and A: `inferred_reliability(R)` = `reliability(C)` × `reliability(A)` ≃ 0.76 × 0.38 ≃ 0.29.

3.4 Example of Results Using TAAABLE with and without MKM

Let Q = GrenadineSyrup ∧ GuavaJuice be the query. Table 1 presents the three first recipes returned by CBR_s and CBR_r with their original ingredients, their reliability scores and their adaptation ids which corresponds to an adaptation id of Table 2, that presents adaptations involved in the retrieved recipes. In this example, adaptations have been computed by generalization-specialization

Table 1. The three first recipes returned by CBR_s and CBR_s, according to Q, with their indexes, their reliability scores and their adaptation ids in Table 2

id	name	$idx(R_i)$	system	reliability	adaptation
R_1	Bora bora	AppleJuice $\wedge$ PineappleJuice $\wedge$ LemonJuice $\wedge$ GrenadineSyrup	CBR_s + CBR_r	0.76	A_2
R_2	Tequila sunrise	Tequila $\wedge$ OrangeJuice $\wedge$ GrenadineSyrup	CBR_r	0.73	A_3
R_3	Bacardi cocktail	Rum $\wedge$ GrenadineSyrup $\wedge$ LemonJuice	CBR_r	0.72	A_4
R_4	Spice shoot	GrenadineSyrup $\wedge$ Tabasco $\wedge$ Vodka	CBR_s	0.73	A_5
R_5	MTS cocktail	Martini $\wedge$ TripleSec $\wedge$ CaneSugarSyrup $\wedge$ LemonJuice	CBR_s	0.4	A_1

Table 2. Adaptations of recipes returned by CBR_s

id	adaptation	reliability
A_1	MultifruitJuice $\rightsquigarrow$ GuavaJuice	0.52
A_2	PineappleJuice $\rightsquigarrow$ GuavaJuice	0.38
A_3	OrangeJuice $\rightsquigarrow$ GuavaJuice	0.35
A_4	LemonJuice $\rightsquigarrow$ GuavaJuice	0.33
A_5	Vodka $\rightsquigarrow$ GuavaJuice	0.11 (filtered)

process of the query to the part of food hierarchy presented in Fig. 3. In this figure, FruitJuice $\sqsupseteq$ Vodka is a KU created by a user and this KU is unreliable since most users have considered that Vodka is not a fruit juice.

The first three results of CBR_s are, in this order:

1. s_{s1}: Spice shoot with adaptation Vodka $\rightsquigarrow$ GuavaJuice;
2. s_{s2}: Bora bora with adaptation PineappleJuice $\rightsquigarrow$ GuavaJuice;
3. s_{s3}: MTS cocktail with adaptation MultifruitJuice $\rightsquigarrow$ GuavaJuice.

The first three results of CBR_r are, in this order:

1. R_{s1}: Bora bora with adaptation PineappleJuice $\rightsquigarrow$ GuavaJuice;
2. R_{s2}: Tequila sunrise with adaptation OrangeJuice $\rightsquigarrow$ GuavaJuice;
3. R_{s1}: Bacardi cocktail with adaptation LemonJuice $\rightsquigarrow$ GuavaJuice.

s_{s1} is not returned by CBR_r because even if the case entitled "Spice shoot" has a reliability score of 0.73, the adaptation knowledge Vodka $\rightsquigarrow$ GuavaJuice has been filtered because of its reliability score of 0.15.

The first result returned by CBR_r is R_{s1} which is only the second result returned by CBR_s. The *inferred reliability* of R_{s1} is 0.29 corresponding to the products of the reliability score of its case and its adaptation. The *inferred reliability* of s_{s2} is 0.26 and the *inferred reliability* of s_{s3} is 0.24. With lower *inferred reliability* s_{s3} is not returned in the first three results of CBR_r.

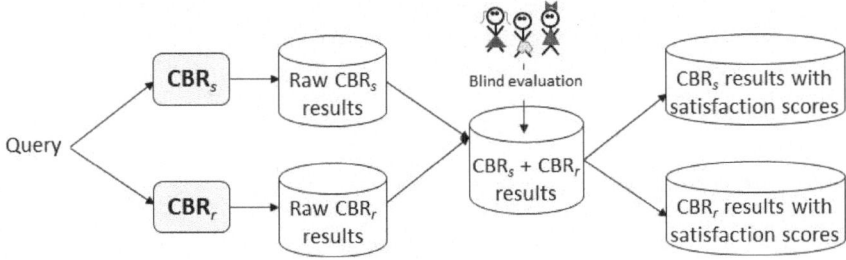

Fig. 4. Evaluation process

This example illustrates that KUs which are not reliable (e.g., `FruitJuice` ⊒ `Vodka`) are not used by CBR_r and that a case with a *good* reliability adapted by a long generalization/specialization path can be better ranked than a case with a *poor* reliability adapted by a shorter generalization/specialization path (e.g., S_{s3} vs. R_{s3}).

4   Evaluation

This section proposes a methodology of evaluation and explains how to analyze the results coming from user tests in order to compare two CBR systems. Section 5 applies this methodology to the TAAABLE use case.

4.1   Evaluation Methodology

As our objective is to demonstrate that managing reliability thanks to MKM in a CBR system improves results returned by the system, the two systems that will be compared are CBR_s and CBR_r and the hypothesis we have to validate is:

> (H) CBR_r returns more satisfactory results than CBR_s.

In order to validate (H), a set of queries is performed on the two systems CBR_s and CBR_r. The results of both systems are presented to users who have to evaluate their relevance using an evaluation scale. Fig. 4 shows that for each query, results of the two systems are computed. The results of these two systems are gathered and displayed to a user in a random order so that the user does not know from which system a result has been computed nor the ranking of this result in the result list.

For each evaluation, the user has to evaluate her satisfaction for each proposed results inferred by the two systems according to the target problem. Many possibilities exist to collect the satisfaction feedback of the user. One of them is the Likert scale [14], which is based on a set of grades, allowing the users to qualify their satisfaction degree. The most frequent scale uses 5 grades, so that users have not too many nor too few options to express their opinions. Among these 5 grades, 2 are positive (very satisfied, satisfied), 1 is neutral (neither satisfied nor unsatisfied) and 2 are negatives (unsatisfied and very unsatisfied). An advantage of this scale is also that it can be easily be turned into a numerical score.

4.2   Result Analysis

In order to verify (H), we have to analyze data from the user evaluation. We draw inspiration from results analysis in recommender systems evaluation. Recommender systems are information filtering systems which predict ratings of users about items (e.g., movies) and propose the best rated item to a user. Recommender systems and CBR systems may be considered as similar both because they match a user query and they return a set of results which may satisfy the user. For this reason, and like it was done in [15], it is possible to analyze the results provided by two CBR systems in the same way that results are analyzed in recommender systems evaluations. In recommender systems, evaluation usually consists in comparing at least two systems that return a set of recommended items in order to measure the performances of the systems. As we are in a similar case, i.e. evaluating the performance of two different systems, the way to compare their performance must be discussed.

Comparing recommender systems. Let A and B be two recommender systems. Testing whether B is better than A consists in testing that the results provided by B are better that those provided by A. For that, if a positive difference between the two systems is observed, this difference has to be significant. A standard measure for that is the p-value which goal is to evaluate the null hypothesis. The null hypothesis assumes that positive results of an experimentation are obtained due to chance. In order to obtain significant results, the null hypothesis must be rejected. If p-value ≤ 0.05 the null hypothesis is rejected and results are significant. To test (H), the results of $\mathrm{CBR_s}$ and $\mathrm{CBR_r}$ will be compared. Data resulting from the user evaluations must be prepared and cleaned before analysis. Moreover, variables of the analysis must be determined in order to establish the appropriate statistical test.

Data preparation and cleaning. Degrees of the Likert scale must be transformed into numerical scores to allow quantitative analysis. We associate a score to each term of the evaluation scale. Very satisfied: 2, satisfied: 1, neither satisfied nor dissatisfied: 0, unsatisfied: -1 and very unsatisfied -2.

Each query is rated by several (at least 4) users, so that the lowest and the highest ratings are excluded in order to limit the impact of *random* ratings.

Variables and tests. The choice of the system is the independent variable (or controlled variable), that is to say, it is the variable that impacts results. The user satisfaction score is the dependent variable (or measured variable) which depends on the independent variable. The measured variable is a non parametric (ordinal) variable which does not have a Gaussian distribution and where possible values are -2, -1, 0, 1 and 2.

The users which evaluates the results of CBR_s and of CBR_r are the same. More precisely, a same user evaluates the set of results returned by both systems on a same query. These two sets are paired samples.

The test requires to evaluate that a significant result must take into account no parametric variables and paired sample. The *Wilcoxon signed-rank test* [16] allows to test that the median of the aggregated satisfaction scores of CBR_r is significantly higher than the median of the added satisfaction scores of CBR_s.

Performance measurement. In recommender systems, precision and recall are performance metrics frequently used [16]. Precision is the proportion of retrieved items which are relevant among the set of items returned by the system while recall is the proportion of relevant items retrieved on the total set of relevant items. For a given query, a retrieved item is relevant for a user if the preference score given by the user is higher or equal than a given threshold. In our evaluation, the threshold is set to 1. Because results to a query in CBR systems is not a finite set, we cannot apply any recall measure.

5   Evaluation of CBR_s and CBR_r in the TAAABLE Use Case

The evaluation aims at validating (H) where the two compared systems have to reason on knowledge coming from the Web. The methodology presented in the previous section has been used to perform this evaluation in the context of the TAAABLE system, using knowledge of ATAAABLE[2], a collaborative web site in which users may interact with KUs used by TAAABLE. ATAAABLE was initially built by translating in French the domain knowledge of WIKITAAABLE[3], the recipes which form the case base (a case is a recipe described by ingredients, dish types, etc.), as well as the adaptation rules have been entered manually by the users of the community. At the time of writing this paper, ATAAABLE contains 2325 classes linked by 2551 hierarchical relations (coming from WIKI-TAAABLE), and 163 recipes (of which 129 cocktail recipes), 11 specific and 25 generic adaptation rules, were entered by 80 users. Entering one of these KU is facilitated by specific interfaces, so that it is very easy to add new ones. Each of the KU may also be easily rated (in one click) by the users of the community using a 5-point scale.

However, the KUs describing the domain knowledge of ATAAABLE are rather consensual, because they were built by a knowledge representation specialist. To simulate knowledge coming from an e-community, the domain knowledge container has been a little bit damaged, by adding, for example, a few bad subsumption relations (e.g. FruitJuice $\sqsupseteq$ Vodka), in order to test if MKM is able to manage erroneous KUs using the evaluation of these KUs by the ATAAABLE users.

---

[2] http://ataaable.loria.fr/

[3] http://wikitaaable.loria.fr/

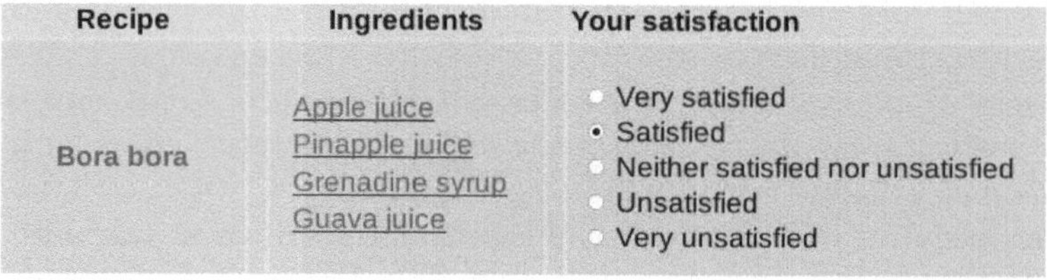

Recipe	Ingredients	Your satisfaction
Bora bora	Apple juice Pinapple juice Grenadine syrup Guava juice	○ Very satisfied • Satisfied ○ Neither satisfied nor unsatisfied ○ Unsatisfied ○ Very unsatisfied

Fig. 5. Evaluation interface for the query GrenadineSyrup ∧ GuavaJuice

5.1 Applying the Evaluation Methodology on TAAABLE

In order to apply the evaluation on TAAABLE, some choices have been made to fix the experimental conditions. In order to limit the number of tests, we focused on a subset of the case base: the ATAAABLE recipes about cocktails. Each query used for the test is composed of 2 required ingredients, each of them appearing at least once in a cocktail recipe of the base. The choice of 2 ingredients is motivated by the expectation of results which will be built by adaptation instead of simply being retrieved as an exact matching recipe. The set of queries has been randomly generated and has been slightly filtered, in order to keep interesting queries, i.e. queries that might be submitted by a real user. Two criteria were taken into account for this filtering:

1. The query must only contain ingredients known from a majority of users, because the users have to evaluate the results. For example, queries containing *"Pisang Ambon"* or *"Angustura"* have been manually removed.
2. The query must look like a real user request. For example, a query like Salt ∧ Pepper has been manually removed.

Subjects were bachelor students who participate to an approximative 20-30 minutes test session organized in a classroom by the authors of this paper. They received an short (5 minutes) oral explanation about the ATAAABLE and TAAABLE systems, the instructions for their participation to the evaluation test being written in the test interface. They register on ATAAABLE to become users and evaluate at least 10 KUs each, randomly proposed but having a link with the adaptation of cocktail (cocktail recipes and hierarchical relations about liquids). The goal was to refine the reliability about these KUs; 396 rates have been collected from 18 users in 5 minutes. Then, each user evaluates the results of about 4 queries, and each query has been evaluated by 4 different users. For the experimentation presented here, 15 queries have been evaluated (thanks to 22 users). For each query, the system results were displayed to users. A result is composed of the title of the original recipe (a link allows the user to access the recipe), and the ingredient list of the adapted recipe is the result of the adaptation. Fig. 5 shows a part of the evaluation interface for one result of the query GrenadineSyrup ∧ GuavaJuice. A user has to enter her satisfaction only according to the composition of the cocktail (so, without testing it).

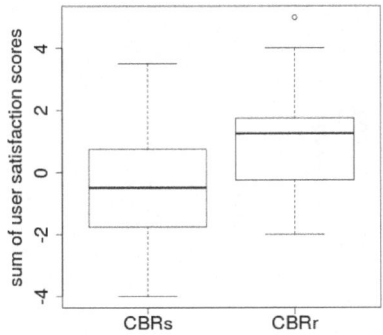

Fig. 6. Comparison of median satisfaction scores of users for each query in CBR_s and CBR_r

Table 3. Average of user satisfaction scores of CBR_s for query #13

result id	user satisfaction score average
S_{s1}	-0.5
S_{s2}	1.5
S_{s3}	1
S_{s4}	-1
S_{s5}	0.5

Table 4. Average of user satisfaction scores of CBR_r for query #13

result id	user satisfaction score average
R_{s1}	1
R_{s2}	1
R_{s3}	1
R_{s4}	1.5
R_{s5}	0.5

5.2   Results Analysis

The results provided in this section show that using reliability on the knowledge of ATAAABLE using by TAAABLE returns more satisfactory results for users than not using it.

Wilcoxon signed-Rank Test. Fig. 6 compares box-plot of CBR_s and CBR_r. We observe that the median of the satisfaction scores attributed to CBR_s is -0.5 and the median of the satisfaction scores attributed to CBR_r is 1.25. The *Wilcoxon signed-rank test* allows to observe a true difference between the results of CBR_r and CBR_s. The p-value is 0.05: the null hypothesis is rejected. Fig. 7 compares CBR_s (black background rectangles) and CBR_r (white background rectangles). A rectangle corresponds to the sum of the average of the satisfaction scores attributed by users on each of the five first results of the query (as those presented in Tables 3 and 4, for the query with the id 13). We observe that the user satisfaction is higher with CBR_r than with CBR_s 11 times out of 15. The query with the id 13 in Fig. 7 is the query GrenadineSyrup $\land$ GuavaJuice. Tables 3 and 4 show the average of the user satisfaction scores of each of the five first results returned by CBR_s and CBR_r for this query. Except for the second result, user satisfaction scores are better with CBR_r.

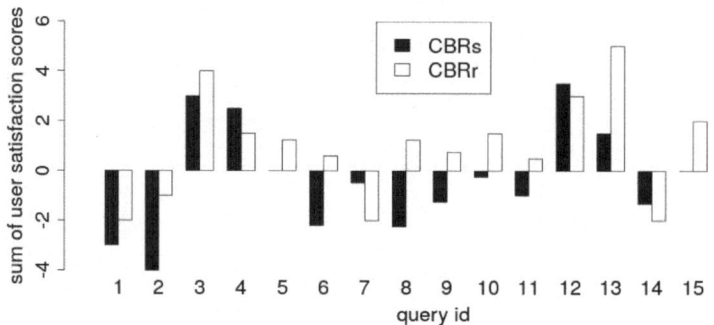

Fig. 7. Comparison of the aggregated satisfaction score averages for each query in CBR_s and CBR_r

Precision. The precision of the CBR_s system is 0.3 while the precision of the CBR_r system is 0.43. Thus, precision of CBR_r system is higher than the one of CBR_s.

Hypothesis Conclusion. Since a significant positive difference has been found between CBR_r and CBR_s and precision of CBR_r is higher than precision of CBR_s, (H) is validated: CBR_r returns better results than CBR_s. Thus, using MKM model in a CBR system improves results returned by the system.

6   Conclusion and Ongoing Work

This paper shows that extending a CBR system reasoning on knowledge coming from the Web, with MKM, a meta-knowledge model which manages KUs reliability, increases user satisfaction. MKM model ensures to reason on the most reliable knowledge and to rank inferred results according to a reliability point of view.

An experimentation validates the hypothesis that the results of CBR_r are more satisfying than those of CBR_s and that there is a significant difference between user satisfaction of CBR_s and CBR_r.

In a short term, we want to demonstrate that personalizing the reliability for users of the e-community, taking into account user preferences in the computation of the *reliability score* of KUs increases user satisfaction. Next, we want to implement filter and ranking functions of MKM in the engine of a CBR system, where retrieval and adaptation processes will be guided by reliability.

References

1. Gaillard, E., Lieber, J., Naudet, Y., Nauer, E.: Case-based reasoning on E-community knowledge. In: Delany, S.J., Ontañón, S. (eds.) ICCBR 2013. LNCS, vol. 7969, pp. 104–118. Springer, Heidelberg (2013)

2. Riesbeck, C.K., Schank, R.C.: Inside Case-Based Reasoning. Lawrence Erlbaum Associates, Inc., Hillsdale (1989)
3. Richter, M.: The knowledge contained in similarity measures. Invited talk at the International Conference on Case-Based Reasoning (1995)
4. Cordier, A., et al.: Taaable: a Case-Based System for personalized Cooking. In: Montani, S., Jain, L.C. (eds.) Successful Case-based Reasoning Applications-2. SCI, vol. 494, pp. 121–162. Springer, Heidelberg (2014)
5. Gaillard, E., Lieber, J., Nauer, E.: Adaptation knowledge discovery for cooking using closed itemset extraction. In: The Eighth International Conference on Concept Lattices and their Applications - CLA 2011 (2011)
6. Naudet, Y., Latour, T., Vidou, G., Djaghloul, Y.: Towards a novel approach for high-stake decision support system based on community contributed knowledge base. In: 10th International Conference on Intelligent Systems Design and Applications (ISDA), pp. 730–736 (2010)
7. Artz, D., Gil, Y.: A survey of trust in computer science and the Semantic Web. Web Semantics: Science, Services and Agents on the World Wide Web 5(2), 58–71 (2007)
8. Golbeck, J.A.: Computing and applying trust in web-based social networks. Ph.D. thesis, University of Maryland (2005)
9. Young, A.K., Muhammad, A.A.: Trust, distrust and lack of confidence of users in online social media-sharing communities. Knowledge-Based Systems 37, 438–450 (2013)
10. Quijano-Sánchez, L., Bridge, D., Díaz-Agudo, B., Recio-García, J.A.: Case-based aggregation of preferences for group recommenders. In: Agudo, B.D., Watson, I. (eds.) ICCBR 2012. LNCS, vol. 7466, pp. 327–341. Springer, Heidelberg (2012)
11. Saaya, Z., Smyth, B., Coyle, M., Briggs, P.: Recommending case bases: Applications in social web search. In: Ram, A., Wiratunga, N. (eds.) ICCBR 2011. LNCS, vol. 6880, pp. 274–288. Springer, Heidelberg (2011)
12. Barry, S., Keane, M.T.: Remembering to forget: A competence-preserving case deletion policy for case-based reasoning systems. In: Proceedings of the 14th International Joint Conference on Artificial Intelligence, JCAI 1995, vol. 1, pp. 377–382. Morgan Kaufmann Publishers Inc., San Francisco (1995)
13. Leake, D.B., Whitehead, M.: Case provenance: The value of remembering case sources. In: Weber, R.O., Richter, M.M. (eds.) ICCBR 2007. LNCS (LNAI), vol. 4626, pp. 194–208. Springer, Heidelberg (2007)
14. Likert, R.: A technique for the measurement of attitudes. Archives of Psychology 22(140), 1–55 (1932)
15. Quijano-Sánchez, L., Recio Garcia, J.A., Díaz-Agudo, B.: Using personality to create alliances in group recommender systems. In: Ram, A., Wiratunga, N. (eds.) ICCBR 2011. LNCS, vol. 6880, pp. 226–240. Springer, Heidelberg (2011)
16. Ricci, F., Rokach, L., Shapira, B., Kantor, P.B.: Recommender Systems Handbook. Springer (2010)

Case-Based Object Placement Planning

Kellen Gillespie, Kalyan Moy Gupta, and Michael Drinkwater

Knexus Research Corporation
Springfield, VA, USA
{kellen.gillespie,kalyan.gupta,michael.drinkwater}@knexusresearch.com

Abstract. Autonomous object placement is a central task in many military and industrial applications. Deciding where to place an object, or placement planning, is a crucial reasoning task within object placement. Placement planning is especially challenging in 3D worlds. In previous work we developed approaches for placement planning that perform well but require manually-engineered knowledge. Consequently, we introduce case-based object placement (COP) planning to reduce the manual engineering burden and improve portability. COP includes the ability to automatically learn cases from problem solving sessions and reuse them to solve previously unseen placement planning problems. We evaluated COP on object placement tasks in virtual office environments. Our findings show that COP can reduce knowledge engineering and increase portability without compromising placement accuracy.

Keywords: Object Placement Planning, Case-Based Reasoning, Spatial Planning, Natural Language Understanding.

1 Introduction

Placing objects in physical or virtual worlds is a routine task in many applications. For instance, unmanned vehicles with manipulators are used to clear mission areas prior to operations. Likewise, game designers carefully place objects in virtual worlds to construct realistic gaming environments [1].

The level of automation in such object placement systems is highly variable. At a low level of automation, a human operator teleoperates an unmanned vehicle using joysticks on an operator console. This manual interaction is time-consuming and requires considerable skill and training. At a high level of automation, the operator can issue natural language commands and the autonomous system responds appropriately. Our long-term goal is to develop autonomous systems that accurately place objects in response to natural language commands. However, this is challenging because natural language can be notoriously vague and difficult to interpret. These problems are exacerbated in placement commands because they involve spatial terms and prepositions that are polysemous [2]. We have developed approaches for language understanding and ambiguity resolution using placement planning [3] [4].

In [3] we introduced the task of object placement planning, its goal to output a placement plan consisting of a final position and orientation. This computational task of placement planning must be performed before a physical or virtual

L. Lamontagne and E. Plaza (Eds.): ICCBR 2014, LNCS 8765, pp. 170–184, 2014.
© Springer International Publishing Switzerland 2014

object can be placed. A challenge with placement planning in 3D worlds is that the space of possible placement plans can be very large. Producing acceptable plans requires substantial real world knowledge about objects and how humans interact with them as well as planning algorithms to leverage such knowledge. In [4] we introduced and evaluated a knowledge-intensive placement planning algorithm called OPOCS. While OPOCS performed well, it requires manually-engineered knowledge which limits its portability. Consequently, in this paper we introduce a case-based object placement (COP) planning algorithm. COP includes approaches for automatically extracting richly structured cases from placement problem solving sessions and reusing them to create placement plans for previously unseen problems. We evaluate COP on a variety of indoor object placement tasks and demonstrate that it can learn cases from problem solving sessions and still perform comparably to OPOCS in placement performance.

We organize the remainder of this paper as follows. In Section 2 we elaborate on the problem of pragmatic ambiguity in language. Section 3 introduces the terminology and the formulation for the placement planning task. In Section 4 we present COP; including its case representation, case learning algorithm, retrieval and reasoning. We describe our empirical evaluation and results in Section 5. Section 6 discusses related work and how COP provides a novel approach to placement planning. Section 7 concludes the paper and suggests possible future work.

2   Pragmatic Ambiguity in Placement Commands

Natural language commands can have multiple interpretations, making them semantically ambiguous. In the context of real world tasks such as object placement, *pragmatic* ambiguities may remain even after these semantic ambiguities have been resolved.

For instance, consider a virtual office environment with typical objects such as a desk, a computer monitor, and a chair (see Figure 1). An agent is given the placement command "Put the printer on the desk." For simplicity, assume the agent is already holding and capable of placing the target object. Let's now assume this agent has access to a suitable semantic representation of the spatial relation 'on' which can be used to select the desktop as a placement surface. Although this interpretation narrows the task's search space, the number of possible placement solutions is still quite large. These solutions could potentially include areas to the left, right, front and back of the monitor. The agent can also consider human factors to constrain the problem space by reasoning that the area behind the monitor is unreachable and that the area in front of the monitor will obscure the human's view of the monitor. The utterance also does not specify a proper orientation of the printer, leaving numerous potential orientations as possible solutions. Without such a specification, the printer could be oriented in numerous ways in relation to the monitor and the chair. However, some potential orientations of the printer, such as upside down or facing the wall, would be invalid. We term this residual ambiguity arising in the context of a situation and a natural language command as "pragmatic ambiguity".

Fig. 1. Office Environment

Computational methods are needed to understand pragmatically ambiguous commands. To this end, we introduce placement planning approaches to understand placement commands and resolve pragmatic ambiguity.

3   Object Placement Planning

Let's revisit our virtual 3D office world in Figure 1 and consider the object placement command "Put the book on the desk". As we explained in Section 2, the spatial term "on" is pragmatically under-specified and covers a continuous space of 3D coordinates and possible orientations for placing the target object "book" in relation to the landmark object "desk". Formally, a spatial plan for placing an object is a tuple of the final position as a 3D coordinate and final orientation as a quaternion: $(position < x, y, z >, orientation < x, y, z, w >)$.

We define a placement command as a tuple including a target object O_t, a landmark object O_l, and a spatial relation or constraint C_p. We combine this command information with all other objects in the world O_w to compute a placement plan.

$$placementPlan \leftarrow computePlacementPlan(O_t, O_l, C_p, O_w) \qquad (1)$$

Searching for a placement plan that satisfies the expectation of the user can be computationally hard due to the inherently large search space. For an ambiguous constraint C_p on a surface area of s square units with an expected positional accuracy of p units and orientation accuracy o degrees for each of yaw, pitch, and roll, the size of the search space can be computed as follows:

$$searchSpace = (s/p) * (360/o)^3 \qquad (2)$$

For instance, for a required positional accuracy within 1 cm and an orientation accuracy within 2 degrees for a typical office desk surface the number of points to evaluate is roughly 4 billion. The problem is worse for larger surfaces and less restrictive placement constraints. However, humans routinely perform this task with ease by utilizing vast amounts of knowledge about objects and their interaction requirements to create acceptable placements.

We previously approached the spatial planning task with an algorithm that utilizes richly encoded domain knowledge about objects and human interactions to effectively prune the search space [4], thereby only evaluating a small fraction of the entire search space where promising solutions may exist. The algorithm uses a knowledge base along with a search algorithm that simulates human interaction to find the best placement plan that meets the command requirements. A limitation of this approach is the difficulty and the effort needed to identify and encode the necessary knowledge. Another limitation is the inherent brittleness of the approach when the algorithm is presented with problems involving objects not covered in its knowledge base.

To overcome the knowledge acquisition and brittleness problems of our prior approach, we have developed case-based placement planning as an alternative. Case-based reasoning (CBR) uses the solutions of similar previous problems to solve the problem at hand [5]. CBR has been successfully applied to design and planning tasks [6]. In one such application CBR was utilized for selecting gaming strategies based on opponent behavior analysis and prediction [7]. CBR planning applications can present numerous challenges [8], largely because the solution space in such problems is large impractical to properly generate case coverage for. Likewise, we contend that the placement planning task is a novel CBR planning task and presents interesting challenges for case representation, learning, retrieval and reuse that we in this paper.

4   Case-Based Object Placement Planning (COP)

Case-based reasoning for object placement planning requires novel approaches to case representation, learning, retrieval, and reuse. Figure 2 shows an overview of the COP learning and reasoning processes. For case learning, we assume the availability of problem solving sessions where decision-makers have manually created placement plans by placing objects in 3D worlds. Learning involves the extraction and representation of cases from these sessions. In Section 4.1 we describe our case representation followed by a detailed approach for case learning in Section 4.2. In addition to extracting cases during case learning, we extract our object taxonomy (see Section 4.3) to be used as similarity knowledge in the case retrieval process.

The input to the reasoning portion of COP is an object placement task and its 3D world state. From this information we perform similarity-based case retrieval and use the retrieved cases to identify and build spatial constraints relevant to the current task. Our case retrieval utilizes our compiled taxonomy of objects for query expansion (see Section 4.4). Additionally, we use the current task to select candidate placement surfaces for the target object to be placed on. We use the spatial constraints combined with our selected placement surfaces as input to the placement planning phase, within which we search for an optimal position using a genetic algorithm (GA) [9]. Given an optimum position we then search for a corresponding orientation, and with both we have our final placement plan. We detail the entirety of the reasoning portion of COP in Section 4.4.

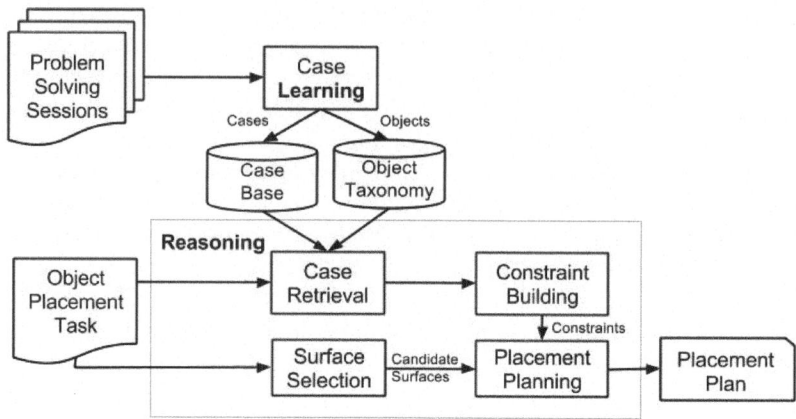

Fig. 2. Learning And Reasoning In COP

4.1 Case Representation

A COP case represents a placement task and a set of applicable spatial constraints and relations $R\{r_1, r_2, ...\}$ that must be solved to generate a spatial plan (see Figure 3). Each relation r in R comprises the related object O_r, the spatial relation constraint C_r, the positional and rotational offsets between O_t and O_r. The inventory of spatial constraints we use in the case base corresponds to linguistic terms such as 'on', 'near', 'in', 'in front of' and so on. The positional and rotational offsets implicitly represent the practical constraints of human interactions with objects. For a "Put the book on the desk" placement planning

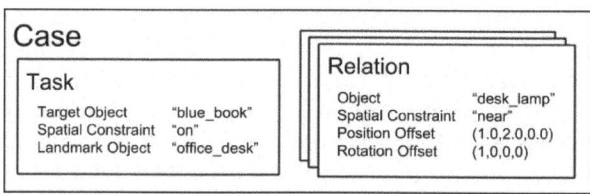

Fig. 3. Episodic Case Structure

task, O_t is the "book", C_p is "on", and O_l is the "desk". With the placement made, R contains all the spatial constraints that apply between the book (O_t) and other objects in the scene.

4.2 Case Learning

To accomplish case learning we make use of object placement problem sessions that record placement tasks. Included in these sessions is the final state of the virtual world including its objects, their positions and orientations, and the corresponding placement plan generated by a human user. Such sessions can be realized within a game design environment, using a 3D scene and its recorded state as input for case learning.

Algorithm 1 shows the case extraction process, which makes use of a set of relation extractors that encode spatial constraints between the target object O_t and each related object O_r in the scene. These constraints also include the landmark object O_l for the given problem to capture the task context.

Algorithm 1. Case Extraction

$O_t \leftarrow TargetObject$
$C \leftarrow \text{GENCASE}()$
for nearby object O_n in scene **do**
    for relation extractor REX **do**
        if $REX(O_t, O_n)$ **then**
            $r \leftarrow \text{GENRELATION}(O_t, O_n, PC_{REX})$ ▷ Use placement constraint of REX
            $C.\text{ADDRELATION}(r)$
        end if
    end for
end for
$\text{STORECASE}(C)$

The case learning thus encodes a problem solving episode into a case for use in placement planning.

A key component of case learning is the encoding of a 3D world state into a collection of relevant spatial constraints. We identify and extract the relations between objects in the world and the target object in the command. This is performed with relation extractors that determine which relations apply between the target object O_t and another object O_r. Relation extractors have a one-to-one correspondence with the placement constraints (e.g. "in", "near"). Algorithm 2 shows an example of a relation extractor for the spatial constraint "near".

Algorithm 2. Relation Extraction

---

function $\text{NEAREXTRACTOR}(O_t, O_r)$ ▷ Take In Target/Related Objects
    $threshold \leftarrow 4.0$ ▷ Predefined Distance Threshold
    $targetBounds \leftarrow O_t.\text{GETBOUNDS}()$
    $volumeDiag \leftarrow targetBounds.\text{GETEXTENT}()$
    $d \leftarrow \text{DISTANCE}(O_t, O_r)$
    return $d \leq (threshold * volumeDiag.\text{LENGTH}())$
end function

In the extractor shown in Algorithm 2, we set a distance threshold of four meters together with the length of the target object's volume diagonal (or space diagonal) derived from its 3D extent. Therefore, when the target object is voluminous (e.g., a building), it is easier to be "near" it. A more sophisticated version of the same extractor could consider the size of both O_t and O_r. Additionally, it could consider other factors such as occlusion or line of sight between the objects to better represent the subtleties of human-object interaction.

4.3   Similarity Knowledge: Object Taxonomy

The object taxonomy is a knowledge artifact that we use for query expansion in case retrieval. The taxonomy is a network of object types with *is-a-sub-type-of* relations. For instance, the entries `blue_book : book` and `book : object` tell us that objects of type `blue_book` are of type `book`, which in turn are of type `object`. Figure 4 shows an example taxonomy hierarchy.

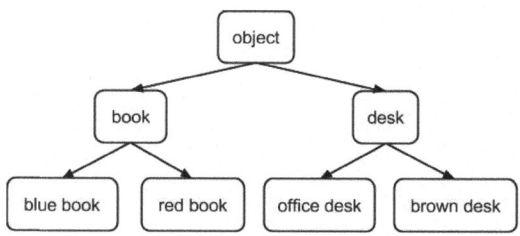

Fig. 4. Example Taxonomy

We automatically compile the object taxonomy as we encode cases from scenarios. This entails obtaining all objects in each 3D scene, getting their type and parent type from their annotations, and encoding them into the hierarchy on-the-fly. Upon completion of case learning we have a full object taxonomy.

4.4   COP Reasoning

Reasoning in COP begins with the identification of candidate placement surfaces and the use of case retrieval to identify a relevant set of cases and their constraints. Together they are used to formulate a placement plan search problem which we solve using a GA. We begin this subsection with a method for selecting candidate surfaces. We follow this with our case retrieval algorithm. Finally we describe the use of the GA to create a placement plan.

Selecting Placement Surfaces. Some objects in the world can have other objects placed on them, depending on the placed object's size and the size of the area it is being placed upon. We refer to these areas in which objects may be placed as surfaces, which are explicitly defined as quadrilaterals made up of four points and a normal vector. For a table, one such surface is the tabletop.

To obtain candidate placement surfaces, we first obtain all the surfaces for all the objects in the world. Then, given a placement constraint C_p and our landmark object O_l, we filter these to only include surfaces that apply to the current task. For instance, if C_p is 'on' and O_l is 'table', we narrow down our surfaces to only those that belong to table objects. Each placement constraint has its own surface filter. The remaining surfaces are our candidate surfaces to search over when selecting the best placement plan.

Case Retrieval. The objective of case retrieval is to obtain cases to assist in identifing a set of relevant and applicable placement constraints to be used for subsequent reasoning. We use a retrieval algorithm that progressively relaxes the exact matching requirements on objects in the query until a pre-specified number of cases is found. We accomplish this with the object taxonomy to search over related objects not explicitly mentioned in the problem task.

Query Levels (QLs) describe the type of matching constraints we impose on the retrieval queries. At first, these constraints apply to the comparison of the current object placement task with those in the cases. Next, they are used to compare objects in the query with those in spatial relations in the cases. As we move up QLs, we expand the query by generalizing the objects up the object taxonomy. The list of QLs for a two-level taxonomy is shown in Table 1.

Table 1. Query Levels

Query Level	Object Expansion	Constraint Set
QL1	None	Task Constraints
QL2	One Parent	Task Constraints
QL3	Both Parents	Task Constraints
QL4	None	Spatial Constraints
QL5	One Parent	Spatial Constraints
QL6	Both Parents	Spatial Constraints
QL7	All	All

For instance, say we are placing an object of type `blue_book` on an object of type `office_desk`. At QL5 we would search for cases in which a `book` happened to be on an `office_desk` or conversely cases in which a `blue_book` happened to be on a `desk`. We say 'happened to' since we are looking at spatial constraints at QL5 rather than the task constraints in the cases.

The QLs described above only use the lowest two levels of the object taxonomy, which we found to be sufficient. More query expansion would increase the number of query levels.

Spatial Constraints Selection and Weighting. Using the retrieved cases we identify and weight the relevant spatial constraints that must be solved to identify the best placement plan. The spatial constraint relations in the retrieved cases provide us with constraints that may be relevant to the current problem. Relevant relations are those that involve similar objects to those in the current scenario, and as such we obtain them by filtering the relations using the objects in the new scene. For instance, if the new problem requires placing a book on a table and there are no lamps in the scene, the relation book:near:lamp is not relevant to our current task and is discarded. Multiple occurrences of the same constraint across several cases increases that constraint's weight. For instance, if the case base contains a large number of cases that contain the relation book:on:bookshelf, it will be a heavily weighted constraint favoring such a placement provided an object of type bookshelf exists in the current scene.

Placement Planning with GA. We search for an optimal placement plan in two steps as follows. For each candidate surface, we first identify the best position and then find the best orientation for the chosen position. Together they form the best plan for the candidate surface. We then choose the highest scoring placement plan across all surfaces to generate our final plan.

To determine the best position on a candidate surface to place O_t, we employ the GA to search through candidate positions (parameterized as u,v coordinates) on that surface. The genetic algorithm is given the constraints C generated in the previous step which it uses to compute fitness as follows:

$$Fitness(u, v) = \sum_{i=0}^{|C|} solved(C[i], C[i] \to weight) \tag{3}$$

For every new generation, using a random walk procedure, the GA creates a new set of points to evaluate. Over several generations, it converges to a candidate position that maximizes the fitness score. We leave the GA with an approximate solution position P_{sol} and the constraints it resolved C_{res}. Next, based on the satisfied constraints, we search for the best orientation O_{sol} as follows. We use the rotational offsets stored in each constraint in C_{res} to retrieve a set of desired orientation values for O_t. We then cluster these orientation values using hierarchical agglomerative clustering. Finally we select the largest cluster and use its centroid as the best orientation O_{sol}.

We run the GA as described above on each selected placement surface, and for each we get the solution (P_{sol}, O_{sol}). Our final solution is the position and orientation (P_{sol}, O_{sol}) from surface with the highest overall constraint satisfaction score or fitness score over the satisfied constraints.

5   Evaluation

Our main goal was to assess whether COP can reduce the knowledge engineering burden by automatically learning spatial constraint knowledge as cases and use them to effectively solve previously unseen placement planning problems. In this section, we present our evaluation approach and results. Our approach consists of gathering test data and associated ground truth for performance evaluation, candidate placement planning algorithms for evaluation, and metrics for evaluation.

5.1   Test Data and Ground Truth

We acquired data for evaluation using the crowd sourcing service Amazon Mechanical Turk [10]. We developed the Integrated 3D Environment Annotation (IDEA) tool to pose object placement tasks for 4 unique 3D virtual offices to hundreds of qualified users and gather placement plan responses. For each world, we posed 74 different object placement tasks covering 10 placement constraints, namely: in, on, near, beside, in front of/behind, to the left/right of, and over/under. We randomly assigned multiple instances of these tasks, each to a different user, until we had 20 placement plans for each task. Table 2 shows our final extraction data from our Mechanical Turk exercise.

Table 2. Learning Results

Environment	Cases	Constraints	Taxonomy
Manager Office	1445	63457	13
Executive Office	1440	31581	13
Home Office	1440	31352	13
Satellite Office	1441	32489	13

Unlike conventional single-class classification tasks that have one acceptable answer, planning problems can have multiple acceptable plans. Therefore, our challenge was to compile multiple acceptable placement plans as ground truths from the raw crowd-sourced data. To this end, we used agglomerative hierarchical clustering of placement plans to create one or more clusters or ground-truths comprising position and orientation for each problem (see Figure 5a). We use the orientation quaternions as input to our orientation clustering (see Figure 5b). These clusters were weighted based on the number of members they contained, an indication of a particular plan preference. As we explain in Section 5.3, these position and orientation clusters along with their weights serve as ground truths for comparing placement planning performance.

The black spheres in Figure 5 represent individual solutions for placement positions and orientations, while the white spheres and cones show the computed position and orientation clusters, respectively.

5.2   Placement Planning Algorithms

We compared the performance of the following three placement algorithms on the same set of test cases.

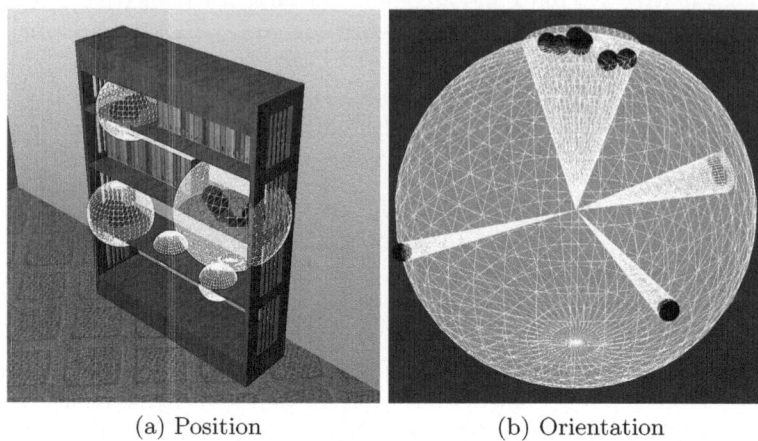

(a) Position   (b) Orientation

Fig. 5. Ground Truth Clusters

Random Placement Planner: This serves as a baseline for evaluating all object placement planning algorithms. It generates placement plans by first collecting all valid placement surfaces in the current scene using an approach similar to the surface selection algorithm described in Section 4.4. Next, it randomly selects a candidate placement surface and places O_t in a random available position on that surface. Finally the algorithm chooses a random cardinal orientation for O_t as its orientation. This algorithm does not require any knowledge engineering.

COP: This is our primary candidate for evaluation. This is the only algorithm capable of automatically learning the knowledge required for reasoning, so no knowledge engineering is required.

OPOCS: This is the heuristic placement planning algorithm from prior research [4], and requires a KB populated with objects and relations covering the objects in the test data. It took roughly 50 person hours to manually encode and validate this KB. Since the KB contains inter-object relations between most objects and all other objects, as the number of objects N grows the KB size grows at roughly N^2 pace. Therefore, the marginal knowledge engineering effort can be substantial as new objects are added to the KB.

5.3   Metrics

For the knowledge engineering and validation effort we use person-hours as a metric. For the placement planning performance, we use a placement plan quality measure (PPQ) as follows:

$$PPQ = PA * OA \qquad (4)$$

PA (Position Accuracy) compares the generated placement plan to ground truth position clusters. If the position in the generated plan lies within one of the clusters then it gets a score in $(0\ 1]$ or 0 otherwise. Given the ground truth cluster (if any) hit by the generated placement plan PC_{hit} and the largest ground truth cluster PC_{max} we calculate our Position Accuracy PA as follows:

$$PA = |PC_{hit}|/|PC_{max}| \tag{5}$$

OA (Orientation Accuracy) is computed the same way with the orientation clusters instead and has the same value range.

$$OA = |OC_{hit}|/|OC_{max}| \tag{6}$$

5.4   Test Regimen and Analysis

Since COP was the only placement planning approach that requires learning prior to reasoning, we used a 4-fold cross validation approach where problem cases for each office world served as a fold. That is, we trained on the data and tasks in three of the four office worlds and tested it on placement planning tasks in the remaining one and repeated the process for all folds. We compared the PPQ of each algorithm using a paired t-test. We present our results next.

5.5   Results

COP case learning for 4 folds of data took approximately 3 computation hours with negligible manual effort to setup and start the execution. In contrast, as described before, OPOCS knowledge engineering took approximately 50 person hours.

Figure 6 presents the PPQ for all the three algorithms across 10 spatial constraints. Both COP and OPOCS outperform the baseline algorithm of Random Placement Planner, which often failed to even score due to the harshness of the metrics and the scoring formula. The average PPQ scores for COP and OPOCS are 0.170 and 0.173 *(p=0.16)*, respectively. Using the PPQ formula we can extrapolate the average position and orientation component (PA/OA) scores for each algorithm. This is equal to the root of each algorithm's PPQ, or 0.412 and 0.418 for COP and OPOCS respectively. Overall, they are not significantly different from each other. Therefore, this demonstrates that COP can effectively ameliorate the knowledge engineering burden without sacrificing performance. Although the overall performance of the two algorithms did not differ significantly, they did so on certain relations. COP outperforms OPOCS on the spatial constraint 'front' roughly 0.429 to 0.160 *(p=0.007)*, nearly tripling its average score. Conversely, OPOCS outperforms COP roughly tripling its score 0.217 to 0.068 *(p=0.022)* on the most ambiguous relation tested 'in'. Likewise, it performs better than COP, although not significantly in this evaluation, on some of the other relations such as behind and beside. Therefore, we believe that there is an opportunity to improve placement planning by combining the performance advantages of the both algorithms. We plan to address in this in our future work among other things (see Section 7).

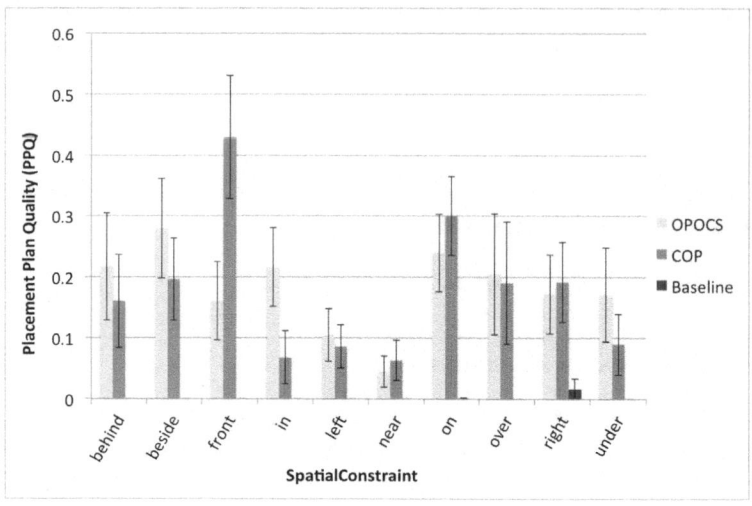

Fig. 6. Placement Results

The PPQ values for COP and OPOCS ranged from 0.05 to 0.43, which appears to be a somewhat weak overall performance. There are a few reasons for this. First, our cluster-based metric was harsh as it penalized near misses just outside of a cluster with a score of 0, despite the fact that it might still be acceptable to a user. Second, the multiplicative PPQ formula means any placement with a PA or OS of 0 will have an overall PPQ of 0. Third, effective placement planning is a hard problem and our system covered a small fraction of the possible human interaction factors that were considered by our crowd-sourced users. We plan to address these limitations in our future work.

6   Related Work

CBR application to placement planning is similar to the application of CBR to layout planning. For instance, in CLAVIER, CBR was applied to the layout for curing ceramic parts of jet aircraft [11]. However, the task of layout planning was not used to resolve natural language commands as we did with COP.

Computational techniques other than CBR have been used for placement planning. For instance, Wordseye [1] performs scene construction (i.e. object placement) via natural language commands. Wordseye includes a knowledge base of functional interactions between human agents and objects using a notion of spatial tags. Its scene construction or depiction process exploits handcrafted rules along with these spatial tags to place objects. Unlike Wordseye, COP does not rely upon handcrafted rules and data regarding objects and instead relies solely upon the use of previous placement episodes and their respective world states. Furthermore, we present an extensive evaluation of our approach across 10 spatial relations.

Cosgun et al. present a planning algorithm for placing objects on a table surface [12]. Unlike COP, their focus is on rearranging existing objects in order to place a new object. They evaluate their push planning approach on 2D maps of objects. In contrast, COP is focused on placing objects with linguistic commands on available spaces in 3D.

Kurup and Cassimatis present a spatial reasoning approach that can be used for spatial planning [13]. However, their approach is purely quantitative and does not address the issue of pragmatic ambiguity discussed here, nor does it handle natural language commands. Furthermore, the approach is limited to 2D grid worlds.

Si et al. use grounded spatial relations in order to probabilistically detect events being performed by agents in a scene [14]. While this work is more focused on determining what semantic actions are being performed, it reasons about objects and how they interact with one another as well as how meaningful these relations are.

While many of COP's disciplines and techniques have been visited in the work described above, we believe it is the first system of its kind to attempt to address all of them at once.

7   Conclusions and Future Work

Placement planning is a novel computational task for CBR. We introduced COP as a CBR alternative to a heuristic algorithm to reduce the knowledge engineering burden. Our evaluation over a representative set of placement planning tasks in virtual office environments demonstrated that COP can indeed reduce this burden without compromising placement planning performance.

Our contributions are two fold. First, a novel combination of CBR and a GA to solve a computationally hard placement planning problem; and second, a novel crowd-sourcing approach to gather placement planning test data for evaluation and subsequent use with metrics for placement planning.

However, our approach is only a first step in solving this problem and is not without limitations. For instance, we identified that the overall performance of OPOCS and COP could be improved by increasing the richness of knowledge and case representation that better captures the subtleties of human-object interaction and also increases the coverage of the KB. We will address these issues in our future work.

We used a simple counting approach for learning constraint weights. Possible improvements to constraint weighting could include the use of information theory measures for weight computation and the inclusion of negative (poor placement) examples. Our evaluation was also limited in scope, and our metric was harsh given the acceptability of a wide range of solutions in practice.

Therefore, we plan to conduct evaluations with larger data sets with a wider range of spatial constraints, and develop soft metrics to address these limitations. We will investigate approaches for combining the planning accuracy and knowledge of OPOCS with the learning and generalization capabilities of COP into a hybrid system to obtain overall better coverage and performance.

References

1. Coyne, R., Sproat, R.: WordsEye: an automatic text-to-scene conversion system. In: 28th Annual Conference on Computer Graphics, pp. 487–496 (2001)
2. Herskovits, A.: Language and Spatial Cognition: An Interdisciplinary Study of the Prepositions in English. Studies in Natural Language Processing. Cambridge University Press (2009)
3. Gupta, K.M., Schneider, A., Klenk, M., Gillespie, K.: Representing and reasoning with functional knowledge for spatial language understanding. In: Proceedings of the Workshop on Computational Models of Spatial Language Interpretation and Generation (2011)
4. Gupta, K.M., Gillespie, K., Karneeb, J., Borck, H.: Spatial planning for placement command understanding (2013)
5. Aamodt, A., Plaza, E.: Case-based reasoning: Foundational issues, methodological variations, and system approaches. AI Communications 7(1), 39–59 (1994)
6. Veloso, M.M., Muñoz-Avila, H., Bergmann, R.: Case-based planning: Selected methods and systems. AI Communications 9(3), 128–137 (1996)
7. Gillespie, K., Karneeb, J., Lee-Urban, S., Muñoz-Avila, H.: Imitating inscrutable enemies: Learning from stochastic policy observation, retrieval and reuse. In: Bichindaritz, I., Montani, S. (eds.) ICCBR 2010. LNCS, vol. 6176, pp. 126–140. Springer, Heidelberg (2010)
8. Lee-Urban, S., Muñoz-Avila, H.: Adaptation versus retrieval trade-off revisited: An analysis of boundary conditions. In: McGinty, L., Wilson, D.C. (eds.) ICCBR 2009. LNCS, vol. 5650, pp. 180–194. Springer, Heidelberg (2009)
9. Mitchell, M.: An Introduction to Genetic Algorithms. MIT Press (1998)
10. Ipeirotis, P.G.: Analyzing the amazon mechanical turk marketplace. XRDS 17(2), 16–21 (2010)
11. Hinkle, D., Toomey, C.: Clavier: Applying case-based reasoning to composite part fabrication. In: Proceedings of the Sixth Annual Conference on Innovative Applications of Artificial Intelligence (1994)
12. Cosgun, A., Hermans, T., Emeli, V., Stilman, M.: Push planning for object placement on cluttered table surfaces. In: International Conference on Intelligent Robots and Systems, pp. 4627–4632 (2011)
13. Kurup, U., Cassimatis, N.L.: Quantitative spatial reasoning for general intelligence. In: Proceedings of the Third Conference on Artificial General Intelligence Conference, pp. 1–6 (2010)
14. Si, Z., Pei, M., Yao, B.Z., Zhu, S.C.: Unsupervised learning of event and-or grammar and semantics from video. In: IEEE International Conference on Computer Vision, pp. 41–48 (2011)

On the Role of Analogy in Resolving Cognitive Dissonance in Collaborative Interdisciplinary Design

Ashok K. Goel and Bryan Wiltgen

Design & Intelligence Laboratory, School of Interactive Computing
Georgia Institute of Technology, Atlanta, Georgia, USA
{ashok.goel,bryan.wiltgen}@cc.gatech.edu

Abstract. Analogies play multiple roles in cognition. In this paper, we explore the roles of analogy in collaborative interdisciplinary design. We describe two analyses of a case study of a design team engaged in biologically inspired design. In the first analysis, we sought to understand the multiple roles of analogy in interdisciplinary design. The goal of the second analysis was to understand the relationship between analogy and collaboration. During this latter analysis, we discovered another, unexpected, role for analogy: resolving cognitive dissonance. Cognitive dissonance typically refers to the mental discomfort a person experiences when simultaneously holding two conflicting goals, values, beliefs, thoughts or feelings. We observed that interdisciplinary design teams too have cognitive dissonance. We also observed that analogies play an important role in helping induce shifts in the perspectives of teammates, align their mental models, and thereby resolve the cognitive dissonance in interdisciplinary design teams. We discuss some implications of our observations for developing case-based systems for collaborative interdisciplinary design.

Keywords: analogical reasoning, cognitive dissonance, collaborative work, interdisciplinary design, design teams.

1   Background, Motivation and Goals

In 1957, Leon Festinger [1] famously introduced the notion of cognitive dissonance into the literature on cognitive psychology. Cognitive dissonance refers to the mental discomfort a person experiences when simultaneously holding two conflicting goals, values, beliefs, thoughts or feelings. Festinger suggested that when a cognitive agent suffers from cognitive dissonance, he tries to reduce the dissonance. Consider, for example, the famous parable of the fox that tried to grasp a bunch of grapes. When the fox failed to reach the grapes, he suffered from a cognitive dissonance between two thoughts: wanting the grapes and the failure to reach them. To reduce the cognitive dissonance, the fox surmised that the grapes must have been sour and therefore not worth reaching! Over the last fifty years, cognitive dissonance has become a classic theory in cognitive science.

Although cognitive science typically views individual humans as cognitive agents, social aggregates of humans, such as teams, organizations, and nations, too may be

L. Lamontagne and E. Plaza (Eds.): ICCBR 2014, LNCS 8765, pp. 185–199, 2014.
© Springer International Publishing Switzerland 2014

viewed as cognitive agents. Consider as an example Ireland, the host country for this conference. Like any cognitive agent, Ireland perceives its environment and acts on it in the pursuit of its goals; it also acquires, organizes, accesses and uses knowledge to decide on appropriate actions; and so on. Indeed, the cognitive science literature is replete with examples of viewing, if only implicitly, social aggregates of cognitive agents themselves as cognitive agents. Sylvan et al. [2] explicitly viewed Japan as a cognitive agent in modeling its decision making related to energy security. Given that social aggregates of humans, such as nations, organizations and teams, can be viewed as cognitive agents, we posit that such social aggregates too may have cognitive dissonance.

This notion of cognitive dissonance in social aggregates of humans seems intimately related to how humans work, play, and live. Consider as an example how humans design. Most design is collaborative in practice: designers typically work in teams. It follows that a design team may suffer from cognitive dissonance due to the conflicting goals, models and thoughts of its members, and that the team will employ cognitive strategies for reducing the dissonance. Further, design often is interdisciplinary in practice: members of a team working on complex system design, for example, may have very different knowledge and expertise. It follows that interdisciplinary design teams in general may suffer from deep cognitive dissonance because of the deep conflicts in the metaphors, models and methods used by its members.

This notion of cognitive dissonance in design cognition raises another question: what cognitive strategies may an interdisciplinary design team use to resolve its cognitive dissonance? Recently, Spitas & Badke-Schaub [3] presented a preliminary information-processing model of cognitive dissonance in interdisciplinary design teams in terms of flow of ideas in "synaptic networks." Our work described below suggests that interdisciplinary design teams use analogies as a mechanism for reducing cognitive dissonance. Informally, this makes sense when one considers "analogy as the core of cognition" [4,5] and especially of creativity [6,7]: Given that analogy is a core process of cognition, we would expect it to appear throughout human behavior; we would especially expect analogy to be prevalent in creative design.

In this paper, we describe two related analyses of a case study of a design team engaged in biologically inspired design. We chose this context for our study because biologically inspired design is intrinsically interdisciplinary (because it entails analogies from biology to a design domain), and collaborative (because most designers are novices in biology and most biologists are naïve about design). In the first analysis, we sought to relate the case study to prior research [8] that described multiple roles of analogy in biologically inspired design. The goal of the second analysis was to understand the relationship between analogy and collaboration. It is during this latter analysis that we unexpectedly discovered the role of analogy in resolving cognitive dissonance. We also discuss the implications of our findings for developing case-based tools for collaborative interdisciplinary design.

2   Biologically Inspired Design

Biologically inspired design (also known as biomimicry) is an important and widespread movement in modern design [9-11]. The paradigm espouses the use of analogies to biological systems for generating conceptual designs for technological systems. One prominent example is the design of windmill turbine blades inspired by the design of humpback whale flippers [12]. The designers adapted the shape of the whale flippers—specifically, the bumps on their leading edges—to turbine blades, creating blades that improve lift and reduce drag, increasing the efficiency of the turbine [13].

The rapidly growing movement of biologically inspired design has led to the development of several case-based interactive tools for supporting its practice, including Biomimicry 3.8 Institute's [14] AskNature, Chakrabarti et al.'s [15] IDEA-INSPIRE, and Vincent et al.'s [16] BioTRIZ. DANE (http://dilab.cc.gatech.edu/dane/) [17], for example, provides access to a digital library of functionally indexed multimodal and structured representations of biological and technological systems for idea generation in conceptual design, and Biologue [18] is an interactive tool for collaborative tagging of biology articles with semantic tags and for accessing biology articles based on the semantic tags. All these case-based tools are based on the fundamental assumption that biological analogies are useful mainly for the task of design idea generation in the conceptual design of technological systems.

However, empirical studies of biologically inspired design indicate that analogies play multiple roles in biologically inspired design. By the "role" of analogy, we mean the "function" of analogy in the design process, i.e., the design task or subtask that the analogy addresses. For example, Christensen & Schunn [19] identified three functional roles of analogy in engineering design: problem identification, problem solving, and concept explanation.

Vattam, Helms & Goel [8] identified multiple roles of analogies in biologically inspired design, including "*solution generation, evaluation,* and *explanation*". Analogy for solution generation occurs when an analogue is used to transfer a mechanism or decompose a design problem. Analogy for evaluation occurs when an analogue is used to assess the candidate design. Analogy for explanation occurs when an analogue is used to explain some part of a design solution.

Helms, Vattam & Goel [20] identified problem formulation as another role analogy in biologically inspired design, and Helms & Goel [21] identified design problem reformulation as yet another role of analogy in biologically inspired design. While analogy for problem formulation occurs when an analogue is used to identify and specify a problem, analogy for problem reformulation occurs when analogy is used to revise a problem formulation.

3   A Case Study of Biologically Inspired Design

Georgia Institute of Technology offers a course on biologically inspired design that provides the context for our case study. ME/ISyE/MSE/PTFe/BIOL 4740 is a yearly, interdisciplinary, project-based undergraduate class taught jointly by biology and engineering faculty in which mostly senior-level design students work in small teams of 4-5 on design projects. The class is composed of students from biology, biomedical engineering, industrial design, industrial engineering, mechanical engineering, and a variety of other disciplines.   The projects involve identification of a design problem and conceptualization of a biologically inspired solution to the problem.   Yen et al. [22] provide details of the project-based learning and teaching in this course.

We performed a participant observation of a design team in the Fall 2010 session of the course.   This entailed one of the authors of this paper (Wiltgen) taking the course and engaging in all activities expected of a student, including attending lectures, doing homework, and participating as a design team member in the observed design team.   Wiltgen's team consisted of five people representing five different majors: architecture, biology, computer science (Wiltgen), mechanical engineering, and industrial engineering.   In this paper, we will focus on what we call the Shark Attack project that was developed by Wiltgen's team.   The goal of this project was to prevent shark attacks off the coast of the United States without harming the sharks. The team designed an underwater sound-based shark repellant device inspired by the snapping shrimp [23,24], a small shrimp with the ability to create loud, underwater sounds using one of its claws, which it uses to hunt prey and communicate with other snapping shrimp.   The final design worked by emitting sounds that are generated by the same cavitation mechanism that the snapping shrimp uses to create sound, but at a frequency that sharks dislike.

4   Roles of Analogy

The first research question in this paper pertains to the roles of analogy in biologically inspired design: for what design tasks does a collaborative, interdisciplinary team use analogies?   Based on our prior research [8], our initial hypothesis was that analogies will address the tasks of what we now call design idea generation, design concept evaluation and design concept explanation.   Our analysis added the role of domain concept explanation.

4.1   Method

Several kinds of data were collected during the participant observation: audio recordings of team discussions, e-mail communications amongst the team, design artifacts (including some class assignments), and field notes of class sessions.   For this paper, we focus strictly on the audio recordings and transcripts derived from those recordings.   The audio recordings cover in-class team activities and out-of-class team meetings.   We transcribed the recordings into written form.   Once transcribed, we analyzed transcripts related to the first design episode, looking for instances where an analogy explicitly occurred.   Phrases such as "like <X>" where used to identify passages.   Once the instances were identified, we then categorized the instance by its role.

4.2   Preliminary Results

Table 1 summarizes our preliminary results pertaining to the roles of analogy in biologically inspired design. The table provides three rows per covered role of analogy: identification of the role, definition of the role, and an example of the role. We identify each of the five team members by their major: CS for computer science (Wiltgen), ME for mechanical engineering, ARCH for architecture, BIO for biology, and IE for industrial engineering.

Table 1. Roles of Analogy

Role: Design Idea Generation
Definition: A problem is defined and then an analogy is used to meet the parameters of that problem.
Example: In one class session, ME used the source analogue of the flying fish to address the problem of drag in maglev trains: *ME: "So, what I was thinking about was using these fins off the flying fish on the side of a maglev train to also create a little bit of lift where you, you actually use the air resistance to help you get over the actual friction and move more efficiently"*
Role: Design Concept Evaluation
Definition: An analogue related to the design solution is used to infer the quality of the design.
Example: In this snippet of a discussion from an out-of-class meeting, only IE and CS are present. As we interpret this snippet, IE brings up an argument for why sharks would not get desensitized to the sound created by the team's design. Here, we present an edited version of the transcript to ease readability. In Table 3, we present the unedited transcript. (Actually, this example was derived from the analogy identified in the second analysis. However, this makes little difference to our results.)   *IE: "the whole thing on like desensitizing the sharks I mean sharks don't eat humans you know"*   *CS: "right"*   *IE: "they're not their food source huma- we're not their food source so does that mean that they've already been I mean it's not like they've been de- desensitized to human sound right now anyways right?"*   *CS: "mmm"*   *IE: "so do we really have to worry about them them being desensitized? Since I mean humans basically are like decoys in a sense already ((CS says "right" overlapping with the ready part of already)) in the fact that they don't get anything out of it"*   *CS: "right"*   *IE: "so if they haven't been desensitized already then"*

Table 1. (*continued*)

Role: Design Concept Explanation

Definition: An analogy is used to explain the entirety or some aspect of the design solution.

Example: In this example, ARCH is describing a component of the design solution: a proximity sensor that would turn on the sound generator only when a shark passes nearby. She makes an analogy to a "fish monitor" to explain her solution component to the team.

ARCH: "Let's say it comes within a certain boundary of the coast like a large organism."

ME: "Yeah"

ARCH: "Cause I mean there's already things that like"

ME: "Let them come off the-"

ARCH: "Can detect large organisms. You know, like a fish monitor if you're going fishing it can-"

CS: "Yeah"

ARCH: "And then so if there's like a large organism that passes through maybe then it triggers something."

Role: Domain Concept Explanation

Definition: An analogy is used as an explanatory tool to help build an understanding about a biological concept.

Examples: In this example, ME had previously presented a magnolia leaf as his found object. A magnolia leaf has two distinct sides, a top, waxy side and a fuzzy brown bottom side. In his explanation of the leaf, he could not explain the bottom side, so the team set forth to generate an explanation. BIO tried to explain the bottom of the leaf by making an analogy to an umbrella, where water collects on the umbrella's underside. He proposes that the fuzzy-like bottom side of the magnolia leaf helps it overcome this problem:

ME: "Yeah, there's not really a lot about the brown on the bottom."

BIO: "What I'm thinking is, uh, you- you just mentioned that kinda good for the umbrella"

ME: "Yeah"

BIO: "Yeah, cause uh, yeah I agree with you. Cause uh, all the umbrellas I think I've kinda experienced that if the water pours out then the water kinda, kinda getting into like underneath of the umbrella."

CS: "Huh"

BIO: "But if it's [[inaudible]] stuff then that kinda prevents the water going underneath, so I think that's [good]"

Thus, our study confirms that analogies play multiple roles in biologically inspired design, including design idea generation, design concept evaluation and design concept explanation. In addition, our study shows that analogies in biologically inspired design also play a role in explaining novel domain concepts. Consistent with Christensen & Schunn [19], we posit that the use of analogies for domain concept explanation is more likely to occur in interdisciplinary design teams because the team members are more likely to begin with different concepts and models of the world.

5   Cognitive Dissonance and Perspective Shifts

The articulation of an analogy by one member of a design team typically results in transfer of the knowledge to other team members and assimilation of the knowledge in the receivers' mental model. We will call this Transfer Analogy. Transfer Analogy appears to be a natural extension of current theories of analogy [e.g., 7]. However, in design concept evaluation, the goal of analogy appears to help a teammate look at the current issue under discussion from a different perspective. We will call this type of analogy Change Analogy. We hypothesize that a teammate may use a Change Analogy as a means of persuading other teammates to view the current problem from a different perspective when there is a cognitive dissonance on a design team.

5.1   Method

We discovered Change Analogy through analysis (conducted after the analysis presented above) of a discussion by the same team in the same Shark Attack project as before. We used the audio file of the discussion and a partial transcript of that audio file in various stages of our method. We analyzed our data in three steps: (1) identify instances of analogical reasoning; (2) for each instance of analogical reasoning that we chose to investigate, understand what changes in knowledge occurred in the speaker of that instance before the occurrence of the instance; and (3) determine if there are any relationships between steps 1 and 2.

We pursued these goals through a variety of techniques. For step 1, we first listened to the audio recording to identify all possible instances of articulated analogies. Then, we selected a specific instance of articulated analogy for further analysis. We coded this instance of articulated analogy (and related passages) using a scheme inspired and informed by Richland et al.'s [25] method for identifying analogies in a discourse. We used these codes to decide whether to accept or reject the instance as a true analogy. For step 2, we used a coding scheme that characterized contents of passages in the transcript in terms of attributes and relationships. Each passage could have zero to many such codes associated with it. We primarily coded only those passages spoken by the speaker of the selected instance of articulated analogy and those passages spoken by others that we felt were important to understanding that speaker's passages. Finally, for step 3, we reviewed our codes, looking for changes in the knowledge of the speaker who articulated the selected instance of analogy. We surmised that the changes in knowledge that occur before the selected instance of articulated analogy might be related to that instance.

5.2   Preliminary Results

Table 2 and Table 3 display sections of the transcript. We use the same speaker identifiers in Table 2 and Table 3 as in Table 1. The discussion in Table 2 occurred before the discussion in Table 3. Table 2 shows an articulation of knowledge by CS (Table 2 Passage Numbers 6 and 8) that we identified as meaningful and some discussion around it. We believe that this articulation of knowledge is at least a contributing factor to IE's identification of a cognitive dissonance. Table 3 shows the analogy that IE articulated. We note that some passages occurred between these two transcript sections.

In both tables, we divide our passages by turns of speaking. We will describe Table 3's columns. From left to right, the columns show the following: a uniquely identifying number for the passage (set to a table-relative number for this paper), the speaker of the passage, what was spoken for the passage, and any analogy-related codes that were created for the passage. Table 2's columns are a subset of Table 3's columns and have the same meaning.

The symbols used in the Spoken column in both tables were derived from Du Bois et al. [26]. <X-UNKNOWN-X> refers to indecipherable speech and that the transcriber could not distinguish the number of syllables in that speech. <X content X> refers to uncertain words or phrases, and <X content X><X2 other-content X2> displays possible alternatives when they exist. X: in the speaker column means an unknown speaker. ((?)) signifies that the transcriber felt the preceding text was a question. ((CONTENT)) is a transcriber comment. [content], [[content]], etc., refer to overlapping speech. We vertically align overlapping speech by the left-most bracket; the right-most brackets always end at the end of a word; and the number of brackets identifies the particular instance of overlapping speech.

The analogy in Table 3 is between the human-shark system and the decoy-shark system. We derived the relationships in both systems from our codes in Table 2 with the exception of Decoy makes Sound. However, we feel this relationship is appropriate because the team was discussing a decoy that would use sound to attract sharks. There is an error in the mapping identified in Table 3 P# 5. The human-shark system part should be "sharks don't get anything out of attacking/eating (?) humans". We put a (?) because we are not exactly sure what precisely IE meant here. Although he does not articulate all of his thought, we conjecture that IE transfers the "not desensitized to" relationship between Sharks and Sound.

Now consider our data as it relates to the process of resolving the cognitive dissonance between IE and CS on the team. IE received CS's articulation of knowledge that revealed to IE that his mental model was different from CS's mental model. At some later point, IE decided to reconcile this difference, i.e., to convince CS to change his mental model to be aligned with IE's mental model. IE attempted to do so by articulating an analogy. Thus, CS's articulation of knowledge provoked analogy making by IE. IE recognized a conflict with his knowledge and sought to fix it not by changing his own knowledge but by attempting to change the knowledge of the articulator (CS).

Table 2. Section of the Transcript Showing CS's Knowledge Articulation. (The Passage Numbers are relative to Table 2 only.)

P#	Speaker	Spoken
1	IE:	huh ((?)) I don't know <X for ((SOUNDS LIKE FUR)) X> but then if <X we're X> just focusing on sound
2	CS:	mhm
3	IE:	do we have to worry about the other two ((?)) like once we've successfully like attracted them or repelled them using sound
4	CS:	mhm
5	IE:	do we really have to worry about what they see or smell afterwards ((?))
6	CS:	we might if they get desensitized like if that if that question is true that it it'll attract them and then they'll go <X well X> there's nothing here and go somewhere else
7	IE:	<X mhm X> ((HARD TO HEAR. IT IS POSSIBLE THIS IS JUST A NOISE AND NOT IE))
8	CS:	then we might want to have in our solution talking about solutions something that uh you know creates smell or creates uh uh something that looks like a <X seal X> or something like[that] you know
9	IE:	[right]
10	IE:	<X yeah X> cause I was thinking like our the purpose of our solution is to like if you're talking about attracting them we're <X trying to attract them X> as far away as needed
11	CS:	mhm
12	IE:	so that they won't be attracted to human sound right ((?))

Table 3. Section of the Transcript Showing IE's Articulated Analogy. (The Passage Numbers are relative to Table 3 only.)

P#	Speaker	Spoken	Analogy Codes
1	IE:	the whole thing on like desensitizing the sharks I mean sharks don't eat humans you know	
2	CS:	right	
3	IE:	they're not their food source <X huma-X> we're not their food source so does that mean that they've already been I mean it's not like they've been de-desensitized to human sound right now anyways right ((?))	Relationship: (human-shark system) sharks don't get anything out of attack-ing/eating (?) humans Relationship: (human-shark system) sharks are not desen-sitized to sounds humans make
4	CS:	<X mhm X><X2 mmm X2>	
5	IE:	<X so X> do we really have to worry about them them being desensitized ((?)) since I mean humans [basically] are like decoys in a sense al[[ready]] in the fact that they don't get anything out of it	Source object: human-shark system Target object: decoy-shark system Relationship: (decoy-shark system) sharks don't get anything out of attack-ing/eating (?) decoys Mapping: (human-shark system) sharks don't get anything out of attack-ing/eating (?) LIKE (decoy-shark system) sharks don't get anything out of attack-ing/eating (?) decoys
6	X:	[<X-UNKNOWN-X>] ((SHOULD BE CS BUT THIS MAY JUST BE A NOISE))	
7	CS:	[[right]]	
8	CS:	right	
9	IE:	so if they haven't been desensitized already then	Transfer: (human-shark sys-tem) sharks are not desensi-tized to sounds humans make THEREFORE (decoy-shark system) sharks are not desen-sitized to sounds decoys make

6   Implications for Case-Based Technology

Our findings have several implications for developing case-based technology to aiding biologically inspired design in particular, and interdisciplinary collaborative design in general. The current generation of case-based tools for aiding biologically inspired design – such as AskNature, IDEA-INSPIRE, DANE, etc. – were designed only for the task of design idea generation. This provokes two questions: (1) To what extent do current case-based tools already serendipitously support other tasks? (2) How may one adapt current case-based technologies to address another task?

We will consider these two questions relative to DANE [17] and Biologue [18]. We briefly described both tools in Section 2: While DANE provides access to a library of conceptual models of biological and technological designs, Biologue provides access to a library of biology articles that contain descriptions of biological systems in natural language. Biologue articles too are annotated with conceptual models, but they are skeletal compared to the detailed models in DANE.

First, let us consider the tasks of design concept explanation and domain concept explanation. We posit that DANE's conceptual models already serendipitously support the task of design concept explanation and also provide partial support for the domain concept explanation task, so one may not need to adapt DANE for it to address these tasks. However, Biologue's skeletal models may be less effective for these tasks, and so it may benefit from adaptation. We conjecture that we could use DANE-like conceptual models to add the two functionalities to Biologue; that is, we could annotate the articles in Biologue with the DANE-like conceptual models of the biological systems and concepts described in the articles. In fact, we are currently developing an AI agent that can automatically derive DANE-like conceptual models for the biological design described in Biologue's articles.

Now let us consider how we may adapt DANE for the task of design concept evaluation, a task it does not serendipitously support. At present, a user interacts with DANE by browsing models or creating/editing models. Let us suppose that a user could also input a design problem and associate it with a conceptual model of the proposed design. We conjecture that an AI agent could (a) use DANE's schema for checking the structure of the conceptual model of the proposed design for consistency, and (b) use DANE's other conceptual models for checking the contents of the conceptual model of the proposed design for correctness.

6.1   Case-Based Techniques for Addressing Cognitive Dissonance

The problem of developing a case-based technique for reducing cognitive dissonance in collaborative interdisciplinary design is more complicated as well as more subtle. Cognitive dissonance on a design team may be healthy (at least to some degree) because it reveals that the team members have different mental models and starts the process of aligning their mental models with one another's. Thus, instead of reducing the occurrence of cognitive dissonance, we want to focus on how to address cognitive dissonance once it is manifested in a design team. Given our finding about analogy as a cognitive strategy for addressing cognitive dissonance, we want to make analogies more effective so that they may more readily help persuade team members to shift their perspectives and align their mental models. But this raises another question: Why are some analogies not effective? This question likely has several answers, but one answer surely is that some analogies are just not very good. This suggests a way in which case-based techniques could help address cognitive dissonance: effective analogy evaluation. It is noteworthy that while there has been a significant amount of work on analogy generation, there has been relatively little research explicitly on analogy evaluation.

Yen et al. [22] found that designers struggle with (a) articulating why a biological analogue is appropriate for a given problem and (b) consistently explaining why one biological analogue is better than another. To support designers' evaluation of biological analogues, our laboratory has developed a simple tool called a T-Chart that enables designers to compare the similarities and differences between a design problem and a biological analogue along multiple dimensions such as function, operating environment, constraints, and performance criteria [27]. Preliminary evidence indicates that T-Charts help designers better evaluate the appropriateness of biological analogues to design problems.

7   Future Work

The current work raises several questions for exploration. The first set of questions pertains to the generalization of our findings. What other empirical evidence exists for cognitive dissonance in social aggregates of humans such as teams, organizations and nations? How can we collect this evidence? Similarly, what other empirical evidence exists for the use of analogies to reduce cognitive dissonance in cognitive agents at various levels of social aggregation from individual humans to nations? How can we collect this evidence?

The second set of questions relates to case-based techniques and tools for addressing cognitive dissonance once it is manifested, and in particular, for effective analogy evaluation. We are presently developing a new technique for automatic evaluation of analogies in biologically inspired design, including the conceptual design that results from such an analogy [28]. This technique uses multiple strategies for analogy evaluation such as requirement checking, model checking, and qualitative analysis and simulation. As part of its evaluation, this technique will generate justifications for why an analogy is good or bad. We hypothesize that this kind of analogy evaluation resulting in justifications could potentially help address cognitive dissonance on a team because it should lead to better and more persuasive analogies.

8   Conclusions

In this paper, we reported on an empirical study of collaborative, interdisciplinary design. It is important to ground the development of case-based theories, techniques and tools in empirical studies of human behavior for at least two reasons. Firstly, empirical grounding of case-based tools increases the likelihood that humans will actually use them. Secondly, observations of human behavior sometimes lead to new inspirations for theory construction.

Our observations of human practices in biologically inspired design seem to present four kinds of opportunities for constructing theories of case-based reasoning. Firstly, our work suggests that analogies are useful for the tasks of design idea generation, design concept evaluation, design concept explanation, and domain concept explanation. Other studies indicate that analogies are also useful for problem formulation as well as problem reformulation. These findings raise questions about how can we repurpose existing case-based tools or build new case-based tools for these design tasks.

Secondly, our empirical studies indicate that cognitive dissonance occurs not only in individual humans as Festinger originally postulated, but also in interdisciplinary design teams. The degree to which this finding can be generalized to other human activities or other human social aggregates is presently unclear. However, it seems clear that yet another role of analogy is to reduce cognitive dissonance in human teams.

Thirdly, and more specifically, analogy is not only a mechanism for transfer of knowledge from a familiar situation to a new situation, but also a cognitive strategy for reducing cognitive dissonance on a team. In particular, analogy is also a strategy for inducing shifts in the perspective of a teammate and alignment of mental models when the mental models of teammates are not well aligned. This raises the hard question of how we can develop techniques for persuading a teammate to shift perspectives, align mental models, and thereby reduce cognitive dissonance on interdisciplinary design teams once the dissonance manifests itself. We expect that articulation of stronger, well-justified analogies would be more persuasive in inducing teammates to change perspectives. Thus, we are exploring a technique that relies on analogy evaluation: it critiques analogies and generates justifications for why they are good or bad.

Finally, and more generally, analogy is not only an internal cognitive process, but it is also situated in the external physical, information and social worlds of cognitive agents. Nersessian & Chandrasekharan [29] found that some analogies are situated in the world. Kokinov & Petrov [30] and Kulinski & Gero [31] found that analogy construction depends in part on external representations of artifacts. At last year's ICCBR conference, we presented a paper that showed that analogies in biologically inspired design typically are situated online [18]: Given a design problem, design teams typically find relevant biological analogies on the Web instead of retrieving them from their long-term memories. In the present study, we found that analogies often are situated in teamwork. This raises the deep question of how can we build new theories of analogy that take into account the affordances and constraints of the physical, information and social worlds in addition to the internal mental worlds of cognitive agents.

Acknowledgments. We are grateful to the instructors and students of the 2010 class of the Georgia Tech ME/ISyE/MSE/BME/BIOL 4740 course, especially Professor Jeannette Yen and Wiltgen's teammates for permission and support for conducting this study. We thank Cristina Weiler for her help in transcribing the audio files, and David Majerich, Swaroop Vattam, and Michael Helms for many discussions about this work.   We are grateful to the US National Science Foundation for its support of this research through a CreativeIT Grant (#0855916) titled "Computational Tools for Enhancing Creativity in Biologically Inspired Design."   We note that in [32] we previously reported on the four functional roles described in Section 4.   Since that paper, we have revised our characterization of the roles.   Finally, we thank the anonymous reviewers of this paper for their helpful comments.

References

1. Festinger, L.: A Theory of Cognitive Dissonance. Stanford University Press (1957)
2. Sylvan, D., Goel, A., Chandrasekaran, B.: Analyzing Political Decision Making from an Information Processing Perspective: JESSE. American Journal of Political Science 34(1), 74–123 (1990)
3. Spitas, C., Badke-Schaub, P.: Transdisciplinary Design and Human Cognition: Exploring Cognitive Dissonance. In: Proc. International Workshop on the Future of Transdisciplinary Design, Luxembourg (June 2013)
4. Hofstadter, D.: Analogy as the core of cognition. In: Gentner, D., Holyoak, K., Kokinov, B. (eds.) The Analogical Mind: Perspectives from Cognitive Science, pp. 499–538 (2001)
5. Hofstadter, D., Sander, E.: Surfaces and Essences: Analogy as the Fuel and Fire of Thinking. Basic Books (2013)
6. Hofstadter, D. (ed.): Fluid concepts & creative analogies: Computer models of the fundamental mechanisms of thought. Basic Books, New York (1995)
7. Holyoak, K., Thagard, P.: Mental Leaps: Analogy in Creative Thought. MIT Press, Cambridge (1995)
8. Vattam, S., Helms, M., Goel, A.: A Content Account of Creative Analogies in Biologically Inspired Design. AIEDAM 24, 467–481 (2010)
9. Benyus, J.: Biomimicry: Innovation Inspired by Nature. William Morrow (1997)
10. French, M.: Invention and evolution: design in nature and engineering, 2nd edn. Cambridge University Press (1994)
11. Vincent, J., Man, D.: Systematic Technology Transfer from Biology to Engineering. Philosophical Transactions of the Royal Society A: Mathematical, Physical and Engineering Sciences 360(1791), 159–173 (2002)
12. Fish, F., Battle, J.: Hydrodynamic design of the humpback whale flipper. Journal of Morphology 225, 51–60 (1995)
13. Fish, F., Weber, P., Murray, M., Howle, L.: The tubercles on humpback whales' flippers: application of bio-inspired technology. Integrative and Comparative Biology 51(1), 203–213 (2011)
14. Biomimicry 3.8 Institute (2008). AskNature, http://www.asknature.org/ (last retrieved on May 29, 2013)
15. Chakrabarti, A., Sarkar, P., Leelavathamma, B., Nataraju, B.: A functional representation for aiding biomimetic and artificial inspiration of new ideas. AIEDAM 19, 113–132 (2005)
16. Vincent, J., Bogatyreva, O., Bogatyrev, N., Bowyer, A., Pahl, A.: Biomimetics: its practice and theory. Journal of the Royal Society Interface 3, 471–482 (2006)
17. Goel, A., Vattam, S., Wiltgen, B., Helms, M.: Cognitive, collaborative, conceptual and creative - Four characteristics of the next generation of knowledge-based CAD systems: A study in biologically inspired design. Computer-Aided Design 44(10), 879–900 (2012)

18. Vattam, S.S., Goel, A.K.: Biological Solutions for Engineering Problems: A Study in Cross-Domain Textual Case-Based Reasoning. In: Delany, S.J., Ontañón, S. (eds.) ICCBR 2013. LNCS, vol. 7969, pp. 343–357. Springer, Heidelberg (2013)
19. Christensen, B., Schunn, C.: The relationship of analogical distance to analogical function: The case of engineering design. Memory and Cognition 35(1), 29–39 (2007)
20. Helms, M., Vattam, S., Goel, A.: Biologically Inspired Design: Process and Products. Design Studies 30(5), 606–622 (2009)
21. Helms, M., Goel, A.: Analogical Problem Evolution in Biologically Inspired Design. In: Proc. Fifth International Conference on Design Computing and Cognition, College Station, Texas. Springer (July 2012)
22. Yen, J., Helms, M., Goel, A., Tovey, C., Weissburg, M.: Adaptive Evolution of Teaching Practices in Biologically Inspired Design. In: Goel, A., McAdams, D., Stone, R. (eds.) Biologically Inspired Design: Computational Methods and Tools, pp. 153–199. Springer, London (2014)
23. Ritzmann, R.: Mechanisms for the Snapping Behavior of Two Alpheid Shrimp. Alpheus Californiensis and Alpheus Heterochelis. Journal of Comparative Physiology 95, 217–236 (1974)
24. Versluis, M., Schmitz, B., von der Heydt, A., Lohse, D.: How Snapping Shrimps Snap: Through Captivating Bubbles. Science 289(5487), 2114–2117 (2000)
25. Richland, L., Holyoak, K., Stigler, J.: Analogy Use in Eighth-Grade Mathematics Classrooms. Cognition and Instruction 22(1), 37–60 (2004)
26. Du Bois, J., Schuetze-Coburn, S., Cumming, S., Paolino, D.: Outline of Discourse Transcription. In: Edwards, J., Lampert, M. (eds.) Talking Data: Transcription and Coding in Discourse Research. Lawrence Erlbaum Associates, Inc. (1993)
27. Helms, M., Goel, A.: The Four Box Method: Problem Specification and Analogue Evaluation in Biologically Inspired Design. To appear in ASME Journal of Mechanical Design (2014)
28. Wiltgen, B.: Interactive Analogy Evaluation. Ph.D. Dissertation Proposal, School of Interactive Computing, Georgia Institute of Technology, Atlanta, USA (2014)
29. Nersessian, N., Chandrasekharan, S.: Hybrid Analogies in Conceptual Innovation in Science. Cognitive Systems Research 10, 178–188 (2009)
30. Kokinov, B., Petrov, A.: Integrating Memory and Reasoning in Analogy Making: The AMBR model. In: Gentner, D., Holyoak, K., Kokinov, B. (eds.) The Analogical Mind: Perspectives from Cognitive Science. MIT Press (2000)
31. Kulinski, J., Gero, J.: Constructive Representation in Situated Analogy in Design. In: Procs. CAAD Futures. Springer (2001)
32. Wiltgen, B., Goel, A.: Case-Based Reasoning All Over the Place: The Multiple Roles of CBR in Biologically Inspired Design. In: Procs. Workshop of Human-Centered and Cognitive Approaches to Case-Based Reasoning at the 19th International Conference on Case Based Reasoning, Greenwich, London, UK (September 12, 2011)

On Retention of Adaptation Rules

Vahid Jalali and David Leake

School of Informatics and Computing, Indiana University
Bloomington IN 47408, USA
{vjalalib,leake}@cs.indiana.edu

Abstract. The difficulty of acquiring case adaptation knowledge is a classic problem for case-based reasoning (CBR). One method for addressing this problem is to use the cases in the case base as data from which to learn adaptation rules. For numeric prediction tasks, adaptation rules have been successfully learned from the case base by using the *case difference heuristic,* which generates rules based on comparisons of pairs of cases. However, because the case difference heuristic could potentially generate a rule for each pair of cases in the case base, controlling growth of adaptation rules is potentially an even more acute problem than controlling case base growth. This raises the question of how to select adaptation rules to retain. The ability to generate adaptation rules from cases also raises questions about the relative benefit of learning cases, learning the adaptation rules generated from them, or learning both. This paper proposes and evaluates a new adaptation rule retention approach and presents a case study assessing the relative benefits of learning cases versus learning adaptation rules derived from the cases, at different points in the growth of the case base.

Keywords: case adaptation learning, case-base maintenance, knowledge containers, rule retention.

1 Introduction

Case adaptation is a classic challenge for case-based reasoning. Because acquiring adaptation knowledge by hand may be difficult or expensive, much research has explored machine learning methods for generating case adaptation knowledge automatically (e.g., [1–11]). The case difference heuristic approach, first proposed by Hanney and Keane [3], is a popular approach to deriving adaptation rules from cases used for numeric prediction (regression) tasks [1, 5, 7, 8, 11], by comparing pairs of cases in the case base. For each pair of cases, it assesses the problem–problem differences between the problems solved by the cases, and also assesses their solution–solution differences. These two differences are used as the basis for generating the antecedent and consequent of a new adaptation rule. The new rule applies to a retrieved case if the retrieved case and new problem have similar problem–problem differences, and it adjusts the solution of the retrieved case according to the previous solution–solution difference. The case difference heuristic approach has received much study, but two fundamental questions it raises have received little attention: How to control growth of the set of adaptation rules, and how the benefit of learning new cases compares to the benefit of learning rules derived from those cases.

L. Lamontagne and E. Plaza (Eds.): ICCBR 2014, LNCS 8765, pp. 200–214, 2014.
© Springer International Publishing Switzerland 2014

The utility issues associated with case base growth are widely recognized in the CBR community, and methods for controlling case base growth are an important CBR research area (e.g., [12]). The utility issues associated with growth of automatically-generated adaptation knowledge may be even more acute. When the case difference heuristic is used to generate cases from the case base, each ordered pair of cases may result in an adaptation rule. For a case base with n cases, the number of possible ordered pairs of cases is $2\binom{n}{2}$, so as cases are added the number of candidate rules grows much more rapidly than the case base. The need to select which rules to retain was noted when the case difference heuristic was first proposed, but since then the problem has received little attention. This paper proposes a new rule retention approach and demonstrates its effectiveness compared to alternative methods.

The ability to use cases as a source for rules also raises an interesting question about the relative benefits of learning cases and/or learning adaptation rules derived from those cases. It is well known that CBR enables system developers to strategically place knowledge in different *knowledge containers*, and that knowledge in one container may compensate for lack of knowledge in another [13]. However, the question of the relative benefit of adding cases directly or applying the case difference heuristic to existing cases, to add rules, is unexplored. This paper provides a case study as a first step towards addressing these questions.

The paper proposes a general strategy for rule retention, based on a test process for credit/blame assignment to automatically-generated adaptation rules. This strategy can be used for any CBR task in which there is a trade-off between retaining adaptation versus case knowledge, regardless of how the adaptation knowledge is generated. The paper presents an evaluation of the strategy in the context of CBR for numerical prediction, a task which has been widely studied for domains such as property value estimation, using the knowledge light "case difference heuristic" method to generate the adaptation rules. Experimental results demonstrate the retention strategy's effectiveness compared to baselines and a previously-proposed frequency-based approach [4]. It also provides an initial experimental exploration of how the knowledge content contribution of adding rules varies at different points in the growth of the case base, illustrating that the knowledge contribution of adaptation rules generated using the case difference heuristic may converge before the knowledge contributions from simply adding new cases, which suggests a knowledge growth strategy of retaining a more limited set of learned adaptations but continuing to add new cases.

2   Related Work

Our research on adaptation rule retention falls within the broad category of *case-based reasoning system maintenance* (e.g., [14]). Case-based reasoning system maintenance extends maintenance considerations beyond the case base, to consider other knowledge containers as well. The work in this paper is motivated by research on adaptation knowledge generation and contributes to the study of adaptation rule ranking and retention, and of the relationship between knowledge in different case-based reasoning knowledge containers. This section briefly highlights relevant work in each of these areas.

Adaptation Knowledge Generation: The CBR literature contains numerous examples of automated adaptation knowledge generation. As a few examples, Craw et al. [15] and Shiu et al. [10] propose building rules by decision tree learning; Leake, Kinley and Wilson [6], Leake and Powell [16], and Minor, Bergmann, and Görg [17] propose capturing cases from prior adaptations; numerous projects have investigated the case difference heuristic for adaptation rule generation (e.g. [3, 5, 7, 8, 11]). D'Aquin et al. [1] combine the case difference approach with data mining methods to improve rule quality and Fuchs et al. [18] present a framework for differential adaptation. In principle, methods from the extensive CBR literature on case-base maintenance could be applied to control retention of adaptation knowledge in the form of cases. However, for systems which learn adaptation rules, methods are needed to guide rule retention.

Filtering and Ranking Adaptation Rules: Li et al. [19] propose filtering learned adaptation rules by removing duplicate rules and merging rules which conflict (*i.e.*, rules which have the same antecedents but different consequents). For rules with numeric consequents, their approach aggregates rules into new rules whose consequents are the average of those of the previous rules; for rules with symbolic features, their approach clusters rules with the same antecedent and ranks them by frequency. This leaves open the question of how to prioritize distinct rules for retention.

Leake and Dial [20] propose selecting the rules to apply to a particular adaptation based on using provenance information to assign blame to the adaptation rules involved in errors, to assess rule quality. They assume that all rules are always present; their focus is only on rule selection from the complete pool to maximize accuracy for a particular problem, rather than determining which rules to retain for best overall performance on future problems.

Hanney and Keane's [3] seminal work on the case difference heuristic proposes retaining the adaptation rules that are generated more frequently by the pair-wise comparison of cases. However, to our knowledge, this maintenance strategy has not been formally evaluated; this paper evaluates it in comparison to our new approach.

We have proposed Adaptation-Guided Case Base Maintenance (AGCBM) [21], a method for rule and case retention aimed at simultaneously controlling the number of cases and adaptation rules generated from them. AGCBM considers the contribution of each case both as (1) a source case to provide initial estimates for input queries, and (2) as a building block for adaptation rules. It selects cases to retain based on a ranking scheme considering both types of contributions. The work presented in this paper differs in three ways. First, AGCBM is computationally expensive ($O(n^3)$ in the initial size of the case base); this paper seeks an approach feasible to apply to large case bases. Second, the approach in this paper focuses solely on rule retention, aiming to minimize the adaptation knowledge container size given a fixed case base. Third, in contrast to AGCBM, and for added efficiency, the method introduced in this paper uses single adaptation rules rather than ensembles of adaptation rules. As discussed in Section 4.4, these changes make the training of the system substantially faster than AGCBM, enabling it to be applied to large case bases which would not be feasible for AGCBM.

Relationships Between CBR Knowledge Containers: Richter [13] observed that the knowledge of CBR systems resides in multiple "knowledge containers:" the case vocabulary, similarity measures, solution transformation knowledge (i.e. adaptation knowledge) and the case base, and that the knowledge containers are overlapping: added knowledge in one may reduce the need for knowledge in another (e.g., adding cases may decrease the need for adaptation knowledge). However, there has been limited research on how learning in different containers interacts.

Leake, Kinley, and Wilson [6] demonstrate that knowledge container interactions during learning may be important, by showing that uncoordinated additions to case and adaptation knowledge may interact negatively, degrading system performance even if adding the same knowledge to one container individually, with the knowledge of the other fixed, would improve performance. The benefits of both types of learning were restored when similarity knowledge was learned as well. Shiu et al. [10] propose that adaptation rules generated from cases can be used to transfer knowledge from the case base container to the adaptation knowledge container, to compact the case base, and demonstrate the approach for adaptation rules learned as decision trees.

The ability to generate adaptation rules by the case difference heuristic raises the question of when knowledge should be retained in the form of cases, when it should be retained in the form of adaptation rules derived from these cases, and when it should be retained in both forms. This paper presents a case study on a facet of this question.

Case Knowledge Maintenance: The CBR community has investigated many methods for retaining/discarding cases, as *case base maintenance*. One prominent trend of case base maintenance is the use of *footprint-based* case base maintenance approaches, originating in work by Smyth and Keane [22]. Such approaches guide maintenance according to Reachability, Coverage and Relative Coverage, which respectively refer to the set of cases that can solve a particular case, the set of cases that can be solved by a particular case, and the set of other cases in the case base that can solve cases in the coverage set of a particular case. Footprint-based methods favor retaining cases with strong competence contributions not duplicated by other cases. We will not further survey the extensive case-base maintenance literature here, but will apply footprint-based methods as our baseline case base maintenance method for a comparative study of effects of rule set and case base growth, in Section 5.2.

3   ARR: A General Approach to Adaptation Rule Retention

As the basis for our approach to adaptation rule retention, we first present a simple framework, ARR (Adaptation Rule Retainer), aimed at automatically generating a compact set of accurate case adaptation rules. ARR involves three steps:

1. Generate a preliminary subset S of possible adaptation rules, based on a given rule generator. The size of the subset may be limited, e.g., to generate x rules.
2. Do leave-one-out testing of the CBR system with S as its adaptation rule set, applying the system to the problem parts of a subset of the cases in the case base. For each test, assess error, assign credit/blame for the error to the adaptation rule used to generate the solution, and retain error data. To enable generating data about multiple rules from each trial, multiple solutions may be generated for each problem, one for each of the k nearest cases to the test problem, for some fixed value of k.

Algorithm 1. ARR's adaptation rule scoring algorithm

Input:
x: number of candidate adaptation rules to generate
k: number of source cases to adapt to solve each query (each generates a different solution)
CB: case base
Output: Scored adaptation rules

$\quad AdaptationRules \leftarrow$ RuleGenerator(CB,x)
$\quad$**for** r in $AdaptationRules$ **do**
$\quad\quad r.score \leftarrow 0$
$\quad$**end for**
$\quad$**for** c in CB **do**
$\quad\quad$**for** i in 1 to k **do**
$\quad\quad\quad RuleToApply \leftarrow$ SelectRule($AdaptationRules$,$Neighbor_i(c)$,$c.problem$)
$\quad\quad\quad ValEst(c) \leftarrow$ Adjust($Neighbor_i(c)$, RuleToApply)
$\quad\quad\quad EstErr(c.problem, Neighbor_i(c), r) \leftarrow |c.value$ - $ValEst(c.problem)|$
$\quad\quad$**end for**
$\quad$**end for**
$\quad$**for** r in $AdaptationRules$ **do**
$\quad\quad r.score \leftarrow ErrorScore(EstErr, r)$
$\quad$**end for**
$\quad$return $AdaptationRules$

---

3. Rank the rules according to a scoring function applied to their error data and retain y highest-ranked rules, for some user-selected y.

The parameters x and k adjust the amount of data considered in the generation process. The parameter y adjusts the final number of rules retained.

Alg. 1 presents ARR's method in more detail. $Neighbor_i(c)$ denotes the ith closest neighbor of c in the case base. $RuleGenerator$ is a method for generating a subset of the possible candidate adaptation rules (e.g., $RuleGenerator$ could apply the case difference heuristic to a desired number of the possible case pairs, randomly selected without replacement). $Adjust(case, rule)$ applies an adaptation rule to a case. $c.problem$ is the problem part of a case, and $c.value$ its stored solution value. $ErrorScore$: $[0, \infty) \rightarrow [0, \infty)$ is a function that maps raw error values to a transformed value reflecting domain-specific error importance characteristics (e.g., $ErrorScore$ could assign the same scores to different error levels if the differences are deemed inconsequential).

4   ARR1: An Instantiation of the ARR Approach

ARR is a general framework. For experimental tests, it is necessary to instantiate the framework with specific choices for the rule generation procedure, rule selection, and error scoring. ARR1 is an instantiation of ARR in which $RuleGenerator$ applies a specific form of the case difference heuristic, $SelectRule$ is based on the rule selection process developed in our previous work on CAAR (Context-Aware Adaptation Retrieval) [23], and $ErrorScore$ is designed to balance observed accuracy of each rule by the amount of evidence acquired about that rule during testing. These functions are described in the following sections.

4.1 ARR1's Rule Generator

ARR1 generates adaptation rules using the case difference heuristic approach proposed by Hanney and Keane [4] and further explored by others (e.g., [5, 7, 8, 11]). The case difference heuristic approach builds new adaptation rules by comparing problem parts and respectively solution parts of cases in the case base. Each rule maps the observed differences in the problem descriptions of a pair of cases to the observed difference in their solutions. For example, in apartment rental domain, if two apartments differ in that one has an additional bedroom, and its price is higher, the case difference heuristic could generate a rule which would increase the rent estimated for a new apartment when estimating based on a previous apartment with one bedroom fewer.

Applying the case difference approach depends on addressing questions such as which pairs of cases will be used to generate adaptation rules, what function will be derived from a given difference between the values estimated by the two cases (e.g., in the rental domain, a $100 difference could prompt a rule to adjust the price by $100, to adjust the price by the same percentage difference reflected by the $100, or any of many other alternatives), and how to select the rule to be applied to a given new problem.

ARR1 is designed for cases whose problem parts are described by numeric feature vectors. ARR1's rule generator generates rules whose antecedents are the vector difference between the problem parts of the two cases from which the rule was generated, and whose consequents are the numerical difference of those two cases' solution values. However, we note that nothing about ARR or the basic ARR1 approach precludes applying alternative methods.

The number of rules generated by ARR1's $RuleGenerator$ is determined by a user-selected parameter k, a small fixed value which determines the number of neighbor cases to which each source case should be compared to generate rules. For each source case s in the selected source cases to consider, if $\{c_i\}_{i=1,...k}$ is the set of the k nearest cases to s, ARR1 generates one adaptation rule to adapt s to each c_i. For a case base with n cases, if each case is compared with its top k neighbors, $n \times k$ adaptation rules are generated.

4.2 ARR1's Value Estimation

ARR1's rule scoring is based on the errors in estimated values resulting from applying the rules. Given a query, ARR1 generates an estimated value by adapting the k nearest cases (for a fixed value of k) and averaging the adapted values, following Alg. 2. $NeighborhoodSelector$ is a function for selecting cases in the neighborhood to adapt. ARR's $SelectRule$ function applies the context-based method for adaptation rule selection developed in our previous work on CAAR [23] (Context-Aware Adaptation Retrieval), which we selected because in previous evaluations it outperformed alternative methods. The specific rule selection procedure used is not significant to the lessons of our experiments, so for reasons on space, we do not describe the rule selection

Algorithm 2. ARR1's Algorithm for Value Estimation

Input:
Rules: input adaptation rules Q: input query
k: number of source cases to adapt to solve query
CB: case base
Output: Estimated solution value for Q

$CasesToAdapt \leftarrow$ NeighborhoodSelector(Q,k,CB)
for c in $CasesToAdapt$ **do**
    RuleToApply $\leftarrow$ SelectRule($Rules,c,Q$)
    $ValEst(c) \leftarrow$ Adjust($c, RuleToApply$)
end for
return Average$_{c \in CasesToAdapt} ValEst(c)$

procedure further here. However, we refer interested readers to previous work on CAAR for the details [23].

4.3 ARR1's Rule Scoring

ARR ranks adaptation rules based on an $ErrorScore$ function assigning blame/credit scores to adaptation rules, based on the error that results when they are used to adapt source case values. ARR1's $ErrorScore$ function is designed to favor rules believed to have low error from more extensive testing. Depending on the problems used in the testing phase, some rules may be used multiple times, while others may be used seldom or never, giving less information for predicting their performance. ARR1's $ErrorScore$ favors rules which showed reasonable accuracy for multiple training cases over a rule which was only applied in a single trial even if it showed excellent accuracy there; the rationale is the expectation that results of multiple tests will be more reliable predictors of future performance.

Specifically, ARR1's $ErrorScore$ function is defined as follows. Let $EstErr$ be an array of estimated error values, with $EstErr(q, c, r)$ the calculated error when case c was adapted by adaptation rule r to solve query q. If M represents the number of times r is applied to different source cases (c_i's) during ARR1's error estimation phase, and q_i and c_i, for $i = 1, ..., M$ represent the queries and source cases used in those applications, ARR1's error score is calculated as follows:

$$ErrorScore(EstErr, r) = \sum_{i=1}^{M} \frac{1}{EstErr(q_i, Neighbor_i(c), r) + \epsilon} \tag{1}$$

where ϵ is a small positive value that determines a non-zero minimum value for perfect predictions. Rules are ranked by their $ErrorScore$ values, in order of ascending $ErrorScore$. Ties are broken arbitrarily.

4.4   Time Complexity of ARR1

Given that ARR1 is aimed at reducing potentially large rule sets, its time complexity is an important consideration. As explained previously, for a case base of n cases and for ARR1's adaptation rule generation method, which considers a neighborhood of m cases (for some small fixed m) around each case in the case base, ARR generates $n \times m$ rules, so the time complexity of its adaptation generation process is $O(n)$.

In Alg. 1, leave-one-out testing used for rule scoring will adapt $n \times k$ source cases, for k a small fixed value. For each adaptation, the most relevant adaptation rule must be selected. ARR1 does this by simply comparing the difference between the pair (input query, source case) with all generated adaptation rules to select the adaptation to apply, so its time complexity for the entire training process will be $O(n^2)$. (We assume that k is set to a small value that is significantly less than n.) However, more efficient rule retrieval methods could reduce rule retrieval complexity.

In addition, we note that the training process is not a routine process which means it could only happen once for the life of the system, and that leave-one-out testing could be replaced by sampling methods which could decrease the processing resources required for ARR1.

5   Experiments

Our experiments address four questions:

1. Performance of ARR1: How does final accuracy compare using (1) adaptation rule retention by ARR1, (2) frequency-based rule retention, (3) random rule retention, and (4) k-NN using value averaging rather than adaptation rules?
2. Performance of $ErrorScore$: How does accuracy using a reduced rule set selected based on ARR1's $ErrorScore$ function compare to accuracy using a reduced rule set selected based on simple averaging of errors?
3. Sensitivity to number of source cases adapted per problem during training: How does the number of cases adapted per problem during training affect final system accuracy?
4. Relative benefit of rule and case learning: How does the knowledge content contribution of adding rules and cases compare, at different points in the knowledge acquisition process?

5.1   Experimental Design

We evaluated ARR1's performance on four sample domains from the UCI repository [24]: Automobile (Auto), Auto MPG (MPG), Housing, and Computer Hardware (Hardware). All records and features were used for Housing (506) and Hardware (209). Auto and MPG contained some records with unknown feature values, which were removed (46 out of 205 for Auto and 6 out of 398 for MPG). However, we note that the value imputation methods used to enable k-NN to handle missing features could equally well be applied to ARR1 to enable it to handle such records. CAAR, part of ARR1's rule selection process, uses locally weighted linear regression for defining context, so requires numeric features. All features of MPG and Housing were numeric, but 10 non-numeric features were removed from Auto and 2 from Hardware, in order for all methods to be provided with the same amount of information. We note that the numeric features are not required by the general ARR method. For each feature, values were standardized by subtracting that feature's mean value from each individual feature value and dividing the result by the standard deviation of that feature. For the Auto, MPG, Housing and Hardware domains, the respective values to estimate are price, MPG (automobile fuel efficiency in miles per gallon), MEDV (median value of owner-occupied homes in $1000's), and PRP (published relative performance).

For all experiments, the set of adaptation rules was generated before query processing, following the ARR process of Alg. 1, and values were estimated following Alg. 2. The value of k, the number of cases to adapt, was set independently for each algorithm to maximize its performance. In Alg. 1, we expect it to be desirable for k to be set to a higher value, so that more adaptations have the chance to participate in the case value estimation process, resulting (on average) in more data being available on the performance of each adaptation rule. In Alg. 2, where k is used for estimating the input query value, we expect better accuracy for a smaller value of k (as well as reducing computation time), by focusing on more similar cases.

In our experiments, we tested a range of k values, 10, 20, 40, 80, 100 and 200 for Alg. 1, and 3, 8 and 13 for Alg. 2. Best performance was achieved for k in Alg. 1 set to 80, 100, 80, and 80 for the Auto, housing, MPG and hardware domains respectively, and in Alg. 2 to 3, 8, 8, and 8, for all domains in the same order. Same process with different k values (ranging from 1 to 10 for all domains) was used to determine the optimal k value for k-NN. The lower value of k for the auto domain compared to the other domains parallels the observation that even for k-NN, using lower numbers of nearest neighbors yields higher accuracy in the Auto domain, suggesting more locality for that domain compared to the other tested domains. The percent of possible rules to generate for Auto, MPG, Housing and Hardware domains was set to 8%, 3%, 4% and 8%, meaning that only a small portion of all possible rules was generated for each domain. For questions 1 and 2, performance was compared for varying numbers of rules retained from the set of possible rules, beginning at 20 rules and increasing to 1990 rules. All experiments used ten-fold cross validation. Adaptation rules were assigned scores by applying Alg. 1.

We compared ARR1 to two alternative retention methods. The first method, *Random*, replaces ARR1's ranking strategy with randomly selecting adaptation rules to retain. The second method, *Frequency-Based Rule Retention* (FBRR), follows Hanney and Keane's [4] proposed approach of retaining adaptation rules based on their frequency of occurrence. Because the case difference heuristic may generate multiple rules with extremely small differences, to have a reasonable indication of frequency we defined a threshold on the distance between the antecedents of the adaptation rules, and treated rules with differences below that threshold as identical. Because the antecedents of each rule are simply vectors of the numeric differences between the corresponding features of the two cases from which the rule was generated, we used Euclidean distance to measure rule similarity.

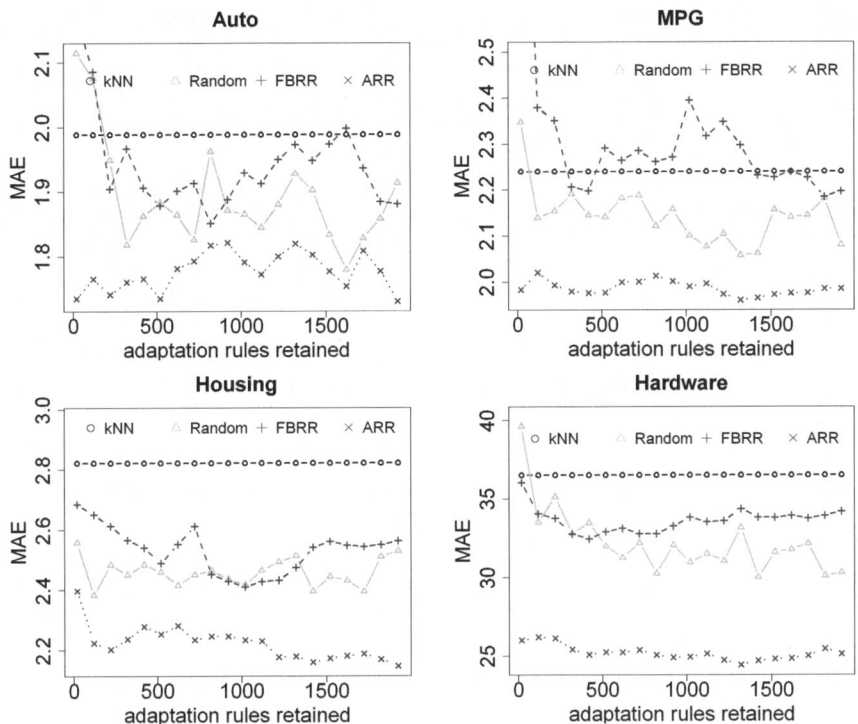

Fig. 1. MAE of the tested methods for different numbers of retained adaptations

5.2   Experimental Results

Adaptation Retention Evaluation: To address question 1, how rule retention by ARR1 affects final system accuracy compared to rule retention by other methods, we compared the Mean Absolute Error (MAE) of our test system using adaptation rules generated and retained by ARR1 to that of k-NN using value averaging rather than adaptation rules, random rule retention, and frequency-based rule retention. Figure 1 shows MAE of all methods in four domains.

In all domains, ARR1 provides the highest accuracy. In all domains except Auto, performance with ARR1 tends to improve as the number of retained rules increases. However, the improvement is not significantly different from the accuracy achieved by retaining a minimum number of adaptation rules. We discuss this further below. For the Hardware and MPG domains, after a certain point increasing the number of adaptation rules slightly degrades the performance. We hypothesize that this deterioration is due to the introduction of less accurate rules, and that this issue could be ameliorated by setting a lower bound threshold on the ranking score for rules to retain.

An interesting observation is that for three of the four domains, regardless of the rule retention method, learning and applying adaptation rules provides substantially better performance than simply averaging results with k-NN, supporting the value of

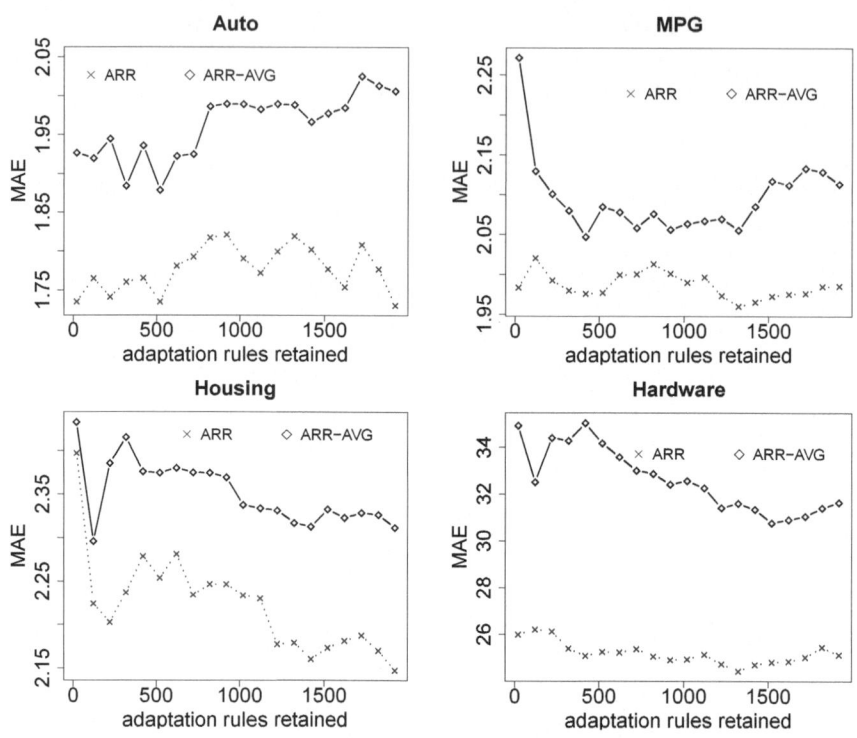

Fig. 2. MAE for retention by ARR1 and ARR-AVG

adaptation rule learning. Even for the exception domain, MPG, two of the three retention methods provide substantially better performance than k-NN.

A surprising result is that frequency-based rule retention for the tested domains usually performs worse than random rule retention. One possible explanation is that the frequency-based approach may sacrifice rule diversity, resulting in poor coverage of parts of the space. However, further study is needed.

Assessing ARR1's Error Scoring for Blame/Credit Assignment: to investigate question 2, on the impact of arr1's error scoring method for blame/credit assignment to rules during the retention process, we compared accuracy using rule sets selected by ARR1 and selected by an ablated version of ARR1, ARR1-AVG, that ranked rules by lowest average error rather than ARR1's *ErrorScore*. ARR1's *ErrorScore* favors rules that are more frequently used during the test phase and at the same time yield accurate estimations; ARR-AVG's scoring favors rules that yield accurate estimations on average, regardless of the number of times the rules were applied in the training phase. Fig. 2 shows that in all test domains, retention by ARR1 resulted in lower MAE than retention by ARR-AVG, supporting the benefit of ARR1's scoring mechanism. The greatest gain over ARR-AVG is observed for the Hardware domain and the gain is least for MPG and Housing domains.

Sensitivity of Resulting Accuracy to Number of Adaptations Per Problem During ARR's Training: As discussed previously, cost of ARR1 depends on the number k of source cases adapted for each test source case. Also, there is a tradeoff in adapting larger numbers of source cases for a given test case: Adapting additional source cases provides more data about rule performance, but because the additional source cases are less similar, the additional adaptations may result in more error, penalizing rules which might have had higher rankings if only applied to more similar cases—and which might only be applied to more similar cases in practice.

To study the effect of k's value (in the training phase) on final system accuracy with the retained rule set, we tested accuracy for 5 k values, 10, 20, 40, 80 and 100. Fig. 3 shows results for the Auto and MPG domains. Best performance for both domains was achieved for a k value of 80. The performance of ARR1 for different k values across MPG and Auto domains shares some general patterns, with some variation. For example, although for the MPG domain the second best performance is achieved when k is set to 100, the same k value in the Auto domain yields lower performance, especially when the number of retained adaptation grows. The results suggest that the performance of ARR1 improves as k's value increases up to a certain point, determined by the domain, and then decreases as k increases further. The housing and MPG domains are not shown for reasons of space, but also show this pattern.

We explain this pattern by the balance between the advantage of judging adaptation rules based on their effectiveness for local cases (because, for a well populated case base, it is more reasonable to select the source cases from the local neighborhood of the input query), versus the previously-noted drawback that using a small number of source cases (lower values for k) in the training phase decreases the number of rules examined by ARR.

Comparative Benefit of Increasing Case or Adaptation Knowledge: To investigate question 4, on the comparative benefit of adding case or adaptation knowledge, we tested how increases in the number of adaptation rules generated by ARR1 affect final performance, versus how increases in the number of cases affect final performance. This experiment explores how sensitive performance is to additions of either type of additional knowledge, which gives an indication of how rapidly the knowledge in that container "converges" to an adequate set. In particular, the experiment measures the gain in accuracy per addition to a target knowledge container (the case base or adaptation knowledge container), compared to the performance of the system for the target knowledge container with minimum size.

Because the effect of case additions depends on the strategy used for selecting cases to add, we tested three methods: (1) Smyth and McKenna's [25] (RelCov), as well as two simplified methods, (2) Cov, which added the new case with the highest competence (Cov) and (3) Reach, which added the new case with the lowest reachability.

Fig. 4 summarizes the results. In the lefthand two graphs, the X axis shows the case base size; in the righthand two graphs, the X axis shows the adaptation rule set size. The Y axis shows the incremental percent improvement from adding another case (in the lefthand graphs) or another rule (in the righthand graphs). The minimum knowledge container size for Auto, MPG, Housing and Hardware domains is 10, 20, 20, and 10 respectively.

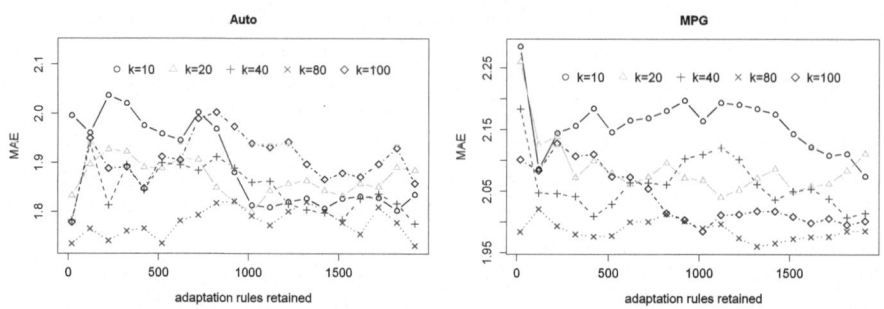

Fig. 3. The effect of the number of source cases used in the training phase on ARR's performance

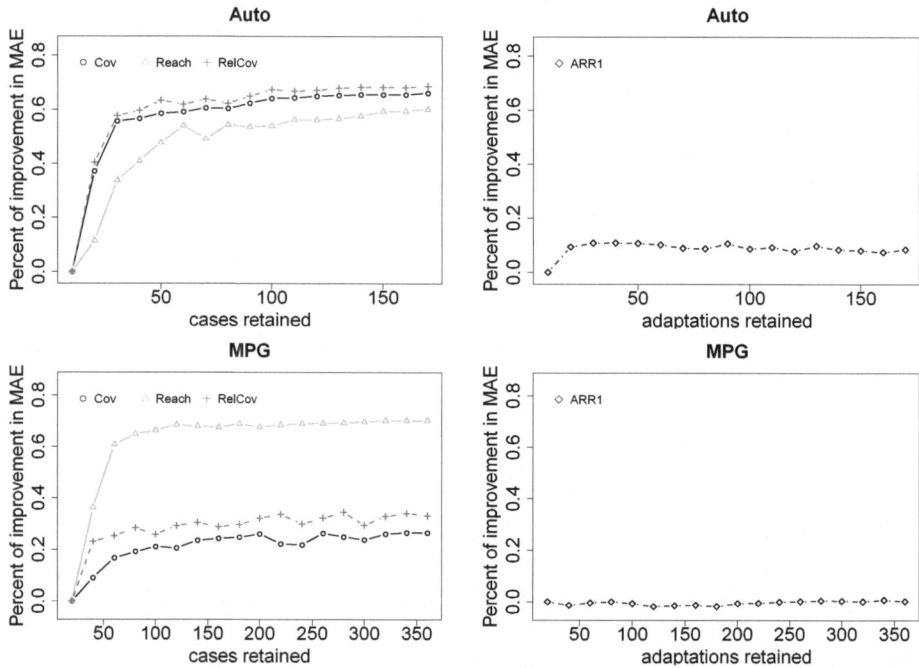

Fig. 4. Percent of improvement in MAE compared to the minimized knowledge container

As expected, increases in size of either target knowledge container generally improve performance. However, ARR1 tends to show less improvement per addition than Cov, Reach, and RelCov. We conclude that once a minimum number of adaptation rules/cases critical for system performance has been retained, performance is less sensitive to additions to the adaptation knowledge container than to additions to the case base.

6   Conclusion and Future Directions

This paper introduced ARR, a general approach to guiding adaptation rule retention. ARR uses a blame/credit assignment mechanism for guiding rule retention based on testing sample adaptation problems. Empirical results for ARR1, a specific instantiation of ARR, showed improvement over k-NN, two other alternative rule retention methods, and an ablated version of ARR with a baseline blame/credit assignment method. Experimental results also showed that ARR1's rule retention process may converge faster than case retention, suggesting a coordinated adaptation and case acquisition strategy of retaining a limited number of rules but continuing case addition.

Future directions for this work include studying ARR's performance for simultaneous maintenance of source case and adaptation knowledge containers (ARR assumes that the source case knowledge container is fixed), and investigating how local coverage characteristics of the case base may affect the choice of whether to retain cases or rules. Another direction is to extend ARR1 by applying ensembles of adaptations and studying the resulting trade-offs in accuracy and time complexity of the algorithm, as well as examining more sophisticated score assignment methods to balance the accuracy and usage frequency of the applied rules in the training phase. Because the applicability of the general ARR method is not restricted to adaptation knowledge generated by any particular method, another interesting avenue would be to explore its use for adaptation knowledge retention in other contexts and for other tasks.

References

1. d'Aquin, M., Badra, F., Lafrogne, S., Lieber, J., Napoli, A., Szathmary, L.: Case base mining for adaptation knowledge acquisition. In: Proceedings of the Twentieth International Joint Conference on Artificial Intelligence (IJCAI 2007), pp. 750–755. Morgan Kaufmann, San Mateo (2007)
2. Wiratunga, N., Craw, S., Rowe, R.: Learning adaptation knowledge to improve case-based reasoning. Artificial Intelligence 170, 1175–1192 (2006)
3. Hanney, K., Keane, M.: Learning adaptation rules from a case-base. In: Smith, I., Faltings, B.V. (eds.) EWCBR 1996. LNCS, vol. 1168, pp. 179–192. Springer, Heidelberg (1996)
4. Hanney, K., Keane, M.: The adaptation knowledge bottleneck: How to ease it by learning from cases. In: Leake, D.B., Plaza, E. (eds.) ICCBR 1997. LNCS, vol. 1266, pp. 359–370. Springer, Heidelberg (1997)
5. Jalali, V., Leake, D.: Extending case adaptation with automatically-generated ensembles of adaptation rules. In: Delany, S.J., Ontañón, S. (eds.) ICCBR 2013. LNCS, vol. 7969, pp. 188–202. Springer, Heidelberg (2013)
6. Leake, D., Kinley, A., Wilson, D.: Learning to integrate multiple knowledge sources for case-based reasoning. In: Proceedings of the Fourteenth International Joint Conference on Artificial Intelligence, pp. 246–251. Morgan Kaufmann (1997)
7. McDonnell, N., Cunningham, P.: A knowledge-light approach to regression using case-based reasoning. In: Roth-Berghofer, T.R., Göker, M.H., Güvenir, H.A. (eds.) ECCBR 2006. LNCS (LNAI), vol. 4106, pp. 91–105. Springer, Heidelberg (2006)
8. McSherry, D.: An adaptation heuristic for case-based estimation. In: Smyth, B., Cunningham, P. (eds.) EWCBR 1998. LNCS (LNAI), vol. 1488, pp. 184–195. Springer, Heidelberg (1998)

9. Patterson, D., Anand, S., Dubitzky, W., Hughes, J.: Towards automated case knowledge discovery in the M^2 case-based reasoning system. Knowledge and Information Systems: An International Journal, 61–82 (1999)

10. Shiu, S., Yeung, D., Sun, C., Wang, X.: Transferring case knowledge to adaptation knowledge: An approach for case-base maintenance. Computational Intelligence 17(2), 295–314 (2001)

11. Wilke, W., Vollrath, I., Althoff, K.D., Bergmann, R.: A framework for learning adaptation knowledge based on knowledge light approaches. In: Proceedings of the Fifth German Workshop on Case-Based Reasoning, pp. 235–242 (1997)

12. Leake, D., Smyth, B., Wilson, D., Yang, Q. (eds.): Maintaining Case-Based Reasoning Systems. Blackwell (2001), Special issue of Computational Intelligence 17(2) (2001)

13. Richter, M.M.: Introduction. In: Lenz, M., Bartsch-Spörl, B., Burkhard, H.-D., Wess, S. (eds.) Case-Based Reasoning Technology. LNCS (LNAI), vol. 1400, pp. 1–15. Springer, Heidelberg (1998)

14. Wilson, D., Leake, D.: Maintaining case-based reasoners: Dimensions and directions. Computational Intelligence 17(2), 196–213 (2001)

15. Craw, S.: Introspective learning to build case-based reasoning (cbr) knowledge containers. In: Perner, P., Rosenfeld, A. (eds.) MLDM 2003. LNCS, vol. 2734, pp. 1–6. Springer, Heidelberg (2003)

16. Leake, D., Powell, J.: Mining large-scale knowledge sources for case adaptation knowledge. In: Weber, R.O., Richter, M.M. (eds.) ICCBR 2007. LNCS (LNAI), vol. 4626, pp. 209–223. Springer, Heidelberg (2007)

17. Minor, M., Bergmann, R., Gorg, S.: Case-based adaptation of workflows. Information Systems 40, 142–152 (2014)

18. Fuchs, B., Lieber, J., Mille, A., Napoli, A.: Differential adaptation: An operational approach to adaptation for solving numerical problems with CBR. Knowledge-Based Systems (in press, 2014)

19. Li, H., Hu, D., Hao, T., Wenyin, L., Chen, X.: Adaptation rule learning for case-based reasoning. In: Third International Conference on Semantics, Knowledge and Grid, pp. 44–49 (2007)

20. Leake, D., Dial, S.: Using case provenance to propagate feedback to cases and adaptations. In: Althoff, K.-D., Bergmann, R., Minor, M., Hanft, A. (eds.) ECCBR 2008. LNCS (LNAI), vol. 5239, pp. 255–268. Springer, Heidelberg (2008)

21. Jalali, V., Leake, D.: Adaptation-guided case base maintenance. In: Proceedings of the Twenty-Eighth Conference on Artificial Intelligence. AAAI Press (in press, 2014)

22. Smyth, B., Keane, M.: Remembering to forget: A competence-preserving case deletion policy for case-based reasoning systems. In: Proceedings of the Thirteenth International Joint Conference on Artificial Intelligence, pp. 377–382. Morgan Kaufmann, San Mateo (1995)

23. Jalali, V., Leake, D.: A context-aware approach to selecting adaptations for case-based reasoning. In: Brézillon, P., Blackburn, P., Dapoigny, R. (eds.) CONTEXT 2013. LNCS, vol. 8175, pp. 101–114. Springer, Heidelberg (2013)

24. Frank, A., Asuncion, A.: UCI machine learning repository (2010), http://archive.ics.uci.edu/ml

25. Smyth, B., McKenna, E.: Building compact competent case-bases. In: Althoff, K.-D., Bergmann, R., Branting, L.K. (eds.) ICCBR 1999. LNCS (LNAI), vol. 1650, pp. 329–342. Springer, Heidelberg (1999)

Case-Based Reasoning for Improving Traffic Flow in Urban Intersections

Anders Kofod-Petersen, Ole Johan Andersen, and Agnar Aamodt

Department of Computer and Information Science (IDI),
Norwegian University of Science and Technology (NTNU),
7491 Trondheim, Norway
{anderpe,agnar}@idi.ntnu.no, olejandersen@gmail.com
http://www.idi.ntnu.no/

Abstract. Congestion in urban areas is a main traffic challenge. Solutions to this problem include building more roads or have more people in each vehicle. Each of these presents their own challenges. However, one inexpensive approach is to improve the traffic flow, in particular by improving signal plans in intersections. We present a prototype case-based reasoning system, which can control traffic lights in an urban intersection. The system uses real historical vehicle counts from an intersection to make new signal plans. jCOLIBRI is used as the framework for the case-based reasoning system and an evolutionary algorithm is used for weighting the cases. Simulations carried out in Aimsun, a simulation tool used by the Norwegian Public Roads Administration, indicates that satisfactory signal plans can be made in a variety of scenarios.

Keywords: Intelligent transportation systems, traffic flow, application of CBR, evolutionary algorithm.

1 Introduction

The traffic situation in urban areas is becoming a large problem. Especially congestion during rush hours is becoming quite visible. Often, this can not be solved by simply building more roads, as most urban areas do not have the available physical space required. One way of solving these problems can be to focus on decreasing the number of vehicles. This can be done by increasing the amount of people traveling in each vehicle, or focusing on alternative transportation such as bicycle or public transport. However, a simpler way is to use the existing infrastructure and improve the control of the traffic flow.

The domain of traffic control is highly complex and many methods exist. Because of this complexity, successful methods in one traffic environment may not give satisfying results in another environment. Road networks vary immensely in their design, which greatly affects the solutions that should be used. Other variables includes traffic culture, weather, day, time of year, season, and so on. Although one solution may be good at a given time, the traffic demand often changes as time passes by, and maintenance is often required to satisfy new

L. Lamontagne and E. Plaza (Eds.): ICCBR 2014, LNCS 8765, pp. 215–229, 2014.
© Springer International Publishing Switzerland 2014

demands. A highly dynamic method, which is able to comprehend complex domains, is desired in order to sufficiently handle this problem.

In the work presented here we want to show that CBR provides the necessary functionality in order to offer a solution to this problem. A CBR-system can relatively easy be programmed to dynamically solve problems, as well as be able to grasp intricate concepts and relationships.

The rest of the paper is organised as follows: Section 2 introduces related work in the area of AI methods for smart intersections; Section 3 details the design and implementation of the system; Section 4 describers the test and evaluation carried out; and the paper ends with a conclusion and outlook on future work.

2 Related Work

Many AI methods have been applied to traffic control in general, and flow through intersections in particular. Frequently applied methods for decision support and learning include neural networks, fuzzy logic, swarm intelligence, evolutionary algorithms, as well as rule-based, model-based and case-based reasoning. A brief overview of some methods and their targeted areas is given in [1] and a discussion of CBR for this type of applications is provided in [2].

We limit our study to case-based methods and start with some general examples in traffic management and control, i.e. systems that analyse traffic flows and incidents in order to provide assistant, advice and automatic control; before moving to closely related work on CBR-methods and intersection traffic flow.

An early system was PLANiTS [3], a planning support system for public transportation problems, e.g. bus priority schemes. It has a rather complex, structured case representation that captures decisions and outcomes. Schutter et al. [4] present a CBR system for assisting traffic operators to improve traffic flow on roads using variable speed limits, dynamic route guidance, opening of shoulder lanes, etc. To increase scalability, large road networks are divided into smaller sub-networks, where each has its own case-base, within an agent architecture.

A prototype CBR system for calculating routing plans was developed by Sadek et al. [5]. Here the initial case base was populated by simulations from a mathematical model. Relying on mathematical models to perform the heavy calculations needed in real-time often turns out to be problematic, but the combination with CBR showed promising results. Real-time performance was enabled by a simple match/retrieve mechanism. Traffic patterns are of a recurrent nature and over time successful solutions are more likely to be reused. The experimental results also showed that it is possible to provide good solutions even without complete information of the traffic situation, e.g. due to sensor failures.

Tightly linked to optimising traffic flow, reducing safety risks is of course always important. Using the CBR tool within the *eGain KnowledgeAgent* tool, Li and Waters [6] built a system for collision analysis and prevention. Cases are question-answer structures grouped into suitable clusters.

Mouncea et al. [7] presents an application for dynamic selection of signal plans based on analysis of the current traffic time series pattern and matching it with past patterns, using a k-NN approach. Experiments show that a similarity measure based on cumulative time series values within a period performs better than similarity based on plain time series.

A CBR method that supports left-turning in signalised intersections was developed by Zohdy and Rakha [8]. A case base was built from more than 9000 observed situations in Christianburg, Virginia. Case parameters include weather condition, light conditions and gap size between opposing cars. Based on the parameters the system advice the driver to make the left turn or not, which is captured in the case together with the advice acceptance by the driver. Evaluation demonstrated a potential for increased safe intersection throughput. An interesting recent development combines controller-to-vehicle advices with vehicle-to-vehicle communication in order to minimise car distance within a platoon through an intersection [9], so far with promising results in terms of increased throughput and reduced fuel consumption.

Another approach to urban intersection control was presented by Zhenlong and Xiaohua [10]. A hierarchical case organisation captures important parameter substructures such as day of the week, time periods within a day and weather conditions in a time period. This approach to was tested against a fixed time control regime that uses a model to minimise intersection delays, which is much in use today. Through cycles of tests and learning, the CBR-system outperformed that system. We found many similarities with this approach and ours. Parts of our system is based on this work. Our system extends their method by adding a genetic algorithm mechanism for learning of feature weights.

Another way of combining CBR with genetic algorithms is done in a system for traffic control in Brazilian cities [11]. Here signal planning is one element in a larger system for traffic control. The focus is to combine the methods in an agent architecture in order to better handle uncertain, incomplete and inconsistent information. Most common situations are modelled by fuzzy rules, which also generates the initial case base. A hybrid GA-CBR method selects a subset of best matching cases and may reuse a specific solution or combine their solutions using standard GA techniques of crossover, mutation, and fitness analysis. If modified in this manner the resulting solution is captured in a new case.

3   Design and Implementation

Our system consists of three main modules: the CBR system, the Simulation Manager and a simulation tool in regular use by the Norwegian Public Roads Administration called Aimsun (see Fig. 1). When running the system, the Simulation Manager will, based on user input, select a scenario to run. The chosen scenario is sent to the CBR-system, which then retrieves a predicted traffic flow based on the features of the scenario. A signal plan is then calculated, using Webster's formula [12] and standard values acquired from the National Public Road Administration [13]. This is returned to the Simulation Manager. The received plan, together with scenario specific information is configured and applied to Aimsun, before running a simulation. The results are displayed in Aimsun.

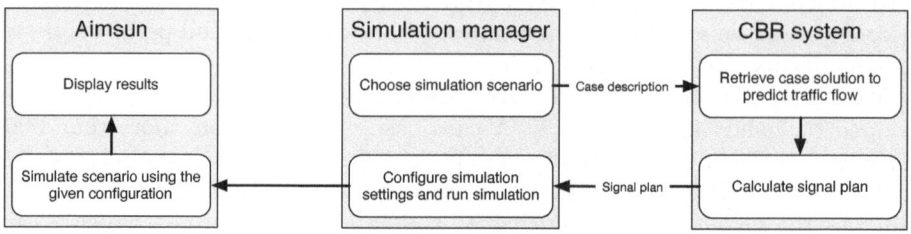

Fig. 1. Functional Overview

Aimsun[1] is a traffic simulation software that support models in all scales. The software is known for its ability to perform demanding simulations at high speed. Simulations can be done on a microscopic, mesoscopic and macroscopic scale, or by combining these to perform hybrid simulations. There are also additional modules that can be applied to enhance the functionality such as adaptive control interfaces, for interfacing with traffic control systems. A programming interface is also available as an additional functionality.

The Simulation Manager is implemented in Python and has access to Aimsun through its Python scripting interface. The CBR-system uses the jCOLIBRI-framework, which is implemented in Java. Inter-process communication between the Simulation Manager and the CBR-system is performed through sockets.

Traffic scenarios are entered by the user into the Simulation Manager. The scenarios are structured as problem descriptions. When the Simulation Manager has retrieved the traffic scenarios, the necessary features are extracted and forwarded to the CBR-system. A new case is constructed and the system will then retrieve similar cases. When a good solution candidate is selected, the CBR-system uses this solution, consisting of a predicted traffic flow, and calculates a signal plan. This plan is then sent back to the Simulation Manager.

The Simulation Manager applies the scenario specific information to Aimsun. This configuration will be reused for each different signal plan that is to be applied. For each new signal plan a simulation is run. After Aimsun has finished a simulation, the results are displayed and stored in its local SQL-database.

3.1   Case Base

The case structure is based on the work presented by Zhenlong and Xiaohue [10], where time (date, weekday and time), weather conditions, temperature, queue lengths for the lanes and special events are used. However, we found that the date generally is less essential than knowing if the day is for example a movable holiday. Information about the given season did also turn out to be important. As in [10], the time of day is important, since commuter traffic influences the signal plans. Finally, non-traffic environmental features are also important. In

---

[1] TSS-Transport Simulation Systems, www.aimsun.com

our case, the weather, temperature and friction (which is pre-calculated based on [14]). Thus, the final case structure contains the features described in Table 1. The solution contains a total traffic flow for each lane in the intersection, which is then used to calculated the actual signal plan.

Table 1. Case structure

Problem description		
Feature	**Values**	**Discrete values**
Season	Winter, Spring, Summer, Fall	Date are discretised into 4 seasons
Weekday	Monday, Tuesday, Wednesday, Thursday, Friday, Saturday, Sunday	Already discrete
Time of day	Night, Morning, Noon, Afternoon, Evening	Night = 00:00-01:00, Morning = 07:00-08:00, Noon = 12:00-13:00, Afternoon = 15:00-16:00, Evening = 21:00- 22:00
Weather	Sunny, Cloudy, Precipitation, Heavy precipitation	Already discrete
Temperature	Hot, Medium, Freezing	Freezing = bellow 0°C, Medium = 0 − 20°C, Hot = above 20°C
Friction	Dry, Wet, Icy	Icy = 0.30, Wet = 0.90, Dry = 1.00
Special event	None, Football match, Holiday	Already discrete
Problem solution		
Total traffic flow for the different lanes	Interger values separated by commas	

3.2 Similarity Functions

All features have multiple possible values. The knowledge of similarities between values is important to incorporate into the system in order to conserve the relationship between concepts from the real world into our model. The similarity functions represent the similarity between the attribute values of the classes.

The similarity values are manually set based on domain knowledge in cooperation with domain experts from the National Public Road Administration.

Table 2 details the local similarities between seasons. Winter and summer are often different when it comes to the traffic conditions. Icy and slippery roads often occur in the winter time and sometimes even in the early spring and late fall. Compared to summer, wet roads are typically the worst road conditions in these seasons. Spring and fall are somewhat similar. In the late fall, snow is a common sight, which is also the case for early spring. During summer, the road conditions are mostly wet or dry.

Table 2. Season similarities

Feature	Winter	Spring	Summer	Fall
Winter	1.00	0.80	0.40	0.60
Spring	0.80	1.00	0.65	0.65
Summer	0.40	0.65	1.00	0.70
Fall	0.60	0.65	0.70	1.00

Table 3. Temperature similarity

Feature	Freezing	Medium	Hot
Freezing	1.00	0.75	0.70
Medium	0.75	1.00	0.95
Hot	0.70	0.95	1.00

When considering temperatures, freezing makes the most different. When a typically wet road turns into a wet and icy road, it may greatly affect the driving conditions. Table 3 summarises the temperature similarities.

Table 4. Day of week similarity

Feature	Mon	Tue	Wed	Thu	Fri	Sat	Sun
Mon	1.0	0.9	0.9	0.9	0.8	0.5	0.4
Tue	0.9	1.0	0.9	0.9	0.8	0.5	0.4
Wed	0.9	0.9	1.0	0.9	0.8	0.5	0.4
Thu	0.9	0.9	0.9	1.0	0.8	0.5	0.4
Fri	0.8	0.8	0.8	0.8	1.0	0.5	0.4
Sat	0.5	0.5	0.5	0.5	0.5	1.0	0.8
Sun	0.4	0.4	0.4	0.4	0.4	0.8	1.0

Table 5. Friction similarity

Feature	Dry	Wet	Icy
Dry	1.0	0.9	0.3
Wet	0.9	1.0	0.4
Icy	0.3	0.4	1.0

When determining the similarity of the days, it is quite clear that the biggest difference is between the weekend and the rest of the weekdays. We have therefore decided to put a similarity of 0.9 between the weekdays, except for Friday that has a similarity of 0.8. The reason is that we assume that Friday differs a little from the other days concerning the working hours. People are more prone to leave at different hours than they normally would, and this day may also be the one day most often skipped when people opt for a "long weekend". Saturday and Sunday stands out, obviously because most people do not work on these days. They also differ because of more traffic in the night time, which both days typically have. Since most stores are closed on Sundays there is generally less traffic than on a Saturday, except when there are football matches.

The similarity between the frictions can be seen in the Table 5, where the friction coefficient determines the similarity.

Table 6. Time of day similarity

Feature	Night	Morning	Noon	Afternoon	Evening
Night	1.00	0.28	0.18	0.20	0.27
Morning	0.28	1.00	0.67	0.42	0.74
Noon	0.18	0.67	1.00	0.62	0.62
Afternoon	0.20	0.42	0.52	1.00	0.42
Evening	0.27	0.74	0.62	0.42	1.00

During weekdays, the difference between night and morning can be considered quite vast. Traffic during morning rush can be as much as 10 times the number of counts during night on a regular day, maybe even more. On Saturday the traffic counts for morning and night are very similar, but on Sundays the night counts are typically the double of what it is in the morning. This is one example concerning the night time, but the general difference between the rest of the time periods, can in most cases be considered high. The similarity between features have been calculated manually by looking at traffic data (see Table 6).

Table 7. Weather situation similarity

Feature	Sunny	Cloudy	Precipitation	Pouring
Sunny	1.00	0.95	0.90	0.80
Cloudy	0.95	1.00	0.95	0.90
Precipitation	0.90	0.95	1.00	0.90
Pouring	0.80	0.90	0.90	1.00

We assume that the different weather conditions are quite similar to each other. Some aspects to consider are that the weather may affect the amount of people driving to work, and also the traffic conditions (see Table 7).

3.3 Evolutionary Algorithm for Weight Setting

Assigning weights using a large amount of non-linear dependent variables is very difficult, even for domain experts. Thus, we employ a genetic algorithm for this optimisation. This algorithm uses a string of 49 bits as genome. For each of our 7 features we have 7 bits, which can give 128 different values. Since our weights need a value between 0 and 1, we calculated the value of each of the bit-strings and divided by 127. The evolutionary operators used are crossover and mutation.

In order to calculate the fitness of each genome, we evaluated the case base using cross-validation. Each genome is applied as weights to the case base, and then a leave-one-out cross validation is executed. The average performance using the weights is calculated by summing the performance of the case base for each removed case, and then dividing by the total amount of cases. This will then return a fitness score between 0 and 1, where 1 is a 100 percent similarity.

The selection of individuals is done by roulette wheel selection. We also use elitism, which ensures that the x most fit individuals are taken to the next generation. This ensures that the population does not lose its fittest individuals, which can save us some time as these do not need to be rediscovered.

As in machine learning in general, overfitting can be a problem here as well. It is therefore important to note than even though the evolutionary algorithm have proposed weights (see Table 8) that give good results when evaluating the case-base, these weights may not give as good results when applied to the test set. This is because the weights are calculated to give optimal results for the known cases, and if new unknown cases appear they might not correspond to the estimated weights. Albeit, we assume that these weights are close to optimal considering the known cases, which we have used for training. In any case, using these weights gives a better performance than when set manually, even by an expert.

Table 8. Feature weights given by the evolutionary algorithm

Feature	Season	Weekday	Time	Weather	Temperature	Friction	Event
Weight	0.008	0.213	0.984	0.039	0.732	0.008	0.291

3.4   Retrieval

When a new problem description is given to the system, the system needs to retrieve the most similar case in order to find the best solution. This process consists of first calculating the similarity of the new case with the other cases in the case base. Weights and similarity functions, as described above, are used in order to determine the similarity between the features, and the total similarity of the cases. The total similarity is calculated by finding the total average similarity by summing all the n similarities and dividing by n. The case that has the highest similarity is then chosen as the solution.

3.5   Modelling the Intersection

The intersection, traffic, and also driver behaviour, need to be modelled in a way that reflects the real world to a highest degree possible, as this greatly affects the correctness of the simulation results.

Fig. 2 depicts the model of the intersection. The lane going from top to bottom corresponds to the lane going from the city centre (north) heading out of the

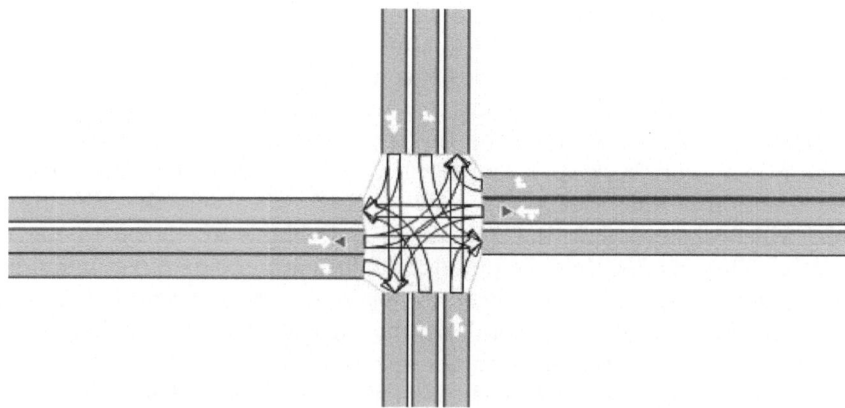

Fig. 2. Model of the intersection

city (south). Comparing to the actual intersection (see Fig. 3), it is clear that one lane is missing in both directions. These lanes are for public transport and priority vehicles that do not affect the static plans currently in use. Thus, we have left them out. Besides this, the model is true to the actual intersection.

Pedestrians are not explicitly considered in this model. In the signal plans in use, the pedestrians receive green light, either at the same time or slightly before cars in the same direction.

4   Test and Evaluation

4.1   Case-base Evaluation

Evaluating the case-base is important before taking on a full simulation. If the case-base has not been evaluated and the whole system performs poorly, it is difficult to say whether it is the coverage of the case-base or other aspects, which are responsible for the lack of performance.

Table 9. Case-base evaluation using different number of cases

Number of cases	55	50	45	35	30	25	20	15
Accuracy	88 %	87 %	85 %	84 %	84 %	83 %	78 %	76 %

To evaluate the CBR-system's case-base, we initially tested the case-base consisting of several number of cases (Table 9 shows the results of size vs. accuracy when classifying cases). Evaluation was carried out using k-fold cross, leave one out validation. Initially, the case-based covered normal traffic-patterns on weekdays and workdays. We then added cases to include football matches and other events that might cause irregular traffic-patters. At 55 cases, most scenarios had multiple cases covering them and by adding more similar cases accuracy would not increase significantly.

4.2   Scenario Descriptions and Settings

The simulation of the system consists of testing five different scenarios: *regular weekday, regular saturday, football match, holiday season* and *slippery roads*. Every simulation is carried out using a *static plan*; a *restricted CBR plan*, being plans that can only have total cycle time of either 60 or 120 seconds; and a *CBR plan*, which is not restricted to a specific cycle time. Each scenario is run three times using different seeds to to determine the arrival of vehicles.

All simulation parameters are set to the default in the simulator. The simulation step is set at 0.1 seconds. Arrival of vehicles is generated from a truncated normal distribution; except where otherwise noted. The distribution of which lanes vehicles chose when leaving the intersection follows the exact distribution as supplied by the NPRA. Table 10 list the distribution, whilst Figure 3 depicts the different lanes. Each scenario is spilt into five different time slots:

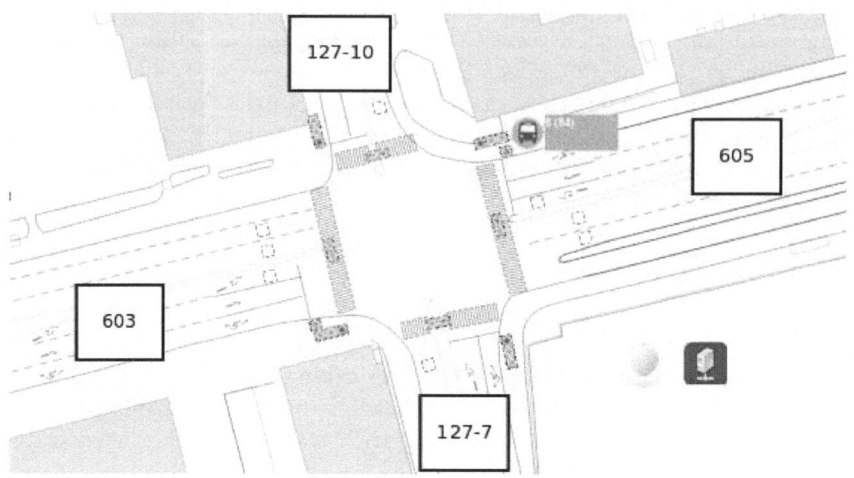

Fig. 3. Lane identification for turning percentages

- Night at 00:00-01:00, when the traffic typically is quite low in weekdays, but may be relative high in the weekends.
- Morning at 07:00-08:00, the time of the day when most people go to work towards downtown in weekdays. This is typically a time period with low traffic in weekends.
- Noon at 12:00-13:00, when the traffic flow is relative low while waiting for the afternoon rush.
- Afternoon at 15:00-16:00, the time of the day when most people are leaving downtown in order to go home after work hours.
- Evening at 21:00-22:00, which is typically a time with relative low traffic.

Scenario 1 – Regular Weekday The night generally has very low traffic, in particular from the east and west going lanes. Morning are significantly more busy than nights, where most of the traffic are going either north or south. Noon has a traffic flow where both south and north going traffic has a high demand. In the afternoon, people are typically leaving the downtown areas, which gives high demands on the south going lane. Finally, in the evening traffic looks much like both morning and noon. Table 11 summaries the descriptions of this day.

Scenario 2 – Regular Saturday The second scenario is similar to the preceding one. This, however, is a Saturday which in distribution is different: there is more traffic in the night time, compared to weekdays; the mornings has very little traffic; noon looks like a normal weekday; in the afternoon traffic primarily flows from the city centre; and evening is similar to night time. Table 12 describes the five time slots for this scenario.

Table 10. Turn distribution for different days (see also Figure 3 for intersection layout)

From → to		Turn percentage		
		Weekdays	Saturday	Sunday
127-7 → 603	E	70 %	60 %	60%
127-7 → 605	E	20 %	38 %	38%
127-7 → 127-10	E	10 %	2 %	2%
127-10 → 603	W	55 %	48 %	45%
127-10 → 605	W	40 %	49 %	53%
127-10 → 127-7	W	5 %	3 %	3%
603 → 605	SN	96 %	96 %	96 %
603 → 127-10	SW	3 %	3 %	3 %
605 → 603	NS	96 %	96 %	96 %
605 → 127-7	NE	2 %	2 %	2 %

Scenario 3 – Football Match Scenario 3 differs from the two preceding ones, as it represents a situation with unusual traffic patterns. This particular scenario is on both a weekday and Sunday, when a football match has just ended[2]. Typically, the traffic just after a football match is quite high in lane from the east (for about 15-20 minutes), whilst the rest of the hour the traffic is normal.

To describe the high variance in traffic distribution in this scenario, we employed both the normal distribution and the ASAP (as-soon-as-possible) distribution for the east lane. Table 13 summarises the description of this scenario.

Scenario 4 – Holiday Season Holiday season obviously affect the traffic pattern. Scenario 4 describes December 25th, which is a day where traffic is most likely to differ the most from a normal day. The morning do not differ so much from a normal Sunday, this is due to the fact that we could only obtain data from a holiday, which was also a Sunday. Afternoon traffic is actually higher than on a normal Sunday. Table 14 describes this scenario.

Scenario 5 – Slippery Roads In order to simulate slippery roads, which affects traffic a lot, we used an existing scenario and changed the friction to "Icy" and temperature to "Freezing". In this case, both Scenario 1 and 2 were used. Table 15 summarises the description of this scenario.

4.3   Results and Discussion

As described in Section 4.2, each simulation is carried out using three different plans: a *static plan*; a *restricted CBR plan*, and an unrestricted *CBR plan*. Each scenario is run three times using different seeds for vehicle arrival.

All plans are compared by *Stop time*, *Travel time*, *Speed* and *Delay* to the existing static plan. Table 16 summarises the results from all simulations. The

---

[2] The real-world intersection is located quite near a local football stadium.

Table 11. Scenario 1 description, Monday, September 17th, 2012

Time slot	Weather	Temperature	Friction	Traffic (NS, SN, E, W, NE, SW)
Night	Cloudy	9.8	Dry	59, 57, 2 ,2, 4, 3
Morning	Precipitating	7.1	Dry	612, 854, 53, 34, 190, 140
Noon	Cloudy	9.0	Wet	733, 755, 35, 32, 53, 41
Afternoon	Cloudy	10.4	Dry	1153, 690, 115, 30, 130, 84, 39
Evening	Cloudy	7.3	Dry	408, 366, 30, 22, 37, 12

Table 12. Scenario 2 description, Saturday, September 22nd, 2012

Time slot	Weather	Temperature	Friction	Traffic (NS, SN, E, W, NE, SW)
Night	Cloudy	6.3	Dry	189, 156, 7, 8, 9, 6
Morning	Cloudy	3.9	Dry	150, 147, 53, 1, 2, 12
Noon	Sunny	7.5	Dry	522, 727, 35, 10, 36, 12
Afternoon	Sunny	9.0	Dry	782, 509, 37, 18, 50, 11
Evening	Cloudy	6.8	Dry	267, 282, 14, 6, 22, 4

Table 13. Scenario 3 description, football match, two days, 2012

Time slot	Weather	Temperature	Friction	Traffic (NS, SN, E, W, NE, SW)
Weekday	Sunny	6.6	Dry	662, 589, 138, 65, 58, 16
Sunday	Cloudy	10,0	Dry	514, 562, 137, 59, 48, 10

Table 14. Scenario 4 description, holiday season, two days, 2011

Time slot	Weather	Temperature	Friction	Traffic (NS, SN, E, W, NE, SW)
Morning	Precipitation	3,0	Wet	142, 131, 2, 0, 7, 7
Afternoon	Precipitation	3.0	Wet	838, 500, 42, 45, 84, 25

Table 15. Scenario 5 description, slippery roads, two days, 2012

Time slot	Weather	Temperature	Friction	Traffic (NS, SN, E, W, NE, SW)
Monday noon	Cloudy	-1.4	Icy	733, 755, 34, 32, 53, 41
Saturday noon	Sunny	-1.4	Icy	522, 727, 35, 10, 36, 12
Monday afternoon	Cloudy	-1.4	Icy	1153, 690, 115, 30, 130, 84, 39
Saturday afternoon	Sunny	-1.4	Icy	782, 509, 37, 18, 50, 11

table lists the best CBR-plan compared to the static plan. Negative numbers indicate that the CBR-plan perform worse than the static plan, whereas positive percentages show how much an improvement has been achieved over static plans.

Early in the simulation process, a weakness appeared in the static plans. The priority for the lanes going from east and west is almost always too high. This results in green time being wasted. As a result of this, the static plans had a bad starting point, and the CBR-plans did in turn give unexpectedly good results.

Table 16. Summary of the results

Scenario	Best plan	Stop time	Travel time	Speed	Delay
Sc1 (Weekday) - Night	Restricted	16%	8%	6%	11%
Sc1 (Weekday) - Morning	Unrestricted	25%	12%	-4%	21%
Sc1 (Weekday) - Noon	Unrestricted	44%	25%	22%	44%
Sc1 (Weekday) - Afternoon	Unrestricted	14%	7%	4%	13%
Sc1 (Weekday) - Evening	Restricted	33%	17%	16%	27%
Sc2 (Weekend) - Night	Restricted	27%	14%	12%	25%
Sc2 (Weekend) - Morning	Restricted	11%	4%	6%	11%
Sc2 (Weekend) - Noon(2p)	Restricted	62%	34%	35%	59%
Sc2 (Weekend) - Noon(3p)	Restricted	37%	17%	-1%	32%
Sc2 (Weekend) - Afternoon	Restricted	49%	28%	39%	50%
Sc2 (Weekend) - Evening	Restricted	32%	16%	13%	30%
Sc3 (Football) - Evening	Restricted	24%	13%	13%	23%
Sc3 (Football) - Evening	Static/Restricted	0%	0%	0%	0%
Sc3 (Football) - Evening	Restricted	23%	13%	14%	22%
Sc3 (Football) - Evening	Static	-15%	-8%	-7%	-12%
Sc4 (Holiday) - Morning	Restricted	49%	26%	22%	48%
Sc4 (Holiday) - Afternoon	Unrestricted	53%	28%	23%	50%
Sc1 (Slippery) - Noon	Restricted	39%	26%	46%	39%
Sc2 (Slippery) - Noon	Restricted	19%	20%	44%	32%
Sc1 (Slippery) - Afternoon	Restricted	19%	8%	0%	13%
Sc2 (Slippery) - Afternoon	Restricted	38%	29%	50%	46%

Especially the noon scenarios would get a big performance boost when applying the CBR-plans. The reason for this seems to be that because of the high traffic flow, the unrestricted CBR was able to make cycles that both lasted longer than 60 seconds, and had a maximum priority to the north and south going lanes. This turned out to be very beneficial under these conditions.

The fact that the CBR plans, which were restricted to a 60 or a 120 second cycle, did actually perform better than the unrestricted one in 14 of the total 19 runs, was a surprise to us. A reason for this would have to be that the restricted plan often has a longer cycle time than the one calculated for the unrestricted one, which turns out to give an advantage, as each cycle iteration gives more red time. The exception is the scenarios where the three phase plan is used.

Even though the general results implies that the plans with the longer cycle times often perform better, there are some aspects of having shorter cycle time that is beneficial. When the waiting time is long, both motorists and pedestrians can be impatient. Especially pedestrians can start to jaywalk when they have waited for a long time, which in turn lowers the safety of the intersection. Therefore, the signal plan having the shorter cycle time should get some kind of benefit during the evaluation.

Another surprise that emerged during the simulations was that the static plans performed similarly, or even better, than the CBR-plans during the simulations of the traffic after a football match. This was the scenario when we decided to use the ASAP arrival model. The high priority which was given to the east and west lanes would in this case turn out to be beneficial. However, the fact that the traffic is high demanding in the east lane for the first 15-20 minutes after the match, is not properly reflected. When using the ASAP-model, the traffic demand will initially be very high, and then when all the cars have left, the traffic going north and south should get maximum priority. As the whole simulation lasts for 1 hour, the plans will need to compromise heavily, resulting in a plan that only gives a decent performance. One thing that could have been done in this situation, is to apply one plan to the first 15 minutes, and then another one to the remaining 45 minutes.

Overall the simulations gave promising results. The system seems to be able to predict the traffic well by retrieving older traffic counts, that does not vary greatly from the current ones. The cross evaluation of the case base shows that most of the cases do get a satisfying solution. Although the results were overall very good, it would have been interesting to evaluate against the plans which is calculated by the currently used system in Trondheim.

Since the restricted plan did perform very well on the two phase plans, compared to the unrestricted one, shows that a lot of improvement can be done to the signal plan calculator.

5   Conclusion and Future Work

The work presented here demonstrates a prototype implementation of a CBR-system that predicts traffic flow and calculates signal plans for urban intersections. The implementation has been evaluated against existing methods with respect to *stop time, travel time, average speed* and *delay*. The evaluation shows promising results for CBR as a method in this domain.

The implementation was, for practical reasons run in off-line mode. To validate this approach, future tests should be carried out in real-time on-line.

Acknowledgements. The authors' would like to thank the following people: Eirik Skjetne, Jo Skjermo and Ørjan Tveit at the Norwegian Public Roads Administration, Tor Wiig from Swarco (developer of Aimsun) and Trond Foss from SINTEF.

References

1. Elkosantini, S., Darmoul, S.: Intelligent public transportation systems: A review of architectures and enabling technologies. In: 2013 International Conference on Advanced Logistics and Transport (ICALT), pp. 233–238 (2013)
2. Sadek, A.W., Demetsky, M.J., Smith, B.L.: Case-based reasoning for real-time traffic management. Computer-Aided Civil and Infrastructure Engineering 14(5), 347–365 (1999)

3. Khattak, A., Kanafani, A.: Case-based reasoning: a planning tool for intelligent transportation systems. Transportation Research Part C: Emerging Technologies 4(5), 267–288 (1996)
4. De Schutter, B., Hoogendoorn, S., Schuurman, H., Stramigioli, S.: A multi-agent case-based traffic control scenario evaluation system. In: Proceedings of the 2003 IEEE Intelligent Transportation Systems, vol. 1, pp. 678–683 (2003)
5. Sadek, A.W., Smith, B.L., Demetsky, M.J.: A prototype case-based reasoning system for real-time freeway traffic routing. Transportation Research Part C: Emerging Technologies 9(5), 353–380 (2001)
6. Li, K., Waters, N.: Transportation networks, case-based reasoning and traffic collision analysis: A methodology for the 21st century. In: Reggiani, A., Schintler, L.A. (eds.) Methods and Models in Transport and Telecommunications. Advances in Spatial Science, pp. 63–92. Springer, Heidelberg (2005)
7. Mounce, R., Hollier, G., Smith, M., Hodge, V.J., Jackson, T., Austin, J.: A metric for pattern-matching applications to traffic management. Transportation Research Part C: Emerging Technologies 29, 148–155 (2013)
8. Zohdy, I., Rakha, H.A.: Framework for intersection decision support in adverse weather conditions. Transportation Research Record: Journal of the Transportation Research Board 2324(1), 20–28 (2012)
9. Zohdy, I.H., Rakha, H.A.: Intersection management via vehicle connectivity: The intersection cooperative adaptive cruise control system concept. Journal of Intelligent Transportation Systems (2014)
10. Zhenlong, L., Xiaohua, Z.: A case-based reasoning approach to urban intersection control. In: 7th World Congress on Intelligent Control and Automation (WCICA 2008), pp. 7113–7118 (2008)
11. Nakamiti, G., da Silva, V.E., Ventura, J.H., da Silva, S.A.: Urban traffic control and monitoring – an approach for the Brazilian intelligent cities project. In: Wang, Y., Li, T. (eds.) Practical Applications of Intelligent Systems. AISC, vol. 124, pp. 543–551. Springer, Heidelberg (2011)
12. Webster, F.V.: Traffic Signal Settings. Road Research Technical Paper No. 39. Great Britain Road Research Laboratory (1958)
13. Vegvesen, S.: Trafikksignalanlegg: planlegging, drift og vedlikehold, Håndbok 142 (2007)
14. Aurstad, J., et al.: Textbook – Road Operation and Maintenance. Traffic Safety, Environment and Technology Department (2011)

Estimating Case Base Complexity
Using Fractal Dimension

K.V.S. Dileep and Sutanu Chakraborti

Department of Computer Science and Engineering
Indian Institute of Technology Madras, Chennai - 600036
{kvsdilip,sutanuc}@cse.iitm.ac.in

Abstract. This paper presents a novel measure of complexity of a case base. The concept of Fractal Dimensions, which is a generalization of the idea of dimensions, is used to estimate complexity. In terms of a classification problem, the idea of Fractal Dimension is used to estimate the ruggedness of the space spanned by instances along the decision boundary. Experiments over collections of varying complexity show that the measure exhibits strong negative correlation with classification accuracies over several classifiers. We also present empirical findings from experiments over non-textual datasets.

Keywords: case base complexity, fractal dimension, correlation dimension, global alignment, local alignment.

1 Introduction

Case Based reasoning (CBR) is a paradigm where problems are solved by using previous experiences of having solved similar problems. The primary assumption of a CBR system is that "similar problems have similar solutions." The extent to which similar problems have similar solutions gives an indication of how well the CBR system will perform. Hence, estimating "complexity" of case bases to solve new problems is an important step in building a CBR system.

Complexity measures can be used to evaluate case base editing algorithms used in case base maintenance [1]. A complementary application to deletion of cases is identification of areas where new cases need to be acquired for the CBR system to perform well [2]. Other applications of complexity measures include finding the right representation for a case base that facilitates effective retrieval. A representation with the least complexity among other representations is more suitable for CBR. Recently, complexity measures have been used to suggest query expansion for retrieval on textual CBR systems [3].

While there are many complexity measures proposed in literature, to the best of our knowledge, there are still some questions that are yet to be answered. Is there an underlying connection between the structure of a case base and its alignment? Does the dimensionality of a case base have any role to play in alignment? Is it better to take complexity on a case-by-case basis or measure the complexity of the collection as a whole? In our work, we attempt to answer these questions by proposing a new measure

L. Lamontagne and E. Plaza (Eds.): ICCBR 2014, LNCS 8765, pp. 230–244, 2014.
© Springer International Publishing Switzerland 2014

based on an intrinsic dimension estimation method. This measure called Fractal Dimension (FD) exploits clustering tendencies present in data. For example, a case base might be represented with 10 features, but only 3 features are independent and rest are dependent on these features. So, the true dimension or intrinsic dimension of the case base is 3, while the embedding/extrinsic dimension is 10.

We will motivate the need to study clustering tendencies in datasets by taking the example of document collections. The more the clustering tendency of documents in the term space, the lesser the complexity. On realistic collections, such as those seen over the web, we often observe pronounced clustering. It is rare for a document to draw words from topics that are totally orthogonal to its theme; a news article on a popular movie may share very few terms (leaving aside function words) with a journal article on medicine. Thus, webpages are not uniformly distributed in the term space. This skew in distribution is exploited by clever metric search algorithms [4] to speed up retrieval. One critical bottleneck in similarity computation is the "curse of dimensionality" [5], which manifests in dramatic slowdown in nearest neighbour search with increasing dimensionality.

Strong clustering tendencies might cause the datapoints to fill up a space which has a fewer number of dimensions compared to the original feature space. Estimates of intrinsic dimensionality thus act as good indicators of collection complexity in unsupervised scenarios. As part of our work, we propose an intrinsic dimensionality indicator for supervised classification. Supervised classification can be considered as a special case in CBR, where the solution side of the case corresponds to class labels. Instead of taking all document instances in training data, we consider those that are hard to classify. Presumably, these are also the ones that are close to the decision boundary. Our hypothesis is that the intrinsic dimensionality of these hard-to-classify points is indicative of the dataset complexity. We report experimental results that support this intuition. Specifically we show that our estimate of classification complexity negatively correlates with accuracies of standard classifiers over datasets of varying difficulty. We also study the performance of the measure over non-textual datasets whose solutions have continuous values instead of class labels. This is more reflective of a generic CBR system. This paper presents our explorations with Fractal Dimension towards this end.

The structure of the paper is as follows. Section 2 discusses related work in alignment measures for CBR. Section 3 introduces Fractal Dimension and an associated concept, the Correlation Dimension, which is a practical way of estimating Fractal Dimension over datasets. Section 4 elaborates our approach of estimating collection complexity in supervised settings. Section 5 details our evaluation methodology, experimental results and key observations. Section 6 discusses the performance of the proposed measure on non-textual datasets and tries to provide some answers to questions raised earlier, followed by concluding remarks.

2   Related Work

Most of the existing complexity measures can be divided into two categories - local and global. Case base complexity is directly related to the concept of alignment. Alignment is the extent to which similar problems in a given case base correspond to similar solutions. Similarity in the Textual CBR context is typically estimated using statistical or linguistic knowledge of lexical relationship. Global alignment is the extent to which problem side clusters correspond to the solution side clusters in a case base. Local alignment looks at a specific case and gives a measure of how the problem side similarity of the case with its neighbours corresponds to the solution side similarity with the same neighbours. Global alignment considers the entire case base while local alignmment takes a case-by-case look.

Consider Fig 1 that gives an illustration to understand global and local alignment better. Fig 1a shows a case base where the cases close to each other on the problem side are close to each other on the solution side as well. The problem side clusters corresponding to cases $\{1, 2\}, \{3, 4, 5\}, \{6, 7\}, \{9, 8\}$ have similar clusters on the solution side as well. So there is a good global alignment in the case base. In contrast, if we look at the case base represented by Fig 1b, the clusters $\{1, 2\}, \{3, 4, 5\}, \{6, 7\}$ do not have corresponding clusters on the solution side. Overall the alignment of the case base is bad. Consider case 9, its problem side is closest to case 8 as well as its solution side. So even if the case base as a whole is complex, case 9 is well-aligned locally.

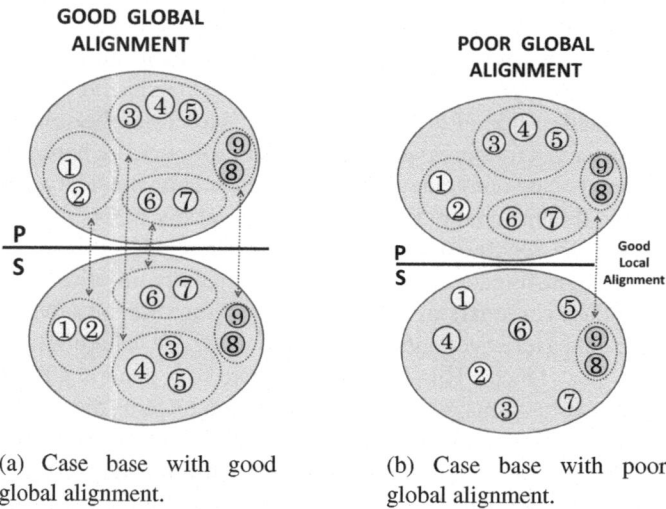

(a) Case base with good global alignment.

(b) Case base with poor global alignment.

Fig. 1. Illustration to understand local and global alignment. P corresponds to problem side and S corresponds to solution side.

Estimating complexity of data collections has been a problem of interest for both the CBR community and the IR community. An approach to estimate the complexity of a collection of unlabelled documents has been proposed in Vinay et al. [4]. First, a random document is generated and its similarity to the closest document, s_{rand} in the collection is measured . For the document obtained, we calculate the similarity to its nearest neighbor s_{nn}. The ratio $\frac{s_{rand}}{s_{nn}}$ is measured multiple times for different random documents and the average is an estimate of the collection's clusterabilty. Higher ratio implies that the data is more clustered.

An interesting application of complexity measures is to evaluate algorithms used for case base maintenance. Cummins et al. [1] evaluated 8 case base editing algorithms by measuring the complexity of case base before and after maintenance along with classification accuracies. Ideally, after maintenance, complexity must decrease and accuracy must increase or remain the same. The authors found counter-examples to this notion(decrease in complexity and decrease in accuracy) and pointed that some case base editing algorithms might be oversimplifying class boundaries.

In case cohesion [6], both the problem side and solution side of a case is taken. The cases that are similar to the problem part of the query case, say RS_{prob} and cases that are similar to the solution part, say RS_{soln} are obtained and compared. The more cases the sets have in common, the more locally aligned the case is. Cohesion does not take into account the distances on either side. Cases closer to the query must contribute more to alignment than cases far from the query. By taking the distances into account, an alignment measure is proposed in Massie et al. [7]. These measures are local, in the sense that they pertain to each query case. The complexity of a case base is calculated by getting the complexity of each case in a leave-one-out scenario and averaged over all the cases.

Global Alignment MEasure (GAME) is proposed in Chakraborti et al. [8]. The main idea is to induce an ordering on the problem side of the case base and compare it with the corresponding solution side ordering. The comparison is made by interpreting the case-feature matrix as an image and find compression ratio of the ordered problem side, the compression ratio of the ordered solution side, and the compression ratio of the solution side ordered w.r.t problem side similarities. If there is good alignment, the compression ratio of the solution side ordered w.r.t problem side will be close to the compression ratio of solution side of case base and this property is used to get an estimate of global alignment.

While most of the existing measures look at the extrinsic dimension based distribution of points to estimate complexity, we feel intrinsic complexity might be a better indicator. A CBR system with cases represented with the optimal number of features will perform significantly better at retrieval that those with extraneous features due to the 'curse of dimensionality'. Thus, information on intrinsic dimensions might give insights for taking a re-look on case representation and drop some features if the difference between the extrinsic and intrinsic dimension is high. Thus, a measure based on estimating the intrinsic dimension of the dataset might give a better estimate of complexity.

3   Fractal Dimension

The term "fractal" has been used w.r.t fractal analogies in recent literature with fractal representation applied in visual analogical reasoning [9]. Fractal Dimension generalizes the concept of dimensionality to include spaces spanned by non-integral number of features. It was first proposed in Mandelbrot [10] while explaining the counter-intuitive notion that the length of Britain's coastline changed with change in scale. This is done based on an elegant observation founded on the concept of self-similarity. As Fig. 2 shows, a line segment can be made of two smaller line segments, a square of four smaller squares and a cube of eight smaller cubes. In each of these cases, the smaller objects look similar to the bigger ones, and the magnification factor (along an edge) is two. Given the number of self-similar objects (C) and the magnification factor (r), the dimensionality is given by $\frac{\log C}{\log r}$ resulting in dimensions 1, 2 and 3 for the line segment, square and cube respectively.

When an object is more self-similar, for the same magnification factor, a lot of self-similar objects are produced. If we reduce a line to half, the new line is 50% of the original line. On the other hand, for a cube it is 12.5% of the original cube. If there are C parts obtained by scaling to a factor of r, each part has a volume of $1/C$. Greater the C, lesser is the volume of each part and more is the loss. More loss implies that the object resides in higher dimension.

Interestingly, the idea of dimension can be generalized to non-integral values of dimensions for objects like Sierpienski triangle which has $C = 3$ and $r = 2$. For the purpose of the current paper, the following intuition is important. An object with Fractal Dimension (FD) close to 1, say 1.1 looks like a line, but an object with FD of 1.9 fills up space like a two dimensional surface. An object with FD of 2.1 is closer to a 2-d surface than it is to a 3-d volume; one with a FD of 2.9 approaches a 3-d object.

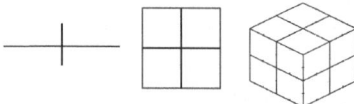

Fig. 2. Illustration of various shapes to measure Fractal Dimension

The idea of Fractal Dimension seems quite intuitive for euclidean objects. But how does it relate to points occupying a space? For example, we can have points in a straight line embedded in a 3 dimensional plane. So, here the extrinsic/embedding dimension is 3 while the intrinsic dimension is 1.

One of the ways to estimate Fractal Dimension is box-counting. In this method, we try to capture the relationship between the number of boxes that cover an object to the size of the box covering it. An illustration of the box-counting approach can be seen in Fig. 3. The smaller the box size, the more the number of boxes that are needed to cover the coastline. The relationship between the number of the boxes C and size of the box r is used to estimate the dimension. For a range of sizes, $r \in [r_1, r_2]$, we calculate the corresponding number of boxes C needed to cover the data. We plot $\log C$ vs $\log r$, and the slope of the line is the fractal dimension of the data.

With the help of an example, the idea of how the notion of Fractal Dimension captures the clustering tendency is demonstrated. Fig. 4 shows both a uniformly distributed data and clustered data along the boxes that cover it. It is seen that uniformly distributed data requires more number of boxes (of same size) to cover it, while the clustered data requires lesser number of boxes. With a uniform distribution, the number of boxes needed to cover the data will gradually decrease until the box is huge enough to cover the data. But in case of clustered data, the number of boxes suddenly decreases with increase in size. A moderate sized box covers more points when data is clustered, so less boxes are needed to cover the data. This can be seen in Fig. 4a and Fig. 4b where

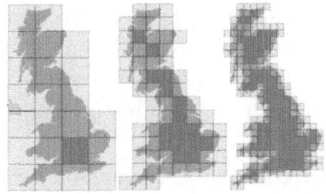

Fig. 3. Illustration of box-counting method to determine Fractal Dimension of the Britain coastline. The light shaded boxes are counted as they cover the coastline *Courtesy: en.wikipedia.org.*

for same box size, the number of covering boxes changes in the two scenarios. It is this change in behavior that we try to exploit when measuring the clustering tendency of data using Fractal Dimension.

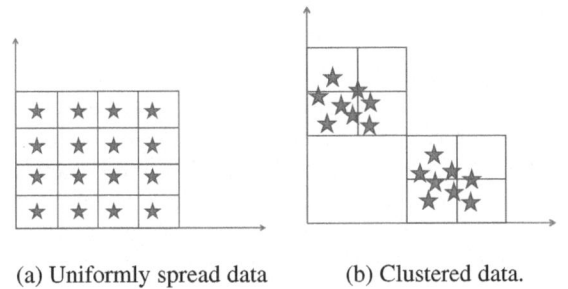

(a) Uniformly spread data   (b) Clustered data.

Fig. 4. Illustration of box-counting method to show clustering tendency in data

For a true self-similar object, the ratio $\log(C)/\log(r)$ is constant. But, it might be the case of real objects hence we try to fit the $\log C - \log(r)$ points to a straight line and then take the slope of the straight line to get an estimate of Fractal Dimension. The FDs of data shown in Fig 4a and 4b were found to be 1.8 and 1.3 respectively. Hence, it is seen that the clustered data has lesser Fractal Dimension than the uniformly distributed one.

For real datasets containing many features, it is hard to extend the notion of Fractal Dimension in a straightforward manner. The best we can do is come up with an estimate. Implementing the box-counting method is a little tedious and the points need to be in a vector space. Hence we use a simpler method which is easy to implement and works even with metric spaces. We use the idea of correlation dimension, where we count pairs of points that fall within a distance threshold. An interesting point to note is that to estimate correlation dimension, we only need pairwise distances of points. Thus, this measure is suitable even for metric spaces.

The correlation dimension is defined as follows:

$$CD = \lim_{r \to 0, N \to \infty} \frac{\log \frac{g_r}{N^2}}{\log r} \tag{1}$$

where, g_r corresponds to the number of pairs of points whose pairwise distance falls below a threshold distance r, and N is the total number of points under consideration. For a given set of points, we take a suitable range of distances $r \in [r_{min}, r_{max}]$, where

r_{min} is the minimum distance and r_{max} is the maximum distance between a pair of points. For each value of r, g_r is computed. Let C be $\frac{g_r}{N^2}$. The correlation dimension is slope of the $\log C$ vs $\log r$ plot. The distance range can be compared to box size and the number of boxes can be compared to the number of points that fall within the pairwise distance range.

In Figure 5 we look at how correlation dimension brings out the difference in the clustering tendencies in the dataset that we have seen earlier. The example clearly shows how for the same distance threshold the number of points differ. Like in box-counting method, the variation in the number of points for same distance threshold for both uniformly distributed and clustered points brings out the difference in correlation dimension. The log-log plots of C vs r are similar as in box-counting case.

While we approximate the log-log plot to a straight line, the actual log-log plot has a sigmoidal shape. At some scale, the ratio $\log(C)/\log(r)$ might be different. This is so because measuring the Fractal Dimension at various scales is like zooming in and zooming out of the structure and estimating the dimensionality at that resolution. When we zoom out of a structure far enough, any set of datapoints will appear like a single point whose dimension is 0. This translates to a flat portion on the right side of the log-log plot where scale is high. Intermediate zoom might give a different Fractal Dimension based on the structure of the data, and this will be reflected in the log-log plot as a piecewise linear structure. Finding the right scale to capture the 'true' structure of underlying data is a challenging task. Therefore, all we can hope for is a decent approximation of Fractal Dimension by fitting the log-log plot to a straight line.

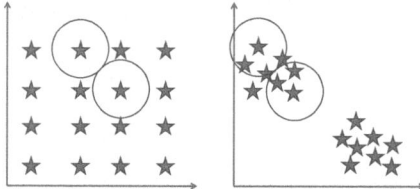

Fig. 5. Illustration of correlation dimension method to show clustering tendency in data. Only a few circles are shown for clarity. The FD behaviour is similar to Fig 4.

4   Our Approach

We are interested in estimating the collection complexity in the context of supervised classification tasks. For simplicity, we consider the case of binary classification. If the dataset is simple, class labels of most instances can be successfully predicted based on the labels of neighbours. This is less so, in a hard problem. A classification problem is hard when documents from the two classes have a large overlap of terms, resulting in complex decision boundaries. In a CBR context, documents correspond to problems and their class labels correspond to solutions. The first step in our approach is to estimate the alignment of the class labels of instances to those of their neighbours. We then focus our attention on those instances that have poor alignment since they are surrounded by points of the opposite class. The correlation dimension of this subset of points is estimated, suitably weighed by their alignments, and used as a measure of classification complexity.

For the same boundary characteristics, a case base with lower proportion of cases near the boundary must be considered less complex than one with a higher proportion of cases near the boundary. In its current form, our approach does not distinguish between the two. But, with the emphasis of CBR systems on competence and case base editing algorithms, cases far from the boundary are usually eliminated. With this observation, we can safely assume a large proportion of case base to lie across the boundary for practical purposes. Fig. 6 shows 2 class data with the boundary marked. If we take

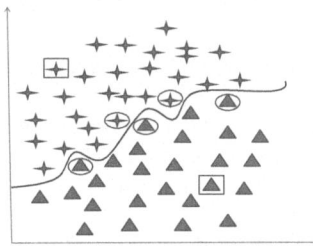

Fig. 6. Classification data with the class boundary marked. Points with low alignment are close to boundary.

the points marked in circle, we observe all of these points have low alignment. In contrast, the points marked in square have high alignment. Hence, if we consider points with low alignment, we get a subset of points close to the class boundary. Thus poorly aligned points are the items of interest when estimating Fractal Dimension. The measure captures intrinsic dimensionality of the poorly aligned datapoints, which is the real number of dimensions in which these points can be embedded while preserving the distances among them. The more uniform the spread of these points in feature space, the higher the intrinsic dimension and harder the classification problem.

The first subproblem, then, is to estimate the local alignment of each of the datapoints. This was previously explored in Massie et al. [7] and is done as follows. For a document instance d, consider its k-nearest neighbours and their labels. The ratio of the number of nearest neighbours having the same class label as the point to the total number of nearest neighbours (that is, k) is indicative of alignment. This measure can be made more granular, by taking into account the distance of d to each neighbour, and allowing neighbours closer to d to have a stronger influence on alignment than ones that are relatively farther. The local alignment of a point d_i with class label $label_i$ is defined as follows:

$$local\,Align(d_i) = \frac{\sum_{k \in nb(d_i)} prob\_sim(d_i, d_k) \times soln\_sim(label_i, label_k)}{\sum_{k \in nb(d_i)} prob\_sim(d_i, d_k)} \quad (2)$$

$$prob\_sim(d_i, d_k) = \cos(d_i, d_k) \quad (3)$$

$$soln\_sim(label_i, label_k) = \begin{cases} 0 : label_i \neq label_k \\ 1 : label_i = label_k \end{cases} \tag{4}$$

where N is the number of documents, d_k a document in neighbourhood of d_i and $label_k$ its class label, $nb(x)$ is the set of nearest neighbours of a document x, $prob\_sim()$ is problem side similarity function which is the cosine similarity ($\cos(d_i, d_k)$ $= \frac{d_i \cdot d_k}{|d_i||d_k|}$) in this case. We use cosine similarity due to the use of textual data. $soln\_sim()$ is solution side similarity as defined in Equation 4. A value close to 1 indicates strong alignment, and a value close to 0 indicated weak alignment. The latter are of interest to us, since they are expected to be closer to decision boundaries.

Algorithm 1 Procedure to calculate Fractal Dimension using alignment threshold

---

1. Given a collection of documents D, and the number of neighbors k.
2. For each document d in the collection, calculate its local alignment using Equations 2 and k neighbours. For our experiments, we used $k = 3$.
3. Fix an alignment threshold, say T.
4. Create a subset S of C, such that $S = \{d \in D | localAlign(d) \geq T\}$.
5. For the subset S of documents, calculate the counts C for a range of distances $r \in [r_{min}, r_{max}]$ where r_{min} is the minimum distance between the points and r_{max} is the maximum distance between the points. For each r, the count is calculated using Equation 1.
6. Find the slope of the log-log plot of counts vs distance, which is the Fractal Dimension of the collection.

Once the local alignments of the points are calculated, we can estimate the collection complexity using one of the following two approaches. The first idea is straightforward. We choose an alignment threshold, and the subset S of all points whose alignments fall below a threshold is considered. Correlation dimension of S is computed as described in Section 3. The detailed procedure is described in Algorithm 1. The measure, however, is sensitive to the choice of thresholds.

The second approach aims at overcoming this limitation. We have seen that estimating Correlation Dimension needs us to count g_r, the number of pairs of points whose pairwise distance falls below a threshold r. For our purpose, it seems reasonable that while counting g_r, a pair of points with local alignments 0.1 and 0.3 should be assigned higher importance than a pair of points with local alignments 0.8 and 0.9. We thus use the product of local alignments of document pairs as a fractional weight assigned to each pair, and aggregate these fractional counts while estimating g_r. This does away with problems related to thresholding, and leads to the following revised definition of weighted correlation dimension (wCD):

$$g'_r = \sum_i \sum_j (1 - localAlign(d_i)) * (1 - localAlign(d_j)) \tag{5}$$

g'_r would be where $distance(d_i, d_j)$ is within r. We use these modified counts to estimate correlation dimension by replacing g_r by g'_r in Equation 1.

Algorithm 2 Procedure for calculating weighed correlation dimension (wCD)

---

1. Given a collection of documents D, and the number of neighbors k.
2. For each document d in the collection, calculate its local alignment using Equations 2 and k neighbours. For our experiments, we used $k = 3$.
3. For the collection D of documents, calculate the counts C using modified counts as given in Equation 5 instead of normal counts, for a range of distances $r \in [r_{min}, r_{max}]$ where r_{min} is the minimum distance between the points and r_{max} is the maximum distance between the points.
4. Find the slope of the log-log plot of modified counts vs distance, which is the weighed correlation dimension of the collection.

The procedure for calculation of wCD is given in Algorithm 2. As it can be seen, the wCD measure does not require any thresholds to be set and gives a fair estimate of the correlation dimension across the boundary.

5   Experiments and Results

For the purpose of our experiments, we consider four binary classification datasets. Two of these were created from the 20 Newsgroups [11] corpus, RELPOL and HARD-WARE, each containing roughly 2000 documents evenly distributed among classes. RELPOL contains 2 groups of discussion related to religion and politics, while HARD-WARE contains 2 groups of discussion related to apple hardware and ibm-pc hardware. Other two are email spam datasets, LINGSPAM [12] (containing around 2400 legitimate and 500 spam messages) and USREMAIL [13] (containing around 1000 mails with 50% spam). The experimental setup is as follows. We create train and test splits using stratified sampling, where each of the sets contains 20% of the original corpus randomly chosen. Feature selection has been done using the Mutual Information (MI) method. Features are represented by binary values indicating presence or absence of a feature in the document. For repeated trials, 15 such train-test splits were created for each of the datasets. So, now we have 15 train-test splits, each containing 4 datasets of varying difficulty. Our aim is to show through experimentation that the proposed measure can predict the complexity of the dataset reasonably well. This is shown by looking at the correlation between the estimated complexity and the accuracy of various classifiers.

The classifier accuracies obtained over the datasets across different classifiers averaged over the 15 train-test splits is shown in Table 1. As outlined in Section 4, we calculate the correlation dimension of the set of hard-to-classify instances, which are identified as those points whose local alignment falls below a threshold. We call this threshold as Local Alignment Threshold (LAT). Without the threshold, if we use all the points in the dataset, we lose the class label information. The mesh plots in Figure 7 show how $\log C - \log(r)$ plot varies as a function of LAT. For creating this plot, we vary LAT from 0.2 through 1 with increments of 0.2. Setting LAT to 1 amounts to considering all points in estimating the correlation dimension. The mesh plots show the sensitivity of C-vs-r curves to the value of threshold. This gives justification to the fact that choosing points near boundary and then calculating correlation dimension gives a difference in complexity and choosing points away from boundary (high threshold) does not give information about complexity.

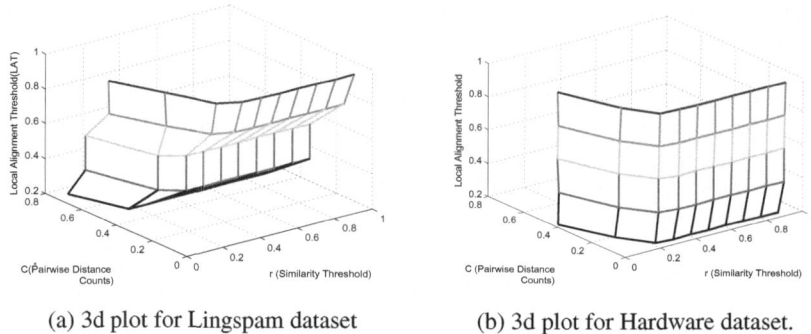

(a) 3d plot for Lingspam dataset   (b) 3d plot for Hardware dataset.

Fig. 7. Comparison of Lingspam and Hardware datasets

Observe that as we relax the threshold, the slope of lingspam curve rises steeply while for hardware it is a gentle rise. In the case of Lingspam, the points that do not belong to boundary, play a role in increase of intrinsic dimension when LAT is relaxed. This shows the need for a proper choice of LAT. The plots of the curves for 2 of the datasets, lingspam and hardware, are shown in Figures 8a and 8b at LAT=0.6. Figure 8a shows the graph of pairwise counts vs similarity threshold. Lingspam data has a smaller slope compared to hardware, and hence smaller correlation dimension.

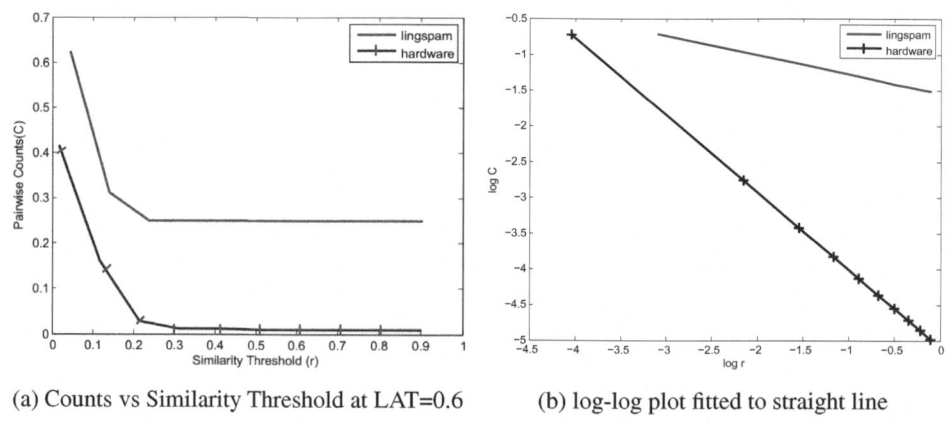

(a) Counts vs Similarity Threshold at LAT=0.6   (b) log-log plot fitted to straight line

Fig. 8. Plots for Lingspam and Hardware Datasets

The average classifier accuracies of datasets over 15 trials is shown in Table 1. This is to give an idea of relative difficulty of datasets to the reader. To examine if correlation dimension acts as a robust measure of complexity, we need to study its correlation against classification accuracies over the 4 datasets. This is compared to Global Alignment MEasure (GAME) which is proposed in Chakraborti et al. [8]. It can be seen in Table 2 that wCD significantly outperforms both correlation dimension with complexity

Table 1. Average classifier accuracies across different datasets for all the trials - Support Vector Machine (svm) with linear kernel, Naive Bayes Classifier (nb), k-Nearest Neighbor (knn)with $k = 3$, Random Forest (randforest)

	svm	nb	knn	randforest
lingspam	97.5	98.1	98.2	96.9
usremail	95.8	95.7	96.6	95.7
relpol	92.3	94.8	93	93.2
hardware	78.1	82.9	74.2	84.4

threshold=0.4 (best correlation performance) and GAME measure. It may be noted that GAME measures simplicity (and not complexity) of datasets and hence shows positive correlation with accuracies while our measures show negative correlation. The magnitude and not sign of correlation is important for comparison; the higher the magnitude, the better the measure is at estimating complexity. We observe here that GAME, unlike our measure does not neatly correspond to intrinsic dimension of hard-to-classify datapoints along the boundary.

Table 2. Correlation values of the complexity scores with the classifiers average over 15 trials

Complexity Measure	knn	svm	nb	randforest
GAME	0.89	0.91	0.91	0.88
correlation dimension with threshold=0.4	-0.92	-0.91	-0.91	-0.89
weighed correlation dimension (wCD)	-0.98	-0.97	-0.96	-0.94

6   Non Textual Datasets

An interesting thread of discussion would be, which of local or global alignment is a better estimate of complexity. What properties of the case base enable global measures to give a better estimate than local and vice-versa? We offer some interesting insights in that direction. We take inspiration from the saying that the whole is greater than the sum of the parts and look at some datasets for some answers.

One of the main steps in estimating complexity through correlation dimension is to find the points close to the boundary. In the previous approaches, we used the concept of alignment with the assumption that points with low alignment are close to the boundary. Another technique that gives the boundary points is the Support Vector Machine (SVM), and the points that lie on either side of the decision boundary and define it are Support Vectors. In this approach, we compute the Support Vectors (SV) for each dataset and for the datapoints corresponding to the SVs, we calculate the correlation dimension. In the same way as we evaluated the wCD, we take the correlation with the accuracies of the classifiers.

The motivation for our line of thinking comes from observing the SVs on synthetic data as shown in Fig 9. In the original space, while the alignment proposed by Massie et.al [7] gave both the datasets to be very well aligned, the correlation dimension of the SVs told a different story. While spiral data had a CD of 1.79, non linearly separable data had a CD of 1.32. This agreed with the classification accuracy in the original space

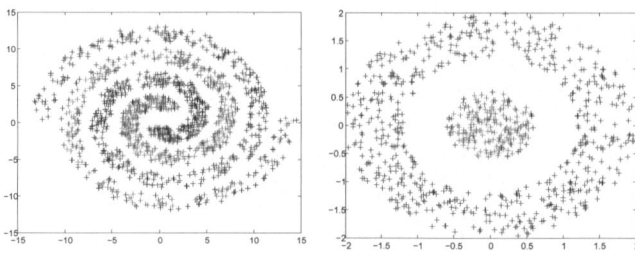

(a) Circular data with support vectors (marked in black).

(b) Circular data with support vectors (marked in black)

Fig. 9. Comparison of Support Vectors for synthetic datasets

which are :- 53.6% and 75% for spiral and non linearly separable data respectively. So there are situations where aggregating local alignments might not give a complete picture. The procedure for SV based correlation dimension is shown in Algorithm 3.

Algorithm 3 Algorithm for calculating fractal dimension using support vectors

---

1. Given data D, run the support vector machine optimization procedure on D to obtain the set of support vectors S.
2. For the subset S of D, calculate the counts C for a range of distances $r \in [r_{min}, r_{max}]$ where r_{min} is the minimum distance between the points and r_{max} is the maximum distance between the points. For each r, the count is calculated using Equation 1.
3. Find the slope of the log-log plot of counts vs distance, which is the fractal dimension of the data.

Having established that some data might have characteristics that cannot be explained as 'sum of its parts', we explore the scenario where we have real-valued solutions to the problems instead of class labels. For any CBR application that has the solution similarity space well defined, we can use the fractal dimension measure. For sake of evaluation, we use regression datasets that come close to a CBR system. We take 4 datasets namely - abalone, dee, plastic and ele-2 and check the correlation of the complexity measure with a performance measure. The datasets were obtained from the KEEL dataset repository [14]. The performance measure we chose for comparison is coefficient of determination $r2$, where $r2 = 1 - \frac{SS_{err}}{SS_{res}}$, where SS_{err} is the sum of squared error $\sum_{i=1}^{n} (y_{pred_i} - y_{true_i})^2$ and SS_{res} is the residual sum of squares $\sum_{i=1}^{n} (y_{mean} - y_{true_i})^2$. y_{pred_i} is the predicted value for i^{th} case, y_{true_i} the true value and y_{mean} the mean of true values. The best fit corresponds to 1 and as the score becomes lower, it implies that the performance is getting worse. The data is split into 5 train-test folds. The regression surface is learned using the training data and error is determined from the test data. We calculate the various complexity scores on the training data and check its correlation to the $r2$ score. The details of the datasets are given in Table 3. The average $r2$ score is computed using k-nn based regression fit on the training data.

Contrary to what we have done in the case of classification, we calculate correlation dimension with the normal counts instead of modified counts. For classification we

Table 3. The details regarding the regression datasets used in our experiments. The r2 score is based on k-nn regression. The data is arranged in order of increasing complexity.

Dataset	No of features	No of cases	Average r2 score
ele-2	4	1056	0.99
dee	6	365	0.85
plastic	2	1650	0.71
abalone	8	4177	0.49

Table 4. Correlation values of the complexity scores with the coefficient of determination r2, knn based regression (knn), support vector regression (svr) for rbf kernel (svr-rbf), svr for polynomial kernel (svr-poly) and linear regression (linear)

Complexity Measure	r2 (knn)	r2 (svr-rbf)	r2 (svr-poly)	r2 (linear)
GAME [8]	0.91	0.65	0.57	0.62
correlation dimension	-0.70	-0.61	-0.34	-0.65
alignment [7]	0.15	0.11	0.18	0.01

estimate the complexity of the classification boundary, while for regression we estimate the complexity of the surface we are trying to fit. We combine the n-features of the training data along with the labels to get a $(n + 1)$-dimensional surface, and we calculate the correlation dimension on this surface. If the problem side of cases is close but the solution is far apart, the combined surface will turn out to be uneven and difficult to fit. The unevenness lends itself to higher intrinsic dimension. So, data that is difficult to fit with regression might have higher correlation dimension. We make use of this hypothesis to estimate the difficulty of regression. The correlation values of the complexity measures to $r2$ score are shown in Table 4. An important step of preprocessing is to normalize the data so that all features lie in the interval $[0, 1]$.

The results show that the global measures of alignment - GAME and correlation dimension significantly outperform the local measure of alignment proposed in Massie et al. [7]. GAME appears to have an edge over correlation dimension in estimating the complexity of the datasets. Except with svr-poly, the correlation with various regression techniques is decent. For linear regression, the correlation dimension measure slightly outperforms GAME.

7 Conclusion

We addressed the problem of measuring the case base complexity in a supervised setting with a new measure based on intrinsic dimensionality. The idea of Fractal Dimension was exploited to estimate the intrinsic dimensionality of the set of points whose class labels are poorly aligned to those of their neighbours. Experiments on binary classification datasets show the effectiveness of this approach in predicting hardness of datasets, as is gauged from accuracies reported by standard classifiers. We studied the measures on synthetic data and regression data and made some interesting observations. We found that aggregating local complexities might make us lose the big picture for certain data. We found that our proposed measure performs comparably with GAME for complexity estimation on regression. Since case base maintenance procedures require local complexity measures to decide which case to retain, a local complexity measure based on intrinsic dimension is worth pursuing. We also plan to extend the idea to multi-class settings, and also take a closer look at sub-regions of the space which contribute the most to complexity. Modifying the measure to consider the proportion of case base that lies across the boundary to give a finer measure of complexity is another interesting problem. We have focused mainly on global measures, it would be interesting to study the data surface complexity as an aggregate of local complexity estimation.

References

1. Cummins, L., Bridge, D.: On Dataset Complexity for Case Base Maintenance. In: Ram, A., Wiratunga, N. (eds.) ICCBR 2011. LNCS, vol. 6880, pp. 47–61. Springer, Heidelberg (2011)
2. Massie, S., Craw, S., Wiratunga, N.: Complexity-guided case discovery for case based reasoning. In: AAAI, pp. 216–221 (2005)
3. Deepak, P., Chakraborti, S., Khemani, D.: Query Suggestions for Textual Problem Solution Repositories. In: Serdyukov, P., Braslavski, P., Kuznetsov, S.O., Kamps, J., Rüger, S., Agichtein, E., Segalovich, I., Yilmaz, E. (eds.) ECIR 2013. LNCS, vol. 7814, pp. 569–581. Springer, Heidelberg (2013)
4. Vinay, V., Cox, I.J., Milic-Frayling, N., Wood, K.: Measuring the complexity of a collection of documents. In: Lalmas, M., MacFarlane, A., Rüger, S.M., Tombros, A., Tsikrika, T., Yavlinsky, A. (eds.) ECIR 2006. LNCS, vol. 3936, pp. 107–118. Springer, Heidelberg (2006)
5. Marimont, R.B., Shapiro, M.B.: Nearest Neighbour Searches and the Curse of Dimensionality. IMA Journal of Applied Mathematics 24(1), 59–70 (1979)
6. Lamontagne, L.: Textual cbr authoring using case cohesion. In: Proceedings of 3rd Textual Case-Based Reasoning Workshop at the 8th European Conf. on CBR (2006)
7. Massie, S., Wiratunga, N., Craw, S., Donati, A., Vicari, E.: From Anomaly Reports to Cases. In: Weber, R.O., Richter, M.M. (eds.) ICCBR 2007. LNCS (LNAI), vol. 4626, pp. 359–373. Springer, Heidelberg (2007)
8. Chakraborti, S., Cerviño Beresi, U., Wiratunga, N., Massie, S., Lothian, R., Khemani, D.: Visualizing and evaluating complexity of textual case bases. In: Althoff, K.-D., Bergmann, R., Minor, M., Hanft, A. (eds.) ECCBR 2008. LNCS (LNAI), vol. 5239, pp. 104–119. Springer, Heidelberg (2008)
9. McGreggor, K., Kunda, M., Goel, A.: Fractals and ravens. Artificial Intelligence 215, 1–23 (2014)
10. Mandelbrot, B.: How Long Is the Coast of Britain? Statistical Self-Similarity and Fractional Dimension. Science 156(3775), 636–638 (1967)
11. Lang, K.: Newsweeder: Learning to filter netnews. In: Proceedings of the Twelfth International Conference on Machine Learning, pp. 331–339 (1995)
12. Sakkis, G., Androutsopoulos, I., Spyropoulos, C.D.: A memory-based approach to anti-spam filtering for mailing lists. Information Retrieval 6, 49–73 (2003)
13. Delany, S.J., Bridge, D.: Feature-Based and Feature-Free Textual CBR: A Comparison in Spam Filtering. In: Bell, D.A., Milligan, P., Sage, P.P. (eds.) Procs. of the 17th Irish Conference on Artificial Intelligence and Cognitive Science, pp. 244–253. Queen's University Belfast (2006)
14. Alcalá-Fdez, J., Fernández, A., Luengo, J., Derrac, J., García, S.: KEEL Data-Mining Software Tool: Data Set Repository, Integration of Algorithms and Experimental Analysis Framework. Multiple-Valued Logic and Soft Computing 17(2-3), 255–287 (2011)

CBR Tagging of Emotions
from Facial Expressions

Paloma Lopez-de-Arenosa, Belén Díaz-Agudo, and Juan A. Recio-García

Department of Software Engineering and Artificial Intelligence
Universidad Complutense de Madrid, Spain
{palomalopezdearenosa,belend}@ucm.es, jareciog@fdi.ucm.es

Abstract. Mobility and context-awareness are two active research directions that open new potential to recommender systems. Usage of dynamically enriched information from the user context leads the system to find better solutions that are adapted to the specific situations. In this paper we focus on the difficult problem of dynamically acquiring the *emotional context* about the user during a recommendation process. We use the fact that emotions are tightly connected with facial expressions and it is difficult for people to hide emotions in facial expressions. We describe PhotoMood, a CBR system that uses gestures to identify emotions in faces, and present preliminary experiments with MadridLive, a mobile and context aware recommender system for leisure activities in Madrid. In the experiments, the momentary emotion of a user is dynamically detected from pictures of the facial expression taken unobtrusively with the front facing camera of the mobile device.

Keywords: context-aware CBR, emotional context, image tagging.

1   Introduction

A case-based reasoning (CBR) system typically relies on knowledge sources that are processed offline, to define the structure of the cases and queries, and to define the similarity of the situations and how to adapt their solutions. However, in a highly interconnected and monitored world around us, there is a great amount of information available about the context surrounding the users that may be relevant to provide a more satisfactory solution. Mobility and context-awareness are two active research directions that open new potential to CBR systems [1–3].

Usage of dynamically enriched information from the user context leads the system to find better solutions that are adapted to the specific situations. This new information coming from the context necessarily alters the CBR processes, that need to be adapted to take into account the dynamic nature of this contextual knowledge which can change in each episode of problem resolution. Our current research intends to explore new ways of performing the CBR processes that take advantage of contextual factors. We focus on contextual *recommender systems* in mobile platforms where the appropriateness of the use of CBR is enhanced by introducing context-sensitive dynamically captured information about: the

L. Lamontagne and E. Plaza (Eds.): ICCBR 2014, LNCS 8765, pp. 245–259, 2014.
© Springer International Publishing Switzerland 2014

social context of the user, who are the user's friends and where are they; the *personal and emotional context*, how the users feel, mood, emotions, personality, preferences, accessibility requirements; the *physical context*, time, weather and position; and the *product context*, dynamic information about the product, like the day events. We are developing a mobile application: MadridLive that makes personalized recommendations for groups of people for leisure activities and events in Madrid. MadridLive uses the smartphone and web services to get the *physical* and *product context* information; uses preference tests and dynamic queries to get the *personal context*; and connects to FaceBook to enrich the *social context* information about the user's friends and the strength of the connections between the people in the group. This social context is taken into account when deciding the group activities following the methods developed in our previous research [4].

In this paper we focus on the difficult problem of dynamically acquiring the *emotional context* about the user during a recommendation process. We use the fact that emotions are tightly connected with facial expressions and it is difficult for people to hide emotions in facial expressions [5]. Research has shown that humans are surprisingly accurate in recognizing basic emotions from facial expressions [6, 7]. Humans can recognize happiness in 96.4% and 89.2% of the times in western and non-western cultures respectively [8]. Our research is a step towards developing an emotion recognition system using a CBR approach which can recognize basic emotions from static pictures taken with the front camera of the mobile device.

There are lots of applications and different approaches to facial recognition that are used to recognize a sample face from a set of given faces by inferring the gestures [9]. Our idea is the use of these features or gestures to identify emotions in the faces. We describe PhotoMood, a CBR system where a case base of manually tagged pictures is used to help dynamically tag new pictures with emotion tags. We propose using these tags to get the *emotional context* in MadridLive, a mobile and context aware recommender system of leisure activities in Madrid. We describe the results on a experiment where the user momentary emotion is dynamically detected from a picture of user's facial expression taken unobtrusively with front facing camera of his mobile device. In this experiment we focus on the lip region of the face. We obtain the emotional state of the user by means of a general case base of pictures and a personal case base where the system learns gestures from the user's self pictures. Preliminary results are promising to define the emotional context with two basic emotions: *I like* or *I dislike* the system recommendation.

The paper runs as follows. Section 2 reviews the related work in the field of tagging emotions from pictures. In Section 3 we describe the recommender system MadridLive and the role played by the emotions in the recommendation process. In Section 4, we detail the PhotoMood CBR module and how it is a integrated with MadridLive to avoid users the task of tagging the product with the like-dislike tags. Section 5 evaluates the results of the experiments and describes some problems found. Section 6 concludes and presents some ideas for future work.

2   Related Work

Facial expression analysis is probably the best non-intrusive way to estimate the emotional state [9]. However the development of an automated system that interprets facial expressions is rather difficult as explained in [10], where authors identify three common problems: the face detection, the extraction of the facial feature information and the expression classification.

Despite its difficulty there are many approaches to face detection, which could be generally classified into the following categories: Template matching methods look for certain patterns in pictures [11]. Feature-based methods use features such as color [12, 13] or shapes [14] to identify expressions and the face contour. Geometry-based techniques use sizes and relative positions of the components of face [15, 16]. Knowledge-based methods try to encode human knowledge of what constitutes a typical face [17, 18]. Appearance based techniques use linear transformation and statistical methods to determine the vectors that represent the face [19]. There are also several machine learning methods that use training samples [20, 21]. And finally, hybrid approaches that combine some of the previous techniques [22].

Regarding face detection applied to tagging of emotions we can find several proposals in the literature [23, 24]. For example, Maglogiannis et al [25] identify the emotions using edge detection and measuring the gradient of eye's or mouth's regions. They consider five major emotions: neutral, happy, sad, surprised and angry. Alternatively, Chawan et al [16] use the nearest matching pattern after applying the bezier curve on eyes and mouth.

These algorithmic techniques for emotion's recognition provide promising approaches in stationary settings [10, 26, 27]. However, mobile settings introduce important complications in capturing facial expressions. Recently some novel solutions have been proposed. For example, in Teeters et al [28] authors present a solution based on a chest mounted self camera detecting 24 feature points and a dynamic bayesian model; Gruebler et al [29] use a interface device to detect facial bioelectrical signals. Not intrusive techniques based on smartphones have been also explored. For example, mobile applications that captures pictures of one's facial expressions throughout the day [30]. There are many other mobile applications using emotion recognition, mostly for personal skills reinforcement or autism disorder therapy.

Next we present our mobile application that integrates the CBR module for emotion recognition: MadridLive.

3   MadridLive

MadridLive is a group recommender for leisure activities and events in Madrid. MadridLive uses the capabilities of mobile devices to enrich the recommendation process by means of contextual knowledge that is acquired dynamically.

This enrichment is provided by four kinds of contextual knowledge: first, the sensors included in the mobile device provide the *physical* context that defines the location, time or weather. Secondly, this knowledge is used to obtain the *product context*. In this case, it is the information about the activities to be recommended (restricted to the location, time and weather). MadridLive loads dynamically from web services the details of different types of activities that are nearby the user: museums, restaurants, parks and walks.

Each retrieved activity is scored using a specialized recommender that can follow a content-based or collaborative strategy depending on the information available. For example, the recommender of museums and restaurants uses a test to obtain the preferences of the user and therefore applies a content-based strategy. On the other hand, the recommender for parks and walks uses a collaborative approach based on the feedback provided by all the users of the system. These preferences of the users obtained through tests or the history of previous consumed items represent their *personal context*. MadridLive also keeps a record of the estimated satisfaction of the user with previous suggestions to bias current recommendations.

Finally, the scores estimated for every activity and user are aggregated to provide a recommendation for the group. The users can be added to the group automatically, both using the locations included in the physical context or the social connections provided by the *social context*. In MadridLive this knowledge is obtained by means of an analysis of the Facebook profiles and is exploited to enhance the group aggregation of individual scorings. To do so, we apply our techniques previously developed in [4] that take into account the social ties and satisfaction of the users.

In this paper we extend these four types of context -product details, physical, personal and social- with an additional context that can potentially improve the global recommendation process: the *emotional context* based on the mood of the user. Concretely, our system uses the emotional context to automatically obtain the immediate satisfaction of the user regarding the activities being recommended. Up to now, this information was obtained using the typical *like-dislike* buttons as shown in Figure 1. Next section describes the Photomood CBR module that is a integrated with MadridLive to avoid users this annoying task. MadridLive automatically takes a picture on the user when the recommended product is presented, and Photomood unobtrusively gets the *like-dislike* feedback from the analysis on his or her facial expression.

4   PhotoMood: A CBR Approach to Face Emotion Recognition

PhotoMood is a module to automatically obtain the user's mood through the analysis of the facial expression. It has been integrated into the MadridLive mobile application to obtain the opinion of the user about the proposed activities. It follows a CBR approach where the case base is composed of tagged user pictures. Therefore, this module implements a CBR classification system where

Fig. 1. Screenshots of MadridLive Mobile App: login screen (left), physical context display (center) and recommendation with like-dislike buttons (right)

tags of the most similar pictures are used to estimate the current user's mood through a majority voting schema. It is important to note that PhotoMood is able to work with several tags although in MadridLive we only use the *like-dislike* classes.

The integration of PhotoMood into MadridLive is graphically described in Figure 2: when an activity is proposed to the user, the MadridLive application takes a picture of his/her facial expression through the front camera of the mobile device. This picture will be the input query of the PhotoMood CBR module that performs the retrieval stage using the similarity metric explained next. Then, the reuse stage consists on a weighted voting schema using the tags/classes associated to every retrieved picture. Finally, the prediction (like or dislike) is sent back to MadridLive to be displayed to the user. Concretely, MadridLive shows the typical like-dislike buttons below the proposed activity but highlighting one of them according to the prediction. This enables the user to modify the prediction of the system and, therefore, revise the proposed solution. Finally, those predictions that were modified by the user are stored in the pictures' case base during the retain stage of the PhotoMood's CBR cycle.

Before performing the CBR cycle itself, PhotoMood needs to preprocess the picture taken by the camera in order to prepare it for the retrieval stage. Following subsections detail both preprocessing and the CBR cycle.

4.1   Picture Preprocessing

PhotoMood's picture preprocessing consists mainly in a face and gesture detection stages. There are many algorithms in the literature that can be used: Principal Component Analysis [31], Independent Component Analysis [32], Kernel

Fig. 2. Interaction between MadridLive Mobile App and PhotoMood

methods [33], Support Vector Machines [34], or Hidden Markov Models [35] are some examples of the approaches being applied. Concretely, we use the Luxand SDK since it offers a high quality according to comparative studies [36]. Concretely, our algorithm follows the steps shown in Figure 3:

1. Face Detection from the picture through Luxand SDK.
2. Cropping and/or scaling Facial Region so that it matches our size specifications.
3. Gesture detection: Luxand SDK provides up to 66 facial feature points (eyes, eyebrows, mouth, nose)
4. Gesture selection of the mouth's points[1].

The result of the preprocessing step is a list of 8 points representing the coordinates of the mouth contour (4 points for the upper lip and another 4 points for the lower one). Next we explain how these points are used as the query of the CBR system.

4.2   The PhotoMood CBR Process

The input of our CBR system is the list of points computed by the preprocessing stage that conform our query $\mathcal{Q}$:

$$\mathcal{Q} = <p_1^q, \ldots, p_n^q> \tag{1}$$

---

[1] Up to now the evaluated algorithm only uses mouth's points although we have done preliminary work with eyes and eyebrows.

Fig. 3. PhotoMood steps (left to right): 1) original picture; 2) Face detection; 3) Face cropping and scaling; 4) Gesture Detection; 5) Lip Points

Therefore the case base is composed of these descriptions plus their corresponding solutions, concretely, the emotion tags associated to each picture.

$$\mathcal{C} = <\mathcal{D}_c, \mathcal{S}_c> \tag{2}$$

where

$$\mathcal{D}_c = <p_1^c, \ldots, p_n^c> \tag{3}$$

$$\mathcal{S}_c = \{t_1, \ldots, t_m\} \tag{4}$$

Given a query $\mathcal{Q}$, the CBR process follows the four standard stages:

Retrieval. A k-Nearest Neighbour algorithm is run configured with a similarity metric based on the angle between points. Following the contour of the mouth, the arctangent between one point and the next one is computed to obtain their angle. Each angle of the query $\mathcal{Q}$ is compared to the corresponding angle of the case $\mathcal{D}_c$ producing n (local) similarity values that are later averaged to obtain the final global similarity:

$$Sim(\mathcal{Q}, \mathcal{D}_c) = \frac{1}{n}\sum_{i=1}^{n} 1 - \sqrt{\arctan(\widehat{p_i^q p_j^q}) - \arctan(\widehat{p_i^c, p_j^c})} \tag{5}$$

$$\text{where } j = (i+1) \bmod n$$

The retrieval stage returns an ordered list with the k cases most similar to the query.

Reuse. To obtain a solution for the query, PhotoMood applies a weighted voting schema according to the similarity of the retrieved cases. Given the scoring function:

$$score(t_i) = \sum sim(\mathcal{Q}, \mathcal{D}_c) \ \forall \ c \, | \, \mathcal{S}_c = t_i \tag{6}$$

The solution assigned to the query is:

$$t_i = arg\,max\{score(t_i),\ i = 1, \ldots, m\} \tag{7}$$

Revise. As it was previously explained, the revision of the proposed solution is directly performed by the user. Thus, it is external to the PhotoMood CBR module. In this case, the MadridLive application allows the user to modify its estimated mood by pushing one of the like or dislike buttons.

Retain. The retain policy of PhotoMood only stores those cases that were revised by the user. That is, solutions incorrectly computed by the CBR process.

Once we have described our CBR system we present its experimental evaluation.

5   Experimental Evaluation

This section describes the experimental evaluation performed in order to validate the CBR process being proposed. We explain the experiment design, its results and some proposals to enhance the results obtained.

5.1   Experimental Setup

We use a general case base (GCB) with 250 pictures of anonymous people taken from web searching processes[2]. Besides, we use 10 tester users with personal case bases (CB_i) of 30 pictures for each $user_i$. These pictures were taken with his/her mobile device and were manually labelled with the *like* or *dislike* tags. This way, the algorithm described in the previous section is configured with two tags $(m = 2)$. It was also parametrized with 8 points describing each mouth $(n = 8)$ and a $k = 3$ for the nearest neighbour method.

We performed five different evaluation tests using alternative case bases to achieve the following goals:

1. $Test_1$: Measure the performance of a generic case base. For each user 10 pictures are randomly chosen as queries. The CBR system contains exclusively pictures from the general case base. This process is repeated 10 times, obtaining an average precision.

2. $Test_2$: Evaluate the performance improvement when using personal pictures. A 3-fold cross-validation test using only the user's personal case base (CB_i). Each validation uses 10 pictures as queries (5 likes and 5 dislikes) and the 20 remaining as the case base.

3. $Test_3$: Analyse the impact of enhancing the general case base with personal pictures. It performs a cross-validation test with pictures from an extended case base (ECB), made of the union between GCB and each personal case base (CB_i). The personal case based is split into 3 folds used as queries against a case base containing the pictures from GCB and the remaining two folds.

---

[2] We used Google Image Search with the queries "Happy face" and "Unhappy face".

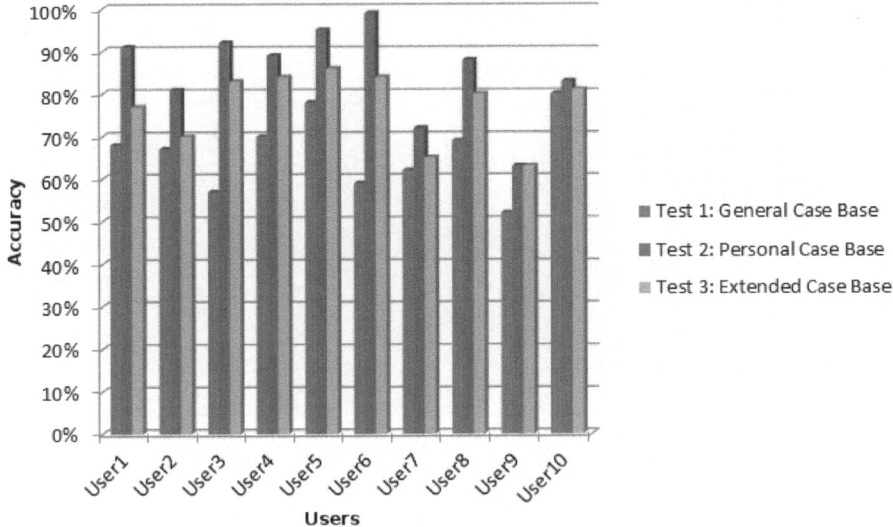

Fig. 4. Summary of the experimental results

4. $Test_4$: Study the evolution of the system regarding to the number of personal pictures learnt by the CBR process. Here, the ECB is progressively evaluated as personal pictures are included.
5. $Test_5$: Discard possible bias depending on the tags *like-dislike*. To do so, the case base is split into both groups of pictures and then re-evaluated.

Additional tests were performed but are not presented here due to their expected –and confirmed– poor performance. For example, a 10-fold cross-validation using GCB exclusively. We also tried several parameter configurations ($n = 6$, $k = 5$ or $k = 7$) but no significant improvements were found. Next we detail the results of our tests.

5.2 Results

Figure 4 summarizes the results of our experimental evaluation comparing the performance of the system when configured according to $Test_1$, $Test_2$ and $Test_3$.

With respect to the accuracy of $Test_1$, it mainly depends on the similarity between the user's facial features and the features of the people in GCB.

$User_9$ is in point for his low precision values. This user is very expressive, with many different happiness expressions, resulting in a wide range of values of similarity when using the mouth's contour obtained from his photos. GCB diversity of cases is not enough to find users with similar happiness expressions. A solution here could be to expand the number of classes/tags (like, dislike, sadness, indifference, surprise,...). We discuss this possibility in Section 6. Another solution could be to increase the number of cases in GCB. However, this doesn't

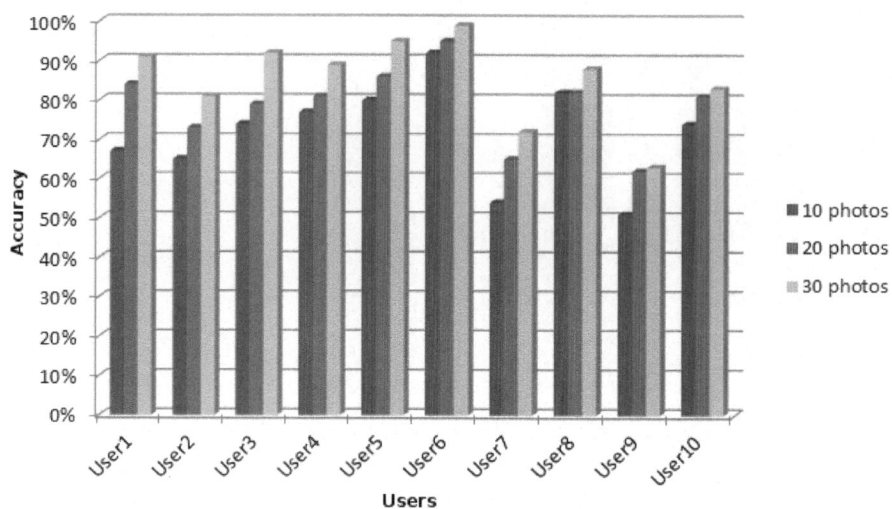

Fig. 5. Analysis of the impact of learning personal pictures

have to work in everyone's favour as it adds noise to the case base. Hence the only plausible solution is the use of the personal case base that is specific for each user. This approach is evaluated in $Test_2$.

For every user, $Test_2$ results using CB_i improve significantly compared to $Test_1$'s, using GCB. It is clearly due to the comparison of the users' pictures to their own actual expressions and facial features. However, we observe that $user_9$ results are still the worst. One reason is that, because of the diversity of gestures, $user_9$'s expressions extracted from his pictures are very different from each other, so it's no easy to find similar cases. Here, a solution is the enlargement of $user_i$'s case base (CB_i) with additional pictures. Another reason –without a clear solution– is that some mouth's contours are ambiguous, even comparing them to the user's own expressions. Here the comparison of other facial features, like eyes and eyebrows, could address this problem.

Another difficult case is $user_7$. Problem with $user_7$ is that she is not expressive at all, and her happiness expressions are too similar to the unlikeness expressions. This problem could be solved by including more pictures with more defined gestures.

$User_3$ and $user_6$ are the ones that are more benefited from the use of the personal case base. This is because his/her expressions are not similar at all to those ones in the GCB.

Figure 4 also shows the results of the third evaluation $Test_3$. The use of ECB, the extended case base mixing pictures from CB_i and GCB, reports better results than using GCB on its own. However, $Test_2$ results, using only the specific user case base CB_i, are still the best. The reason is that this case base minimizes noise, and therefore, improves accuracy.

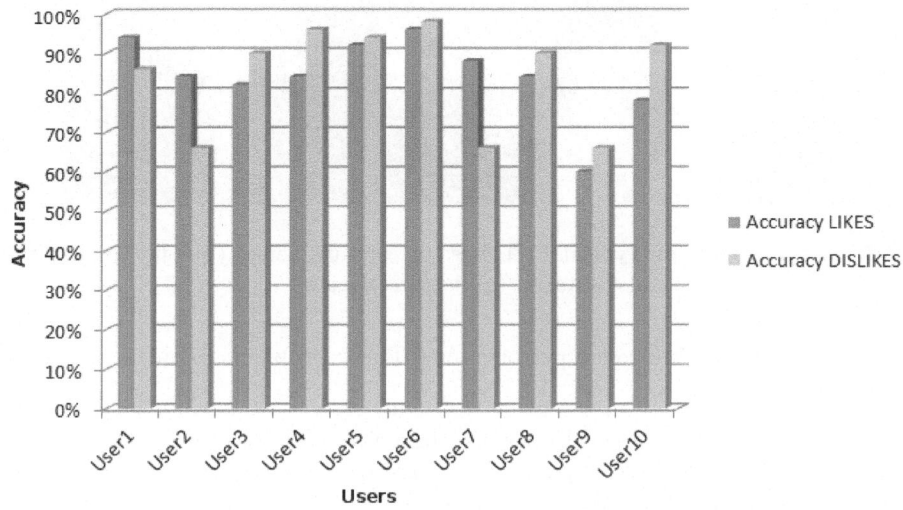

Fig. 6. Analysis of accuracy for each class/tag

We have also analysed the impact of the learning capabilities of our CBR approach in $Test_4$. As we have previously explained, PhotoMood performs pretty good and obtains higher precision values when using user's self pictures. However, when a user starts using the application the CBR system only includes those pictures in the general case base GCB. Personal pictures will be progressively included into the case base as the user interacts with the application. This is the typical cold-start problem. Therefore we have measured the evolution of the system performance depending on the personal pictures being included into the case base. To do so, we have performed an additional test where personal pictures are progressively included (10, 20 or all the 30 personal pictures). This test uses 5 random personal pictures as queries repeated 10 times. Figure 5 clearly shows that the system performs better as more cases/pictures are learnt.

Finally, in $Test_5$ we have studied the possible bias of the system depending on the tags/classes being considered. Results shown in Figure 6 prove that there are no significant differences, and that accuracy only depends on the specific user's gestures and expressiveness.

5.3 Case Base Enhancement

When working with pictures there are some typical problems, like blurriness, lighting, angles, etc. that affect the precision of any image processing algorithm. In our experiments we have identified these irregularities in order to remove these pictures/cases from the case base. Nevertheless, results previously reported were computed using the original set of pictures sent by the users to reflect a real scenario. But a more elaborated preprocessing of the pictures in the case

Fig. 7. Face Recognition Typical Problems: (a) picture angle, (b) blur, (c) lighting conditions, (d) proportional distances, (e) inexpressiveness

base shall lead to an improved performance. These problems found in the users' pictures are:

1. Viewing angle of the face: the application performs pretty good at full frontal faces and around 15/20 degrees off but as soon as you go towards profile the results are much worse (see Figure 7a).
2. Blurry pictures or low resolution leads to problems on recognition the mouth reference points. (see Figure 7b)
3. Lighting conditions specially pictures against the light (see Figure 7c).
4. Difference on proportions due to closeness to the camera (see Figure 7 d).
5. Inexpressive people, showing less emotions or without difference in the likes or dislikes face expressions will obtain bad results.

Addressing these problems will obviously enhance the performance of the system. However it is out of our future work presented next as it is not related to CBR research.

6   Conclusions and Future Work

In this paper we have described PhotoMood, a CBR system that uses gestures to identify emotions in the faces, and we have presented preliminary experiments with MadridLive, a mobile and context aware recommender system of leisure activities in Madrid. When the user gets a recommendation with MadridLive, his/her momentary emotion is automatically tagged from a picture of user's facial expression taken unobtrusively with the front facing camera of his/her mobile device. PhotoMood is a CBR system that uses a case base with previously annotated pictures to identify the user momentary emotion from his/her picture (the query). In the experiments, presented in Section 5, PhotoMood performs pretty good and obtains high precision values in tagging basic emotions. Results are better when using a personal case base made of user pictures than general pictures from anonymous people. The system learning capability offers the possibility of using a general case base in the cold start situation, before the user has acquired his/her own trained case base of tagged self pictures. The advantage of the CBR approach is the capability of training the system with better quality self pictures to improve precision results. Besides, the user could also improve the results by refining the tags associated to a picture, in case of mistakes of annotation, or adding more pictures to his/her personal case base to adjust personal particular expressions.

Although the experiments in this paper use only two basic emotions, "like" and "dislike", PhotoMood works with the number of emotion tags that are used to annotate the user pictures. The algorithm retrieves the most similar neighbours and assigns the emotion tags to the unlabelled query user picture, whatever these emotions are. In the recommender application MadridLive, we only use the two mentioned tags and the mobile interface of the recommender application (see Figure 1) allows to annotate the user pictures. However we have planned to extend the experiments by manually tagging the user pictures with additional emotion tags and see how the use of a diversity of emotions affects the results. As we discuss in Section 5 there are users, like $user_9$ that would benefit from this extension. Emotions we are considering that could be useful to recommender systems are like, dislike, sadness, indifference and surprise; being surprise a sign of likeness, and sadness a sign of unlikeness.

As future work we are also extending our approach to other facial features, namely we have been doing preliminary experiments with eyes and eyebrows. However, initial results have not been good with 5 points in each eyebrow, as the software was not able to distinguish the emotion and always tagged the expression as "neutral". Besides, next step in our research is experimenting about the exploitation of the acquired emotional context and how it can be used to improve the recommendation results. Once the results of the automatic emotional tagging system PhotoMood are validated, we are using these results to better personalize the products that are recommended to this user and the groups of friends of this user. We are measuring how the use of the emotional context affect the average user satisfaction.

Acknowledgements. Supported by UCM (Group 910494) and Spanish Ministry of Economy and Competitiveness under grant IPT-2011-1890-430000.

References

1. Adomavicius, G., Mobasher, B., Ricci, F., Tuzhilin, A.: Context-aware recommender systems. AI Magazine 32, 67–80 (2011)
2. Braunhofer, M., Kaminskas, M., Ricci, F.: Location-aware music recommendation. IJMIR 2, 31–44 (2013)
3. Benou, P., Bitos, V.: Context-aware query processing in ad-hoc environments of peers. JECO 6, 38–62 (2008)
4. Quijano-Sánchez, L., Recio-García, J.A., Díaz-Agudo, B., Jiménez-Díaz, G.: Social factors in group recommender systems. ACM Transactions on Intelligent Systems and Technology 4, Article 8 (2013)
5. Cohn, J.F.: Foundations of human computing: Facial expression and emotion. In: Huang, T.S., Nijholt, A., Pantic, M., Pentland, A. (eds.) AI for Human Computing. LNCS (LNAI), vol. 4451, pp. 1–16. Springer, Heidelberg (2007)

6. Ortony, A., Turner, T.J.: What's basic about basic emotions? Psychological Review 97(3), 315–331 (1990)
7. Russell, J.A.: Is there universal recognition of emotion from facial expressions? A review of the cross-cultural studies. Psychological Bulletin 115, 102–141 (1994)
8. Scollon, C.N., Kim-Prieto, C., Diener, E.: Experience sampling: Promises and pitfalls, strengths and weaknesses. Journal of Happiness Studies 4 (2003)
9. Eckman, P.: Facial expression and emotion. American Psychologist 48, 384–392 (1993)
10. Pantic, M., Rothkrantz, L.J.M.: Automatic analysis of facial expressions: The state of the art. IEEE Trans. Pattern Anal. Mach. Intell. 22(12), 1424–1445 (2000)
11. Feris, R.S., de Campos, T.E., Cesar Jr., R.M.: Detection and tracking of facial features in video sequences. In: Cairó, O., Sucar, L.E., Cantu, F.J. (eds.) MICAI 2000. LNCS, vol. 1793, pp. 127–135. Springer, Heidelberg (2000)
12. Solina, F., Peer, P., Batagelj, B., Juvan, S., Kovac, J.: Color-based face detection in the "15 seconds of fame" art installation. In: Proceedings of Mirage 2003 (INRIA Rocquencourt), pp. 38–47 (2003)
13. Kovac, J., Peer, P., Solina, F.: Human skin color clustering for face detection. In: EUROCON 2003, Computer as a Tool. The IEEE Region 8, vol. 2, pp. 144–148. IEEE (2003)
14. Wang, J., Tan, T.: A new face detection method based on shape information. Pattern Recognition Letters 21, 463–471 (2000)
15. Paul, S.K., Uddin, M.S., Bouakaz, S.: Extraction of facial feature points using cumulative histogram. CoRR abs/1203.3270 (2012)
16. Chawan, P.M., Jadhav, M.M.C., Mashruwala, J.B., Nehete, A.K., Panjari, P.A.: Real time emotion recognition through facial expressions for desktop devices. International Journal of Emerging Science and Engineering (IJESE) 1, 104–108 (2013)
17. Lin, C., Fan, K.-C.: Triangle-based approach to the detection of human face. Pattern Recognition 34, 1271–1284 (2001)
18. Yang, G., Huang, T.S.: Human face detection in a complex background. Pattern Recognition 27, 53–63 (1994)
19. Draper, B.A., Baek, K., Bartlett, M.S., Beveridge, J.: Recognizing faces with PCA and ICA. Computer Vision and Image Understanding 91, 115–137 (2003) (Special Issue on Face Recognition)
20. Rowley, H.A., Baluja, S., Kanade, T.: Neural network-based face detection. IEEE Trans. Pattern Anal. Mach. Intell. 20, 23–38 (1998)
21. Sung, K.-K., Poggio, T.: Example-based learning for view-based human face detection. IEEE Trans. Pattern Anal. Mach. Intell. 20, 39–51 (1998)
22. Lin, H.J., Yen, S.H., Yeh, J.P., Lin, M.J.: Face detection based on skin color segmentation and svm classification. In: SSIRI, pp. 230–231. IEEE Computer Society (2008)
23. Fragopanagos, N., Taylor, J.: Emotion recognition in human computer interaction. Neural Networks 18, 389–405 (2005) (Emotion and Brain)
24. Pantic, M., Rothkrantz, L.J.: Facial action recognition for facial expression analysis from static face images. Trans. Sys. Man Cyber. Part B 34, 1449–1461 (2004)
25. Maglogiannis, I., Vouyioukas, D., Aggelopoulos, C.: Face detection and recognition of natural human emotion using markov random fields. Personal and Ubiquitous Computing 13, 95–101 (2009)
26. Anderson, S., Conway, M.: Investigating the structure of autobiographical memories. Journal of Experimental Psychology: Learning, Memory, and Cognition 19, 1178–1196 (1993)

27. Cohen, I., Sebe, N., Garg, A., Chen, L.S., Huang, T.S.: Facial expression recognition from video sequences: temporal and static modeling. Computer Vision and Image Understanding 91, 160–187 (2003) (Special Issue on Face Recognition)
28. Teeters, A., El Kaliouby, R., Picard, R.: Self-cam: Feedback from what would be your social partner. In: ACM SIGGRAPH 2006 Research Posters. SIGGRAPH 2006. ACM, New York (2006)
29. Gruebler, A., Suzuki, K.: Analysis of social smile sharing using a wearable device that captures distal electromyographic signals. In: Stoica, A., Zarzhitsky, D., Howells, G., Frowd, C.D., McDonald-Maier, K.D., Erdogan, A.T., Arslan, T. (eds.) EST, pp. 178–181. IEEE Computer Society (2012)
30. Karapanos, E.: Modeling Users' Experiences with Interactive Systems. SCI, vol. 436. Springer, Heidelberg (2013)
31. Turk, M., Pentland, A.: Eigenfaces for recognition. J. Cognitive Neuroscience 3, 71–86 (1991)
32. Hyvärinen, A., Oja, E.: Independent component analysis: Algorithms and applications. Neural Netw. 13, 411–430 (2000)
33. Liu, Q., Huang, R., Lu, H., Ma, S.: Face recognition using kernel-based fisher discriminant analysis. In: Proceedings of the Fifth IEEE International Conference on Automatic Face and Gesture Recognition, pp. 197–201 (2002)
34. Guo, G., Li, S.Z., Chan, K.L.: Support vector machines for face recognition. Image and Vision Computing 19, 631–638 (2001)
35. Nefian, A.V., Hayess, M.H.: Hidden Markov Models for Face Recognition. In: Proc. International Conf. on Acoustics, Speech and Signal Processing (ICASSP 1998), vol. 5, pp. 2721–2724 (1998)
36. Degtyarev, N., Seredin, O.: Comparative testing of face detection algorithms. In: Elmoataz, A., Lezoray, O., Nouboud, F., Mammass, D., Meunier, J. (eds.) ICISP 2010. LNCS, vol. 6134, pp. 200–209. Springer, Heidelberg (2010)

A Proposal of Temporal Case-Base Maintenance Algorithms

Eduardo Lupiani, Jose M. Juarez, and Jose Palma

Computer Science Faculty, Universidad de Murcia, Spain
{elupiani,jmjuarez,jtpalma}@um.es

Abstract. Time plays a key role in describing stories and cases. In the last decades, great efforts have been done to develop Case-Based Reasoning (CBR) systems that can cope with the temporal dimension. Despite this kind of system finds it difficult to maintain their case-base, little attention has been paid to maintain temporal case-bases. Case-Base Maintenance (CBM) algorithms are useful tools to build efficient and reliable CBR systems. In this work, we propose the extension of different CBM approaches to deal with this problem. Five temporal CBM algorithms are proposed (t-CNN, t-RENN, t-DROP1, t-ICF and t-RC-FP). These algorithms make the maintenance of case-bases possible when cases are temporal event sequences.

Keywords: Case-Base Maintenance, Temporal CBR framework, Algorithms.

1 Introduction

There are myriads of domains in which the temporal dimension is essential to describe a problem to be solved, such as the observation of the evolution of variables through time or activity planning descriptions. In Case-Based Reasoning (CBR), temporal case representation mainly rely on time series, episodes, workflows and event sequences [2, 6, 14]. As in classical CBR systems, the performance of temporal CBR systems can be reduced if the number and size of cases in the case-base are increased.

Case-Base Maintenance (CBM) can reduce the size of a case-base keeping its problem solving correctness and, therefore, improving its performance in most of the situations. CBM implements policies for revising the organization or contents (e.g. representations, domain content or accounting information) of the case-base in order to simplify the future problem-solving process subjected to a particular set of performance objectives [8]. In particular, according to [22], CBM algorithms aim to reduce the case-base size or to improve the quality of the solutions provided by the CBR system.

Different CBM algorithms have been proposed such as CNN, RENN, DROP1 or ICF among others [1, 12, 18, 22]. Whereas CBM algorithms typically use a particular heuristic to remove (or select) cases from the case-base, the resulting maintained case-base relies on the proportion of redundant and noisy cases that

L. Lamontagne and E. Plaza (Eds.): ICCBR 2014, LNCS 8765, pp. 260–273, 2014.
© Springer International Publishing Switzerland 2014

are present in the case-base, among other factors. That is, each type of CBM algorithm is suitable for certain types of case-bases that share some indicators, such as redundancy and noise levels.

Case bases containing event sequences often require automatic and semi-automatic actions to be maintained. In our opinion, CBM is a suitable approach to avoid underperformance in such case-bases. However, as far as we know, little attention has been paid to the analysis of the effect of the maintenance in temporal cases.

In this work, we propose to adapt different CBM algorithms to maintain temporal case-bases and to analyse its suitability for event sequences. Section 2 surveys related work in both temporal CBR systems and CBM algorithms. In Section 3, a formalization of event sequences and temporal distances is given. Our proposal of temporal CBM algorithms is described in Section 4. Then, we present experimental results and discuss in Section 5. Finally, in Section 6 we end with conclusions and future works.

2   Background

In the last decade, the inclusion of temporal features within the problems descriptions has gained a relevant role in the CBR systems, as is highlighted by the amount of publications related with CBR systems working with time dependent features [4, 7, 15]. In particular, the most used temporal structures are time series, workflows and event sequences. This work is focused on event sequences.

Event sequences consists of a collection of ordered heterogeneous events, where each event is composed of an event type and a time-stamp that represent when the event occurs [7, 11]. While event types could be either categorical or numeric values, time-stamps could be a date or just simply a number representing a time instant. Furthermore, time series are restricted to the observation of one temporal feature while event sequences could involve two or more different temporal features. Moreover, event sequences could also be used to represent the execution of workflows, by defining every existing task in the workflow and their begin time as an event type and a time-stamp respectively.

Heterogeneous event sequences cannot use time series similarity techniques and new proposals are required to compare this kind of cases [17]. The distance between event sequences is usually based on the temporal edit distance [11, 13], an adaptation of the Levenshtein distance [9]. The Match & Mismatch (M&M) algorithm splits each event sequence into several event sub-sequences, one list for each event type [23].

Essentially, for every event type the algorithm obtains an edit-distance matrix. From all the matrices built, four different measures are calculated: the time difference between the events (TD), the missing events (NM), the extra event (NE) and the number of swamping events (SN). These measures are the basis of one overall score that quantifies the similarity between two event sequences. Other approaches are based on heuristics [6, 24], vector theory [4] or possibilistic temporal constraint networks [7].

There is a wealth of fields in which CBR has been successfully applied to solve problems where cases are heterogeneous event sequences. For example, in [19] a CBR system for the oil well drilling domain is presented. This system is based on *episodes* of cases, where a case is designed as a static snapshot to record a set of parameters values that characterised an oil drilling process. The complete operation process is built by linking different cases through the time. Since episodes always have the same number of cases, then the similarity between cases is done using sequence pattern matching. In [3], a case contains a sequence of complex events, where these complex events could represent information related with well oil drilling operations. Since the event sequences are transformed into vectors of fixed length, the similarity between cases can be done by calculating the distance between vectors, for example, using the cosine similarity.

For predicting the length of stay of patients in a hospital a system is proposed in [6]. In this work, the event types comprise different clinical events, such as admission or discharge. They are arranged into patient traces, which follow the structure of an event sequence. In order to compute the similarity between two event sequences, a temporal edition distance is proposed.

According to [16, 20], there are different dimensions in which CBM algorithms can be classified: *case search*, *direction*, *order sensitive* and *type of cases* to retain, as well as the different CBM algorithm families: *NN*, *DROP* and *Competence*. On top of those families and dimensions, a myriad of CBM algorithms have been proposed in the literature [1, 5, 12, 18, 20–22]. For the sake of simplicity, table 1 shows the different dimensions and algorithms, as well as some CBM algorithms of each of their combinations. Each cell represents a CBM algorithm of a given family algorithm: *NN*, *DROP* and *Competence*, and a given dimension: *case search direction*, *order sensitive* and *type of cases* to retain. For instance, the algorithm CNN is a member of the *NN* family, which uses an incremental case search direction, is sensitive to the case ordering and retains cases in the borders of case clusters.

Table 1. Outline of CBM algorithms by their dimension and family

Dimensions vs. Family		NN	DROP	Competence
Case search direction	Incremental	CNN	-	RC-FP
	Decremental	RENN	DROP1,2,3	ICF
Order sensitive	Yes	CNN,RENN	DROP1	ICF
	No	-	DROP2,3	RC-FP
Type of Cases to Retain	Border	CNN	DROP1	ICF,RC-FP
	Central	RENN	DROP2,3	COV-FP

Despite of the interest of the management of temporal case-bases, up to our knowledge, there is no study related with CBM algorithms working with event sequences.

3   Temporal CBR Framework

An event sequence represents different observations of temporal dependent variables through the time.

Definition 1. *(**Event Sequence**) Given a set of event types E, an event sequence s_l is a total ordered set of events, and is represented as:*

$$s_l = \langle \left(e_1^l, t_1^l\right), \ldots, \left(e_n^l, t_n^l\right) \rangle,$$

where e_1^l and t_i^l are the i-th event type and time-stamp of the sequence s_l. Furthermore, it is always true that $t_i \leq t_{i+1}$ for each pair of event types e_i and e_{i+1}.

In the following, the main concepts related with temporal CBR systems are defined, such as temporal case, temporal problem, solution and temporal case-base.

Definition 2. *(**Temporal Problem Domain**) Given a set of event sequences S, a set of non temporal dependent variables $V_1, \ldots, V_n$, a temporal problem domain Π is defined as:*

$$\Pi = V_1 \times \ldots \times V_n \times S$$
$$\pi = (v_1, \ldots, v_n, s)$$

Definition 3. *(**Solution Set**) The solution set Ω is the set of all the existing solutions to a given temporal problem domain Π.*

$$\Omega = \{\omega_i \mid \exists \pi_i \in \Pi \ : \omega_i \text{ solves the problem } \pi_i\}$$

Definition 4. *(**Temporal Case**) Given a temporal problem domain Π, and a set of solutions Ω, a temporal case is defined formally as follows:*

$$c = (\pi, \omega),$$

where $\pi = (v_1, \ldots, v_n, s) \in \Pi$, and $\omega \in \Omega$.

In order to ease the understanding, Figure 1 shows a graphical representation of a temporal case.

Definition 5. *(**Temporal Case-Base**) Given a set of event sequences S, a set of non temporal dependent variables $V_1, \ldots, V_n$, and a set of solutions Ω, a Temporal Case-Base (TC) is a set of temporal cases:*

$$TC \in \wp(V_1 \times \ldots V_n \times S \times \Omega)$$
$$c = (v_1, \ldots, v_n, s, \omega) \in TC$$

Note that TC could be the empty set, representing a temporal case-base with no cases. For the sake of simplicity, hereafter, the terms case and temporal cases are used interchangeably.

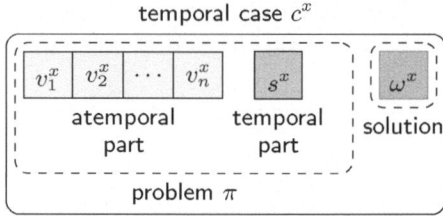

Fig. 1. Temporal case c. Regarding the problem description, the v_1 to v_n are values of the non temporal dependent variables V_1 to V_n, respectively, and s^x is an event sequence. The problem solution is represented by ω^x.

Definition 6. *(Global Distance between two Temporal Problems) Given the temporal problem domain Π, the global distance function δ_{global} is defined as:*

$$\delta_{global}^{temp} : \Pi \times \Pi \to \mathbb{R}^+$$

In particular, we consider an adaptation of the Minkowski distance:

$$\delta_{global}^{temp}(\pi^x, \pi^y, r) = \left(\sum_{i=1}^{n} \left(\delta_{local}^i \left(v_i^x, v_i^y \right)^r \right) + \delta_{temp} \left(s^x, s^y \right)^r \right)^{\frac{1}{r}},$$

where $r \in \mathbb{N}^+$, δ_{local} and δ_{temp} are local distances and temporal local distances respectively. There is a wide range of δ_{temp} function descriptions [11, 13, 24]. In this work we adopted the temporal edit distance and the M&M distance [23].

4   Temporal Case-Base Maintenance

In this section, we present temporal CBM algorithms to manage temporal case-bases (see definition 5). In particular we adapt different types of CBM algorithms according to their family approach and reduction strategy. Table 2 depicts the CBM algorithms that have been adapted from the table 1.

CNN, an incremental maintenance algorithm, is adapted in algorithm 1. In line 10, *correctlyClassified* function checks the outcome correctness of the temporal CBR system when the input case c^x is provided, using $\delta_{global}^{temporal}$ temporal similarity function (see definition 6) and considering k neighbours.

Algorithm 2 describes t-RENN, a decremental maintenance algorithm based on RENN for temporal case-bases.

t-DROP algorithm is described in algorithm 3. In line 24, $t-NearestNeighbour$ function calculates the k-most similar cases of c^x in TC using the $\delta_{global}^{temporal}$ function and considering $k + 1$ neighbours.

Table 2. Outline of proposed temporal CBM algorithms by their dimension and family

Dimensions vs. Family		NN	DROP	Competence
Case search direction	Incremental	t-CNN	-	t-RC-FP
	Decremental	t-RENN	t-DROP1,2,3	t-ICF
Order sensitive	Yes	t-CNN,t-RENN	t-DROP1	t-ICF
	No	-	t-DROP2,3	t-RC-FP
Type of Cases to Retain	Border	t-CNN	t-DROP1	t-ICF,t-RC-FP
	Central	t-RENN	t-DROP2,3	t-COV-FP

Algorithm 1. t-CNN

Require: A temporal case-base TC, $k \in \mathbb{Z}$ as the number of neighbours.
Ensure: A maintained temporal case-base TC'
1. $TC' \leftarrow \emptyset$
2. **for all** $\omega \in \Omega$ **do**
3. $c^x \leftarrow$ random case of TC with solution ω
4. $TC \leftarrow TC - \{c^x\}$
5. $TC' \leftarrow TC' \cup \{c^x\}$
6. **end for**
7. **repeat**
8.   **for all** $c^x = (\pi^x, \omega^x) \in TC$ **do**
9. $TC - \{c^x\}$
10.   **if not** $\boxed{correctlyClassified(c^x, TC, \delta_{global}^{temporal}, k)}$ **then**
11. $TC' \leftarrow TC' \cup \{c^x\}$
12.   **else**
13. $TC \leftarrow TC \cup \{c^x\}$
14.   **end if**
15.   **end for**
16. **until** TC without changes
17. **return** TC'

Algorithm 2. t-RENN

Require: A temporal case-base TC
Ensure: A maintained temporal case-base TC'
1. $TC' \leftarrow TC$
2. **repeat**
3.   **for all** $c^x \in TC'$ **do**
4.   **if not** $correctlyClassified(c^x, TC, \delta_{global}^{temporal}, 3)$ **then**
5. $TC' \leftarrow TC' - \{c^x\}$
6.   **end if**
7.   **end for**
8. **until** TC' without changes
9. **return** TC'

Algorithm 3. t-DROP1

Require: A temporal case-base TC, a number of neighbours $k \in \mathbb{Z}$
Ensure: A maintained temporal case-base TC'

1. $TC' \leftarrow TC$
2. **for all** $c^x \in TC'$ **do**
3. $NEIGHBOUR^x \leftarrow$ t-NearestNeighbours$(c^x, TC', \delta_{global}^{temporal}, k+1)$
4.   **for all** $c^y \in NEIGHBOUR^x$ **do**
5. $ASSOCIATE^y \leftarrow ASSOCIATE^y \cup \{c^x\}$
6.   **end for**
7. **end for**
8. **for all** $c^x \in TC'$ **do**
9. $with \leftarrow 0$
10. $without \leftarrow 0$
11. $TC'' \leftarrow TC' - \{c^x\}$
12.   **for all** $c^y \in ASSOCIATE^x$ **do**
13.   **if** $correctlyClassified(c^y, TC')$ **then**
14. $with \leftarrow with + 1$
15.   **end if**
16.   **if** $correctlyClassified(c^y, TC'')$ **then**
17. $without \leftarrow without + 1$
18.   **end if**
19.   **end for**
20.   **if** $without \geq with$ **then**
21. $TC' \leftarrow TC''$
22.   **for all** $c^y \in associate^x$ **do**
23. $NEIGHBOUR^y \leftarrow NEIGHBOUR^y - \{c^x\}$
24. $c^z \leftarrow$ t-NearestNeighbour$(c^y, TC, \delta_{global}^{temporal}, 1)$
25. $NEIGHBOUR^y \leftarrow NEIGHBOUR^y \cup \{c^z\}$
26.   **end for**
27.   **for** $c^y \in NEIGHBOUR^x$ **do**
28. $ASSOCIATE^y \leftarrow ASSOCIATE^y - \{c^x\}$
29.   **end for**
30.   **end if**
31. **end for**
32. **return** TC'

---

t-ICF is an order sensitive and decremental temporal CBM algorithm that adapts ICF algorithm for event sequences. Algorithm 4 describes t-ICF key details.

Algorithm 4. t-ICF

Require: A temporal case-base TC, a number of neighbours $k \in \mathbb{Z}$
Ensure: A maintained case-base TC

    $TOREMOVE \leftarrow \emptyset$
    for all $c^x \in TC$ **do**
      if not $correctlyClassified(c^x, TC, \delta_{global}^{temporal}, k)$ **then**
        $TOREMOVE \leftarrow TOREMOVE \cup \{c^x\}$
      end if
    end for
    $TC \leftarrow TC - TOREMOVE$
    repeat
      for all $c^x = (\pi^x, \omega^x) \in TC$ **do**
        for all $c^y = (\pi^y, \omega^y) \in$ t-NearestNeighbour$(c^x, TC, \delta_{global}^{temporal}, k)$ **do**
          if $(\omega^x = \omega^y)$ **then**
            $COVERAGE^x \leftarrow COVERAGE^x \cup \{c^y\}$
          end if
        end for
        for all $c^y \in TC - c^x$ **do**
          for all $c^z \in$ t-NearestNeighbour$(c^y, TC, \delta_{global}^{temporal}, k)$ **do**
            if $\delta_{global}^{temporal}(c^x, c^z) = 0$ **then**
              $REACHABLE^x \leftarrow REACHABLE^x \cup \{c^y\}$
            end if
          end for
        end for
      end for
      $TOREMOVE \leftarrow \emptyset$
      for all $c^x \in TC$ **do**
        if $|REACHABLE^x| > |COVERAGE^x|$ **then**
          $TOREMOVE \leftarrow TOREMOVE \cup \{c^x\}$
        end if
      end for
      $TC \leftarrow TC - TOREMOVE$
    until not progress in TC
    return TC

Finally, t-RC-FP, as in previous algorithms, adapts key elements of RC-FP algorithm for temporal case-bases as shown in algorithm 5.

5   Experiments and Results

The objective of the following experiments is to study the capacity of the temporal CBM algorithms to obtain a good temporal case-base considering the characteristics of an original case-base (repository of event sequences).

Algorithm 5. t-RC-FP

Require: A temporal case-base TC, a value $k \in \mathbb{Z}$ as the number of neighbours.
Ensure: A maintained temporal case-base TC'

$\quad R \leftarrow$ t-RENN(TC)
$\quad TC' \leftarrow \emptyset$
$\quad$ **while** $R \neq \emptyset$ **do**
$\quad\quad$ **for all** $c^x = (\pi^x, \omega^x) \in R$ **do**
$\quad\quad\quad COVERAGE^x \leftarrow \emptyset$
$\quad\quad\quad REACHABLE^x \leftarrow \emptyset$
$\quad\quad\quad$ **for all** $c^y = (\pi^y, \omega^y) \in$ t-NearestNeighbour($c^x, R, \delta_{global}^{temporal}, k$) **do**
$\quad\quad\quad\quad$ **if** $(\omega^x = \omega^y)$ **then**
$\quad\quad\quad\quad\quad COVERAGE^x \leftarrow COVERAGE^x \cup \{c^y\}$
$\quad\quad\quad\quad$ **end if**
$\quad\quad\quad$ **end for**
$\quad\quad\quad$ **for all** $c^y \in TC - \{c^x\}$ **do**
$\quad\quad\quad\quad$ **for all** $c^z \in$ t-NearestNeighbour($c^y, TC, \delta_{global}^{temporal}, k$) **do**
$\quad\quad\quad\quad\quad$ **if** $\delta_{global}^{temporal}(c^x, c^z) = 0$ **then**
$\quad\quad\quad\quad\quad\quad REACHABLE^x \leftarrow REACHABLE^x \cup \{c^y\}$
$\quad\quad\quad\quad\quad$ **end if**
$\quad\quad\quad\quad$ **end for**
$\quad\quad\quad$ **end for**
$\quad\quad$ **end for**
$\quad\quad max \leftarrow -1$
$\quad\quad$ **for all** $c^x \in R$ **do**
$\quad\quad\quad cov \leftarrow 0$
$\quad\quad\quad$ **for all** $c^y \in COVERAGE^y$ **do**
$\quad\quad\quad\quad cov \leftarrow cov + |REACHABILITY^y|$
$\quad\quad\quad$ **end for**
$\quad\quad\quad$ **if** $\frac{1}{cov} > max$ **then**
$\quad\quad\quad\quad cov \leftarrow \frac{1}{cov}$
$\quad\quad\quad\quad c^{next} \leftarrow c^x$
$\quad\quad\quad$ **end if**
$\quad\quad$ **end for**
$\quad\quad TC' \leftarrow TC' \cup \{c^{next}\}$
$\quad\quad R \leftarrow R - COVERAGE^{next}$
$\quad$ **end while**
$\quad$ **return** TC'

5.1   Experiment Protocol

In this experiment, the temporal CBM algorithms are evaluated considering the accuracy and the reduction rate.

- The accuracy of the temporal CBM measures how accurate the temporal CBR system is using the temporal case-base obtained by a temporal CBM algorithm.
- The reduction rate is the division of the number of cases removed from the temporal CBM by the total number of cases of the temporal case-base.

Regarding the characteristics of the case-base we consider:

- The *redundancy* of the case-base: the proportion of cases with the same solution.
- The *case-base size*: the number of event sequences of the case-base.

Finally, we also analyse how temporal distance function affects the tCBM algorithms. To this end, temporal edition and the M&M distance are considered in the experiments.

The evaluation methodology used is the $\alpha\beta$ [10], an extension of the Cross-Validation for evaluating CBM algorithms using a simple CBR system. $\alpha\beta$ method extracts from the original temporal case-base the training and test sets β times, and every temporal CBM algorithm is executed α times using the training set. At the end of each experiment, there will be up to β accuracy and reduction rate results, which are the evaluation scores of the experiment. $\alpha\beta$ method has shown to be a reliable with values $\alpha = 10$ and $\beta = 10$ when the CBR system configuration is a K-NN classifier with $K = 3$.

In our experiments, the temporal case-base size can be one of the following values $\{60, 300, 600\}$. The level of redundancy is analysed for $\{75\%, 80\%, 85\%\}$.

5.2   Results

In the following we show the results of the experiments carried out according to the goals and factors stated previously.

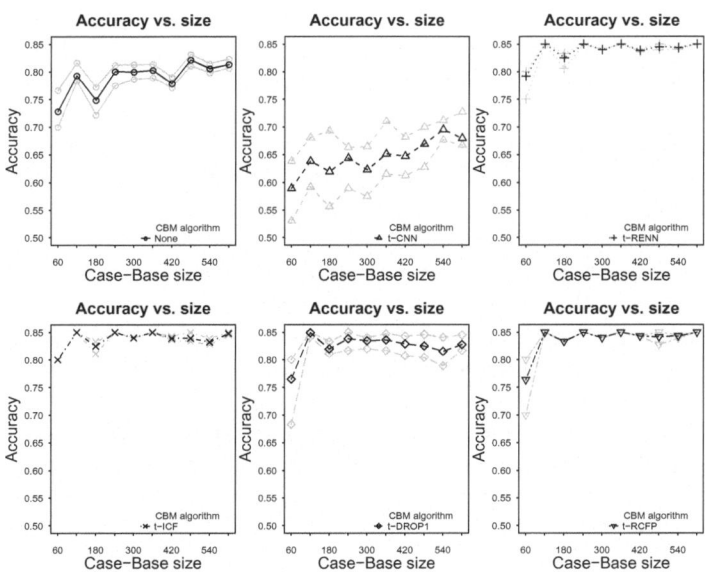

Fig. 2. Accuracy and reduction rate with different synthetic temporal case-base with different sizes for temporal edit distance

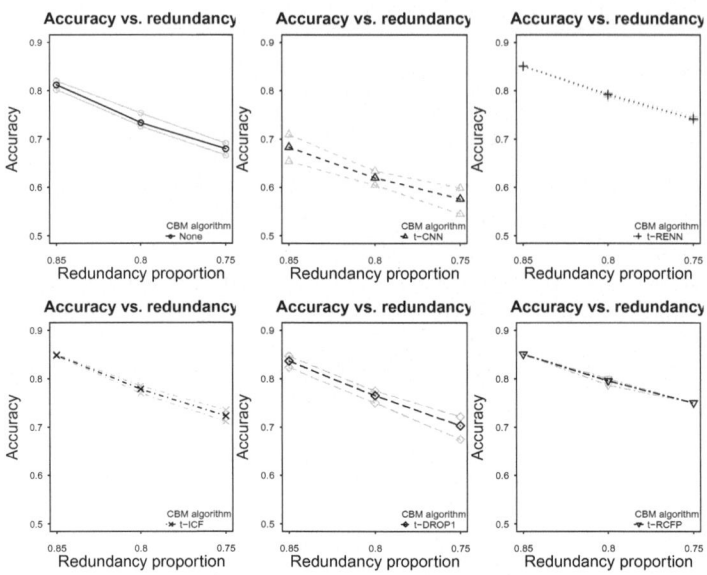

Fig. 3. Accuracy with different case-bases and different redundancy levels for temporal edit distance

Figure 2 shows how accurate the temporal CBR system is when the original temporal case-base is maintained using temporal CBM algorithms and varying the number of cases from the original case-base. Dotted grey lines show the minimum and maximum accuracy in the evaluation while dark lines are the average. *None* expresses the response of the system when no temporal CBM are used.

The relation between the accuracy of the temporal CBR system and the redundancy of the temporal case-base is depicted in Figure 3. As before, dotted grey lines expresses the minimum and maximum accuracy values and dark lines the accuracy average of the evaluation.

Figures 2 and 3 show the results considering the temporal edit distance. However, a key issue in this study is to analyse how the accuracy is affected by the temporal similarity function used. To this end: (1) the experiments has been replicated using M&M distance;(2) a t-test has been carried out to analyse the statistical difference between edit and M&M distance.

Table 4 shows the reduction rate for the different temporal CBM algorithms using edit and M&M distances. As previously, a t-test is carried out to analyse how the reduction rate is affected by the temporal similarity function used.

Table 3. Analysis of the accuracy of the system considering different similarity functions of the temporal TCM algorithm

	Avg. accuracy		
	p-value	Edit distance	M&M
None	1.00	0.79	0.77
t-CNN	0.03	0.65	0.66
t-RENN	1.00	0.84	0.79
t-ICF	1.00	0.84	0.79
t-DROP1	1.00	0.82	0.76
t-RC-FP	1.00	0.84	0.78

Table 4. Analysis of reduction rate considering different similarity functions of the temporal TCM algorithm

	Reduction rate average		
	p-value	Edit distance	M&M
t-CNN	1.00	0.65	0.62
t-RENN	$< 10^{-5}$	0.18	0.22
t-ICF	1.00	0.86	0.75
t-DROP1	1.00	0.81	0.74
t-RC-FP	1.00	0.78	0.78

5.3   Discussion

According to the previous results, we would like to highlight:

1. All temporal CBM algorithms obtain acceptable results when the number of cases increases. Note that, according to figure 2, most of the algorithms obtain similar results to the original CBR system (none). While *t-CNN* obtains worst accuracy and a high dispersion, *t-ENN* and t-ICF outperform the original CBR system.
2. As shown in figure 3, when the redundancy decreases, the accuracy also descends. Moreover, the results obtained using temporal CBM are similar or outperform the original temporal CBR system.
3. Table 3 shows the average accuracy and the p-value of the statistical test. According to the results obtained, the accuracy obtained does not depend on the temporal distance function.
4. As shown in table 4, reduction rates are independent from the temporal distance function used, except for t-RENN.

6   Conclusions

There is an increasing interest of the CBR community in the management of cases described by temporal event sequences. However, up to our knowledge, very little attention has been paid to the reduction of the number of temporal cases. In this work, we extended some of the well-known CBM algorithms to reduce the number of sequences in a case-base. Due to the wealth of approaches to develop CBM algorithms, firstly we classify the algorithms in families (NN, DROP and Competence-Based) and dimensions (search direction, order sensitivity, cases retained); and secondly we extended one CBM algorithm of each category to deal with event sequences.

In this paper, we present a formal framework for describing temporal CBM when cases contain event sequences. We also propose 5 new algorithms: t-CNN, t-RENN, t-DROP1, t-ICF and t-RC-FP. Finally, we carry out some experiments to evaluate empirically our proposal using temporal case-bases from a synthetic sequence generator.

The experiments carried out show that most of the temporal CBM algorithms obtain promising results, similarly as classical CBM do in atemporal case-bases. For both, accuracy and reduction rate, most temporal CBM algorithms outperform the original temporal CBR system. The algorithm t-CNN is the only one that underperform the maintenance task. In particular, t-CNN builds case-bases with lower accuracy than the original. In this way, more experiments need to be carried out with this algorithm in order to gather more evidence and to provide a final conclusion. These limitations are being addressed in an ongoing work.

According to the results of the experiments, the temporal CBM proposed seem not to depend on the temporal distance function used. Therefore, it seems feasible to adapt any existing CBM algorithm in order to work with temporal case-bases.

Among future works, we plan to integrate temporal CBM algorithms in CBR systems for home monitoring and healthcare domains.

Acknowledgement. This study was funded by Seneca Foundation, the Regional Research Office of the Region of Murcia, Spain, under project no. 15277/PI/10.

References

1. Brighton, H., Mellish, C.: On the consistency of information filters for lazy learning algorithms. In: Żytkow, J.M., Rauch, J. (eds.) PKDD 1999. LNCS (LNAI), vol. 1704, pp. 283–288. Springer, Heidelberg (1999)
2. Cordier, A., Lefevre, M., Champin, P.A., Georgeon, O.L., Mille, A.: Trace-based reasoning - modeling interaction traces for reasoning on experiences. In: FLAIRS Conference (2013)
3. Gundersen, O.E., Sørmo, F., Aamodt, A., Skalle, P.: A real-time decision support system for high cost oil-well drilling operations. AI Magazine 34(1), 21–32 (2013)
4. Gundersen, O.E.: Toward measuring the similarity of complex event sequences in real-time. In: Díaz Agudo, B., Watson, I. (eds.) ICCBR 2012. LNCS, vol. 7466, pp. 107–121. Springer, Heidelberg (2012)
5. Hart, P.: Condensed nearest neighbor rule. IEEE Transactions on Information Theory 14(3), 515+ (1968)

6. Huang, Z., Juarez, J.M., Duan, H., Li, H.: Length of stay prediction for clinical treatment process using temporal similarity. Expert Systems with Applications 40(16), 6330–6339 (2013)
7. Juarez, J.M., Guil, F., Palma, J., Marin, R.: Temporal similarity by measuring possibilistic uncertainty in CBR. Fuzzy Sets and Systems 160(2), 214–230 (2009)
8. Leake, D., Wilson, D.: Categorizing case-base maintenance: Dimensions and directions. In: Smyth, B., Cunningham, P. (eds.) EWCBR 1998. LNCS (LNAI), vol. 1488, pp. 196–207. Springer, Heidelberg (1998)
9. Levenshtein, V.I.: Binary codes capable of correcting deletions, insertions, and reversals. Soviet Physics Doklady 10, 707–710 (1966)
10. Lupiani, E., Juarez, J.M., Jimenez, F., Palma, J.: Evaluating case selection algorithms for analogical reasoning systems. In: Ferrández, J.M., Álvarez Sánchez, J.R., de la Paz, F., Toledo, F.J. (eds.) IWINAC 2011, Part I. LNCS, vol. 6686, pp. 344–353. Springer, Heidelberg (2011)
11. Mannila, H., Moen, P.: Similarity between event types in sequences. In: Mohania, M., Tjoa, A.M. (eds.) DaWaK 1999. LNCS, vol. 1676, pp. 271–280. Springer, Heidelberg (1999)
12. McKenna, E., Smyth, B.: Competence-guided case-base editing techniques. In: Blanzieri, E., Portinale, L. (eds.) EWCBR 2000. LNCS (LNAI), vol. 1898, pp. 186–197. Springer, Heidelberg (2000)
13. Moen, P.: Attribute, Event Sequence, and Event Type Similarity Notions for Data Mining. Ph.D. thesis, University of Helsinki (2000)
14. Montani, S., Bottrighi, A., Leonardi, G., Portinale, L., Terenziani, P.: Multi-level abstractions and multi-dimensional retrieval of cases with time series features. In: McGinty, L., Wilson, D.C. (eds.) ICCBR 2009. LNCS, vol. 5650, pp. 225–239. Springer, Heidelberg (2009)
15. Montani, S., Leonardi, G.: Retrieval and clustering for supporting business process adjustment and analysis. Information Systems 40, 128–141 (2014)
16. Pan, R., Yang, Q., Pan, S.J.: Mining competent case bases for case-based reasoning. Artificial Intelligence 171(16-17), 1039–1068 (2007)
17. Roddick, J., Spiliopoulou, M.: A survey of temporal knowledge discovery paradigms and methods. IEEE Transactions on Knowledge and Data Engineering 14(4), 750–767 (2002)
18. Smyth, B., McKenna, E.: Competence models and the maintenance problem. Computational Intelligence 17(2), 235–249 (2001)
19. Valipour Shokouhi, S., Skalle, P., Aamodt, A., Sormo, F., et al.: Integration of real-time data and past experiences for reducing operational problems. In: International Petroleum Technology Conference (2009)
20. Wilson, D.R., Martinez, T.R.: Instance pruning techniques. In: Machine Learning: Proceedings of the Fourteenth International Conference (ICML 1997), pp. 404–411. Morgan Kaufmann (1997)
21. Wilson, D.R.: Asymptotic properties of nearest neighbor rules using edited data. IEEE Transactions on Systems Man and Cybernetics SMC 2(3), 408–421 (1972)
22. Wilson, D.R., Martinez, T.R.: Reduction techniques for instance-based learning algorithms. Machine Learning 38(3), 257–286 (2000)
23. Wongsuphasawat, K., Shneiderman, B.: Finding comparable temporal categorical records: A similarity measure with an interactive visualization. In: IEEE Symposium on Visual Analytics Science and Technology, VAST 2009 (2009)
24. Wongsuphasawat, K., Plaisant, C., Taieb-Maimon, M., Shneiderman, B.: Querying event sequences by exact match or similarity search: Design and empirical evaluation. Interact. Comput. 24, 55–68 (2012)

Using Case-Based Reasoning
to Detect Risk Scenarios
of Elderly People Living Alone at Home

Eduardo Lupiani[1], Jose M. Juarez[1], Jose Palma[1],
Christian Serverin Sauer[2], and Thomas Roth-Berghofer[2]

[1] University of Murcia, Spain
{elupiani,jmjuarez,jtpalma}@um.es
[2] The University of West London, UK
{thomas.roth-berghofer,christian.sauer}@uwl.ac.uk

Abstract. In today's ageing societies, the proportion of elderly people living alone in their own homes is dramatically increasing. Smart homes provide the appropriate environment for keeping them independent and, therefore, enhancing their quality of life. One of the most important requirements of these systems is that they have to provide a pervasive environment without disrupting elderly people's daily activities. The present paper introduces a CBR agent used within a commercial Smart Home system, designed for detecting domestic accidents that may lead to serious complications if the elderly resident is not attended quickly. The approach is based on cases composed of event sequences. Each event sequence represents the different locations visited by the resident during his/her daily activities. Using this approach, the system can decide whether the current sequence represent an unsafe scenario or not. It does so by comparing the current sequence with previously stored sequences. Several experiments have been conducted with different CBR agent configurations in order to test this approach. Results from these experiments show that the proposed approach is able to detect unsafe scenarios.

Keywords: Elderly monitoring, Temporal case-base, Temporal case similarity, Implementation.

1 Introduction

According to the World Health Organization and the US National Institute of Ageing/Health [25], industrialized countries are facing the problem that the population's average age is drastically increasing. One of the side effects of an increased aged population is the rising number of elderly people living alone at home. Information Technology and Artificial Intelligence (AI) may play a relevant role in looking after people living alone at home, as well as in providing care assistance. Examples of this key role are some policies such as the Ambient Assisted Living initiative promoted by the European Union. In all of these initiatives, Smart Homes are encouraged as a tool to detect unsafe scenarios at home

L. Lamontagne and E. Plaza (Eds.): ICCBR 2014, LNCS 8765, pp. 274–288, 2014.
© Springer International Publishing Switzerland 2014

[5], as for instance falls, which are one of the major causes of serious accidents for the elderly people living alone.

Smart Home systems are usually based on agent architectures [8], where each agent is responsible of one particular task, such as the control of the home environment [20], the assistance of home inhabitants [13,14,24] or monitoring of residents' health status [6]. In order to fulfill their purpose, Smart Homes require a set of sensors to gather home data and deliver them to agents. To this end, different types of sensors have been considered, ranging from intrusive devices such as cameras or wearable sensors [7,24], to pervasive approaches such as movement or pressures devices connected by wireless sensor networks [2,21]. Once data is collected and processed, AI techniques can carry out some inference processes in order to interpret the scenario based on the processed data.

The use of Case-Based Reasoning (CBR) as a reasoning agent in Smart Home systems has several advantages [8,15]: First, the learning process is implicit in the CBR cycle so CBR agents can learn from concrete situations as time goes by, making it possible for the CBR agents to adapt themselves to the resident's specific needs. Second, the system response time can be reduced because CBR avoids resolving already solved problems, which may involve a great amount of information and computation in a Smart Home environment. Third, an expert can define personalised cases to represent a customised problem and its solutions. Finally, as CBR systems use similar past solved cases to solve a current problem, these cases can be used to provide explanations on why a concrete solution is proposed. [9,22].

Some authors have already proposed CBR agents to solve problems in Smart Homes [3,15]. However, to our understanding, little attention has been paid to the temporal dimension in the development of these CBR systems, since the use of the time dimension is limited to determine the context of the case [17].

In this work, we propose a CBR agent able to detect potential unsafe scenarios in a Smart Home, as for instance falls, using a spatial-temporal approach. The agent is based on the retrieval of previous cases which represent the different locations visited by the elderly resident during one of his/her daily activities. These cases are represented by event sequences where each event consists of a location and a time-stamp. The proposed agent has been integrated in the *proDIA* monitoring system [4]. This system consists of a wireless sensor network which uses pervasive sensors, such as motion detection infra-red sensors, pressure sensors (located in bed, chairs, sofas, etc) and magnetic sensors to detect door opening and closing. A prototype of this system has been placed in 100 houses in the province of Murcia, Spain. Furthermore, the CBR agent has been implemented using *myCBR* software development kit [23].

The remainder of this paper is organized as follows. In section 2 we review the background of this work. In section 3 we describe in detail the proposed system to detect unsafe scenarios with a CBR agent. Section 4 describes our experiments which we performed using a synthetic case-base with cases that represent daily activities at home. Finally, in section 5 we present our conclusions and describe planned future work.

2   Related Work

CBR has been used in Smart Home systems in different approaches and with different purposes. In [26], a CBR system is proposed as a decision support system to place the sensors in a Smart Home. The decision is done according to the resident's physical disabilities, such as their cognitive abilities, mobility, dexterity and other personal details. However, the system does not make use of the time dimension to reflect the change in a resident's physical state.

A CBR architecture for Smart Home in order to enhance the inhabitants comfort at home is proposed in [17]. The cases in this approach are representations of what actions are occurring at home and how the Smart Home should react to them to enhance the comfort at home, as for instance lowering the AC temperature or adjusting the light brightness in a room. The case structure is a frame with slots for representing the user information, data gathered from the sensors, and a time-stamp to represent when the observation of the house was done.

In [15], the authors introduced how a CBR system may be used in order to detect problems at a home, so as to propose actions to amend them, however, most of the description are not related with the spatial-temporal representation of the actions taken at home. According to this work, the success of the system relies on the quality of the retained cases in the case-base. Thus, the use of good case engineering practices are recommended not only to create the first set of cases, but also to create cases personalised to a particular user. Furthermore, since the case learning task is included in the CBR cycle, the case-base size increases with time, making the inclusion of case-base maintenance policies necessary to maintain the quality and performance of the case-base and thus the CBR system.

The core of the Smart Home proposed in [8] is a CBR system. This system is placed in a residential ward where nursery staff takes care of patients. The purpose of the CBR system is to plan the future tasks to be done by the staff. The case representation keeps a record of one task already done by a staff member, along with information related to the time in which the task started and ended as well as its priority and providing the next task to do as the solution of the case. Furthermore information related to the patients' health status is retained in the cases as well, since this data is relevant for the planning of future tasks.

AmICREEK uses a CBR system to detect the situation taking place within the system's environment [11,12]. AmICREEK is based upon a three layer architecture [4]: the *perception* layer as the middle-ware gathering the information from the sensor network, the *awareness* layer as a CBR system that detects the context in which the action is taking place and the goal of the action. Lastly, there is the *sensitivity* layer, in which a sequence of tasks is built in order to satisfy the goal given by the CBR system according to the system's context. The authors tested this approach in a hospital ward but the cases only contained one time-stamp to represent the time in which they were created.

3   A Smart Home for Alarm Detection

The architecture of the *proDIA* system is built upon three levels: *sensor level*, the *communication level* and the *data processing level*. Figure 1 depicts the system levels, as well as an example of the distribution of sensors in a house, located in room, kitchen, bathroom, bedroom and the corridor.

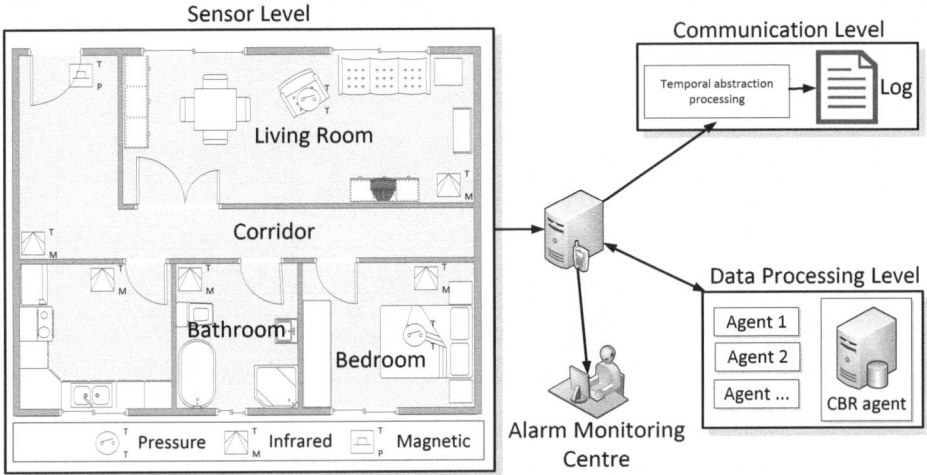

Fig. 1. The three levels of the system, as well as an example of the distribution of sensors in the house in order to monitor the person at home

The *sensor level* is the first level of the system's architecture, it manages the sensor-data acquisition from the wireless sensor network. This network uses three types of sensors: infrared motion detection sensors, pressure sensors and magnetic sensors to detect whether the main door is opened or closed. The basic configuration of the system implies the placement of motion sensors in every location, and one single magnetic sensor to detect whether the main door is opened or closed. Furthermore, pressure sensors may be located in places such as sofas and beds to detect whether the person is resting or lying on one of them. With these pressure sensors it is possible to detect the location of the person, even if the person is not moving.

The second architecture level is the *communication level*, where the data provided by the sensor level is recorded. Using the IEEE-802.15.4 communication standard, the data gathered by the sensors is kept in a home-station (mini-PC), which synchronises the data sent by the sensors according to the timestamps in which they are received. The communication level creates a log of the data with a given frequency, which is used by the *data processing level*. The *log* is a comma separated values (CSV) file, where each line contains information related to the readings sent by the sensors, such as the identifier of the sensor, the time-stamp in which the sensor is activated, the location of the sensor and the content of the sensor's reading. Every newly created log starts empty, so the frequency in which the log is created will determine the amount of data stored in it. Therefore, if long observation periods of the home are required, the log must be created with a low frequency. On the contrary, log files created with a high frequency contain less data since they represent short observation periods.

For instance, the sensor level may produce a log file as shown in table 1. In this example, the log records the data sent by the sensors when the user arrives at home.

Table 1. Example of activity log

Time-stamp	sensor id.	Location	Message	Rationale/Justification
7932	sensor197680	Corridor	MOVE: true	getting house
7932	sensor197683	Door	isOPENDOOR: true	
7940	sensor197683	Corridor	isOPENDOOR: false	
7959	sensor347050	Bedroom	MOVE: true	going to the bedroom
7972	sensor530111	Bedroom	PRESS: true	sitting down on the bed
7980	sensor197680	Corridor	MOVE: false	
8054	sensor530111	Bedroom	PRESS: off	standing up
8113	sensor197680	Corridor	MOVE: true	going to the toilet
8121	sensor197680	Bedroom	MOVE: false	
8122	sensor536770	Bathroom	MOVE: true	getting in the bathroom

Starting from the information provided by the communication level, the *data processing level* attempts to infer the state in which the elderly is, that is, the complete situation in accordance to the situation context is described. The system relies on the assumption that the location of the resident, the activity or absence of it, and the moment of the day in which these facts are registered are enough to detect possible emergency situations. For example, if the attendee has fallen and lost conscience or broke a bone in such a way that prevents him or her from moving, detection of this situation is based on an excessive time of inactivity being measured in a context in which this is abnormal (i.e. the attendee is in the house and she is not supposed to be resting or sleeping). To this end a behavioural model was developed, based on a finite state automaton. Once an abnormal situation is detected, the system sends an alarm to the Alarm Monitoring Centre using UMTS telecommunication technology, where a specific predefined protocol is fired. For instance, if a fall is detected the monitoring centre may call an ambulance, or if the system detects frequently that the person is having problems at night, then the alarm monitoring centre may contact by phone with the elderly person to show interest in their health status.

3.1   Case-Based Reasoning Agent

In this work we propose to include a CBR agent, which tries to check whether the daily activity at home is normal or abnormal, indicating an unsafe scenario is taking place. To this end, the approach followed is to keep a record of the movement of the resident at home within given time-frame. The CBR agent checks whether a current activity, or event sequence, is similar to previously recorded activities/event sequences.

According to the definition of case given in [1], a case consists of a problem and a solution. In order to classify and detect unsafe situations at home, the cases represent a daily activity and its type. Thus, whereas the problem represents the visited locations during one daily activity, the solution is a label describing the type activity or scenario. The set of valid values for the solution is not limited, being possible the inclusion of new *solution* labels on demand when the CBR is running.

The event sequences consist of ordered heterogeneous events in time, where each event is composed of an event type and a time-stamp that represents when the event occurs [18,19]. Thus, each event in the sequence is a tuple of the location visited by the person and a time-stamp.

The following expressions are shown to detail the case representation (see expression 1) and the event sequence (see expression 2).

$$case\ c = (sequence, solution) \tag{1}$$
$$solution \in \{normal, scenario_1, scenario_2, \ldots\}$$

$$\begin{aligned} sequence = \ &\langle (loc_1, t_1), \ldots, (loc_2, t_2), (loc_n, t_n) \rangle \mid \\ &\mid \forall loc \in \{Corridor, Bedroom, \ldots\} \wedge \\ &\wedge\ t \in \mathbb{N}^+ \wedge \forall_{i=1}^{n-1} t_i \le t_{i+1} \end{aligned} \tag{2}$$

Given a log, generated by the communication level, the operation of the CBR agent is the following:

1) The CBR agent reads the log when a new one is created by the communication level. This log contains the data from sensors chronologically ordered according to their time-stamps.
2) An event sequence is built from the collected sensor data. Later, this event sequence is used as a *input* query to the retrieval step.
3) The CBR agent retrieves from its case-base those cases with the most similar event sequence to the *input*.
4) Based on the retrieved cases, the system infers the type of the activity best matched to the *input*.
5) When the activity is classified as abnormal (according to the defined solution labels), the system sends a message to the Alarm Monitoring Centre, where the expert decides which is the most suitable action for the detected scenario.

Finally, a new case is retained in the case-base when an *abnormal* scenario is classified correctly

3.2   Computing the Similarity between Cases

The CBR agent uses the edit distance between event sequences proposed in [18,19]. This distance measure computes the *cost* of transforming an event sequence into another. This cost is represented as the number of operations needed to perform the transformation. The operations are applied on the query event sequence, until it matches with the retrieved event sequence, which is known as pattern. Therefore, a high number of transformation operations stands for two not very similar event sequences. On the contrary, two more similar event sequences need fewer transformation operations. The set of available operations are *insertion, deletion* and *displacement*. While the insertion is used when the query does not contain an event that is present in the pattern, the deletion is used if the query contains an event not appearing in the pattern. The displacement operation is used when two events in both the query and pattern match with the same location. Whereas the operations insertion and deletion have a cost for their application, the operation displacement has a cost based on the difference between the timestamps of the two events. In order to ensure the correct functioning of the edit distance, the displacement costs has to be lower than the insertion and deletion operations. Thus, the difference between the timestamps is normalized between 0 and 1, and the cost of the insertion and deletion operations is set to values higher or equal to 1. Algorithm 1 presents a dynamic programming approach for searching the minimum number of operations to transform a query into a pattern.

Algorithm 1. Edit distance between two event sequences x, y [18,19]

Input: Two event sequences $x = \langle (loc_1^x, t_1^x), \ldots, (loc_n^x, t_n^x) \rangle$ and $y = \langle (loc_1^y, t_1^y), \ldots, (loc_m^y, t_m^y) \rangle$, with $loc \in \{Bedroom, Corridor, \ldots\}$, the costs $w(loc^x), w(loc^y)$ of the *insertion* and *deletion* operations.

Output: Edit distance between the two given sequences.

1. $r \leftarrow$ matrix of $n \times m$ dimensions
2. $r(0,0) \leftarrow 0$
3. **for** $i \leftarrow 0$ **to** m **do**
4. $r(i,0) \leftarrow r(i-1,0) + w(loc^x)$
5. **end for**
6. **for** $j \leftarrow 0$ **to** n **do**
7. $r(0,j) \leftarrow r(0,j-1) + w(loc^y)$
8. **end for**
9. **for** $i \leftarrow 1$ **to** m **do**
10.   **for** $j \leftarrow 1$ **to** n **do**
11. $update_x \leftarrow r(i-1,j) + w(loc^x)$
12. $update_y \leftarrow r(i,j-1) + w(loc^y)$
13. $align \leftarrow r(i-1,j-1)$
14.   **if** $loc^x = loc^y$ **then**
15. $align \leftarrow align + (\frac{|t_i^x - t_j^y|}{max(t) - min(t)})$
16.   **else**
17. $align \leftarrow align + w(loc^x) + w(loc^y)$
18.   **end if**
19. $r(i,j) \leftarrow min(update_x, update_y, align)$
20.   **end for**
21. **end for**
22. **return** $r(n,m)$

3.3   Study of Alarm Scenarios

Four different scenarios are considered based on different common scenarios that usually occur at home: a *normal* daily activity, a *bad night* were the person wake at night one more times for voiding, a fall resulting in a conscious status and a fall with an unconsciousness status. These particular scenarios are chosen because they are frequent in elderly people living alone. On the one hand, these two types of falls are common in elderly people at home [16]. On the other hand, elderly people may suffer from nocturia (the medical term for excessive urination at night) [10]. Next, an example of a case is given for each type of scenario, where each location used by the event sequences is one of the following

$$loc = \{ \; Corridor = 0, Kitchen = 1, LivingRoom = 2,$$
$$Toilet = 3, Bedroom = 4, Out = 5\}$$

The normal daily behaviour represents the locations visited by the resident during one of his/her daily activity taken at home, such as having a shower in the bathroom, watching the TV in the living room or sleeping on the bed in the bedroom.

$$c = (\quad \langle (4, 539), (3, 29303), (1, 29439), (4, 30737), (2, 31420),$$
$$(0, 35352), (1, 49882), (3, 53750), (2, 54011), (0, 62753),$$
$$(1, 74758), (2, 76114), (1, 82977), (3, 85593)\rangle, \text{'normal'})$$

The *bad night* template represents the locations visited by the resident when he or she has not been able to sleep due to nocturia status. In this scenario, the event sequence represents regular visits to the bathroom during the night.

$$c = (\quad \langle (4, 447), (3, 7685), (3, 28669), (1, 29196), (4, 30618),$$
$$(2, 31049), (1, 49726), (3, 53542), (2, 54109), (0, 61434),$$
$$(1, 73228), (2, 77069), (1, 83260), (3, 85484)\rangle, \text{'bad night'})$$

The two fall scenarios represent two types of falls: a fall where the person stays motionless after losing consciousness and another where the person stays conscious and may crawl on the floor. Both scenarios usually occur in the bathroom as a consequence of a fall in the bathtub and the difference between them is the activity of the location visited. Whereas an unconscious person after a fall does not activate any movement sensor, a conscious person try to move or crawl to other different location to call for help. The following are examples of cases of both fall scenarios:

$$c = (\; \langle (3, 29040), (1, 29985), (4, 30871), (2, 31343), (1, 49764),$$
$$(3, 53915), (3, 53960), (3, 54393), (3, 54482), (0, 54628),$$
$$(3, 54663), (0, 54892), (0, 54968), (2, 55115)\rangle,$$
$$\text{'fallen with consciousness'})$$

$$c = (\ \langle (3, 29229), (3, 29997), (3, 30055), (3, 30119),$$
$$(3, 30178), (3, 30235), (3, 30290), (3, 30350) \rangle,$$
$$\text{'fallen with unconsciousness'})$$

3.4 *myCBR* Extension for Temporal Similarity

A key goal in the developing of the open source software *myCBR* is the aim to provide a compact and easy-to-use tool for rapidly prototyping CBR applications. The *myCBR* tool is especially intended to be used in the contexts of research and teaching as well as to allow businesses to allow for the timely implementation of CBR systems with low initial development effort. *myCBR* allows for the integration of the developed CBR systems into other applications. A key factor of the SDK is to allow also for the implementation of extensions into the SDK. These extensions can, for example, be based on specific requirements such as additional similarity calculations, as it is demonstrated in this paper.

As described in this paper the need for providing a similarity measure for event sequence was met with the integration of this similarity measure into the *myCBR* SDK. This extension of the SDK was eased by the high modularity of the existing *myCBR* SDK's code and its extensive documentation. As *myCBR* is implemented in Java it follows the object oriented approach, easing the extension of the SDK. The ability to extend the SDK allows for researchers and businesses to quickly integrate their innovations or experimental features into a robust existing CBR system development tool. Thus a CBR developer can test run their own innovative feature within a robust existing CBR system without having to implement the whole CBR system from scratch [23].

The extension of *myCBR* that was implemented to provide the similarity measure for event sequences was based on the creation of a new attribute. This new attribute represents event sequences. To define this new attribute we adopt the event sequence definition of expression 2.

Figure 2 shows the new classes that were implemented as an extension to the existing *myCBR* SDK. The extension allows us to represent an event sequence as an attribute of a case, which allows *myCBR* to compute the similarity between two different attribute values, e.g. event sequences.

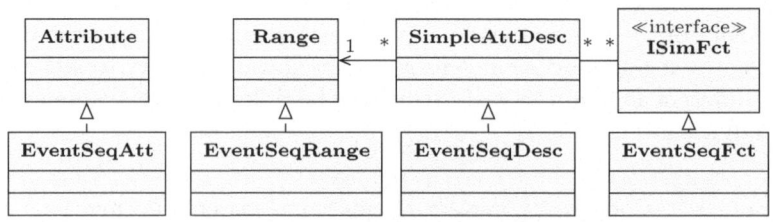

Fig. 2. Sequence of operations to compute the similarity between two cases

EventSeqAtt: Contains the information of a particular event sequence.

EventSeqRange: Contains the possible values of event sequences. Also allows for the creation of new *EventSeqAtt*.

EventSeqDesc: Contains the details about how the event sequences are implemented.

EventSeqFct: This corresponds to the implementation of the similarity measures that we have worked on in the present paper. When an object of this class is created it is mandatory to specify which type of similarity measure it should employ.

Next to the generation of CBR systems and the extension of these systems by experimental features, *myCBR* also allows for the generation of cases from CSV files. This form of rapid case acquisition is highly desirable for the use of sensor based systems as these systems' data output can be transformed into CSV data quite easily. Aside from the project focused on in this paper, the authors are currently taking part in research work on the effectiveness and accuracy of large numbers of cases from CSV-formatted sensor data.

4   Experiments

The goal of the experiments was to study whether the CBR agent is able to detect the type of a scenario occurring at a residents home, particularly unsafe scenarios. The CBR agent used in the experiments was configured with regard to its case-base, which was created synthetically, and the length of the home observation, that is, the length of the queries generated from the log. The CBR agent retrieved the most similar cases using an 1-NN global similarity function. As only one case was retrieved by each query, no adaptation process was performed. So the solution of the retrieved case was returned as the inferred type of scenario taking place at the residents home. Every experiment has been carried out with two case-bases. The first case-base is the query case-base, which are solved using the second one. The query case-base is the same for all the experiments, and represents the activity at home during one year (being each case an event sequences of one day duration). Besides, in the query case-base the 95% are labelled as normal activities and the remaining are labelled as non normal situations (equally distributed among the different unsafe scenarios).

4.1   Generating a Synthetic Case-Base

The evaluation is based on the creation of synthetic case-bases. The main reason to do so is to keep control of the different existing scenarios at home, and being able to study if the CBR agent is able to detect them.

From the proposed scenarios in subsection 3.3, new cases are created to populate the case-base. Every created case contains an event sequence which represents the visited locations for one behavioural template. That is, the templates may be understood as workflows, and the event sequences as their executions.

Furthermore, the case-base has a 90% of its cases representing normal daily activities, since abnormal scenarios should appear less frequently than normal scenarios. In order not to discriminate any abnormal scenario, the 10% of the remaining cases is equally shared by the abnormal cases. In particular, the normal behaviour is created using as template a workflow given in figure 3.

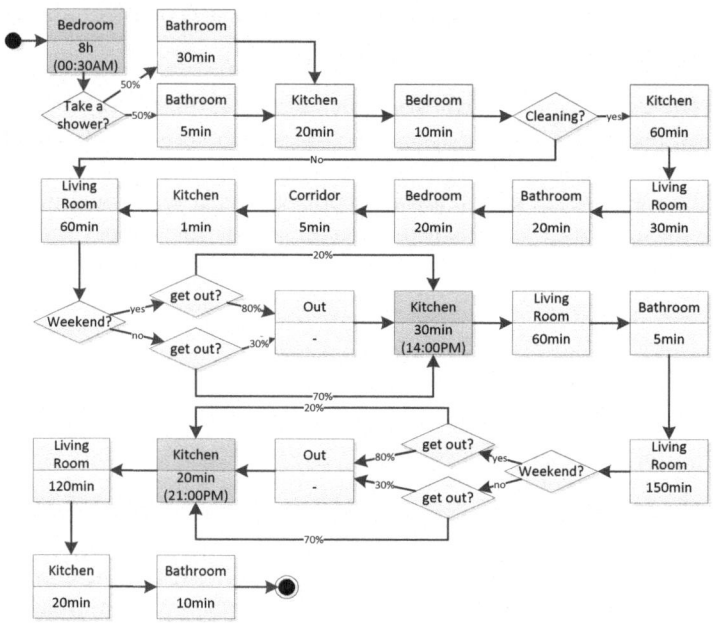

Fig. 3. Workflow of activities for a normal behaviour at home

Regarding the similarity between the different types of scenarios, while the bad night activities are almost similar to normal activities, the fallen activities are similar between themselves but not to normal and bad night activities.

4.2   Experiments Results

Several experiments have been run in order to evaluate the effects of the case-base size and the frequency in which the log is created by the communication level. That is, the CBR agent is evaluated for different case-base sizes in different log scenarios. The proposed log creation frequency are 6, 12, 18 and 24 hours. The cases in the case-base representing normal and bad night activities cover up to 24 hours of movements at home, and the rest of the cases represent normal activities until the unsafe scenario occurs.

Figure 4 shows the evaluation results from the performed experiments. For each experiment, the accuracy of the system is observed, as well as the false positive rate regarding the normal behaviour. Additionally, the true positives rates for the different behaviours are recorded to study if the CBR agent succeeds to detect them. Note that the false positive rate means the proportion of unsafe scenarios that were classified as normal, which must be avoided because this type of misclassification is the most dangerous one for the residents safety.

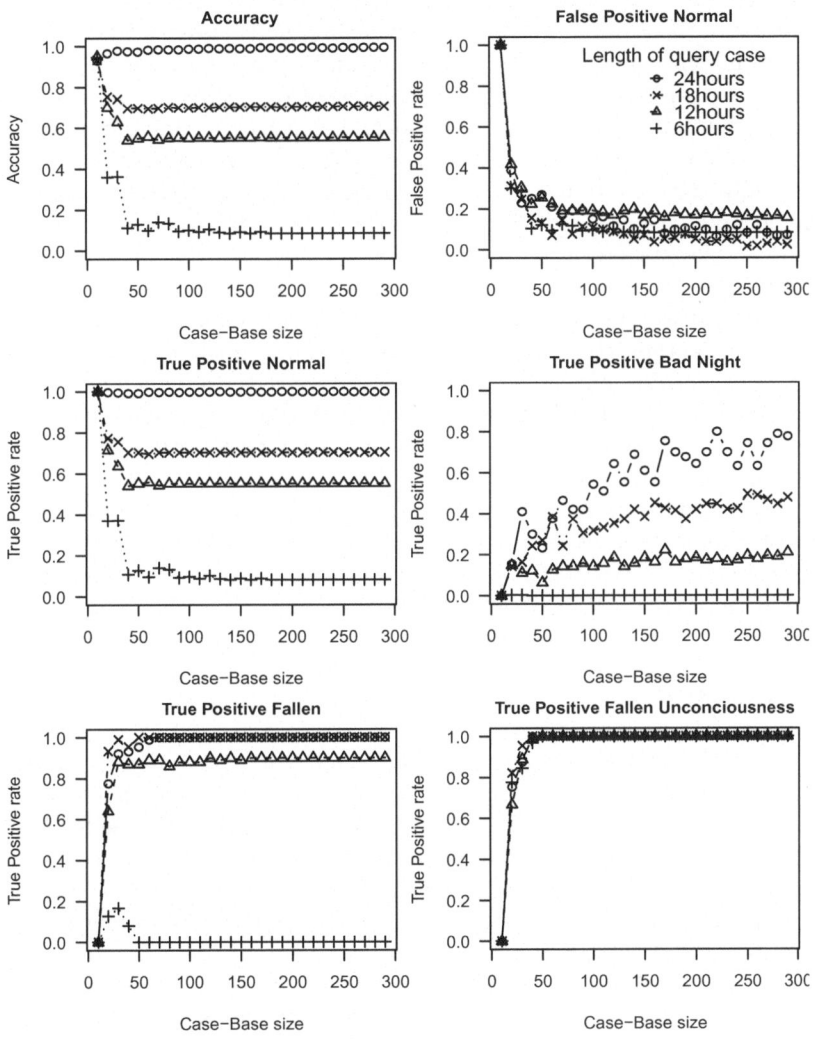

Fig. 4. Evaluating the CBR agent varying the case-base size

4.3   Discussion

Based on the results shown in figure 4, the following observations could be made:

- The accuracy is stable in all the scenarios when the case-base size is over 50 cases. However, a CBR agent using short observation periods has a lower accuracy than a CBR agent using longer observation periods. In fact, the experiments where the observation period spanned up to 24 hours gets an accuracy value close to 1, even with small case-bases.
- Retaining a large amount of cases decreases the ratio of false positives for the normal behaviour. That is, increasing the number of stored cases means a decrease of the number of times in which the agent miss-classifies a behaviour as normal when in fact there is an abnormal situation at the residents home.
- The true positives rate for the normal activities is directly correlated to the accuracy of the system, which is normal due to the predominance of this type in the case-base (90% of the cases). Nonetheless, when the observations are frequently acquired, then the CBR agent is increasingly unable to detect the normal activities, which increases the ratio of false alarms of the system. That is, the system is permanently classifying the abnormal situations as normal.
- Regarding the true positive rate of bad night scenarios, in all the experiments an increment of the case-base size will ease the correct classification of these scenarios, except for the shortest observations, which never classifies this activity correctly. In fact, CBR agents using longer observation periods classify the bad night scenarios correctly more frequently than those using shorter observation periods.
- The detection of falls is possible with most observation lengths, except for the shortest observation periods, which fail to detect falls without loss of consciousness. Notwithstanding, longer observation periods are not suitable for the needed quick response in the fall scenario.

5   Conclusions

In the present work, we propose a CBR agent for a Smart Home system to detect potential dangerous scenarios for elderly people living alone at home. The CBR agent gathers data on the activity of the monitored resident (in the form of event sequence) and uses temporal similarity to retrieve previously stored activities. The agent is implemented using the *myCBR* SDK. We evaluate the suitability of this approach using a simulation of activities at home representing normal and different dangerous scenarios.

Unlike other work [11,12,15,17], our proposal is based on a spatial-temporal representation using event sequences. Furthermore, our system relies on a low-cost pervasive sensor network. That is, the available information is limited and the temporal dimension plays an essential role. Essentially the CBR agent is a temporal case retrieval system using a temporal edit distance.

Our experiments are based on synthetic case-bases using a simulator in order to analyse the responses provided by the CBR agent. Results show that the CBR agent is able to detect the proposed unsafe scenarios, although this detection is limited by the amount of data within each observation of the house. Thus, the accuracy is affected by the frequency in which the house status is observed and the amount of data in each observation. On the one hand, long observations of the house ease the identification and classification of all the activities. However, low frequent observations can increase the response time of the system when a fall occurs. On the other hand, high frequency observations make possible for the system to response quickly to falls, but they make very difficult to detect long activities such as having a bad night. What is more, if the observation frequency is very high then the system is not able to detect falls neither.

We can also conclude that a CBR agent can detect unsafe scenarios at home, although some considerations need to be addressed before using a CBR agent in a real deployment. The observation of the environment needs to be done with short and long intervals in order to get a quick response to dangerous situations, such as falls, and to detect long activities. Finally, since new cases are added constantly to the case-base, there is a risk of a decreasing performance of the CBR agent. This problem may be solved with the application of an appropriate Case-Base Maintenance task. Our future steps will be focused on exploring available maintenance approaches, the adoption of other temporal similarity measures, analyse other experiment parameters and the practical evaluation of the CBR agent in a real world test environment. Finally, real data from operational *proDIA* will be used in order to study the performance of the proposed CBR agent.

Acknowledgements. This work was funded by the Seneca Research Foundation of the Region of Murcia under project 15277/PI/10.

References

1. Aamodt, A., Plaza, E.: Case-based reasoning: Foundational issues, methodological variations, and system approaches. AI Communications 7, 39–59 (1994)
2. Akyildiz, I., Su, W., Sankarasubramaniam, Y., Cayirci, E.: Wireless sensor networks: a survey. Computer Networks 38(4), 393–422 (2002)
3. Augusto, J.C., Nugent, C.D.: Smart homes can be smarter. In: Augusto, J.C., Nugent, C.D. (eds.) Designing Smart Homes. LNCS (LNAI), vol. 4008, pp. 1–15. Springer, Heidelberg (2006)
4. Botía, J.A., Villa, A., Palma, J.T.: Ambient assisted living system for in-home monitoring of healthy independent elders. Expert Systems with Applications 39(9), 8136–8148 (2012)
5. Broek, G.V.D., Cavallo, F., Wehrmann, C.: AALiance Ambient Assisted Living Roadmap. IOS Press (2010)
6. Copetti, A., Loques, O., Leite, J.C.B., Barbosa, T., da Nóbrega, A.: Intelligent context-aware monitoring of hypertensive patients. In: Conference on Pervasive Computing Technologies for Healthcare, pp. 1–6 (2009)
7. Corchado, J.M., Bajo, J., de Paz, Y., Tapia, D.I.: Intelligent environment for monitoring alzheimer patients, agent technology for health care. Decision Support Systems 44(2), 382–396 (2008)
8. Corchado, J., Bajo, J., Paz, Y.: A CBR system: The core of an ambient intelligence health care application. In: Prasad, B. (ed.) Soft Computing Applications in Industry. STUDFUZZ, vol. 226, pp. 311–330. Springer, Heidelberg (2008)

9. Cunningham, P., Doyle, D., Loughrey, J.: An evaluation of the usefulness of case-based explanation. In: Ashley, K.D., Bridge, D.G. (eds.) ICCBR 2003. LNCS, vol. 2689, pp. 122–130. Springer, Heidelberg (2003)
10. van Kerrebroeck, P., Abrams, P., Chaikin, D., Donovan, J., Fonda, D., Jackson, S., Jennum, P., Johnson, T., Lose, G., Mattiasson, A., Robertson, G., Weiss, J.: The standardisation of terminology in nocturia: Report from the standardisation sub-committee of the international continence society. Neurourology and Urodynamics 21(2), 179–183 (2002)
11. Kofod-Petersen, A., Aamodt, A.: Contextualised ambient intelligence through case-based reasoning. In: Roth-Berghofer, T.R., Göker, M.H., Güvenir, H.A. (eds.) ECCBR 2006. LNCS (LNAI), vol. 4106, pp. 211–225. Springer, Heidelberg (2006)
12. Kofod-Petersen, A., Aamodt, A.: Case-based reasoning for situation-aware ambient intelligence: A hospital ward evaluation study. In: McGinty, L., Wilson, D.C. (eds.) ICCBR 2009. LNCS, vol. 5650, pp. 450–464. Springer, Heidelberg (2009)
13. Krishnan, N.C., Cook, D.J.: Activity recognition on streaming sensor data. Pervasive and Mobile Computing 10, Part B, 138–154 (2014)
14. Krose, B., Kasteren, T.V., Gibson, C., Dool, T.V.D.: Care: Context awareness in residences for elderly. In: ISG 2008 (2008)
15. Leake, D., Maguitman, A., Reichherzer, T.: Cases, context, and comfort: Opportunities for case-based reasoning in smart homes. In: Augusto, J.C., Nugent, C.D. (eds.) Designing Smart Homes. LNCS (LNAI), vol. 4008, pp. 109–131. Springer, Heidelberg (2006)
16. Lehtola, S., Koistinen, P., Luukinen, H.: Falls and injurious falls late in home-dwelling life. Archives of Gerontology and Geriatrics 42(2), 217–224 (2006)
17. Ma, T., Kim, Y.D., Ma, Q., Tang, M., Zhou, W.: Context-aware implementation based on cbr for smart home. In: WiMob 2005, vol. 4, pp. 112–115 (2005)
18. Mannila, H., Moen, P.: Similarity between event types in sequences. In: Mohania, M., Tjoa, A.M. (eds.) DaWaK 1999. LNCS, vol. 1676, pp. 271–280. Springer, Heidelberg (1999)
19. Moen, P.: Attribute, Event Sequence, and Event Type Similarity Notions for Data Mining. Ph.D. thesis, University of Helsinki (2000)
20. Rawi, M., Al-Anbuky, A.: Passive house sensor networks: Human centric thermal comfort concept. In: ISSNIP 2009, pp. 255–260 (2009)
21. Safavi, A., Keshavarz-Haddad, A., Khoubani, S., Mosharraf-Dehkordi, S., Dehghani-Pilehvarani, A., Tabei, F.: A remote elderly monitoring system with localizing based on wireless sensor network. In: ICCDA 2010, vol. 2, pp. 553–557 (2010)
22. Sormo, F., Cassens, J., Aamodt, A.: Explanation in case-based reasoning - perspectives and goals. Artificial Intelligence Review 24(2), 109–143 (2005)
23. Stahl, A., Roth-Berghofer, T.R.: Rapid prototyping of CBR applications with the open source tool myCBR. In: Althoff, K.-D., Bergmann, R., Minor, M., Hanft, A. (eds.) ECCBR 2008. LNCS (LNAI), vol. 5239, pp. 615–629. Springer, Heidelberg (2008)
24. Tabar, A.M., Keshavarz, A., Aghajan, H.: Smart home care network using sensor fusion and distributed vision-based reasoning. In: MM 2006, pp. 145–154 (2006)
25. WHO, NIA, NIH: Global health and ageing. NIH Publications, WHO (2011)
26. Wiratunga, N., Craw, S., Taylor, B., Davis, G.: Case-based reasoning for matching smarthouse technology to people's needs. Knowledge-Based Systems 17, 139–146 (2004)

An Algorithm for Conversational Case-Based Reasoning in Classification Tasks

David McSherry

School of Computing and Information Engineering
University of Ulster, Coleraine BT52 1SA, Northern Ireland
dmg.mcsherry@ulster.ac.uk

Abstract. An important benefit of conversational case-based reasoning (CCBR) in applications such as customer help-desk support is the ability to solve problems by asking a small number of well-selected questions. However, there have been few investigations of the effectiveness of CCBR in classification problem solving, or its ability to compete with k-NN and other machine learning algorithms in terms of accuracy. We present a CCBR algorithm for classification tasks and demonstrate its ability to achieve high levels of problem-solving efficiency, while often equaling or exceeding the accuracy of k-NN and C4.5, a widely used algorithm for decision tree learning.

Keywords: conversational case-based reasoning, classification, accuracy, efficiency, transparency.

1   Introduction

In contrast to traditional case-based reasoning (CBR), conversational CBR (CCBR) makes no assumption that a description of the problem to be solved is available in advance. Instead, problem features are incrementally and selectively elicited in an interactive dialog with the user, usually with the aim of minimizing the number of questions that need to be asked to solve the problem. As shown in applications such as customer help-desk support and interactive fault diagnosis, guiding the selection of features (or questions) that are most useful for solving a given problem is an important advantage of CCBR [1–7].

Although CCBR has been successfully adapted for use in planning and recommendation tasks (e.g., [2, 8–11]), there have been few investigations of its effectiveness in classification problem solving or ability to compete with k-NN and other machine learning algorithms in terms of classification accuracy. In previous work, we presented a CCBR algorithm called iNN(k) that achieved good levels of accuracy in a variety of classification tasks related to medicine and health care, while often requiring the user to provide only a small subset of the features in a complete problem description [12–13].

Question selection is guided in iNN(k) by the goal of confirming a target class. A feature's discriminating power is assessed in terms of its relative frequencies among

L. Lamontagne and E. Plaza (Eds.): ICCBR 2014, LNCS 8765, pp. 289–304, 2014.
© Springer International Publishing Switzerland 2014

cases with and without the target class as their solution, and also takes account of the number of values of a selected attribute. The k parameter in iNN(k) plays a similar role in the retrieval of similar cases as in k-NN. Another parameter that controls the algorithm's behavior is a choice between local and global feature selection; only features that occur in one or more of the most similar cases are considered in local feature selection, whereas global feature selection relaxes this constraint.

While iNN(k) has been shown to be effective when applied to classification datasets with only nominal or discrete attributes with small numbers of values, the need for a feature selection measure that is equally suitable for nominal and numeric (i.e., discrete or continuous) attributes, and avoids any possible bias in favor of one attribute type or the other, is one of the issues addressed in iNN$_2$(k), a new algorithm for CCBR in classification tasks that we present in this paper. The new algorithm also requires only a single parameter (k) to be selected for a given classification task.

Related work is briefly discussed in Section 2. In Sections 3 and 4, we describe our approach to CCBR in iNN$_2$(k) and show how the algorithm's goal-driven approach to question selection, a feature it shares with iNN(k), contributes to problem-solving transparency in a CCBR system called CBR-*Converse*. In Section 5, we present empirical results that demonstrate the ability of iNN$_2$(k) to achieve high levels of problem-solving efficiency on a selection of classification datasets, while often equaling or exceeding the accuracy of k-NN and C4.5, a widely used algorithm for decision tree learning [14]. Our conclusions are presented in Section 6.

2   Related Work

Inference Corporation's CBR3 was one of the first commercial CBR tools for developing help-desk applications of CCBR [1], while CCBR tools for interactive fault diagnosis include CaseAdvisor [3], ExpertGuide [6], NaCoDAE [2], and SHRIEK [5]. A common feature of CCBR systems for help-desk and similar applications is that every case has a unique solution [2, 15–16], which means that classification accuracy cannot be assessed by conventional methods. Although some CCBR algorithms have also been applied to traditional classification datasets, they have been evaluated mainly in terms of dialog efficiency (e.g., [17–20]).

Gu and Aamodt [18] propose an evaluation strategy for CBR systems that combines both domain-independent and domain-dependent goals. Assessment of domain-independent evaluation goals is exemplified by the use of multiple classification datasets to investigate the effects of similarity calculation methods and other factors on dialog efficiency in CCBR [19, 17]. Domain-dependent evaluation goals are discussed in the context of a knowledge-intensive CCBR (KI-CCBR) system for retrieval of image processing software components. The evaluation reported includes an analysis of features such as dialog relevance and explanation of reasoning and an ablation study of the effects of KI-CCBR techniques on dialog efficiency.

Bogaerts and Leake [21] note that performance measures like dialog efficiency and precision do not reflect important dialog properties such as the quality of cases that the user is shown in the early stages of a CCBR dialog. They propose a measure of rank quality for assessing how well the list of cases that the user is shown in each cycle of a CCBR dialog approximates to the list of cases that would be generated for a complete problem description. However, the performance measures on which we focus in our initial evaluation of $iNN_2(k)$ are classification accuracy and problem-solving efficiency assessed in terms of the average length of CCBR dialogs.

3   Conversational CBR in $iNN_2(k)$

In this section, we describe the key features of $iNN_2(k)$ as a CCBR algorithm, such as the similarity measures used for retrieval, the method used to select the most useful questions, and the conditions for terminating a CCBR dialog.

3.1   Cases, Attributes, and Queries

We assume the structure of the case base to be similar to that of a traditional classification dataset (i.e., the same attributes are used to describe each case, and each class is typically represented by several cases). Both nominal and numeric attributes are supported by $iNN_2(k)$, and there may be missing attribute values in the case base and/or attributes with unknown values in the elicited problem description. Both missing and unknown values are denoted by question marks (?) in the following discussion. We denote by A the set of attributes, other than the class attribute, used to describe each case C, and by $\pi_a(C)$ the value of a in C for each $a \in A$. A case's problem description is thus a list of features $a = v$, one for each $a \in A$, such that $v \in values(a)$ $\cup \{?\}$. The solution stored in a case C is a class label, which we denote by $class(C)$.

Supporting Cases. An important role in $iNN_2(k)$ is played by the set of cases that support a given class [12–13]. For any class G and case C, we say that C supports G if $class(C) = G$.

A query Q describing the problem to be solved is incrementally elicited in $iNN_2(k)$ by asking the user for the values of attributes that are most useful for confirming a target class G. The user can answer *unknown* to any question, in which case they will be asked the next most useful question. After one or more questions have been answered in a CCBR dialog, the current query in $iNN_2(k)$ is a list of problem features $Q = \{a_1 = v_1, ..., a_r = v_r\}$, where $v_i \in values(a_i) \cup \{?\}$ for $1 \leq i \leq r$. We denote by A_Q the set of attributes in Q, and by $\pi_a(Q)$ the value of a in Q for each $a \in A_Q$.

3.2   Similarity in $iNN_2(k)$

For any case C and query Q we define:

$$Sim(C, Q) = 0 \text{ if } Q = \varnothing$$

$$Sim(C, Q) = \sum_{a \in A_Q} w_a \times sim_a(C, Q) \text{ otherwise}$$

An importance weight w_a may be assigned to each $a \in A$, although the case attributes are equally weighted in our evaluation of $iNN_2(k)$ in Section 5. For each $a \in A_Q$, $sim_a(C, Q)$ is a local measure of the similarity between the values of a in C and Q. For any case C, query Q, and nominal attribute $a \in A_Q$, we define:

$$sim_a(C, Q) = 0 \text{ if } \pi_a(C) \neq \pi_a(Q)$$
$$sim_a(C, Q) = 0 \text{ if } \pi_a(C) = \pi_a(Q) = ?$$
$$sim_a(C, Q) = 1 \text{ otherwise}$$

For any case C, query Q, and numeric attribute $a \in A_Q$, we define:

$$sim_a(C, Q) = 0 \text{ if } \pi_a(C) = ? \text{ or } \pi_a(Q) = ?$$
$$sim_a(C, Q) = 0 \text{ if } \pi_a(Q) > \max(a) \text{ or } \pi_a(Q) < \min(a)$$
$$sim_a(C, Q) = 1 - \frac{|\pi_a(C) - \pi_a(Q)|}{\max(a) - \min(a)} \text{ otherwise}$$

where, for example, $\max(a)$ is the maximum value of a in the case base.

3.3 The $iNN_2(k)$ Retrieval Set

As in $iNN(k)$ [12–13], the cases retrieved by $iNN_2(k)$ in each cycle of a CCBR dialog are those for which there are less than k more similar cases. More formally, the $iNN_2(k)$ retrieval set for a given query Q, integer $k \geq 1$, and case base CB is:

$$r_k(Q) = \{C \in \text{CB} : \text{card}(more\text{-}similar(C, Q)) < k\} \tag{1}$$

where $more\text{-}similar(C, Q)$ is the set of cases that are more similar to Q than C. In common with versions of k-NN that include all ties for the kth most similar case in the retrieval set (e.g., [22–23]), the $iNN_2(k)$ retrieval set for a given query may contain more than k cases. Also as in $iNN(k)$, the $iNN_2(k)$ retrieval set for the empty query at the start of a CCBR dialog is the set of all cases in the case base, as no case is more similar to an empty query than another case.

3.4 The Target Class in $iNN_2(k)$

The target class used to guide question selection in $iNN_2(k)$ is the class that is supported by most cases in $r_k(Q)$, the $iNN_2(k)$ retrieval set for the current query Q (or equivalently, the majority class in the retrieval set). If there is a tie for the class supported by most cases in the $iNN_2(k)$ retrieval set, then the tied class with the most similar supporting case is selected as the target class. Fig. 1 shows an example in which $k = 5$ and two of the 3 classes in the $iNN_2(k)$ retrieval set are supported by the same number (2) of retrieved cases. As the tied class with the most similar supporting case is Class 2, Class 2 is selected as the target class in this example.

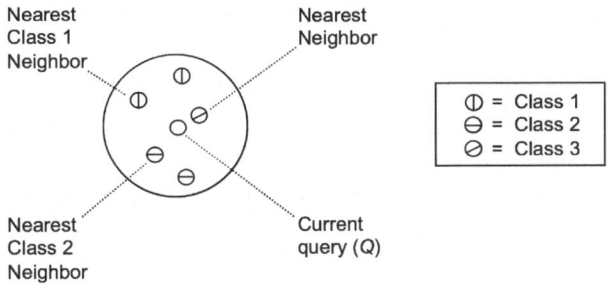

Fig. 1. Breaking a tie for the class supported by most cases in the $iNN_2(k)$ retrieval set

If there is also a tie for the class with the most similar supporting case, then the tied class that is supported by most cases in the case base as a whole is selected as the target class. A class tie that remains after this step is broken alphabetically. As noted in Section 3.3, the $iNN_2(k)$ retrieval set at the start of a CCBR dialog consists of all cases in the case base, and so the target class is initially the class that is supported by most cases in the case base. However, the target class is subject to change at any time as the problem description is elicited.

3.5   Relative Frequencies in $iNN_2(k)$

The relative frequency of the target class in the $iNN_2(k)$ retrieval set plays an important role in the selection of questions that are most useful for confirming the target class. For any class G and subset S of the case base, the relative frequency of G in S is:

$$p(G, S) = \frac{card(\{C \in S : class(C) = G\})}{card(S)} \tag{2}$$

For example, if the target class G in a CCBR dialog is supported by 6 out of 9 cases in $r_k(Q)$, the $iNN_2(k)$ retrieval set for the current query Q, then $p(G, r_k(Q)) = 0.67$.

3.6   Selecting the Most Useful Question

In each cycle of a CCBR dialog, $iNN_2(k)$ selects the feature $a^* = v^*$ that it predicts to be most useful for confirming the target class and asks the user for the attribute's value. Specifically, the feature $a^* = v^*$ predicted to be most useful for confirming the target class G is the one that maximizes:

$$p(G, r_k(Q \cup \{a = v\})) \tag{3}$$

i.e., the relative frequency of G in the $iNN_2(k)$ retrieval set for the query $Q \cup \{a = v\}$ created by adding $a = v$ to the current query Q.

However, only features $a = v$ that occur in at least one case in the $iNN_2(k)$ retrieval set that supports the target class are considered by the algorithm. The selected feature must also be one that increases the relative frequency of the target class in the $iNN_2(k)$ retrieval set i.e.,

$$p(G, r_k(Q \cup \{a = v\})) > p(G, r_k(Q)) \tag{4}$$

Finally, any attribute for which the user has already been asked to provide a value is excluded from the selection process. If there is a tie for the feature that maximizes the relative frequency of the target class, then the tied feature that maximizes:

$$\text{card}(r_k(Q \cup \{a = v\})) \tag{5}$$

is selected. A tie that remains after this step is broken alphabetically by attribute name. We say that a feature $a = v$ confirms a target class G if:

$$p(G, r_k(Q \cup \{a = v\})) = 1 \tag{6}$$

3.7   The $iNN_2(k)$ Solution

A CCBR dialog based on $iNN_2(k)$ typically continues until all cases in the retrieval set have the same class label. However, additional termination criteria are needed in practice, for example to avoid prolonging a dialog that appears to be making no progress towards confirming the target class. Specifically, a CCBR dialog based on $iNN_2(k)$ continues until one of the following conditions is satisfied:

- All cases in the $iNN_2(k)$ retrieval set have the same class label
- The user has already been asked for the values of all the case attributes but there are still cases in the $iNN_2(k)$ retrieval set with different class labels
- None of the cases in the $iNN_2(k)$ retrieval set that support the target class has a feature which, if added to the current query, will increase the relative frequency of the target class

At this point, the dialog is terminated and the class G that is supported by most cases in the $iNN_2(k)$ retrieval set is presented as a solution to the problem described by the user. Any ties for the class supported by most cases in the $iNN_2(k)$ retrieval set are broken as described in Section 3.4.

3.8   Algorithm Overview

The $iNN_2(k)$ algorithm is outlined in Fig. 2. The class supported by most cases in the $iNN_2(k)$ retrieval set is selected as the target class (G) in Line 6. Two of the conditions for terminating the CCBR dialog (i.e., target class confirmed, no more attributes) are tested in Line 7. The feature $a^* = v^*$ that maximizes the relative frequency of G over all features that occur in at least one case in the $iNN_2(k)$ retrieval set that supports G is selected in Lines 10–12. The third and final condition for terminating the CCBR dialog (i.e., no increase in the relative frequency of the target class) is tested in Line 13. Only if none of the termination conditions are satisfied, the user is asked for the value of a^* in Line 16 and the user's answer is used to extend the current query (Q) in Lines

17–18. In Line 19, $a*$ is removed from the set of attributes available for selection. The cycle is then repeated until one of the conditions for terminating the CCBR dialog is satisfied. Finally, G is returned in Line 23 as the $iNN_2(k)$ solution to the problem described by the user.

4   Conversational CBR in CBR-*Converse*

In a CCBR dialog based on $iNN_2(k)$, a description of the problem to be solved is incrementally elicited by asking the user questions with the goal of confirming a target class. In this section, we demonstrate the approach in CBR-*Converse*, a CCBR system that uses a similar approach to explanation as its predecessor CBR-*Confirm* [12–13], but with $iNN_2(k)$ replacing $iNN(k)$ as the algorithm used to guide the CCBR process. Explanation in CBR-*Converse* is discussed in Sections 4.1 and 4.2, and a CCBR dialog based on the Labor Relations dataset from the UCI Machine Learning Repository [24] is presented in Section 4.3.

Algorithm: $iNN_2(k)$
Inputs: integer (k), case base, attributes (A)
Output: problem solution (G)

```
 1  begin
 2      Q ← ∅
 3      solved ← false
 4      while not solved do
 5      begin
 6          G ← class supported by most cases in r_k(Q)
 7          if class(C) = G for all C ∈ r_k(Q) or A = ∅
 8          then solved ← true
 9          else begin
10              select the feature a* = v* that maximizes p(G, r_k(Q ∪ {a = v}))
11              over all features a = v such that a ∈ A and π_a(C) = v for at least
12              one C ∈ r_k(Q) such that class(C) = G
13              if p(G, r_k(Q ∪ {a* = v*})) ≤ p(G, r_k(Q))
14              then solved ← true
15              else begin
16                  ask user for value of a*
17                  v ← value of a* reported by user
18                  Q ← Q ∪ {a* = v}
19                  A ← A − {a*}
20              end
21          end
22      end
23      return G
24  end
```

Fig. 2. Algorithm for CCBR in classification tasks

4.1   Explaining Why a Question Is Relevant

Before answering any question in CBR-*Converse*, the user can ask why it is relevant. In response, CBR-*Converse* uses one of the following explanation templates to explain the relevance of the feature $a = v$ it has selected as most useful for confirming the target class:

E_1: if $a = v$ this will confirm G_1
E_2: if $a = v$ this will decrease the likelihood of G_2
E_3: if $a = v$ this will increase the likelihood of G_1

where G_1 is the target class and G_2 is a competing class in $r_k(Q)$, the iNN$_2(k)$ retrieval set for the current query Q.

The explanation template used by CBR-*Converse* depends on how the class distribution in the iNN$_2(k)$ retrieval set will be affected if the selected feature $a = v$ is added to the current query Q. If all cases in $r_k(Q \cup \{a = v\})$ support the target class G_1, then CBR-*Converse* uses template E_1 to explain that $a = v$ will confirm G_1. Another possible effect of a selected feature $a = v$ is to eliminate a competing class G_2 (though perhaps only temporarily) from the iNN$_2(k)$ retrieval set. In this case, CBR-*Converse* uses template E_2 to explain the relevance of the selected feature. If a selected feature neither confirms the target class nor eliminates a competing class from the iNN$_2(k)$ retrieval set, then CBR-*Converse* uses template E_3 to explain its relevance.

4.2   Explaining the iNN$_2(k)$ Solution

A common approach to explaining CBR solutions is to show the user a similar case with the same solution [25]. However, the value of the most similar case as an explanation case has been questioned by several authors (e.g., [26–28]). For example, Doyle *et al.* [27] show that a case closer to the decision boundary may provide a more convincing explanation than the most similar case in nearest neighbor classification. A related issue in classification based on k-NN is that the CBR solution may differ from the solution for the most similar case. Perhaps the simplest strategy in this situation is to use the most similar case with the same solution as an explanation case [28]. A similar approach to explaining iNN$_2(k)$ solutions, which may also differ from the solution for the most similar case, is used in CBR-*Converse*.

As discussed in Section 3.7, the iNN$_2(k)$ solution to the problem described by the user is the class G that is supported by most cases in the iNN$_2(k)$ retrieval set at the end of the CCBR dialog. When presenting the iNN$_2(k)$ solution (G) to the user, CBR-*Converse* also shows the user the most similar case that supports G, which by definition is one of the cases in the iNN$_2(k)$ retrieval set. If there is a tie for the most similar supporting case, then the tied case that appears first in the case base is used to explain the solution.

4.3   Example CCBR Dialog

Fig. 3 shows a CCBR dialog in CBR-*Converse* based on the Labor Relations dataset [24] with question selection based on iNN$_2(k)$ with $k = 5$. The dataset contains descriptions of 57 employment contracts and outcomes (good or bad) of the contract negotiations. The 16 non-class attributes in the dataset include both nominal and numeric attributes. The example dialog was created by removing one case (Case 22) from the dataset and using its description to simulate the user's answers to questions asked by CBR-*Converse* when provided with a case base containing the other 56 cases in the dataset. Because there are missing values for several attributes in the left-out case, only some of the problem features are known to the simulated user.

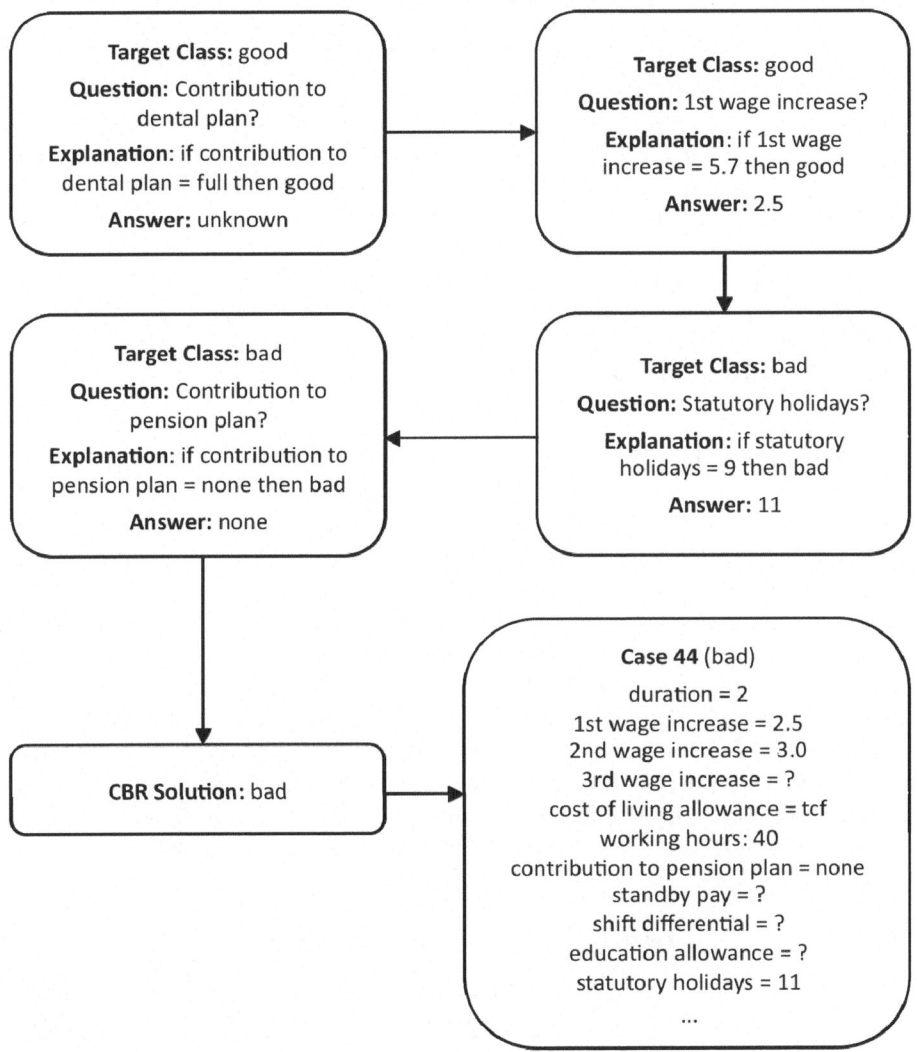

Fig. 3. Example dialog in CBR-*Converse* based on the Labor Relations dataset, with question selection based on $iNN_2(k)$ with $k = 5$

At the start of the example dialog, the majority class in the dataset (good) is selected by CBR-*Converse* as the target class and the user is asked about the employer contribution to the dental plan. As the value of this attribute is unknown to the user, the dialog moves on to the next most useful question. When the 1st wage increase is reported to be 2.5 in the second cycle, the target class changes from good to bad and the user is asked for the number of statutory holidays. The target class (bad) remains unchanged in light of the user's answer, and is confirmed by the user's answer to the next question.

The example dialog also shows how CBR-*Converse* can explain the relevance of each question it asks the user in terms of the target class it is trying to confirm. For example, when the user asks about the relevance of the 1st wage increase, CBR-*Converse* explains that if the 1st wage increase is 5.7 this will confirm the target class (good). Note that the only explanation template needed by CBR-*Converse* in the example dialog is E_1 (with the wording simplified in Fig. 3 to *if a = v then G*).

In this example, CBR-*Converse* is able to reach a solution after asking the user only 4 out of 16 possible questions and in spite of the employer's contribution to the dental plan being unknown to the user. The $iNN_2(k)$ solution (bad) is also the same as in the left-out case. The case used by CBR-*Converse* to explain the $iNN_2(k)$ solution (Case 44) is partly shown in Fig. 3.

5   Empirical Evaluation

In this section, we evaluate $iNN_2(k)$ on a selection of classification datasets from the UCI Machine Learning Repository [24]. The performance criteria of interest are classification accuracy and problem-solving efficiency assessed in terms of the average length of CCBR dialogs (i.e., number of questions required to reach a solution). We also examine the hypothesis that selecting a preferred k value for a given classification task may involve a trade-off between classification accuracy and problem-solving efficiency in $iNN_2(k)$. The number of cases in the selected datasets (Table 1) ranges from 57 to 435, the number of classes from 2 to 6, the number of non-class attributes from 4 to 16, and the percentage of missing values from 0% to 37%. Two of the datasets (Iris and Glass) have only numeric attributes, while Voting Records has only nominal attributes. Zoo and Labor Relations have both nominal and numeric attributes.

Table 1. UCI datasets used in the evaluation

Dataset	Cases	Classes	Non-Class Attributes	Nominal Attributes	Numeric Attributes	Missing Values
Iris	150	3	4	0	4	0%
Zoo	101	6	16	15	1	0%
Labor Relations	57	2	16	8	8	37%
Glass	214	6	9	0	9	0%
Voting Records	435	2	16	16	0	6%

5.1   Experimental Methods

We assess the accuracy of $iNN_2(k)$ in comparison with k-NN and C4.5, a widely used algorithm for decision-tree learning developed by J.R. Quinlan [14]. The evaluated version of k-NN uses the same similarity measures and tie-breaking methods as $iNN_2(k)$, and the attributes in each dataset are equally weighted in both algorithms. Our results for C4.5 are based on J48, an open source implementation of C4.5 in the Weka data mining toolkit [29]. The algorithm is used with the default settings for the parameters that control its pruning behavior. We also compare $iNN_2(k)$ with a simplified version of the algorithm called $iNNR(k)$ in which there is no target class, questions are selected randomly, and dialog length is limited to a randomly selected number of questions. As only 2 of the 3 conditions for terminating a CCBR dialog in $iNN_2(k)$ (Section 3.7) are applicable in the absence of a target class, dialog length is randomly constrained in $iNNR(k)$ to make it more competitive in terms of problem-solving efficiency.

Leave-one-out cross validation is used to assess the accuracy of $iNN_2(k)$ on the selected datasets as well as the average lengths of CCBR dialogs. For each dataset, we temporarily remove one case at a time and use its description to simulate the user's answers to questions asked by a CCBR system based on $iNN_2(k)$ with a case base containing all the other cases in the dataset. For any attribute with a missing value in the left-out case, the simulated user answers *unknown* when asked for its value. All questions asked contribute to dialog length whether or not the attribute's value is known to the simulated user. Average dialog lengths for the two datasets with missing values may thus be lower when a complete problem description is always known to the user.

At the end of each dialog, we record the number of questions asked and whether or not the $iNN_2(k)$ solution is correct (i.e., the same as in the left-out case). We repeat the process for all values of k from 1 to $n - 1$, where n is the size of the dataset. The performance of $iNNR(k)$ on each dataset is similarly assessed in the evaluation, in this case with the results for each k value averaged over 3 repeated experiments. We also use leave-one-out cross validation to assess the accuracy of k-NN and C4.5 on each dataset, and in the case of C4.5 using the tools provided in Weka [29].

5.2   Overall Accuracy

Table 2 shows the highest levels of accuracy achieved by $iNN_2(k)$, $iNNR(k)$, and k-NN on the selected datasets over all values of k. The default accuracy of C4.5 is also shown for each dataset, and the best accuracy results are highlighted in bold. $iNN_2(k)$ equaled or exceeded the best accuracy of k-NN on 4 of the 5 datasets, but was outperformed by k-NN on Labor Relations. $iNN_2(k)$ also gave higher levels of accuracy than C4.5 on 4 of the 5 datasets, but was unable to match the accuracy of C4.5 on Voting Records. $iNNR(k)$ was least accurate on 4 of the 5 datasets, though did better than C4.5 on Labor Relations. The k values that maximized the accuracy of $iNN_2(k)$ were in the range from 1 to 10 for all but one of the datasets (Iris).

Table 2. Best accuracy achieved by $iNN_2(k)$, $iNNR(k)$, and k-NN on each dataset over all k values. The default accuracy of C4.5 (as implemented in Weka) on each dataset is also shown.

Dataset	Cases	Best $iNN_2(k)$		Best $iNNR(k)$		Best k-NN		C4.5
		Acc.	k	Acc.	k	Acc.	k	Acc.
Iris	150	**96.0**	42	91.8	27	**96.0**	39	95.3
Zoo	101	**96.0**	1	83.5	1	**96.0**	1	92.1
Labor Relations	57	89.5	5	87.1	7	**91.2**	14	77.2
Glass	214	**73.8**	10	64.6	19	72.9	2	66.8
Voting Records	435	95.9	4	90.5	3	93.1	4	**96.8**

5.3   Accuracy *vs.* Dialog Length

Table 3 shows the accuracy (A) and average lengths of $iNN_2(k)$ dialogs (D) for $k = 1$ to 10 on the selected datasets. The highest level of accuracy achieved by $iNN_2(k)$ for the specified range of k values is shown in bold for each dataset. Average lengths of $iNN_2(k)$ dialogs can be seen to increase or remain unchanged for 3 of the 5 datasets as k increases from 1 to 10. The accuracy of $iNN_2(k)$ varies considerably over the range of k values for most of the datasets, but with no predictable pattern.

A trade-off between accuracy and average dialog length can be seen in the $iNN_2(k)$ results for Glass. An increase in accuracy from 50.0% to 73.4% is gained at the cost of an increase in average dialog length from 1.8 to 3.6 as k increases from 1 to 6. A further increase in average dialog length from 3.6 to 4.3 can be seen as k continues to increase until the maximum accuracy of 73.8% on this dataset is reached for $k = 10$. Similar trade-offs are evident in the results for Labor Relations and Iris, and to a lesser extent Voting Records. These observations support the hypothesis that selecting a preferred k value for a given classification task may involve a trade-off between classification accuracy and problem-solving efficiency in $iNN_2(k)$. However, it can be seen from the results for Zoo that such a trade-off does not always exist. For this dataset, there are two k values (1 and 2) that maximize accuracy while also minimizing average dialog length.

Table 3. Accuracy (A) of $iNN_2(k)$ related to average dialog length (D) for $k = 1$ to 10 on the selected datasets

k		1	2	3	4	5	6	7	8	9	10
Iris	A	94.7	92.7	94.0	94.7	94.7	94.7	94.7	**95.3**	94.7	94.7
	D	1.2	1.3	1.3	1.4	1.4	1.5	1.5	1.5	1.5	1.5
Zoo	A	**96.0**	**96.0**	94.1	96.0	92.1	90.1	88.1	88.1	84.2	84.2
	D	2.6	2.6	2.7	2.8	2.6	2.7	2.7	3.1	2.8	2.9
Labor Relations	A	84.2	80.7	82.5	86.0	**89.5**	84.2	82.5	82.5	82.5	75.4
	D	2.6	3.1	3.1	3.3	3.7	4.7	4.9	4.9	5.2	5.3
Glass	A	50.0	58.9	65.4	71.0	69.6	73.4	70.1	73.4	70.6	**73.8**
	D	1.8	2.0	2.6	2.7	3.0	3.6	3.7	4.1	4.2	4.3
Voting Records	A	95.2	95.6	95.4	**95.9**	95.4	95.6	95.2	94.7	94.3	94.9
	D	3.2	3.2	3.3	3.3	3.4	3.4	3.4	3.5	3.5	3.4

Table 4 shows the accuracy (A) and average lengths of iNNR(k) dialogs (D) for k = 1 to 10 on the selected datasets. In spite of dialog length being randomly constrained in iNNR(k), average dialog lengths are generally higher than in iNN$_2$(k), though to a lesser extent on Iris and Glass. iNNR(k) came close to matching the accuracy of iNN$_2$(k) on Labor Relations. However, an average dialog length of 6.4 is required for accuracy of 87.1% in iNNR(k) compared to 3.7 for accuracy of 89.5% in iNN$_2$(k).

Table 4. Accuracy (A) of iNNR(k) related to average dialog length (D) for k = 1 to 10 on the selected datasets

k		1	2	3	4	5	6	7	8	9	10
Iris	A	83.6	86.9	88.7	88.4	88.9	88.4	**90.9**	87.3	88.0	89.3
	D	1.4	1.5	1.5	1.5	1.7	1.6	1.6	1.7	1.7	1.7
Zoo	A	**83.5**	83.2	82.2	81.5	78.9	76.9	82.8	78.9	81.2	77.6
	D	4.3	4.8	4.6	4.8	4.9	5.3	5.4	5.3	5.8	5.6
Labor Relations	A	80.1	78.9	83.0	83.6	82.5	82.5	**87.1**	84.2	81.3	83.6
	D	3.2	3.6	3.6	4.7	5.2	5.4	6.4	5.8	7.2	7.6
Glass	A	43.1	59.8	60.9	**64.3**	60.7	63.4	64.0	61.1	62.6	60.4
	D	1.6	2.4	3.2	3.6	3.9	4.0	4.4	4.2	4.5	4.5
Voting Records	A	87.3	90.2	**90.5**	89.3	88.8	90.0	89.3	90.2	88.8	89.1
	D	5.4	5.5	5.8	5.8	5.7	5.9	6.0	5.8	6.1	6.1

5.4   Problem-Solving Efficiency

Table 5 shows the minimum, average and maximum lengths of iNN$_2$(k) dialogs for the k values from 1 to 10 that maximized the accuracy of iNN$_2$(k) on the selected datasets. The maximum possible length of a CCBR dialog (i.e., the number of non-class attributes in a given dataset) is reached for only 3 of the 5 datasets. Average dialog length is also shown as a percentage of the number of non-class attributes in each dataset. A lower value for this measure indicates a higher level of problem-solving efficiency.

For example, average dialog length on the Labor Relations dataset (3.7) amounts to only 23% of the number of features (16) in a complete problem description. iNN$_2$(k) can also be seen to have achieved high levels of problem-solving efficiency on Zoo and Voting Records, though to a lesser extent on Iris and Glass. A relatively low level of problem-solving efficiency for the k value (10) that maximizes the accuracy of iNN$_2$(k) on Glass is perhaps not surprising given the strong trade-off between accuracy and average dialog length on this dataset (Section 5.3).

Table 5. Minimum, average and maximum lengths of $iNN_2(k)$ dialogs for the k values from 1 to 10 that maximized the accuracy of $iNN_2(k)$ on the selected datasets. Average dialog length is also shown for each dataset as a percentage of the number of non-class attributes in the dataset.

Dataset	k	Dialog Length			Non-Class Attributes	Avg. Dialog Length (%)
		Min.	Avg.	Max.		
Iris	8	1	1.5	4	4	38%
Zoo	1	1	2.6	7	16	16%
Labor Relations	5	1	3.7	12	16	23%
Glass	10	1	4.3	9	9	48%
Voting Records	4	2	3.3	16	16	21%

6 Conclusions

We presented an algorithm for CCBR called $iNN_2(k)$ and demonstrated its ability to achieve high levels of accuracy and problem-solving efficiency in a range of classification tasks. For example, only 16%, 21%, and 23% on average of the features in a complete problem description were considered in CCBR dialogs for the k values that maximized the accuracy of $iNN_2(k)$ on Zoo (96.0%), Voting Records (95.9%), and Labor Relations (89.5%). Moreover, $iNN_2(k)$ equaled or exceeded the accuracy of k-NN in 4 of the 5 classification tasks studied, and exceeded the default accuracy of C4.5 in 4 of the 5 tasks. We also showed that selecting a preferred k value for a given classification task may involve a trade-off between accuracy and problem-solving efficiency in $iNN_2(k)$, for example as seen in the algorithm's behavior on Glass.

As in its predecessor $iNN(k)$, question selection is guided in $iNN_2(k)$ by the goal of confirming a target class, a feature that contributes to problem-solving transparency as shown in CBR-*Converse*. Ranking of candidate features is based in $iNN_2(k)$ on their impact on the relative frequency of the target class, an approach that is equally suitable for nominal and numeric attributes. Future research will include an in-depth evaluation of $iNN_2(k)$ in comparison with $iNN(k)$, which has been shown to perform well on classification datasets containing only nominal or discrete attributes with small numbers of values [12–13]. Adapting $iNN_2(k)$ as an algorithm for CCBR in regression tasks is another interesting direction for future work.

References

1. Acorn, T.L., Walden, S.H.: SMART: Support Management Automated Reasoning Technology for Compaq Customer Service. In: 4th Annual Conference on Innovative Applications of Artificial Intelligence, pp. 3–18. AAAI Press, Menlo Park (1992)
2. Aha, D.W., Breslow, L.A., Muñoz-Avila, H.: Conversational Case-Based Reasoning. Appl. Intell. 14, 9–32 (2001)
3. Carrick, C., Yang, Q., Abi-Zeid, I., Lamontagne, L.: Activating CBR Systems through Autonomous Information Gathering. In: Althoff, K.-D., Bergmann, R., Branting, L.K. (eds.) ICCBR 1999. LNCS (LNAI), vol. 1650, pp. 74–88. Springer, Heidelberg (1999)

4. Göker, M., Roth-Berghofer, T., Bergmann, R., Pantleon, T., Traphöner, R., Wess, S., Wilke, W.: The Development of HOMER: A Case-Based CAD/CAM Help-Desk Support Tool. In: Smyth, B., Cunningham, P. (eds.) EWCBR 1998. LNCS (LNAI), vol. 1488, pp. 346–357. Springer, Heidelberg (1998)
5. McSherry, D., Hassan, S., Bustard, D.W.: Conversational Case-Based Reasoning in Self-healing and Recovery. In: Althoff, K.-D., Bergmann, R., Minor, M., Hanft, A. (eds.) ECCBR 2008. LNCS (LNAI), vol. 5239, pp. 340–354. Springer, Heidelberg (2008)
6. Shimazu, H., Shibata, A., Nihei, K.: ExpertGuide: A Conversational Case-Based Reasoning Tool for Developing Mentors in Knowledge Spaces. Appl. Intell. 14, 33–48 (2001)
7. Watson, I.: Applying Case-Based Reasoning: Techniques for Enterprise Systems. Morgan Kaufmann, San Francisco (1997)
8. Bridge, D.G., Göker, M., McGinty, L., Smyth, B.: Case-Based Recommender Systems. Knowl. Eng. Rev. 20, 315–320 (2005)
9. McSherry, D.: Explanation in Recommender Systems. Artif. Intell. Rev. 24, 179–197 (2005)
10. Shimazu, H.: ExpertClerk: A Conversational Case-Based Reasoning Tool for Developing Salesclerk Agents in E-Commerce Webshops. Artif. Intell. Rev. 18, 223–244 (2002)
11. Thompson, C.A., Göker, M., Langley, P.: A Personalized System for Conversational Recommendations. Journal of Artif. Intell. Res. 21, 393–428 (2004)
12. McSherry, D.: Conversational Case-Based Reasoning in Medical Classification and Diagnosis. In: Combi, C., Shahar, Y., Abu-Hanna, A. (eds.) AIME 2009. LNCS, vol. 5651, pp. 116–125. Springer, Heidelberg (2009)
13. McSherry, D.: Conversational Case-Based Reasoning in Medical Decision Making. Artif. Intell. Med. 52, 59–66 (2011)
14. Quinlan, J.R.: C4.5: Programs for Machine Learning. Morgan Kaufmann, San Mateo (1993)
15. Aha, D.W., McSherry, D., Yang, Q.: Advances in Conversational Case-Based Reasoning. Knowl. Eng. Rev. 20, 247–254 (2005)
16. McSherry, D.: Minimizing Dialog Length in Interactive Case-Based Reasoning. In: 17th International Joint Conference on Artificial Intelligence, pp. 993–998. Morgan Kaufmann, San Francisco (2001)
17. Gu, M., Aamodt, A.: Dialog Learning in Conversational CBR. In: 19th International Florida Artificial Intelligence Research Society Conference, pp. 358–363. AAAI Press, Menlo Park (2006)
18. Gu, M., Aamodt, A.: Evaluating CBR Systems Using Different Data Sources: A Case Study. In: Roth-Berghofer, T.R., Göker, M.H., Güvenir, H.A. (eds.) ECCBR 2006. LNCS (LNAI), vol. 4106, pp. 121–135. Springer, Heidelberg (2006)
19. Gu, M., Tong, X., Aamodt, A.: Comparing Similarity Calculation Methods in Conversational CBR. In: 2005 IEEE International Conference on Information Reuse and Integration, pp. 427–432. IEEE (2005)
20. Jalali, V., Leake, D.: Custom Accessibility-Based CCBR Question Selection by Ongoing User Classification. In: Díaz Agudo, B., Watson, I. (eds.) ICCBR 2012. LNCS, vol. 7466, pp. 196–210. Springer, Heidelberg (2012)
21. Bogaerts, S., Leake, D.: What Evaluation Criteria are Right for CCBR? Considering Rank Quality. In: Roth-Berghofer, T.R., Göker, M.H., Güvenir, H.A. (eds.) ECCBR 2006. LNCS (LNAI), vol. 4106, pp. 385–399. Springer, Heidelberg (2006)
22. Cover, T.M., Hart, P.E.: Nearest Neighbor Pattern Classification. IEEE T. Inform. Theory 1, 21–27 (1967)

23. Ripley, B.D.: Pattern Classification and Neural Networks. Cambridge University Press, Cambridge (1996)
24. Bache, K., Lichman, M.: UCI Machine Learning Repository. University of California, Irvine, School of Information and Computer Sciences (2013)
25. Cunningham, P., Doyle, D., Loughrey, J.: An Evaluation of the Usefulness of Case-Based Explanation. In: Ashley, K.D., Bridge, D.G. (eds.) ICCBR 2003. LNCS (LNAI), vol. 2689, pp. 122–130. Springer, Heidelberg (2003)
26. Cummins, L., Bridge, D.: KLEOR: A Knowledge Lite Approach to Explanation Oriented Retrieval. Comput. Inform. 25, 173–193 (2006)
27. Doyle, D., Cunningham, P., Bridge, D., Rahman, Y.: Explanation Oriented Retrieval. In: Funk, P., González Calero, P.A. (eds.) ECCBR 2004. LNCS (LNAI), vol. 3155, pp. 157–168. Springer, Heidelberg (2004)
28. McSherry, D.: A Lazy Learning Approach to Explaining Case-Based Reasoning Solutions. In: Díaz Agudo, B., Watson, I. (eds.) ICCBR 2012. LNCS (LNAI), vol. 7466, pp. 241–254. Springer, Heidelberg (2012)
29. Hall, M., Frank, E., Holmes, G., Pfahringer, B., Reutemann, P., Witten, I.: The WEKA Data Mining Software: An Update. SIGKDD Explorations 11, 10–18 (2009)

Towards Process-Oriented Cloud Management with Case-Based Reasoning

Mirjam Minor and Eric Schulte-Zurhausen

Wirtschaftsinformatik, Goethe University Frankfurt,
60325 Frankfurt am Main, Germany
{minor,eschulte}@cs.uni-frankfurt.de

Abstract. The paper is on a novel cloud management model based on Case-based reasoning. Cloud resources are monitored and (re-)configured according to cloud management experience stored in a case-based system. We introduce a process-oriented, multi-tier cloud management model. We propose a case representation for cloud management cases, define similarity functions and sketch adaptation and revise issues. A proof-of-concept of this ongoing work is given by a sample application scenario from the field of video ingest.

Keywords: Process-oriented CBR, Cloud Computing, Cloud Management, Configuration Problems.

1 Introduction

Cloud management deals with management methods for provisioning and use of cloud services [2]. It is of vital importance to achieve rapid scalability of cloud services which is one of the main characteristics of cloud computing according to the NIST definition [5]. Cloud management addresses monitoring and configuration methods for cloud systems considering technical, organizational and legal aspects. The monitoring methods include measuring technical parameters like the utilization of physical resources, observing the quality of service in compliance *Service Level Agreements* (SLAs), and predicting the system behavior. SLAs are specifications of the terms of use of a service. The configuration methods include traditional network management methods like switching on and off the physical resources, managing virtual resources like *Virtual Machines* (VMs) or managing virtual network facilities.

From a technical point of view, cloud management is a resource management problem that can be solved by a multi-dimensional optimization approach [20] balancing resource consumption with other optimization criteria like performance or costs for SLA violations. However, decisions in cloud management have to be taken immediately. Solutions with a lower computational complexity than multi-dimensional optimization are advisable. *Case-based Reasoning* (CBR) has been considered for intelligent cloud management recently in the literature [11]. The work of Maurer et al. applies CBR to implement automatic cloud management following the MAPE reference model (Monitor - Analyse - Plan -

L. Lamontagne and E. Plaza (Eds.): ICCBR 2014, LNCS 8765, pp. 305–314, 2014.
© Springer International Publishing Switzerland 2014

Execute) [6], which originates in autonomic computing. A case in cloud management records a cloud configuration with current workloads to be processed as a problem situation. A solution describes the optimal distribution of work on the optimal number and configuration of cloud resources while maintaining SLAs. Maurer et al. use a bag of workloads to schedule the work, which makes it difficult to predict future workloads and to achieve stable configurations for a few future time steps.

In this paper, we make use of the workflow paradigm to address this gap by a case-based cloud configuration approach that takes into consideration the workloads to be scheduled next. The bag of workloads is replaced by the set of ongoing workflow instances. A workflow is "the automation of a business process, in whole or part, during which documents, information or tasks are passed from one participant to another for action, according to a set of procedural rules" [4]. The currently active tasks represent the workload of the cloud system. This process-oriented cloud management approach provides further benefits in addition to better prediction capabilities: It allows to use process-oriented modelling and monitoring tools instead of conventional cloud management tools, which are usually not aware of business processes and which are fairly frequently command-line oriented. The use of modelling tools also for cloud management tasks is more convenient for administrators who prefer graphical tools. Workflow reasoning supports the configuration task of cloud management by well-informed suggestions or provides even an automated cloud management approach based on previous experience. This paper extends previous work on a very early version of the proposed cloud management solution [15] by introducing the workload concept, elaborating a running sample in a video ingest application and refining some other definitions.

The paper is organized as follows: First, we discuss some related work in Section 2. In Section 3, we present on a multi-tier model for process-oriented cloud management implementing different cloud layers. In Section 4, we introduce a case-based approach for task placement on cloud resources. Section 5 provides a proof-of-concept for the approach by means of an application scenario on video ingest workflows. Finally, we conclude the paper in Section 6.

2   Related Work

Many commercial cloud systems still use quite straight-forward algorithms for cloud management. Frequently, cloud management activities are chosen following simple rules based on observations on the number of open connections [8] or on the CPU utilization. In contrast, our approach considers the characteristics of workloads like CPU intensive, storage intensive, memory intensive or network intensive tasks.

Some work has already been done on the automated placement of VMs on physical resources. Experiments have shown that VM placement decisions should consider the characteristics of the tasks [10]. In the All4green project [1] , energy consumption profiles are investigated to design Green-SLAs between data centers and end users considering the conservation of resources via SLAs. Both approaches are not aware of future workloads. In CloudBay [20], software agents negotiate on available resources like VMs with other software agents. The agents lease and configure resources to orchestrate virtual appliances automatically. The act of sale is not in the scope of our work which is restricted to manage resources in a technical sense.

The placement of jobs on VMs has been studied in the field of High Performance Computing [16,9]. Similar to the job placement problem is the application placement problem [17] which is the problem to to allocate applications with dynamically changing demands to VMs and how to decide how many VMs must be started. In contrast to such bag of work approaches, workflow tasks are executed in a given order.

The problem to assign workflow tasks to computing resources has been investigated in Grid Computing [19]. Wu et al. [18] describe a cloud computing approach inspired by Grid Computing to solve the task placement problem for scientific workflows with meta-heuristics. This Task-to-VM assignment is implemented by SwinDeW-C, a cloud workflow system. A unified resource layer is defined that is not reusable for our approach since we aim to involve heterogeneous cloud environments which are required in recent hybrid cloud solutions, for instance, to combine private and public cloud resources.

3   A Multi-Tier Cloud Management Model

Cloud management has to deal with complex topologies of resources. A multi-tier model separates physical from virtual resources in different layers. It allows to cascade configuration activities from layer to layer. The model is an extension of the cloud management model from Maurer at al. [11], which consists of three layers for hierarchical configuration activities. We place a workflow tier on top of the three tiers from Maurer et al.'s model in order to achieve a process-oriented perspective (see Fig. 1). The *physical machine tier* at the bottom is manipulated by configuration activities like to add and remove compute nodes. The *virtual machine tier* allows activities like to increase or decrease incoming and outgoing bandwith of a VM, its memory, its CPU share, or to add or remove allocated storage by x%. Further, VMs can be migrated to a different physical machine or moved to and from other clouds in case of outsourcing/insourcing capabilities. The *application tier* is dedicated to management activities for individual applications. The same set of activities as for VMs can be applied but with an application-specific scope. Obviously, the migration and insourcing/outsourcing activities refer to the placement of applications on VMs at the application tier. The *workflow tier* adresses the placement of workflow tasks on VMs called *task placement*. Two management activities can be conducted at the workflow tier:

- Task migration
- Task tailoring

Task migration means that a workflow task is scheduled on a different VM for execution. The initial placement of a task on a VM is a special case of migration.

The tailoring of tasks addresses the adaptation of tasks of an ongoing workflow as follows: Tasks can be replicated by splitting the corresponding input and output data in case the monitoring and prediction values advise to do so. This may require an additional task that aggregates the output data resulting from the replicated tasks. The task tailoring is a shallow form of agile workflow technology [12] which allows to structurally adapt ongoing workflow instances.

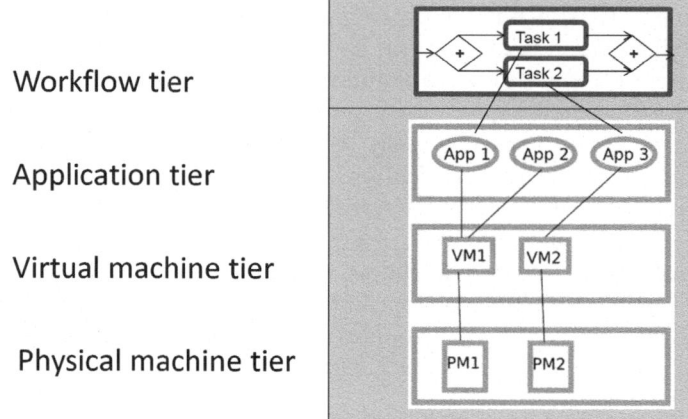

Fig. 1. Process-oriented model for cloud management extending Maurer et al. [11]

4   CBR for Task Placement

A case records an experience of a solved problem [13]. In cloud management, a case refers to solving a cloud management problem like to avoid an impending SLA violation or to schedule workflow tasks to be triggered next.

4.1   Representation of a Cloud Management Case

The *problem part* of a cloud management case describes a state of the cloud system where an action is required. The state of the system comprises of:

- the cloud configuration,
- the task placement, i.e. the set of ongoing workflow tasks (workloads) and their distribution on the cloud resources
- the actual utilization values measured for the virtual and physical resources
- the utilization values agreed on in the SLAs

The cloud configuration CC is a tuple $\langle PM, VM, VMP \rangle$. PM denotes a set of physical machines. Each physical machine is characterized by a tuple $(bandwidth, CPU, memory, storage, costs\ per\ time\ unit)$ describing the hardware parameters of the physical machine. The costs per time unit specifies an average value for the operating costs, particularly the energy consumed while the machine is running. VM stands for a set of virtual machines. Each virtual machine is characterized by a tuple $(bandwidth, CPU, memory, storage)$ describing the maximum share of the physical resources consumed by the virtual machine running under full workload. $VMP : VM \rightarrow PM$ is the placement function for virtual machines on physical machines. Each VM $vm_i \in VM$ is assigned to exactly one $pm_j \in PM$ but a PM can contain more than one VM, for example $\{(vm_1, pm_1), (vm_2, pm_1)\}$.

The task placement TP is described by a tuple $\langle AT, VM, TPF \rangle$. AT is the set of currently active tasks. Tasks join this set at the moment of being triggered for execution by the workflow engine. A workflow task $wt \in AT$ is described by a tuple $(task\ name, input\ data, execution\ time\ per\ unit, cpu\ usage, memory\ usage, storage\ usage, bandwidth\ usage)$. The task name is a string; input data provides a link to the input data to be processed by the task. The execution time per unit provides an average value for the duration of the task per data unit of a standardized size. The cpu usage is given in Million Instructions Per Second (MIPS), memory usage in gigabytes, storage usage in gigabytes and bandwith usage in megabytes per second. For example, a task "render image" might be described by (render image, 10GB, 27.5 GB/h, 100 MIPS, 4GB, 1GB, 0Mbit/s). VM is a set of virtual machines. $TPF : AT \rightarrow VM$ is the placement function for tasks on virtual machines.

The actual utilization values and the utilization values agreed on in the SLAs are expressed in percentage of the values provided by the cloud configuration.

Further, the event for which an action is required is part of the problem description, for instance the event of observing the CPU utilization of a physical server exceeding a threshold.

The *solution part* of a cloud management case describes the action taken in the past as well as the action recommended by a post mortem analysis for the past situation. The post mortem analysis has been inspired by the work of Gundersen et al. [7] on real-time analysis of events. The recommended action can be determined by an optimization approach, for instance.

4.2   Retrieval of a Cloud Management Case

A query is a partial case describing a recent situation of a cloud system with an event causing a cloud management problem. The retrieval of cases uses a similarity function for cloud management cases following the local-global-pinciple [13]. In this early stage of the work, we are planning to use straight-forward similarity functions, which might be replaced by more sophisticated functions such as object-oriented functions considering classes of tasks. Both, cloud configuration and task placement can be denoted as a bi-partite graph with the nodes of the one graph being mapped to the nodes of the other graph. Hence, graph edit distances [3] are used for the local similarity functions for cloud configurations and for task placements. The edit distance seems a natural approach since it imitates the actual configuration steps. The utilization values are compared by weighted sums for numerical distances. The events are compared by a simple structured similarity function based on a taxonomy of events.

4.3   Reuse of a Cloud Management Case

The reuse of a cloud management case requires an adaptation of the retrieved case to the current situation. Further, it has to be investigated whether the recommended management activities actually solve the problem. The latter can be done by a short-time simulation of the behaviour of the cloud system making use of the workflow execution logic. Based on the estimated execution time of the particular current workloads, a prediction of the workloads in the following n time steps is computed by simulating the tasks to be scheduled next for each ongoing workflow instance. For this sequence of n time steps, the prospective utilization values are simulated and assessed. A rough approximation of such a short-time simulation could be to consider only one future step with the set of all tasks that are subsequent to ongoing tasks. This set can be easily derived from the control flow structure of the workflow instances. The workload of this set of tasks can be approximated based on the estimated size of the output data of the previous task. The characteristics of the application domain might have an impact on the number of future steps to be considered best by the workload approximation.

5   Video Ingest as a Sample Application Scenario

We have chosen a sample application scenario on video ingest processes in order to provide a first proof-of-concept for our novel, case-based, process-oriented cloud management approach. Fig. 2 depicts a sample workflow for a video ingest process that is inspired by the work of Rickert and Eibl [14]. Video content on VHS video tapes is digitized and stored in different formats on a tape in the more recent LTO standard. Different transformation steps like creating a preview file, creating a legacy proxy for archiving purposes, and creating an analysis proxy for further processing, for instance, for face detection, are executed as workflow tasks in parallel. Part of the workflow tasks include algorithms with a high computational complexity. A cloud computing solution allows to accelerate the workflow execution by running tasks on different computing resources in parallel.

The cloud configuration that we have chosen for this illustrating example is CC = {PM, VM, VMP}. Let us assume a 3GHz CPU that can handle around 9000MIPS, a 2GHz CPU that is able to handle 6000MIPS and a 1GHz CPU with about 3000MIPS. Further, we assume that we have two PMs $pm_1, pm_2 \in$ PM and a configuration of both PMs pm_i = (3GHz, 4GB, 1000GB, 100MBit/s, 3\$ /day). In addition, we assume to have 3 identical VMs $vm_i \in$ VM with the configuration (1GHz, 2GB, 500GB, 50MBit/s). The actual placement of the virtual machines is VMP = {$(pm_1, vm_1),(pm_2, vm_2),(pm_2, vm_3)$}.

Let the sample tasks shown in Fig. 3 have a task placement TP = {AT, VM, TPF} with *create preview file, create legacy proxy* and *create analysis proxy* being active tasks. Table 1 describes the properties of the tasks. Let the VMs as described in CC be assigned as follows by TPF = {$(vm_1$, create preview file),$(vm_2$,create legacy proxy),$(vm_3$, create analysis proxy)}.

The set of the tasks to be scheduled next would be {detect faces, create QR code} according to Fig. 2.

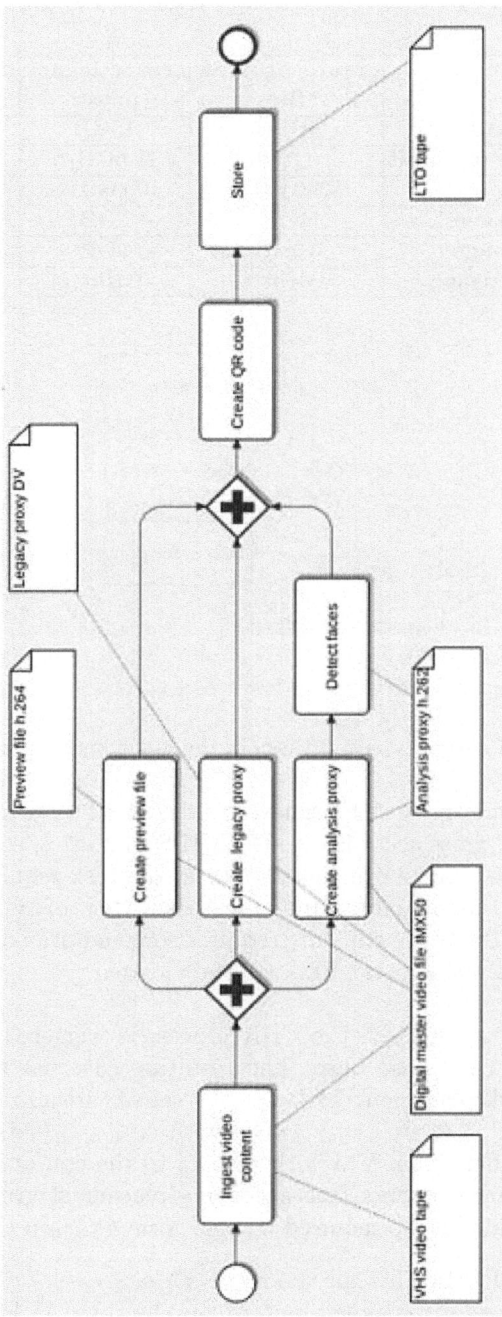

Fig. 2. Example for a video ingest workflow derived from Rickert and Eibl [14]

Table 1. Properties of the sample tasks used in the scenario

task name	create preview file	create legacy proxy	create analysis proxy
input data	10GB	10GB	10GB
execution time per unit	1GB/h	12.6GB/h	3.15GB/h
cpu usage	3000MIPS	1000MIPS	1000MIPS
memory usage	4GB	2GB	2GB
storage usage	10GB	2GB	2GB
bandwidth usage	0MBit/s	0MBit/s	0MBit/s

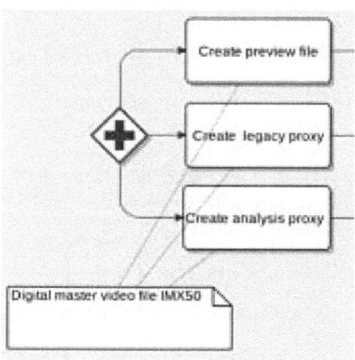

Fig. 3. Example of currently active tasks of the video ingest workflow from Fig. 2

Let us assume that one SLA has guaranteed that all tasks of a user are always provided with CPU times according to the MIPS that are specified in the sample tasks. In case of the "create preview file" task, the task requires 3000MIPS, for instance. However, the virtual machine vm_1 has a CPU of 1GHz only and, thus, can handle 3000MIPS. Since the vm_1 requires some additional CPU time for its operating system, obviously, the SLA will be violated.

If such an event is detected, the CBR process is triggered to retrieve a best matching case from the case base. The resulting case has a cloud configuration CC' and a task placement TP'. Let the set of virtual machines in the retrieved case be VM' = $\{vm_1', vm_2', vm_3'\}$ with vm_i' = (0.5GHz, 1GB, 500GB, 50MBit/s). VM' differs from VM with respect to the cpu and the memory values. The edit distance between the according placement graphs is 6 since the following edit operations are required to transform CC into CC':

− Decrease the CPU by 50% for the three VMs.
− Decrease the memory assigned by 50% for the three VMs.

Let the set of tasks from the retrieved case be the same as in the current situation but with half the MIPS numbers. The local similarity function for the task placement makes use of a table of similarity values specified in Table 2.

Table 2. Local similarity function for MIPSs

	3000	6000	9000
3000	1	0.7	0.3
6000	0.7	1	0.7
9000	0.3	0.7	1

A sample solution suggested by the case could be to increase the CPU share for the virtual machine vm_1. This solution can be transfered to the recent situation namely to increase the CPU share for vm_1'.

6   Conclusion

In this paper we have introduced a novel cloud management model with a process-oriented perspective and a CBR approach for experience-based cloud management. A cloud management case describes the state of a cloud system including an event causing a problem. The solution part of the case provides a solution for the problem by cloud management activities. The novel model has been illustrated by a sample application with a video ingest workflow. Our workflow reasoning approach aims at improving the automated management of computational resources of a cloud system for the different video analysis tasks. It takes into consideration cases reporting previous configuration decisions in similar situations, for instance to switch on an additional virtual machine in case of a high utilization of CPU's. The cloud management model is still in an early phase of ongoing work. In future work, we will further specify the similarity functions and adaptation methods, develop an implementation of the proposed model and conduct some lab experiments.

We believe that cloud computing and especially cloud management and cloud configuration issues are an intriguing, novel application field for CBR. The cloud management model provided in this paper is a first step towards case-based solutions for cloud management and will hopefully stipulate further work in future.

References

1. Basmadjian, R., Niedermeier, F., de Meer, H.: Modelling and analysing the power consumption of idle servers. In: Sustainable Internet and ICT for Sustainability, SustainIT 2012, October 4-5, Pisa, Italy, Sponsored by the IFIP TC6 WG 6.3 "Performance of Communication Systems", pp. 1–9. IEEE (2012)
2. Baun, C., Kunze, M., Nimis, J., Tai, S.: Cloud Computing - Web-Based Dynamic IT Services. Springer (2011)
3. Bunke, H., Messmer, B.T.: Similarity measures for structured representations. In: Wess, S., Althoff, K.D., Richter, M. (eds.) EWCBR 1993. LNCS, vol. 837, pp. 106–118. Springer, Heidelberg (1994)
4. Workflow Management Coalition: Workflow management coalition glossary & terminology (1999), http://www.wfmc.org/standars/docs/TC-1011_term_glossary_v3.pdf (last access May 23, 2007)

5. NIST US Department of Commerce: Final version of NIST cloud computing definition published, http://www.nist.gov/itl/csd/cloud-102511.cfm (last access September 17, 2013)
6. IBM Corporation: An architectural blueprint for autonomic computing (2006), http://www-03.ibm.com/autonomic/pdfs/AC%20Blueprint%20White %20Paper%20V7.pdf (last accessed: April 17, 2014)
7. Gundersen, O.E., Sørmo, F., Aamodt, A., Skalle, P.: A real-time decision support system for high cost oil-well drilling operations. In: Twenty-Fourth IAAI Conference (2012)
8. Hammond, B.W.: Getting Started with OpenShift. Google Patents (2003), US Patent 6,637,020
9. Le, K., Zhang, J., Meng, J., Bianchini, R., Jaluria, Y., Nguyen, T.: Reducing electricity cost through virtual machine placement in high performance computing clouds. In: 2011 International Conference for High Performance Computing, Networking, Storage and Analysis (SC), pp. 1–12 (November 2011)
10. Mahmood, Z.: Cloud Computing: Methods and Practical Approaches Springer (May 2013)
11. Maurer, M., Brandic, I., Sakellariou, R.: Adaptive resource configuration for cloud infrastructure management. Future Generation Computer Systems 29(2), 472–487 (2013)
12. Minor, M., Tartakovski, A., Schmalen, D., Bergmann, R.: Agile workflow technology and case-based change reuse for long-term processes. International Journal of Intelligent Information Technologies 4(1), 80–98 (2008)
13. Richter, M.M., Weber, R.: Case-Based Reasoning: A Textbook. Springer (2013)
14. Rickert, M., Eibl, M.: Evaluation of media analysis and information retrieval solutions for audio-visual content through their integration in realistic workflows of the broadcast industry. In: Proceedings of the 2013 Research in Adaptive and Convergent Systems, RACS 2013, pp. 118–121. ACM, New York (2013)
15. Schulte-Zurhausen, E., Minor, M.: Task placement in a cloud with case based reasoning. In: 4th International Conference on Cloud Computing and Services Science, Barcelona, Spain (2014)
16. Sharma, B., Chudnovsky, V., Hellerstein, J.L., Rifaat, R., Das, C.R.: Modeling and synthesizing task placement constraints in google compute clusters. In: Proceedings of the 2nd ACM Symposium on Cloud Computing, SOCC 2011, pp. 3:1–3:14. ACM, New York (2011)
17. Tang, C., Steinder, M., Spreitzer, M., Pacifici, G.: A scalable application placement controller for enterprise data centers. In: Proceedings of the 16th International Conference on World Wide Web, WWW 2007, pp. 331–340. ACM, New York (2007)
18. Wu, Z., Liu, X., Ni, Z., Yuan, D., Yang, Y.: A market-oriented hierarchical scheduling strategy in cloud workflow systems. The Journal of Supercomputing 63(1), 256–293 (2013)
19. Yu, J., Buyya, R., Ramamohanarao, K.: Workflow scheduling algorithms for grid computing. In: Xhafa, F., Abraham, A. (eds.) Metaheuristics for Scheduling in Distributed Computing Environments. SCI, vol. 146, pp. 173–214. Springer, Heidelberg (2008)
20. Zhao, H., Li, X.: Resource Management in Utility and Cloud Computing, 1st edn. Springer, Dordrecht (2013)

Workflow Streams: A Means for Compositional Adaptation in Process-Oriented CBR

Gilbert Müller and Ralph Bergmann

Business Information Systems II
University of Trier
54286 Trier, Germany
{muellerg,bergmann}@uni-trier.de
http://www.wi2.uni-trier.de

Abstract. This paper presents a novel approach to compositional adaptation of workflows, thus addressing the adaptation step in process-oriented case-based reasoning. Unlike previous approaches to adaptation, the proposed approach does not require additional adaptation knowledge. Instead, the available case base of workflows is analyzed and each case is decomposed into meaningful subcomponents, called workflow streams. During adaptation, deficiencies in the retrieved case are incrementally compensated by replacing fragments of the retrieved case by appropriate workflow streams. An empirical evaluation in the domain of cooking workflows demonstrates the feasibility of the approach and shows that the quality of adapted cases is very close to the quality of the original cases in the case base.

Keywords: process-oriented case-based reasoning, compositional adaptation, workflows.

1 Introduction

Adaptation in Case-Based Reasoning (CBR) is still a very important research field, even after more than 30 years of research in CBR. One obstacle that prevents comprehensive adaptation capabilities is the knowledge acquisition bottleneck for adaptation knowledge, which is required by many adaptation methods. Therefore, most commercial applications of CBR are developed for application domains in which adaptation is not a primary concern, such as in knowledge management or service support. However, in application domains involving complex cases with highly structured solutions, adaptation cannot be disregarded as it is an ability that could lead to significant benefits for the users. One such area is process-oriented case-based reasoning (POCBR) [11], which deals with CBR applications for process-oriented information systems (POIS). POCBR has the potential of enabling POIS to support domain experts in defining, executing, monitoring, or adapting workflows. Thus, workflow adaptation is an important field, in which still little research exist so far.

Existing methods to adaptation in CBR can be roughly classified into transformational and generative adaptation [18]. Transformational adaptation relies

L. Lamontagne and E. Plaza (Eds.): ICCBR 2014, LNCS 8765, pp. 315–329, 2014.
© Springer International Publishing Switzerland 2014

on adaptation knowledge representing links between differences in the problem description and resulting modifications of the solution. For POCBR we investigated a transformational adaptation approach using adaptation cases as adaption knowledge [9], but the acquisition of adaptation cases, although addressed in [10], is still a difficult issue. Methods for learning adaptation knowledge from a case base have been proposed by various authors [7,18,8,3], but no approaches yet exist that are appropriate for complex case representations such as used in POCBR.

Generative adaptation, on the other hand, demands general domain knowledge appropriate for an automated (knowledge-based) problem solver to solve problems from scratch. While this approach is quite successful in planning because knowledge about actions need to be present anyway, it is not appropriate for POIS as the task descriptions of workflows may describe human activities, which cannot be formalized in sufficient detail.

To circumvent the problems with these approaches, we investigate the idea of compositional adaptation for POIS. While compositional adaptation usually means that several cases are used during adaptation, still incorporating transformational or generative adaptation methods involving adaptation knowledge (such as in COMPOSER [12], *DéjàVu* [15] or PRODIGY/ANOLOGY [17] – to cite some classic examples), we propose a pure compositional adaptation approach waiving of any adaptation knowledge (see [1] p.232ff and [16]). Instead, the available case base of workflows is analyzed and each case is decomposed into meaningful subcomponents, called *workflow streams*. During adaptation, deficiencies in the retrieved case are incrementally compensated by replacing fragments of the retrieved case by appropriate workflow streams. The proposed method depends on cases being represented as block-oriented workflows, as the workflow structure is exploited to define the borders of the workflow streams, ensuring their reusability. Hence, the next section introduces the workflow representation as well as certain formal notations being further used. In Section 3, the workflow streams are formally defined as a prerequisite for the compositional adaptation method described in Section 4. Finally, we report on the results of an empirical evaluation of the proposed method in the domain of cooking workflows and wrap-up by discussing related and potential future work.

2   Foundations

2.1   Workflows

A workflow is "the automation of a business process, in whole or part, during which documents, information or tasks are passed from one participant to another for action, according to a set of procedural rules" [19]. Broadly speaking, workflows consist of a set of *activities* (also called *tasks*) combined with *control-flow structures* like sequences, parallel (AND) or alternative (XOR) branches, as well as repeated execution (LOOPs). Tasks and control-flow structures form the *control-flow*. In addition, tasks exchange certain *data items*, which can also be of physical matter, depending on the workflow domain. Tasks, data items, and relationships between the two of them form the *data flow*. We illustrate our approach in the domain of cooking recipes. A cooking recipe is represented as a workflow describing the instructions for cooking a particular dish [14]. Here, the tasks represent the cooking steps and the data items refer to the ingredients being processed by the cooking steps. An example cooking workflow for a pizza recipe is illustrated in Figure 1 also showing the commonly used graph representation, formally defined below.

Definition 1. *A workflow is a graph $W = (N, E)$ with a set of nodes N and edges $E \subseteq N \times N$. The nodes $N = D \cup T \cup C$ can be of different types, namely data nodes D, task nodes T, and control-flow nodes C. The control-flow nodes $C = C_* \cup C^*$ construct blocks of sequences and are either opening $C_* = C_A \cup C_X \cup C_L$ or closing control-flow nodes $C^* = C^A \cup C^X \cup C^L$ representing AND, XOR and LOOP nodes. The set of sequence nodes is defined as $S = T \cup C$. Edges $E = CE \cup DE$ can be control-flow edges $CE \subseteq S \times S$, which define the order of the sequence nodes or data-flow edges $DE \subseteq (D \times S) \cup (S \times D)$, which define how the data is shared between the tasks.*

Figure 1 shows an opening AND control-flow node A_* and a related closing AND control-flow node A^*. Further, it is important to note that the control-flow edges CE induce a strict partial order on the sequence nodes S. Thus, we define $s_1 < s_2$ for two sequence nodes $s_1, s_2 \in S$ as a transitive relation that expresses that s_1 is executed prior to s_2 in W. We further define $n \in [x_1, x_2]$ iff $x_1 < n < x_2$, describing that node n is located between x_1 and x_2 in W w.r.t. the control-flow edges.

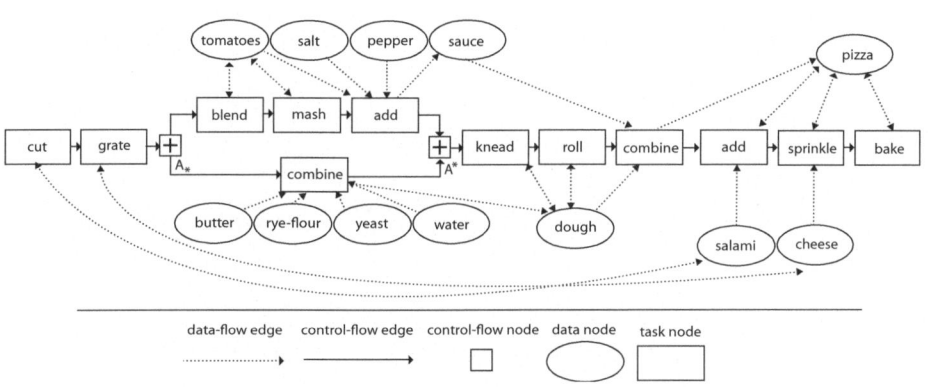

Fig. 1. Example of a block-oriented cooking workflow

2.2   Block-Oriented Workflows

Block-oriented workflows are workflows that are constructed from blocks of workflow elements as defined in Definition 2.

Definition 2. *A block-oriented workflow* $W = (N, E)$ *is a workflow with a block-oriented graph structure according to the following rules:*

a) *A block element is a workflow subgraph of the Graph* $(N \setminus D, CE)$ *which contains either:*
- *a task node*
- *a sequence of block elements*
- *a LOOP block containing an opening loop node, 1 block element, and a closing loop node*
- *a XOR/AND block containing an opening XOR/AND node, 2 branches containing a block element, and a matching closing XOR/AND node[1]*
- *a XOR block containing an opening XOR node, 1 branch with a block element, an empty branch and a closing XOR node*

b) *Each block element must either contain one task or another block element*

c) *The workflow* W *is a single block*

The workflow shown in Figure 1 is a block-oriented workflow. Figure 2 illustrates the workflow blocks, which are marked with dashed rectangles. Workflow block elements may be nested (e.g. see loop block), but not interleaved. The construction of block-oriented workflows restricts the usage of control-flow edges w.r.t. block elements by the correctness-by-construction principle [13,4]. Such a construction ensures the syntactic correctness of the workflow, e.g., that the workflow has one start node (node without ingoing control-flow edges) and one end node (node without outgoing control-flow edges). Such workflows are referred to as consistent workflows. In contrast to the control-flow, the data-flow is not restricted by the block-oriented workflow construction.

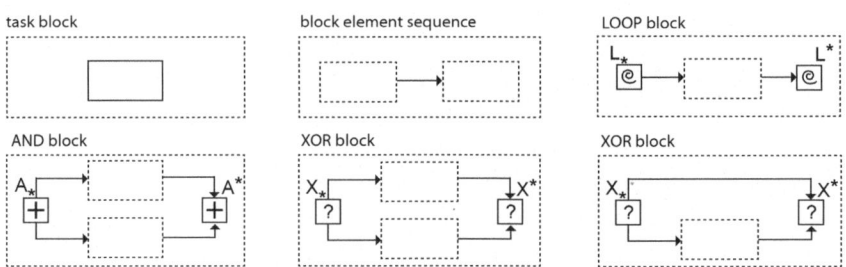

Fig. 2. Illustration of workflow block elements

2.3   Partial Workflows

As we aim at reusing meaningful parts of workflows, we define the notion of a partial workflow as follows.

---

[1] For the sake of simplicity only 2 branches are contained in an XOR/AND block, as multiple branches can be easily be constructed by nesting other XOR/AND blocks.

Definition 3. *For a subset of tasks $T' \subseteq T$, a partial workflow W' of a block-oriented workflow $W = (N, E)$ is a block-oriented workflow $W' = (N', E' \cup CE'_+)$, with a subset of nodes $N' = T' \cup C' \cup D' \subseteq N$. $D' \subseteq D$ is defined as the set of data nodes that are linked to any task in T', i.e., $D' = \{d \in D | \exists t \in T' : ((d, t) \in DE \vee (t, d) \in DE)\}$. W' contains a subset of edges $E' = E \cap (N' \times N')$ connecting two nodes of N' supplemented by a set CE'_+ of additional control-flow edges that retain the execution order of the sequence nodes, i.e., $CE'_+ = \{(n_1, n_2) \in S' \times S' | n_1 < n_2 \wedge \nexists n \in S' : ((n_1, n) \in CE' \vee (n, n_2) \in CE' \vee n \in [n_1, n_2])\}$*

In general, control-flow nodes are part of a partial workflow if they construct a workflow w.r.t. the block-oriented workflow structure. This basically means that each block nested in a control-flow block element must either contain one task or some child block element containing a task. The additional edges CE'_+ are required, to retain the execution order $s_1 < s_3$ of two sequence nodes if for $s_1, s_2, s_3 \in S$ holds $s_2 \in [s_1, s_3]$ but $s_2 \notin N'$. Figure 3 illustrates a partial workflow W' of the workflow W given in Figure 1. One additional edge is required in this example, depicted by the double-line arrow since "grate" and "add" are not linked in W.

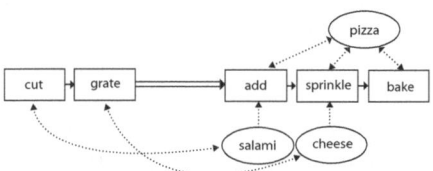

Fig. 3. Example of a partial workflow W'

2.4   Semantic Workflows and Semantic Workflow Similarity

To support retrieval and adaptation of workflows, the individual workflow elements are annotated with ontological information, thus leading to a *semantic workflow* [2]. In particular, all task and data items occurring in a domain are organized in taxonomy, which enables the assessment of similarity among them. We deploy a taxonomy of cooking ingredients and cooking steps for this purpose. In our previous work, we developed a semantic similarity measure for workflows that enables the similarity assessment of a case workflow w.r.t. a query workflow [2]. The similarity of task or data items reflects the closeness in the taxonomy, and further regards the level of the taxonomic elements. In particular, if a more general query element such as "meat" is compared with a specific element below it, such as "pork", the similarity value is 1. This ensures that if the query asks for a recipe containing meat, any recipe workflow from the case base containing any kind of meat are considered highly similar.

The similarity measure performs a kind of inexact subgraph matching, optimizing the overall similarity between the matched workflow elements. It is used during case retrieval in order to find workflows which best match a certain query.

Within the adaptation method described in this paper, the same similarity based retrieval method is used to identify reusable workflow streams, as we will show in Section 4.

3   Workflow Streams

According to Davenport, "[...] a process is simply a structured, measured set of activities designed to produce a specific output for a particular customer on the market" [5]. We define this specific output as the goal of a workflow. This goal is reached by producing partial outputs and combining them to the specific output. Thus, a workflow that, for example, prepares a pizza dish, also prepares the dough, the pizza sauce, and the toppings. Hence, these partial outputs represent partial goals of a workflow. While the entire workflow is designed to fulfill the overall goal of a workflow, particular parts of the workflow attain partial goals (see Fig. 4). Tasks in a workflow that fulfill a partial goal, produce a data node that is not consumed by the same task, i.e., task having at least one outgoing data edge but no ingoing data edge referring to the same data node d. We call such tasks *creator tasks*. For a workflow W the set of creator tasks is defined as follows:

$$CT = \{t \in T | \exists d \in D : ((t,d) \in DE \land (d,t) \notin DE)\} \qquad (1)$$

The creator tasks of the workflow illustrated in Figure 4 are marked with ⊙ symbols. They create the dough, the sauce or combine the dough, and sauce to create the pizza.

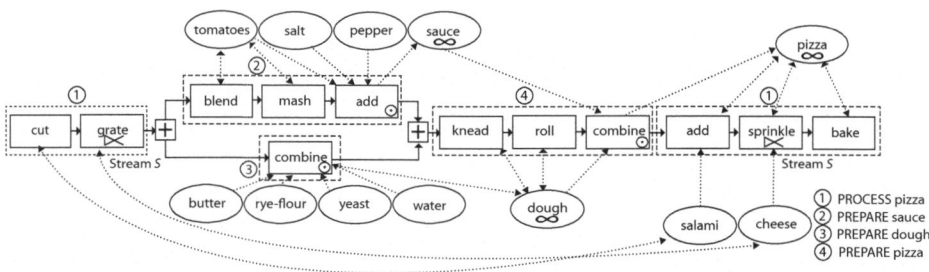

Fig. 4. Wokflow and workflow streams

Furthermore, not only the creator task is sufficient to produce new data. For example, the preparation steps to blend and mash the tomatoes have to be completed before the pizza sauce can be produced. Thus, all tasks that are required, before the creator task can be completed have to be identified. Such tasks can be recognized by aid of the data-flow. We define that two tasks $t_1, t_2 \in T$ are data-flow connected $t_1 \bowtie t_2$ if $t_1 < t_2$ and t_2 consumes a data node produced by t_1. Thus, t_1 has to be completed before task t_2 can be executed as otherwise the data wouldn't be processed in the correct order. Hence, if the data-flow connectedness is regarded and retained when modifying the workflow, the semantic correctness of the workflow is ensured. We define:

$$t_1 \ltimes t_2, \text{iff } t_1 < t_2 \wedge \exists d \in D : ((t_1, d) \in DE \wedge (d, t_2) \in DE) \qquad (2)$$

Further, let $t_1 \overset{d}{\ltimes} t_2$ denote that two tasks $t_1, t_2 \in T$ are data-flow connected via the data object d. Figure 4 shows that the tasks "grate cheese" and "sprinkle cheese over the pizza" (marked with $\ltimes$) are data-flow connected via the data node "cheese". Additionally, we define $t_1 \ltimes^* t_2$ to express that two tasks $t_1, t_2 \in T$ are transitively data-flow connected:

$$t_1 \ltimes^* t_2, \text{iff } t_1 \ltimes t_2 \vee \exists t \in T : (t_1 \ltimes t \wedge t \ltimes^* t_2) \qquad (3)$$

We use the definition of creator tasks and data-flow connectedness to decompose a workflow W into partial workflows each of which represent disjoint partial goals. Each workflow W can be partitioned by the definition given below.

Definition 4. *A partition $\mathcal{S}^W = \{\mathcal{S}_1^W, \ldots, \mathcal{S}_n^W\}$ of a block-oriented workflow W is a set of partial workflows S_i^W, such that each task $t \in T$ is contained in a partial workflow S_i^W and such that the tasks in each S_i^W are transitively data-flow connected and not contained in any other partial workflow $S_{j \neq i}^W$. All creator tasks $y \in CT$ are end nodes of any partial workflow in $\mathcal{S}^W$. Each partition S_i^W is called workflow stream.*

In order to partition the workflow W into streams, for each creator task $y \in CT$ a set of transitively data-flow connected and disjoint tasks $T_\mathcal{S}$ is identified. $T_\mathcal{S}$ is constructed, by adding all data-flow connected tasks that are not data-flow connected to any predecessor creator task of CT (as the creator tasks represent end nodes of partial workflows):

$$T_\mathcal{S}(y) := \{t \in T | t \ltimes^* y \wedge t \notin CT \wedge \not\exists x \in CT : (t \ltimes^* x \wedge y \not< x)\} \cup \{y\} \qquad (4)$$

Based on the set of tasks $T_\mathcal{S}$ that belong to a workflow stream $\mathcal{S}$, a partial workflow can be constructed according to Definition 3, which is then finally referred to as workflow stream. For each of the creator tasks illustrated in Figure 4 (marked with $\odot$) the dashed lines represent the tasks assigned to the referring stream (see stream 2,3,4). Regarding task disjointness, the task "mash tomatoes", for example, does not belong to the workflow stream 4, although it is transitively data-flow connected, as it is separated though a data-flow connected creator task (see "combine" task of stream 4).

According to Definition 4, each task is contained in a partial workflow S_i^W. Thus, each set of transitively data-flow connected tasks not already contained in any stream S_i^W derived from a creator task, is assigned to a new stream. Thus, not only streams that produce new data nodes are regarded, but also streams processing a data node. As an example, see stream 1 in Fig. 4 for putting the toppings on the pizza.

To summarize, the workflow is partitioned in such a manner that each task is assigned to exactly one stream of the workflow. The extracted workflow streams are themselves consistent workflows as they are block-oriented. Furthermore, due to the data-flow connectedness, the streams maintain their "semantic correctness" as they represent meaningful connected subcomponents of the original workflow.

The basic idea for compositional adaptation is, to adapt a workflow by using the workflow streams of other workflows that fulfill the same partial goal in a different manner, e.g., with other tasks or data. In the pizza domain, for example, toppings or preparation steps, can be replaced. Therefore it is required to identify whether some streams can be substituted. We require that substitutable streams must produce the same data and that they must consume the identical data nodes. This ensures that replacing an arbitrary stream doesn't violate the semantic correctness of the workflow, e.g. that a data object is never produced or consumed (except of those never produced or consumed by the entire workflow). The following notion of *anchors* is used to define the relevant data notes to be considered to decide whether two streams are substitutable.

Definition 5. *For a stream $\mathcal{S}$, a set of anchors is defined as $A_{\mathcal{S}} = \{d \in D_S | (\exists t \in T \setminus T_S \wedge \exists t_S \in T_S) : (t \overset{d}{\ltimes} t_S \vee t_S \overset{d}{\ltimes} t)\}$*

Thus, anchors of a stream $\mathcal{S}_i^W$ are data nodes of this stream that are either produced or consumed by a task of another stream $S_{j \neq i}^W$ of the same workflow W. Two streams are substitutable if they have identical anchor sets. The anchor nodes of all streams are marked with ∞ symbols in Figure 4, i.e., sauce, dough, and pizza.

4   Compositional Adaptation Using Workflow Streams

In the following, we assume that each workflow in the case base has been decomposed into workflow streams according to Section 3. These streams are linked to each workflow in the case base and stored in a separate workflow stream repository. The adaptation procedure is now presented and explained by an example scenario. After the retrieval of a most similar workflow W the user might want to adapt this workflow to his or her preferences. Let us assume that the user retrieved the workflow W given in Figure 4.

4.1   Change Request

Following the retrieval, a change request is defined by specifying lists of task or data nodes that should be added (ADD list) or removed (DELETE list) from workflow W. The change request can be either manually acquired from the user after the workflow is presented or it can be automatically derived based on the difference between the query and the retrieval case. As the tasks and data are taxonomically ordered (see Section 2.4), the change request can also be defined by a higher level concept of the taxonomy in order to define a more general change of the workflow. For example, a change request specified as "DELETE meat" ensures that the adapted recipe is a vegetarian dish.

4.2   Search of Substitute Stream Candidates

To perform a workflow adaptation, for each stream S in the retrieved workflow W a substitute stream candidate S' is searched in the constructed stream pool repository. If a candidate can be found, the stream S is substituted with stream S'. The identification of a substitute stream candidate S' requires to identify those workflows that can be replaced with a stream S. As already mentioned, stream S and S' must have the identical set of anchor elements. This condition ensures that the adapted workflow is semantically correct, i.e., that the workflow doesn't contain any data nodes that are never produced or consumed by an other workflow stream. Additionally, the substitute stream candidate must regard the change request. Hence, streams are searched that do not contain any node given in the DELETE list and that contain the highest number of nodes from the ADD list of the change request. For all streams that fulfill these conditions, a retrieval of the most similar stream to stream S is executed, considering the semantic similarity measure sketched in Section 2.4. The retrieved most similar stream is the substitute stream candidate. This approach ensures that a substitute stream is selected that only changes the workflow as much as required w.r.t. the change request.

The following example illustrates this approach. Assume that for a certain query, the workflow from Fig. 4 is retrieved and that the change request is to "DELETE salami" and to "ADD ground beef". Lets assume that stream 1 of Fig. 4 is stream S and the stream illustrated in Figure 5 is the substitute stream candidate S'. Both streams have an identical set of anchor elements (see ∞) which only contains the "pizza" node. Furthermore, workflow stream S' regards the change request as it contains the node given in the ADD list ("ground beef") and no node defined in the DELETE list ("salami").

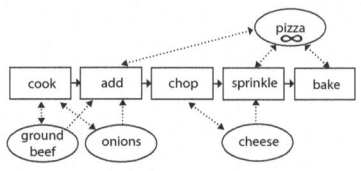

Fig. 5. Substitute stream S'

4.3   Replacing Workflow Streams

To replace stream S by the substitute stream candidate S', S is first removed from the workflow. This means that a partial workflow (see Sec. 2.3) is constructed, containing all tasks of W except of those contained in $\mathcal{S}$. The removal of the work-flow stream in the given scenario is illustrated in Figure 6.

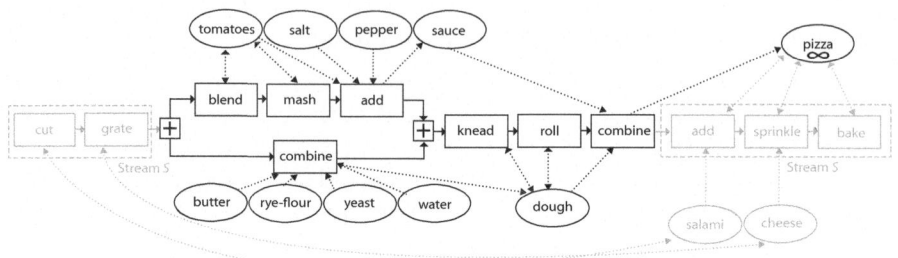

Fig. 6. Stream $\mathcal{S}$ removed from W

Then, the new stream $\mathcal{S}'$ is inserted at the position of the last sequence node of the workflow stream $\mathcal{S}$ in W. This means that all edges, tasks, and data nodes (if not already present) of $\mathcal{S}'$ are inserted into the workflow W. Then, the inserted stream $\mathcal{S}'$ is connected with an additional control-flow edge that links the tasks of the stream at the old position to the last sequence node of the removed stream in W. In the illustrated scenario, the stream $\mathcal{S}'$ is inserted behind the last "combine" task (see Fig. 7). In the special case that the last sequence node of $\mathcal{S}$ is a control-flow node that is still a part of the partial workflow after removing workflow stream $\mathcal{S}$, the stream $\mathcal{S}'$ is inserted behind this control-flow node.

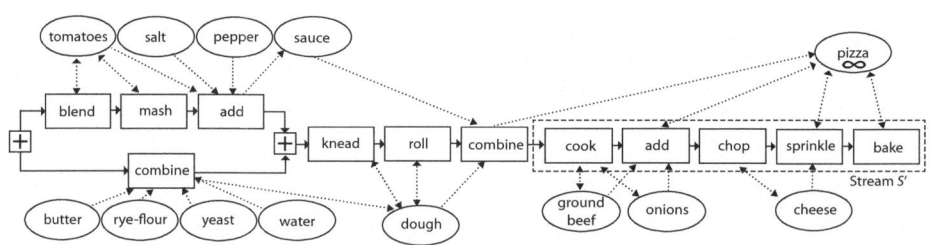

Fig. 7. Stream $\mathcal{S}'$ added to W

Inserting the stream $\mathcal{S}'$ may lead to a shift of tasks. In the given scenario, for example, the start node of the original workflow and the adapted workflow differ (compare Fig. 4 and Fig. 7). However, this does not violate the semantic correctness of the workflow, i.e., the data is still being processed in the right order. This is ensured by the workflow stream definition as all tasks that are transitively data-flow connected are included in the workflow stream as long as they are not separated by a creator task. Additionally, the anchors ensure that only streams are replaced representing an identical partial goal.

5   Evaluation

The described approach to compositional adaptation of workflows has been fully implemented as part of the CAKE framework[2] that already includes a process-oriented case-based reasoning component for similarity-based retrieval of workflows. To demonstrate its usefulness, the approach is experimentally evaluated focussing on two hypotheses explained below.

In Section 3 it was already stated that workflows adapted by the workflow stream method are consistent and semantically correct. Hence, this property will not be subject to the empirical evaluation. However in the experiments the consistency and semantic correctness of the constructed workflows was checked (and confirmed) in order to validate the correctness of the implementation. We conducted an empirical evaluation to analyze whether change requests are fulfilled by the adapted workflows (Hypothesis H1) and to validate whether the adapted workflows are of an acceptable quality (Hypothesis H2).

H1. The compositional adaptation approach is able to produce workflows that fulfill the change requests.

H2. The compositional adaptation produces workflows, whose quality is comparable to the quality of the workflows in the case base.

5.1   Experimental Setup

We developed an automatic workflow generator that enables us to produce workflow repositories for experimental purposes with predefined properties. The generator uses the fast downward planner[3] to produce sequences of tasks, which are further organized into a (partly parallelized) workflow structure. The generated workflows are block-oriented and semantically correct w.r.t. the definition of the planning domain. For the experiments, we generated 240 different cooking workflows, each of which produces a pizza. Overall, 17 different ingredients and 8 different tasks occur in the workflows. Due to the limitation of the workflow generator, the workflows do not contain any XOR or LOOP structures, but AND blocks occur frequently.

The experimental setup to validate the hypotheses defined above is illustrated in Figure 8. The repository of 240 workflows was split into two data sets: One repository containing 10 arbitrary workflows (referred to as test workflows) and a case base containing the remaining 230 workflows. A stream repository was computed for adaptation based on the workflows contained in the case base. In

---

[2] cake.wi2.uni-trier.de
[3] www.fast-downward.org

total 920 workflow streams were identified and stored in the stream repository. For each test workflow TW_i we retrieved a random workflow from the case base (referred to as retrieved workflow) and constructed a pair of workflows (TW_i, RW_i). If the retrieved workflow RW_i was already paired with another test workflow $TW_{i \neq j}$ or if the set of data and task nodes was identical, a different workflow RW_i was selected from the case base. Each pair (TW_i, RW_i) was used to automatically generate a change request for RW_i by determining the set of nodes to be added and deleted in order to arrive at TW_i. A change request "DELETE salami", for example, means that the retrieved workflow RW_i uses a salami topping while the test workflow TW_i does not. We executed the proposed compositional adaptation method for each of the 10 workflows RW_i using the corresponding change request. Thus, 10 adapted workflows are computed.

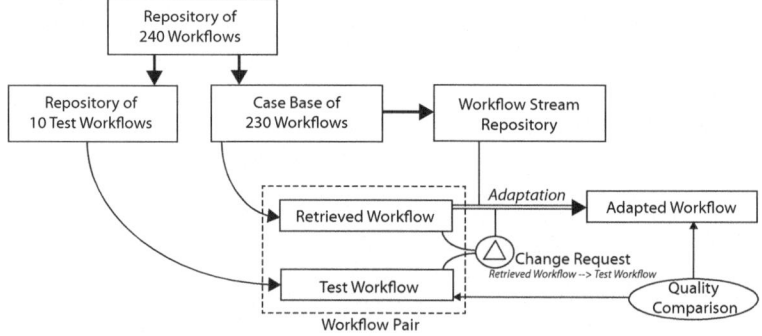

Fig. 8. Experimental Setup

5.2   Experimental Evaluation and Results

To verify hypothesis H1 we analyzed the resulting adapted workflows whether they fulfilled the change request, i.e., whether they did not contain any node from the DELETE list but all nodes from the ADD list. It turned out that each of the 10 adapted workflows fulfilled their change request, thus Hypothesis H1 was confirmed. This outcome was not very surprising, as the workflows in the case base all represent pizza recipes and are thus relatively similar to each other. Due to the large number of 920 workflow streams generated, there was a high chance that streams could be found that perfectly matches the change request.

To evaluate Hypothesis H2 a blinded experiment was performed involving 5 human experts. The experts rated the quality of the 10 test worklows and the 10 corresponding adapted workflows. These 20 workflows were presented in random order, without any additional information. Thus the experts did not know whether the workflow was an original workflow from the case base or an adapted workflow. The experts were asked to assess the quality of each workflow based on 5 rating items on a 5 point Lickert scale (from 1=very low to 5=very high). The rating items are comprised of the intelligibility of the entire preparation process,

Table 1. Item rating assessment

	better test workflows	better adapted workflows	equal
intelligibility of recipe	24	10	16
correctness of recipe	14	7	29
level of process optimization	20	15	15
plausibility of preparation steps	19	11	20
overall quality	17	12	21
aggregated quality	28	15	7

Table 2. Average differences on item ratings

intelligibility of recipe	0.26
correctness of recipe	0.42
level of process optimization	0.20
plausibility of preparation steps	0.18
overall quality	0.28
average per item	0.27
average per workflow	1.34

the correctness of the recipe (w.r.t. order of tasks and data-flow), the level of process optimization (w.r.t. parallel execution and task order), the plausibility of the preparation steps, and the overall quality of the recipe.

The ratings from the 5 experts of all 10 workflow pairs were compared, leading to 50 ratings. We define that one item was rated better for a workflow if it was scored with a higher value than the corresponding item of the compared workflow. Based on this, we conclude that a workflow has a higher aggregated quality, if more of its items were rated better than those of the compared workflow.

Table 1 illustrates the results for each rating item in isolation as well as for the aggregated quality assessment. It shows the number of workflows for which the test workflow or the adapted workflow is better, as well as the number of workflows which were equally rated. In 22 out of 50 rated workflow pairs, the adapted workflow was rated of higher or equal quality (concerning the aggregated quality), whereas 28 test workflows were rated higher. Thus in 44% of the assessments, the adaptation produced workflows with at least the same quality as the workflow from the case base from which they were assembled. When only the rating of the overall quality is regarded, even 66% of the assessments indicate that the adaptation produces a workflow with at least the same quality as the workflow from the case base. Additionally, table 2 illustrates the average rating difference on the items of all 50 workflow pairs. In total, the items of each test workflow are rated 1.34 higher than those of the adapted workflow, which means that each item was rated about 0.27 times better than the corresponding item of the adapted workflow. Thus, the experts rated the items and hence the quality of the test workflows only slightly higher. Altogether, Hypothesis H2 is mostly confirmed.

6   Conclusions and Related Work

Adaptation is a major challenge in case-based reasoning. However, most research only considers cases represented by attribute-values [7,3,8] and there is only little work addressing workflow adaptation. The presented approach of Minor et al. [9], for example, executes workflow adaptations by transforming workflows in a kind of a rule-based manner. Dufour-Lussier et al. [6] presented a compositional adaptation approach for processes, which differs as in their work processes are represented as trees and additional adaptation knowledge is required.

We presented a novel approach to compositional adaptation of workflows by decomposing them into reusable workflow parts (workflow streams). We investigated how to identify and to replace workflow streams and developed an approach to adapt entire workflows regarding a change request. A major advantage is that no manually acquired adaptation knowledge is needed, thus expensive knowledge acquisition is avoided. Instead, the available knowledge contained in the case base is accessed. The proposed approach ensures that the adapted workflows are consistent and semantically correct. Our evaluation indicates that the adaptation process does not significantly decrease the quality of the resulting workflows. However, the presented approach is limited, as it requires similarly structured workflows and workflow streams to be present. Further, we employ the domain knowledge in the ontology during the similarity assessment and for the representation of change requests. Additional adaptation knowledge is yet not used, but could be considered in future extension of the proposed approach.

Future work will extend the existing evaluation towards other domains and varying characteristics of the case base. The developed workflow generator now makes such evaluations feasible. Furthermore, we will explore various ways to improve the proposed method by weakening the equality condition on anchors in the substitutability of streams, and by generalizing the streams by using the taxonomy of tasks and data items. To further increase the flexibility of workflow reuse, workflow streams might be a means to construct abstract workflows, to identify subworkflows, or to optimize workflows (w.r.t. parallel executions).

Acknowledgements. This work was funded by the German Research Foundation (DFG), project number BE 1373/3-1.

References

1. Bergmann, R.: Experience Management. LNCS (LNAI), vol. 2432. Springer, Heidelberg (2002)
2. Bergmann, R., Gil, Y.: Similarity assessment and efficient retrieval of semantic workflows. Inf. Syst. 40, 115–127 (2014)
3. Craw, S., Jarmulak, J., Rowe, R.: Learning and applying case-based adaptation knowledge. In: Aha, D.W., Watson, I. (eds.) ICCBR 2001. LNCS (LNAI), vol. 2080, pp. 131–145. Springer, Heidelberg (2001)

4. Dadam, P., Reichert, M., Rinderle-Ma, S., Göser, K., Kreher, U., Jurisch, M.: Von ADEPT zur AristaFlow BPM Suite-Eine Vision wird Realität: "Correctness by Construction" und flexible, robuste Ausführung von Unternehmensprozessen. University of Ulm (2009), http://dbis.eprints.uni-ulm.de/489/

5. Davenport, T.: Process Innovation: Reengineering Work Through Information Technology. Harvard Business Review Press (2013)

6. Dufour-Lussier, V., Lieber, J., Nauer, E., Toussaint, Y.: Text adaptation using formal concept analysis. In: Bichindaritz, I., Montani, S. (eds.) ICCBR 2010. LNCS, vol. 6176, pp. 96–110. Springer, Heidelberg (2010)

7. Hanney, K., Keane, M.T.: Learning adaptation rules from a case-base. In: Smith, I., Faltings, B. (eds.) EWCBR 1996. LNCS, vol. 1168, pp. 179–192. Springer, Heidelberg (1996)

8. McSherry, D.: Demand-driven discovery of adaptation knowledge. In: Dean, T. (ed.) IJCAI, pp. 222–227. Morgan Kaufmann (1999)

9. Minor, M., Bergmann, R., Görg, S., Walter, K.: Towards case-based adaptation of workflows. In: Bichindaritz, I., Montani, S. (eds.) ICCBR 2010. LNCS, vol. 6176, pp. 421–435. Springer, Heidelberg (2010)

10. Minor, M., Görg, S.: Acquiring adaptation cases for scientific workflows. In: Ram, A., Wiratunga, N. (eds.) ICCBR 2011. LNCS, vol. 6880, pp. 166–180. Springer, Heidelberg (2011)

11. Minor, M., Montani, S., Recio-Garca, J.A.: Process-oriented case-based reasoning. Information Systems 40, 103–105 (2014)

12. Purvis, L., Pu, P.: Adaptation using constraint satisfaction techniques. In: Veloso, M.M., Aamodt, A. (eds.) ICCBR 1995. LNCS, vol. 1010, pp. 289–300. Springer, Heidelberg (1995)

13. Reichert, M.: Dynamische Ablaufänderungen in Workflow-Management-Systemen. University of Ulm (2000), http://dbis.eprints.uni-ulm.de/433/

14. Schumacher, P., Minor, M., Walter, K., Bergmann, R.: Extraction of procedural knowledge from the web. In: Workshop Proc.: WWW 2012, Lyon, France (2012)

15. Smyth, B., Cunningham, P.: Deja vu: A hierarchical case-based reasoning system for software design. In: Neumann, B. (ed.) Proc. ECAI 1992, pp. 587–589 (1992)

16. Stahl, A., Bergmann, R.: Applying recursive CBR for the customization of structured products in an electronic shop. In: Blanzieri, E., Portinale, L. (eds.) EWCBR 2000. LNCS (LNAI), vol. 1898, pp. 297–308. Springer, Heidelberg (2000)

17. Veloso, M.M.: Planning and Learning by Analogical Reasoning. LNCS, vol. 886. Springer, Heidelberg (1994)

18. Wilke, W., Bergmann, R.: Techniques and knowledge used for adaptation during case-based problem solving. In: Mira, J., Moonis, A., de Pobil, A.P. (eds.) IEA/AIE 1998. LNCS, vol. 1416, pp. 497–506. Springer, Heidelberg (1998)

19. Workflow Management Coalition: Workflow management coalition glossary & terminology (1999), http://www.wfmc.org/docs/TC-1011_term_glossary_v3.pdf (last access on April 04, 2014)

Collective Classification
of Posts to Internet Forums

Pádraig Ó Duinn and Derek Bridge*

Insight Centre for Data Analytics,
School of Computer Science and Information Technology,
University College Cork, Ireland
{padraig.oduinn,derek.bridge}@insight-centre.org

Abstract. We investigate automatic classification of posts to Internet forums. We use collective classification methods, which simultaneously classify related objects — in our case, the posts in a thread. Specifically, we compare the Iterative Classification Algorithm (ICA) with Conditional Random Fields and with conventional classifiers (k-Nearest Neighbours and Support Vector Machines). The ICA algorithm invokes a local classifier, for which we use the kNN classifier. Our main contributions are two-fold. First, we define experimental protocols that we believe are suitable for offline evaluation in this domain. Second, by using these protocols to run experiments on two datasets, we show that ICA with kNN has significantly higher accuracy across most of the experimental conditions.

Keywords: collective classification, Internet forums, incremental.

1 Introduction

Internet forums are places for debate and discussion, for the sharing of experience, and for collaborative problem-solving. Users assume a variety of roles. They be *consumers*, seeking information from existing posts; as *contributors*, they might initiate new threads or add posts to existing threads; and as *moderators*, they might monitor discussions and intervene to enforce forum policies.

Forums display the posts within a thread in a linear order, most commonly in chronological order. But a thread's argument structure may not be linear. A post may respond to a post other than the one that immediately precedes it. As the thread becomes longer, contributors, who may be unwilling to read all previous posts, may repeat previous posts, insert unrelated posts, or even start new threads that replicate existing threads. This makes it more difficult for the different categories of users to achieve their goals.

We are interested in forum management software that infers the argument structure and uses it to assist users. For example, for consumers, the software

---

* The authors gratefully acknowledge the financial support of the Irish Research Council. This publication has emanated from research supported in part by a research grant from Science Foundation Ireland (SFI) under Grant Number SFI/12/RC/2289.

might summarise a whole thread, or present it in a hierarchical fashion, or suppress redundant or irrelevant posts. For contributors, it might provide guidance for increasing the usefulness of a contribution. For moderators, it might help identify such things as: spam posts, misplaced posts, flame wars, troll users, or threads that need locking, merging or deleting.

In this paper we are investigating post classification. We wish to classify each post using a label that denotes the post's role within its thread. In a trouble-shooting forum, the labels might include 'question', 'answer', 'clarification', and so on. We make the simplifying assumption in this work that each post has a single label. This is fairly reasonable for a trouble-shooting forum, but more questionable for forums where threads are more discursive. We may lift this assumption in future work. We believe that classifying posts by their role in the thread is a useful precursor to the kinds of tasks we mentioned in the previous paragraph, for example inferring the argument structure so that threads can be presented hierarchically.

In traditional classification, objects are considered in isolation, and classified independently. However, it may be the case that objects are inter-related. In this case, classification accuracy can often be improved by taking into account information from the related objects. When this information includes the *predicted labels* of the related objects, the term *collective classification* is used.

We compare two collective classifiers. One is a Conditional Random Fields (CRF) classifier, and this approach has already appeared in the literature [9]. The other is the Iterative Classification Algorithm (ICA) [14]. ICA has not to our knowledge been previously used for post classification. We expect collective classifiers to obtain higher classification accuracy than traditional classifiers.

The Iterative Classification Algorithm, which we explain fully in Section 2.1, repeatedly invokes another, 'local' classifier [14]. In principle, the local classifier can be any classification algorithm. In practice, the local classifier must be cheap to train because it is re-trained repeatedly on a changing training set. The training set is updated on each iteration with the latest predicted labels of the related objects. For our local classifier, we use a k-nearest neighbours case-based classifier. Case-based classifiers are lazy learners and, as such, have negligible training costs, making them ideal for use within ICA.

In Section 2, we describe the task of post classification in general and we describe the two collective classifiers that we are comparing in this paper (ICA and CRF). Section 3 describes the two datasets which we use in our experiments. Section 4 defines our experimental protocols. Section 5 presents the results of some of the experiments that we have run. The final sections present related work (Section 6) and conclusions (Section 7).

2 Post Classification

A forum comprises a set of threads T. Each thread $t \in T$ comprises a sequence of posts, $t = \langle p_1, p_2, \ldots, p_n \rangle$. Each post has a label $c(p_i, t)$ which is drawn from a finite set of labels C. Note that the label may depend on both the post, p_i, and the thread in which it appears, t: in principle, an identical post in a different thread might receive a different label. The task of the classifier is to predict the label, $\hat{c}(p_i, t)$. It might seem more natural to assume that the label and predicted label of a post p_i both depend only on the preceding posts in the thread, $\{p_j | j < i\}$, rather than the whole thread, t. Certainly, in the experiments we report in this paper, we only use the preceding posts. But, a formalism that allows the label to depend on the whole thread leaves open the possibility that classification can depend on reactions to a post — which come later in the thread. This is something we may experiment with in the future.

Each post p_i will be described by attributes. Three categories of attributes may be used:

- Attributes derived from observations of p_i itself, e.g. author information and lexical attributes such as word counts and counts of the number of punctuation symbols.
- Attributes derived from observations of other posts in the thread, $p_j \in t, j \neq i$, e.g. p_j's author information and lexical attributes, or the degree of similarity between the titles of p_i and p_j. In some cases, it may be possible to *observe* the *labels* of other posts in the thread, e.g. if they are given as part of training data, in which case attributes derived from these would also fall into this category. But, for most posts we cannot observe their labels; we can only *predict* them, and so their labels fall into the third category.
- Attributes derived from the *predicted labels* of other posts in the thread, $\hat{c}(p_j, t)$ where $p_j \in t, j \neq i$, e.g. the predicted label of the previous post, or a count of the number of previous posts that are predicted to have a certain label.

Attributes in the first two categories are called *observed attributes*, whereas attributes in the third category are called *unobserved attributes*. Alternatively, attributes in the first category are *non-relational attributes*, whereas attributes in the last two categories are *relational attributes*.

A classifier that uses only the first of these categories is said to be a *non-relational classifier*; one that uses the second of these is said to be a *relational classifier*; and one that uses the third is said to be a *collective classifier* [14].

Collective classifiers are more usually formalised for the case of a graph. The job is to predict labels for a subset of the nodes in the graph — those whose labels are not already known. In addition to a node's attributes, the classifier may use relational attributes, i.e. the attributes and labels of other nodes. Usually, there is a definition of neighbourhood within the graph, and a node's relational attributes come only from the other nodes in its neighbourhood. In this paper, we are taking the special case where the graph is a chain — a linear arrangement of posts.

Collective classifiers fall into two main types: those that make use of a local classifier, and those that define and optimize a set of weights in a global objective function [15]. We will explain one representative of the first of these two approaches (ICA) and one representative of the second approach (CRF).

Algorithm 1 Iterative Classification of Forum Posts

---

t = thread of posts $\langle p_1, p_2, \ldots, p_n \rangle$
m = the index of the first unlabelled post in t, $m \in [1, n+1]$
M_O = local classifier that uses observed attributes only
M_U = local classifier that uses observed and unobserved attributes

// Step 1: Initial classification
// Step 1a: Derive observed attributes
for all $i \leftarrow 1$ to n **do**
   $o_i \leftarrow p_i$'s observed attributes
end for
// Step 1b: Classify unlabelled posts using observed attributes only
for all $i \leftarrow m$ to n **do**
   $c_i \leftarrow M_O(o_i)$
end for
// Step 2: Iterative classification
repeat
   // Step 2a: Derive unobserved attributes
   for all $i \leftarrow m$ to n **do**
      $u_i \leftarrow p_i$'s unobserved attributes
   end for
   // Step 2b: Re-classify unlabelled posts using all attributes
   for all $i \leftarrow m$ to n **do**
      $c_i \leftarrow M_U(o_i, u_i)$
   end for
until class labels have stabilized or a certain number of iterations has elapsed
return c_i for each p_i, $i \geq m$

2.1 Iterative Classification Algorithm

The Iterative Classification Algorithm (ICA) is probably the simplest collective classifier. Algorithm 1 presents the version that we use for forum post classification. We allow for the possibility that we already know the labels of some of the posts in thread t; for example, where they are supplied in a training set. We use index $m \in [1, n+1]$ to designate the first thread whose label is not known. In other words, we know the labels of posts p_i for $i < m$; and we must predict the labels of posts p_j for $j \geq m$. If $m = 1$, then we have not been given any true labels in this thread; if $m = n + 1$; all posts are already labelled, and there is no work to be done.

Step 1 of ICA is to establish initial labels for the unlabelled posts. This is done using a local classifier which only considers the post's observed attributes. Step 2 continually re-classifies posts, but now using all attributes, until predictions have stabilized or, as in our experiments, a certain number of iterations (in this paper, 10) has been performed. Specifically, Step 2a computes the unobserved attributes. These may simply be the currently predicted labels of the other posts, or attributes that are derived from these. In Step 2b, posts are re-classified using both observed and unobserved attributes.

Note that we make the assumption that posts are processed by the algorithm in chronological order (for-loops run *up* to n). More sophisticated versions of ICA include steps for choosing the order of processing, e.g. based on the degree of uncertainty of the currently predicted label (e.g. [14]).

For both local classifiers, M_O and M_U, we use a k-nearest-neighbours case-based classifier [12]. Its case base contains posts, described by a full set of attributes. Each post is stored with its label. To classify a new post, we retrieve its k-nearest neighbours. The local distance measure we use to compare one attribute-value with another is the range-normalised absolute difference [20]; the global distance measure is simply an unweighted average of the values of the local distance measures for each attribute. M_O, however, only computes distances over the new post's observed attributes, whereas M_U computes distances over all attributes. The predicted class is obtained by a distance-weighted vote of the nearest neighbours. In Section 1, we explained why a lazy learner, such as a case-based classifier, is especially well-suited for use in ICA.

2.2   Conditional Random Fields Classification

The main alternative way to carry out collective classification seeks to optimize a global objective function, e.g., one that weights individual feature functions. In collective classification, techniques include Loopy Belief Propagation (LBP) [18] and Mean-Field Relaxation Labelling (MF) [19]. Although rarely presented as collective classification, Conditional Random Fields (CRFs) are similar [10]; indeed, LBPs are partly inspired by CRFs [18].

However, post classification does not require the full generality of these algorithms, since the posts form a chain, and not an arbitrary graph. The problem simplifies to a sequence labelling problem. For sequence labelling, classifiers include Maximum Entropy Markov Models (MEMMs) [11] and Linear Chain Conditional Random Fields. We will use a Linear Chain CRF, since it overcomes the label bias problem exhibited by MEMMs [10]. Specifically, we use CRF++, which is an open source Linear Chain CRF.[1]

There is another reason for choosing to use Linear Chain CRFs: they were the best-performing approach in an earlier post classification study by Kim et al. [9], where they were compared with a conventional Maximum Entropy learner (ME) and a Structural Support Vector Machine (HMM-SVM) [8]. Here, we are able to compare CRFs to a more competitive approach (ICA) and to do so using the experimental protocols that we define in Section 4, which we believe give more reliable results.

3   Datasets

Our goal is to compare post classification using ICA with the Linear Chain CRF that was used in [9] and with conventional classifiers, namely k-nearest-neighbours (kNN) and Support Vector Machines (SVM). We use two datasets, which we describe here.

---

[1] http://crfpp.sourceforge.net/

The first, the CNET dataset, is the one created by Kim et al. for the work they describe in [9]. It comprises 320 threads but five of them include some unlabelled posts, so we use only the remaining 315 threads; these contain a total of 1332 posts from CNET forums that deal with trouble-shooting for operating systems, software, hardware and web development.[2]

For the CNET dataset, the set C contains twelve labels. The two main labels, *Question* and *Answer*, are sub-divided into sub-categories, having double-barrelled names. For example, a post whose label is *Question-question* contains a new question. A post whose label is *Question-add* contains information which supplements some other question, e.g. providing additional information or asking a follow-up. Similarly, *Answer-answer* offers solutions to problems or answers to questions posed in question posts, and *Answer-add* provides information that supplements some other answer. The other labels are *Question-confirmation, Question-correction, Answer- confirmation, Answer-correction, Answer-objection, Resolution, Reproduction* and *Other*. Note that one label (*Answer-correction*) never occurs.

The true labels of each post were assigned by two human annotators, with adjudication in the case of disagreement. In the original dataset, 65 posts were assigned multiple labels. It is not clear how [9] handles these. For each of these, we made a judgement about which label was best and discarded other labels.

The second dataset, the FIRE dataset, is also from a forum dealing with trouble-shooting for computers[3], which was used by Bhatia et al. in [1]. It comprises 100 threads, containing a total of 556 posts. Bhatia et al. use a set C of just eight labels: *Question, Repeat Question, Clarification, Further Details, Solution, Positive Feedback, Negative Feedback* and *Junk*.

We extracted a set of attributes from each post in the two datasets. We based the attributes on those described in [9]. All attributes are numeric-valued. The observed non-relational attributes are:

- **Punctuation:** Punctuation marks can help identify the nature of a post. The number of explanation marks in a post and the number of question marks in a post are used as two attributes.
- **URLs:** Posts which contain URLs are more likely to be in one of the *Answer* classes. The number of URLs in a post is used as an attribute.
- **Thread Initiator:** The user who starts a thread is less likely to post answers in that thread. We use a binary feature which indicates whether the post's author is also the person who initiated the thread.
- **Post Position:** The position of a post in a thread is important, not least since the first post is always a question. This attribute records the relative position of the current post in its thread, i.e. the ratio of its position to the number of posts in the thread.
- **Author Profile:** Some users ask more questions than they answer, and vice versa. We use the class priors for that user. In other words, we have $|C|$ attributes, one for each class label, where each is the probability that this

---

[2] http://forums.cnet.com/
[3] http://www.isical.ac.in/~fire/

user creates posts of this class, calculated from the labels of the user's posts in the training set.

The observed relational attributes are:

- **Title Similarity:** A post whose title is similar to the title of a previous post is often some form of reply to that previous post. We compute the cosine similarity between the post's title and the titles of all previous posts. We find the previous post whose title is most similar to that of the current post. The relative position for that post is used as the attribute. (This attribute is not used in the experiments with the FIRE dataset because its posts do not have titles.)
- **Post Similarity:** Similarity of content is also a sign of one post responding to another. This attribute is the same as the previous one, but this time based on cosine similarity of post content, rather than post title.

The unobserved relational attributes are:

- **Previous Post**: This attribute records the label of the previous post in the thread. To make it numeric-valued, we use one-hot encoding, i.e. we use $|C|$ attributes, one for each class label, setting them all to zero except for the one that corresponds to the class of the previous post, which is set to one.
- **Previous Post from Same User:** This attribute records the label of the previous post in the thread by the author of the current post. Here again we used one-hot encoding.
- **Full History:** This attribute records the labels of all previous posts in the thread. Again there are $|C|$ attributes, but this time they are counts: the number of previous posts having each label.
- **User History:** This attribute records the labels of all previous posts of the author of the current post. Like Full History, there are $|C|$ attributes, represented as counts: the number of posts having each label.

Notice that we have no lexical attributes (such as unigram and bigram frequencies). In their experiments, Kim et al. did not run the CRF with lexical attributes because there were too many such attributes. They claim the CRF does not scale well to large numbers of attributes. We confirmed this: CRF++ crashed on 1000 attributes. To keep the comparison fair, we do not use such attributes for the ICA even though the local classifiers that we are using cope reasonably well with high-dimensional data. In any case, Kim et al. found that their other classifiers (ME and HMM-SVM) were not very accurate when trained on lexical attributes, possibly because the amount of training data (just 1332 posts of fewer than 100 words each) was too small to allow useful generalization.

Our experimental protocols (next section) work in a chronological fashion, thus requiring that each post have a timestamp. These were missing from the CNET dataset, so we scraped them from the original forum.

4   Experiments

Both Kim et al. and Bhattia et al. use a 10-fold cross-validation methodology. While this gives robust accuracy estimates, it fails to model the way that threads grow over time and the fact that a later post must be classified using the possibly erroneous class labels of previous posts in the same thread. We define *incremental* protocols instead, giving a more realistic setting for the experiments.

In each protocol, we train classifiers on all posts with their true labels from the first 100 days of the dataset. We then ask the classifier to predict labels for all of the next day's posts, and we measure accuracy against the true labels. We repeat this on a daily basis until the dataset is exhausted. Thus we have an accuracy figure for each day subsequent to the 100th day.

Where the protocols differ is whether and how they retrain the classifiers at the end of each day.

Static (S): In this protocol, classifiers are not re-trained. The classifiers that are built from the first 100 days of the training set are used unchanged to classify each of the remaining days' posts.

Supervised Incremental (SI): In this protocol, at the end of each simulated day (after the classifiers have made predictions for that day's posts), the classifiers are re-trained on a dataset that is extended by all of that day's posts with their *true labels*. In other words, when classifying posts submitted to the forum on day $i + 1$, the classifiers will have been trained on all posts up to day i. Re-training includes updating the priors for the Author Profile attribute so that they reflect the distribution of labels in the extended training set.

Semi-Supervised Incremental (SSI): This protocol is similar to the previous one except that the training sets are extended each day with that day's posts but with their *predicted labels*, rather than their true labels. In other words, when classifying posts submitted on day $i+1$, the classifiers have been trained on posts from the first 100 days with their true labels and posts from the next $i - 100$ days with their predicted labels.

Careful Semi-Supervised Incremental (CSSI): This protocol is similar to SSI, however not all posts are included when retraining. Only when the classifier's confidence in its prediction exceeds a threshold does that post and its predicted label get added to the training set. Posts from day i for which this threshold is not met are reclassified on day $i + 1$ along with the new posts from day $i+1$. Each of the classifiers can output a probability that its prediction is correct, and it is this that we compare with the threshold to decide what will happen with the post.

The SI protocol models the situation where forum moderators make decisions about each new post on a daily basis: a classifier might advise the moderator but the moderator always makes the final decision. Hence, the training set only ever contains posts with true labels. By contrast, the SSI protocol models the situation where there is no on-going moderator intervention: the classifier's decision is always final. Hence, training set posts up to day 100 have their true labels but subsequent posts have their predicted labels. The CSSI protocol attempts to reduce the misclassification rate of SSI by taking a more cautious approach.

It is informative to investigate the results of these somewhat extreme protocols. In practice, on a real forum performance will lie somewhere between the

Table 1. Unobserved relational attributes used by ICA in different experiments

Protocol	Dataset	Unobserved attributes used
S	CNET	Previous Post from Same User
S	FIRE	Previous Post from Same User
SI	CNET	Previous Post, Full History, User History
SI	FIRE	Previous Post, Previous Post from Same User, User History
SSI	CNET	Previous Post, Full History, User History
SSI	FIRE	User History
CSSI	CNET	Previous Post, Full History, User History
CSSI	FIRE	Previous Post

extremes. For example, there might be a triage system: if the classifier has low confidence in its decision, a moderator may be asked to review it.

One other aspect of the methodology should be mentioned. Before running these protocols, we use a model selection step. For the conventional kNN classifier, model selection chooses a good value for k (from $k = 5, 7, 11, 15$); for ICA using kNN as its local classifier, model selection chooses a good value for k and also decides which (non-empty) subset of the unobserved attributes to use; for CRF, model selection chooses between two optimization methods and parameters for optimization and regularization; and for the SVM, model selection chooses a penalty coefficient and a coefficient for the kernel (for which we use the Radial Basis Function). For all classifiers, model selection also chooses the best confidence threshold to use in the CSSI protocol.

Model selection uses the first 100 days' posts (with their true labels). We train different versions of the classifiers (e.g. kNN with different k) on approximately the first 70% of these days' posts, we measure accuracy of predictions on the remaining posts, and then we select the best version of the classifier for use in the protocols described above.

Table 1 shows, for each experiment (protocol plus dataset), which unobserved attributes model selection chooses for ICA.

5   Results

Figure 1 shows the results for the Static protocol — where there is no re-training. On the CNET dataset, ICA outperforms all other classifiers tested; in the case of the FIRE dataset, ICA performs best, but in this instance, its performance is matched by kNN.

By way of comparison, we calculated the accuracy of a majority-class classifier, taking the majority class from the 100 days of training data and seeing how often this would be the correct prediction for the remaining posts. The majority classifier has an accuracy of 40.74% for the CNET dataset and 36.87% for the FIRE dataset. ICA is about 70% and 60% accurate on these datasets respectively.

Figure 2 presents the results from experiments with the Supervised Incremental protocol — where daily re-training uses the true labels of the new posts. On

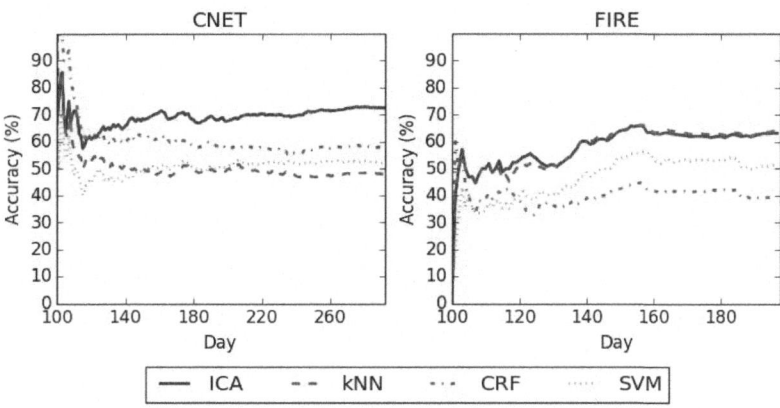

Fig. 1. Accuracy for the Static protocol

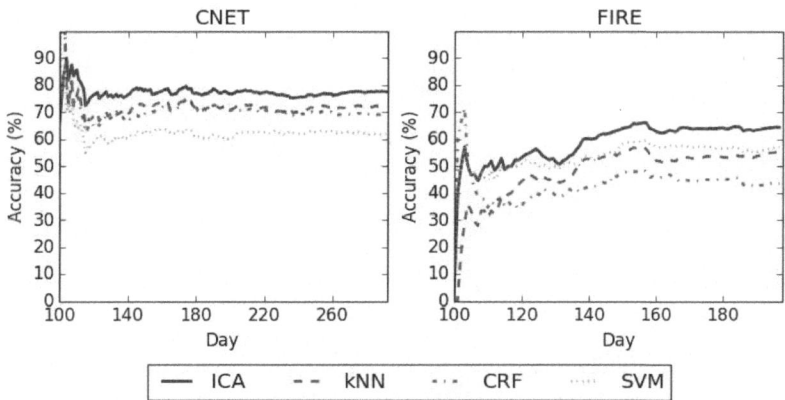

Fig. 2. Accuracy for the Supervised Incremental protocol

the CNET dataset, ICA is again the best performing technique. However, with this protocol there is an increase in the accuracy of all the classifiers, and the difference in performance is smaller than that seen for the CNET dataset using the S protocol. This is to be expected as the classifiers have access to more properly labelled training data for each day. We can make similar observations about the FIRE dataset and this time kNN is not matching ICA.

Figure 3 presents the results for the Semi-Supervised Incremental protocol — where daily re-training uses the predicted labels of the new posts. For the CNET dataset, ICA continues to achieve the best performance. In this case, kNN is also able to achieve good accuracy, but it still falls short of ICA's. The results for the FIRE dataset are similar, with ICA performing best, followed by kNN and, in this case, SVM close behind.

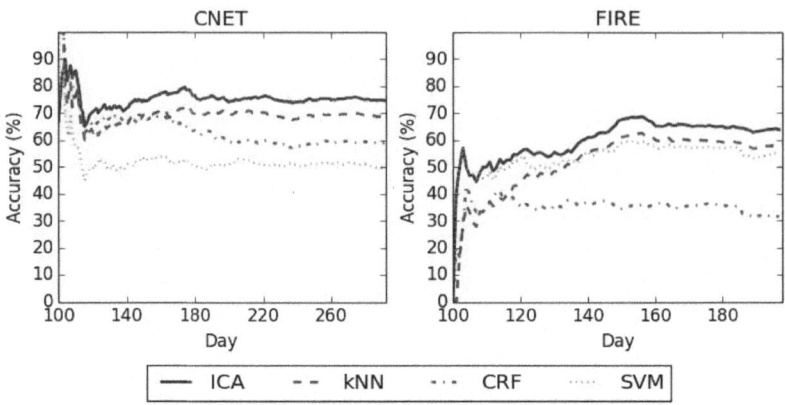

Fig. 3. Accuracy for the Semi-Supervised Incremental protocol

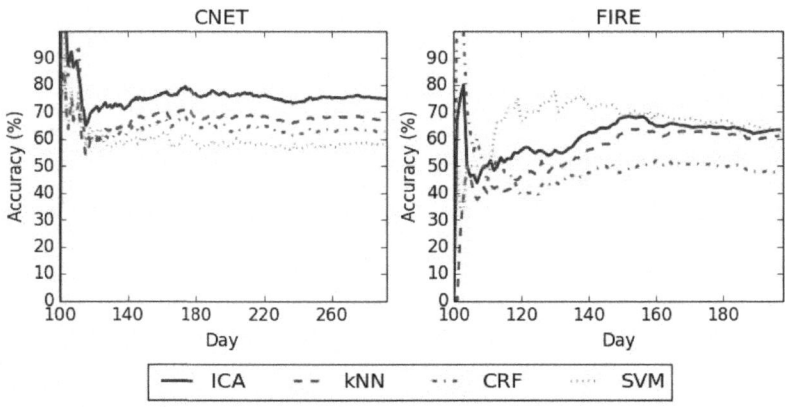

Fig. 4. Accuracy for the Careful Semi-Supervised Incremental protocol

As we would expect, accuracies with this protocol (SSI) are lower than the figures we saw with the SI protocol. Although the training data grows over time, there is no guarantee that the labels in the new training instances are correct: they are predicted labels. It is also the case that, with the exception of kNN, the accuracies of the classifiers see little if any improvement over the accuracies obtained in the S protocol experiment, which implies that the benefits of the extra training data are largely outweighed by the error in the predicted labels.

Finally, the results for the Careful Semi-Supervised Incremental protocol are in Figure 4. Once again, ICA achieves the best results on the CNET dataset. The results for the other classifiers are a little improved relative to the SSI protocol experiment. The results for the FIRE set are more unusual. In this case, SVM performs very well initially. However towards the end, ICA, kNN and SVM all achieve similar accuracy.

Table 2. Average accuracy across all test instances

CNET	ICA	kNN	CRF	SVM	FIRE	ICA	kNN	CRF	SVM
S	**72.47**	47.82	57.83	52.09	S	63.20	63.80	39.47	51.34
SI	**77.57**	72.01	68.95	61.72	SI	**64.39**	55.19	43.62	57.27
SSI	**75.81**	68.67	58.94	49.58	SSI	**63.80**	57.27	31.45	54.90
CSSI	**74.91**	67.01	62.76	57.97	CSSI	63.28	60.91	47.86	60.30

Looking across the experiments, we see that ICA is generally the most accurate classifier for forum posts. We tested for significance using McNemar's Test [7] with a value of 1 for the continuity correction term [5] and $p = 0.05$. We did this on just the final day's test data. The tests confirm that ICA is significantly better than the other classifiers for all experiments except one: there is no significant difference between ICA and kNN on the FIRE dataset for the S protocol.

ICA is also the most robust classifier: variations in the experimental protocol have little effect on its accuracy. The other classifiers have room to improve their accuracy when given more training data and, as we would expect, in general being given new training instances with true labels (SI) is better than being given new instances with confidently-predicted labels (CSSI) which, in turn, is better than being given new instances with predicted labels without regard to prediction confidence (SSI).

We have also summarized the results by computing average accuracy across *all* test instances: see Table 2. Again we tested for significance using McNemar's Test. In the Table, figures in bold are significantly better than all others on that protocol.

From the Table, we observe that, in nearly all cases, ICA is significantly better than the other classifiers. The exceptions are found on the FIRE dataset using the S protocol, where ICA is not significantly better than kNN, and on the FIRE dataset using the CSSI protocol, where there is no significant difference between ICA, kNN and SVM.

It is interesting too to note from the graphs and table the relatively poor performance of CRFs across our experiments. In [9], Kim et al. used a 10-fold cross-validation methodology to compare CRFs with a Maximum Entropy learner and a Structural SVM on their dataset (CNET). They found CRFs to have the highest accuracy. They did not compare with ICA. We repeated their experiment, using their methodology, to compare CRF with ICA: in this experiment, we found ICA and CRF accuracy to be very similar. However, we believe that Kim et al.'s methodology lacks sufficient fidelity to the way that forum post classification happens in practice. Their test set contains randomly-chosen *threads*. This means that entire thread structure is preserved also in the training set. This enables the CRF to build a good model. For the more realistic chronological protocols that we have described in this paper, CRF loses its advantage and ICA becomes the most accurate.

We should add a word of warning about detailed comparisons of classifiers across different graphs in this paper. Although each graph plots accuracy for, e.g., a kNN classifier, the kNN classifier in one graph is not strictly comparable with the kNN classifier in another graph since they may be using different values for k. This is due to the model selection phase in our experimental methodology (described in Section 4). The same applies to all the other classifiers, e.g. ICA may be using a different value for k but also different unobserved relational attributes (Table 1).

6   Related Work

The most closely related piece of work is by Kim et al. [9], since we set out to compare collective classifiers that use local classifiers (ICA) with CRFs, which were the best-performing classifiers in their experiments. We also use their dataset and, as best we could, the same attributes that they used. Their paper also applies CRF to link classification, i.e. the task of predicting thread structure.

Bhatia et al. also report work on post classification [1], and we use the same dataset that they use (the FIRE dataset). However, they do not use collective classification: all their attributes are observed attributes. They experiment with several classifiers from the Weka machine learning tool-kit [6] and report the results of just the logit model classifier (72.02% accuracy).

There has been work on classification for online debates and spoken debates. For example, Somasundaran & Wiebe classify stance (pro or con) in online debates [16,17]. In [2], transcripts from the US congress are classified based on support or opposition to a topic. This paper makes use of collective classification techniques.

A related topic is the classification of speech acts in email conversation. This is explored in [4] and then revisited in [3] using collective classification techniques. The collective classifiers were able to achieve better results.

More generally, there are several overviews and comparisons using collective classification, e.g. [15,12]. The accuracy of the collective classifiers that use local classifiers can often be improved by using 'cautious' variants of the algorithms, which take account of the uncertainty in the predicted labels. These cautious collective classifiers are presented in, e.g., [14,13].

Accordingly, we too used a cautious variant of ICA (the one that [14,13] designates ICA_C) in our experiments. However, in our experimental protocols, we did not find ICA_C to be better than ICA. We omitted the results from this paper to reduce clutter in our graphs.

7   Conclusions

In this paper, we have investigated the task of classifying posts to Internet forums. We compared ICA —a collective classifier that uses a local classifier (in our case, kNN)— with Linear Chain CRFs —a classifier that optimizes global probability functions— and with some conventional classifiers (kNN and SVM). We compared them on two datasets, and we defined a number of experimental protocols for robust testing of these classifiers. ICA out-performed all other classification techniques tested across all the protocols for the CNET dataset. It out-performed all other classification techniques for all but two of the protocols for the FIRE dataset: it was matched by kNN in one protocol; and in another, there was a period during which SVM out-performed ICA, but their end-of-period accuracies were the same.

There are many ways in which we can improve ICA. We can try other attributes, including lexical attributes and also relational attributes that are based on later posts in the thread as well as those that are based on earlier posts in the thread. We can improve the kNN local classifier; for example, we can use attribute weights in the distance computation [12]. We can try other local classifiers, such as Naïve Bayes, or even ensembles of classifiers. Additionally, Gibbs Sampling is another collective classification technique which uses local classifiers. It is possible that this will further improve the accuracy seen in our experiments. However, Gibbs Sampling comes with a considerable time cost compared to ICA, and it remains to be seen whether any improvements would be worth the added expense. We can also trying collective classifiers that do not use local classifiers (other than CRF), such as Loopy Belief Propagation.

More generally, we plan to look at identifying argument structure in threads and then techniques that can be incorporated into the kind of forum management software that we described in Section 1.

References

1. Bhatia, S., Biyani, P., Mitra, P.: Classifying user messages for managing web forum data. In: Ives, Z.G., Velegrakis, Y. (eds.) Procs. of the 15th International Workshop on the Web and Databases, pp. 13–18 (2012)
2. Burfoot, C., Bird, S., Baldwin, T.: Collective classification of congressional floor-debate transcripts. In: Lin, D., Matsumoto, Y., Mihalcea, R. (eds.) Procs. of the 49th Annual Meeting of the Association for Computational Linguistics: Human Language Technologies, pp. 1506–1515. ACL (2011)
3. Carvalho, V.R.: On the collective classification of email speech acts. In: Baeza-Yates, R.A., et al. (eds.) Procs. of the 28th Annual International ACM SIGIR Conference on Research and Development in Information Retrieval, pp. 345–352. ACM Press (2005)
4. Cohen, W.W., Carvalho, V.R., Mitchell, T.M.: Learning to classify email into "speech acts". In: Procs. of the Conference on Empirical Methods in Natural Language Processing, pp. 309–316. ACL (2004)
5. Dietterich, T.G.: Approximate statistical tests for comparing supervised classification learning algorithms. Neural Computation 10(7), 1895–1923 (1998)
6. Hall, M., Frank, E., Holmes, G., Pfahringer, B., Reutemann, P., Witten, I.H.: The WEKA data mining software: An update. SIGKDD Explorations 11(1), 10–18 (2009)
7. Japkowicz, N., Shah, M.: Evaluating Learning Algorithms. Cambridge University Press (2011)
8. Joachims, T., Finley, T., Yu, C.-N.J.: Cutting-plane training of structural SVMs. Machine Learning Journal 77(1), 27–59 (2009)

9. Kim, S.N., Wang, L., Baldwin, T.: Tagging and linking web forum posts. In: Procs. of the Fourteenth Conference on Computational Natural Language Learning, pp. 192–202. ACL (2010)
10. Lafferty, J.D., McCallum, A., Pereira, F.C.N.: Conditional random fields: Probabilistic models for segmenting and labeling sequence data. In: Brodley, C.E., Danyluk, A.P. (eds.) Procs. of the 18th International Conference on Machine Learning, pp. 282–289 (2001)
11. McCallum, A., Freitag, D., Pereira, F.C.N.: Maximum entropy markov models for information extraction and segmentation. In: Langley, P. (ed.) Procs. of the 17th International Conference on Machine Learning, pp. 591–598 (2000)
12. McDowell, L., Gupta, K.M., Aha, D.W.: Case-based collective classification. In: Wilson, D., Sutcliffe, G. (eds.) Proceedings of the Twentieth International Florida Artificial Intelligence Research Society Conference, pp. 399–404. AAAI Press (2007)
13. McDowell, L., Gupta, K.M., Aha, D.W.: Cautious inference in collective classification. In: Proceedings of the Twenty-Second AAAI Conference on Artificial Intelligence, pp. 596–601. AAAI Press (2007)
14. McDowell, L., Gupta, K.M., Aha, D.W.: Cautious collective classification. Journal of Machine Learning Research 10, 2777–2836 (2009)
15. Sen, P., Namata, G., Bilgic, M., Getoor, L., Gallagher, B., Eliassi-Rad, T.: Collective classification in network data. AI Magazine 29(3), 93–106 (2008)
16. Somasundaran, S., Wiebe, J.: Recognizing stances in online debates. In: Su, K.-Y., Su, J., Wiebe, J. (eds.) Procs. of the 47th Annual Meeting of the Association for Computational Linguistics and the 4th International Joint Conference on Natural Language Processing of the AFNLP, pp. 226–234. ACL (2009)
17. Somasundaran, S., Wiebe, J.: Recognizing stances in ideological on-line debates. In: Procs. of the NAACL HLT 2010 Workshop on Computational Approaches to Analysis and Generation of Emotion in Text, pp. 116–124. ACL (2010)
18. Taskar, B., Abbeel, P., Koller, D.: Discriminative probabilistic models for relational data. In: Darwiche, A., Friedman, N. (eds.) Procs. of the 18th Annual Conference on Uncertainty in Artificial Intelligence, pp. 485–492 (2002)
19. Weiss, Y.: Comparing the mean field method and belief propagation for approximate inference in MRFs. In: Opper, M., Saad, D. (eds.) Advanced Mean Field Methods. The MIT Press (2001)
20. Randall Wilson, D., Martinez, T.R.: Improved heterogeneous distance functions. Journal of Artificial Intelligence Research 6, 1–34 (1997)

NudgeAlong: A Case Based Approach to Changing User Behaviour

Eoghan O'Shea[1], Sarah Jane Delany[1], Rob Lane[2], and Brian Mac Namee[1]

[1] Applied Intelligence Research Centre,
Dublin Institute of Technology (DIT), Ireland
[2] Climote, Ireland
{eoghan.oshea,brian.macnamee,sarahjane.delany}@dit.ie, rlane@climote.ie

Abstract. Companies want to change the way that users interact with their services. One of the main ways to do this is through messaging. It is well known that different users are likely to respond to different types of messages. Targeting the right message type at the right user is key to achieving successful behaviour change. This paper frames this as a case based reasoning problem. The case representation captures a summary of a user's interactions with a company's services over time. The case solution represents a message type that resulted in a desired change in the user's behaviour. This paper describes this framework, how it has been tested using simulation and a short description of a test deployment.

Keywords: Case Based Reasoning, Recommender Systems, Simulation.

1 Introduction

In almost all industries businesses make attempts to change the behaviours of their customers. In some cases this is driven by a selfish motivation on behalf of the business - for example, encouraging customers to use a product or service in a particular way so as to earn maximum profit for the business - but it can also be driven by more socially aware motivations - for example, utility companies encouraging customers to use resources more efficiently. Fogg [1] proposes that in order for someone to be successfully convinced to change their behaviour to a new target behaviour that person must (1) have the ability to perform the target behaviour, (2) be sufficiently motivated, and (3) be triggered to perform the new behaviour. Fogg [2] uses the term *captology* to refer to the use of technology to persuade people to change their behaviours.

Many digital consumer scenarios (such as home utilities and the use of online services) are particularly attractive for the application of captology for two reasons. Firstly, at almost all times users have the ability to perform a target behaviour - for example a user can almost always visit a website or interact with their home utility service. Secondly, the current behaviour of users can be constantly and accurately monitored. This monitoring can offer insights into

L. Lamontagne and E. Plaza (Eds.): ICCBR 2014, LNCS 8765, pp. 345–359, 2014.
© Springer International Publishing Switzerland 2014

users' motivations and how open they are to behavioural change triggers. Furthermore, constantly monitoring user behaviour allows a system to determine when behaviour has changed.

To actually effect changes in behaviour, particularly in digital consumer scenarios in which large numbers of users are involved and personal communication is not possible, mass communication with users via SMS, email or other channels is frequently used. This introduces a tension between motivating users and providing them with the right triggers for behaviour change. Different types of communication are likely to motivate and trigger different users. For example, if the goal is to encourage users of a home heating control system to reduce the amount of heating that they use, some users might respond best to a message that focuses on the environmental impact of using excessive home heating, while others are likely to respond better to a message describing the financial benefits of using less home heating energy. This suggests that a degree of personalisation is required so that each user can be sent personalised messages that are most likely to motivate and trigger them to change their behaviour to a desired target.

In this paper we describe the *NudgeAlong* system which uses techniques from *case based reasoning* (CBR) to automatically send users personalised messages to motivate them to use a product or service in a particular way. To do this the system first builds a profile of each user based on information about their usage of the relevant service or product, as well as any available personal details (for example age or gender). When the system determines that a message should be sent to a user this profile is used to retrieve a set of most similar users from a case base containing details of user profiles and previous messaging attempts. The message type that was most effective at encouraging behaviour change for these retrieved users is then used for the current user. Monitoring the user's subsequent behaviour allows the system to determine whether or not the message effectively encouraged behavioural change and this result can be added to the case base. This is a typical CBR cycle and makes CBR an attractive solution to this problem.

Evaluating systems like NudgeAlong is notoriously difficult as actual user interactions are required to determine whether or not behaviour has changed. To address this issue we use a simulation approach to evaluation. Users of a service are simulated to be more or less responsive to different types of motivating messages, and we measure how effectively NudgeAlong can personalise messages so as to take this into account. While an evaluation based on this type of simulation is not as compelling as a real deployment it does provide useful results that can help in preparing for a real deployment. We also describe an initial pilot deployment that has been undertaken with a partner company in the smart-connnected-home space, Climote (`www.climote.com`).

The remainder of this paper proceeds as follows: in Section 2 we give a general overview of the approach taken in NudgeAlong; in Section 3 we present an evaluation of NudgeAlong using simulated data and briefly describe a test deployment; in Section 4 we discuss some related work; and, finally, in Section 5 we present our conclusions.

2   The NudgeAlong System

Figure 1 shows the key components of the NudgeAlong system. User data (coming, for example, from a company's monitoring of a user's interaction with their product) is collected to build a profile of each user and these profiles are fed into a Recommendation Engine. The Recommendation Engine, which is built on a case based reasoning system, uses these profiles and past history about the success or failure of attempts to change users' behaviours, to recommend which type of message will be most likely to encourage a particular user to change their behaviour. Message types are broad categories of messages such as financial, environmental, gamification or motivational.

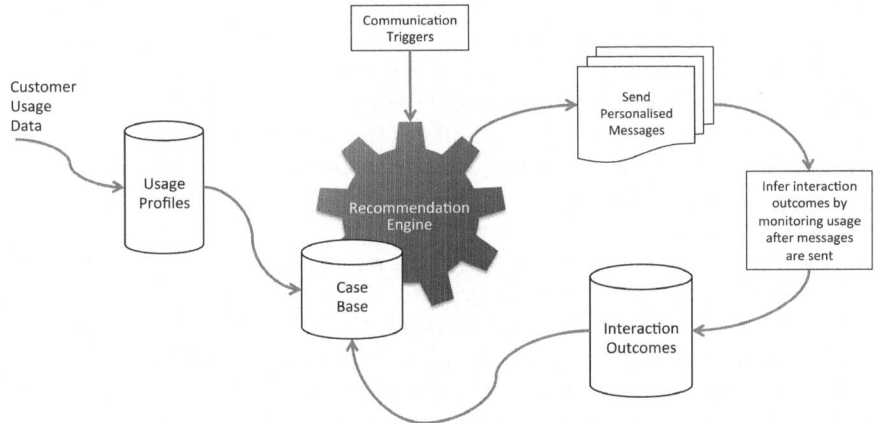

Fig. 1. The key components of the NudgeAlong system

Each case in NudgeAlong represents an instance of a message, of a particular message type, being sent to a specific user and successfully leading to behaviour change. In the current version only details of messages that successfully changed a user's behaviour are stored in the case base. The case representation in NudgeAlong consists of two components: a user profile, containing information on a user's interaction with the system in question and any available personal information, and the type of message sent. The user profile includes features that describe how a user has been interacting with a particular service in the time leading up to the message being sent and varies based on application. Typically the user profile will include features such as how many times a user has interacted with a system in the recent past and how many of the different interaction types that a system allows that a user has recently performed.

For case retrieval the similarity measure we use is a simple Euclidean distance. We choose cases to reuse from the case base by selecting the k nearest neighbours, where, following [3], k is chosen as being the square root of the number of complete cases. Within the k nearest neighbours we use majority voting to choose the most commonly occurring message type. Before being reused, however, we perform a test to see if this message type has recently failed to convince a user to change their behaviour. If this is the case an alternative message type is randomly chosen instead. An alternative would be to use a different message type from the k most similar cases. However, the use of random messages ensures that the system is more dynamic, ensuring users are more likely to receive a variety of message types, particularly in the early stages, so leading more quickly to a situation where users are receiving the most appropriate message type.

The case based system just described determines what type of message to send to a user when the NudgeAlong system has determined that a message should be sent. The triggers shown in Figure 1, are responsible for determining when a message should be sent to a user. These triggers are simple rules that fire if a user's behaviour has diverged sufficiently from expected target behaviour. For example, a message might be sent to a user if they are not using a service a sufficient number of times per week, or if the distribution of a user's use of the different services offered by a company is not optimal. The triggers also include a restriction that imposes an upper limit on the number of messages that users can be sent per week so as to avoid users being bombarded with messages. The trigger rules are evaluated on a regular basis and users who should be sent a message are selected.

Once a user has been deemed to require a message the Recommendation Engine is used to select the message type most likely to encourage that user's behaviour to change. After this message has been sent, the user's subsequent behaviour is monitored over a period of time to determine whether their behaviour has changed or not. The definition of message success or failure is dependent on the targeted behavioural change and is therefore application dependent. Typically, however, a successful behavioural change is signified by a user changing the frequency of their interactions with a particular aspect of a service within a given time window.

3   Evaluation

Evaluating systems like NudgeAlong is notoriously difficult as real user interactions are required to determine whether or not behaviour has changed and, if so, to what extent. To address this issue we primarily use a simulation approach to evaluation. In this section we first outline the simulation framework that we have designed to perform these evaluations and then present our evaluation results. Finally, we provide a short description of a small, live deployment experiment that we have performed with Climote.

3.1   Simulation Framework

Our simulation framework generates daily simulated interactions with a service for a group of simulated users. We assume that any service that NudgeAlong will work with will offer a number of different interaction types that users can potentially perform each day. We simulate a number of different user segments, each of which is characterised by varying likelihoods of performing each of the available interactions each day. These likelihoods are defined by normal distributions of different means and standard deviations. To simulate a user's behaviour we draw a random number of interactions from the appropriate distribution each day for each user. These interactions are then summarised to populate the user profiles, that in turn populate the case base. The number of user segments simulated and the likelihood of users in each segment performing each of the possible interactions each day are set as part of the design of the simulation and vary for different application scenarios.

For each user segment we also predefine the likelihood that a particular message type will result in a user changing their behaviour. Each message type defined in NudgeAlong is assigned a different probability of producing a behavioural change for users in each user segment. When a user is sent a message in the simulation we randomly determine whether or not that message will result in their behaviour changing based on these probabilities. To simulate that a user's behaviour has changed we simply adjust the distributions that determine the likelihoods of a user performing different interactions.

In order to make the simulation more realistic, we also simulate a degree of *satiation*, an important issue in captology research. Satiation refers to the tendency of some users to revert back to their original behaviours after their behaviour has been successfully changed to a new target. In our simulation, we examine two different scenarios: (1) a situation where *every* user whose behaviour has been successfully changed is eventually assumed to revert back to their pre-existing behaviour; and (2) a situation where each user is allocated a satiation probability and only those users with a probability greater or equal to 50% revert back to their pre-existing behaviour. To simulate satiation, a user that is assumed to have gone back to their old behaviour has their interaction likelihoods reset to their original distributions after a pre-set period of time.

In the next section we will illustrate how this simulation framework has been used to evaluate the NudgeAlong system.

3.2   Simulation Evaluation

In order to test the viability of NudgeAlong, we evaluated it using the simulation framework described in the previous section. The simulation was modelled using data collected over a one year period from the remote-controlled home-heating system developed by Climote[1].

Climote allows a user to control their home heating system at a distance, either online, through a smartphone app, or by SMS messaging. In our simulation we allow users 5 separate interactions with the Climote system: *Program* (programming the heating to come on at a particular time), *Boost* (a short-term ad-hoc request for heating, e.g. for 30 mins, 1 hour at a time), *Schedule* (a more

---

[1] http://www.climote.com/

comprehensive, overarching heating schedule covering larger periods such as Spring, Summer, Autumn or Winter), *Temp* (an adjustment of the thermostat) and *Hold* (a suspension of heating for a set period). This is a slight simplification of the interactions that Climote actually offers users.

Corresponding to their use of these 5 interactions, each user has a usage profile built for them that forms the basis of the NudgeAlong case base. This profile stores the number of times a user has used each of the 5 different interactions over the last 7 days as well as the average number of times per week that the user has used each of these interactions since the simulation began. We experimented with a number of other metrics in the profile, e.g., average weekly values, weekend averages, etc. but it was found that this small set of features was sufficient in order to differentiate users in different segments.

The behaviour change sought in this example is to increase users' overall usage of the Climote system and so the total number of interactions performed by each user per week is used as the basis for the message trigger in the simulation. This, relatively simple, behaviour change was of interest to Climote as user engagement with the Climote system is one of their key metrics. If the total number of interactions that a user performs decreases or remains static, then a message is sent out to *nudge* a user back to the desired behaviour. In this simulation, we define a sent message as being successful if it is followed by an increase in a user's total number of interactions with the system that is maintained for a period of time.

In this evaluation we simulated 100 users. Each of the 100 simulated users were assigned to one of 5 different pre-defined user segments (20 users were assigned to each segment). To define these segments we performed a k-means clustering of user profiles generated from real Climote usage data that covered a period of one year. This segmentation revealed 5 distinct segments: *Power Boosters* (PB), *Power Programmers* (PP), *Casual Boosters* (CB), *Casual Programmers* (CP) and *Others*. Power Boosters are considered to use a larger number of Boosts compared to Program interactions, while Power Programmers are considered to use a greater number of Program interactions compared to Boost interactions. In both cases, the user's use of either Program or Boost far outstrips their use of any other command type and overall system usage is high. Casual Boosters and Casual Programmers are considered to use more Boosts and Programs, respectively, than other interactions, but the difference is not as stark as for the Power Boosters/Programmers and overall usage of the system is lower. The Others segment is defined as those users that do not fall neatly into the other 4 segments, e.g. those that use more Schedule and Hold interactions compared to the other types.

Depending on which of the 5 segments a user is in, values, corresponding to command use in our simulated system, were randomly sampled from normal distributions with means corresponding to that segment. For example, for the Power Programmer segment, values were sampled from 5 normal distributions, each with a different mean value, representing Boost, Program, Schedule, Temp and Hold command usage, respectively. For the Power Programmer segment, we gave a higher value mean to the normal distribution representing the values for the Program command, indicating the preference for users in this segment for this command type. In the same vein, for the Power Booster segment, we gave a higher value mean to the normal distribution representing the values for the Boost command. The distributions for the other segments are determined in a similar way.

For our simulation we chose four different message types: Email1, Email2, Text1 and Text2, corresponding to two different types of email messages and two different types of SMS messages. For example, the Email1 message type could be financial messages, giving details of offers, savings, price reductions, etc., while the Email2 message type could be motivational messages such as "Use Climote more!", etc. We note that in this simulation, by using the names Email1, Email2, etc., we are simulating not only the message-type, i.e., financial-type, etc. but also the effectiveness of using a different medium to deliver that message, i.e., email or SMS messaging. This is intentional.

In the simulation, each message type has a probability of success allocated to it for each user segment. These probabilities were not based on the actual Climote data, but were chosen to ensure that a higher value (above 50%) was allocated to a preferred message type for each segment. For example, we choose that for Power Programmers the preferred message type Email2 will have a 57.5% probability of causing a change in behaviour, with Email1, Text1 and Text2 having lesser probabilities of causing a change at 25%, 25% and 10%, respectively. Table 1 shows the probabilities chosen for each segment.

Table 1. Probabilities that a message type will succeed in producing a behavioural change

Segment	Email1	Email2	Text1	Text2
PP	25%	57.5%	25%	10%
PB	25%	15%	20%	60%
CP	70%	10%	12.5%	25%
CB	67%	22%	25%	20%
Other	22%	6%	60%	20%

As mentioned in Section 2, once a message is sent out, the system waits for a specified amount of time before determining whether that message was a success or a failure; in this case has the total number of Climote interactions increased. In this simulation we expect behaviour change to happen at most within 7 days of a message being sent. Within the simulation we also ensure that users will at most receive a message once every 7 days to avoid overloading users. As described in Section 3.1, the simulation is initially set up so that every simulated user experiences satiation (scenario 1). Each user is randomly assigned a satiation value in the range of 7–14 days (following a uniform distribution).

In order to test the simulation, it was set to run through 6 months' worth of simulated user interactions with the Climote home-heating system, one day at a time. The first 7 days of the simulation are considered a bootstrapping phase during which the user profiles are populated and no messages are actually sent out. On the 7th day, a random message was sent to all users in order to initially *seed* the simulation. These initial message types are chosen randomly in the expectation that some of the messages will cause a positive behaviour change. If a message type produces a successful behaviour change, a case is generated for it combining the user profile at the time the message was sent and the successful message type. The case is then added to the case base. Subsequently, as the simulation moves on in time, and users are found to be not behaving as desired, the most appropriate message type for each user in each segment can then be selected using the case base. Eventually the system will begin to choose the *correct* or optimum message types for users in each segment. From Table 1, the optimum message types for each segment are those with the highest probabilities.

As well as simulating the behaviour of users who received messages from NudgeAlong, we also included a control group of users who received randomly selected messages. Whenever the trigger fired, indicating that a user required a message, they were sent a message randomly selected from the 4 message types (following a uniform distribution).

In the top panel of Figure 2, we show the overall result of the simulation, in terms of the average number of interactions all users have with the system over time. There are three results shown here. The solid line shows the interactions performed by users receiving recommended messages from NudgeAlong, the dashed line shows the interactions performed by users receiving random messages (the control group), while the dotted line shows the results for a situation where no user received a message over the course of the simulation.

If we consider the recommended messages first, it can be seen in this plot that for the first 7 days the average number of interactions of all users is about 12. Following the messages that are sent to every user on the 7th day, however, this rises dramatically to approximately 19 (as all users are spurred to act) before decreasing again as satiation takes hold. The system eventually settles down to a steady state that varies between about 16 and 17.5. If we compare this with the initial value of approximately 12, this is a noticeable increase in use of the system. It is also noticeable that there is an increasing trend with time in the average number of interactions. That is, as NudgeAlong is running, it is becoming more efficient at nudging users to a higher usage, as expected (as it is more consistently picking the correct message type for them).

If we compare these results to the control condition (dashed line), we can see some similarities and some important differences. For the control condition, we again see a large increase following the initial sending of messages to users, followed by a decrease. However, here the system settles down to a steady state at a significantly lower level than for the recommended message example, occupying a range of between about 14 and 16. Also no increasing trend over time is evident. We find that, between the 1st November 2013 and the end of the simulation, the improvement in the average number of interactions as a result of using recommended messages as opposed to random messages is of the order of 10%.

If we look at the situation where no user receives a message (the dotted line), we can see that this benchmark level remains fixed at ~12 for the duration of the simulation, as expected.

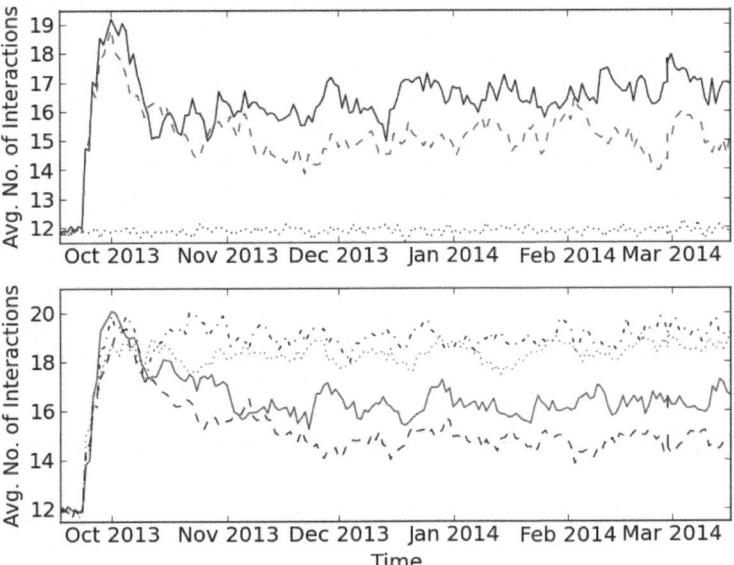

Fig. 2. (upper panel) Variation of the average number of interactions users have with the simulated home heating system. Results from recommended messages are shown as the solid line; results from random messages are shown as the dashed line; results from the situation where no user received a message are shown as the dotted line. (lower panel) The same results for a situation where only users with a satiation probability of greater than or equal to 50% experience satiation (see text for details).

As mentioned Section 3.1, we examined two satiation scenarios in the simulation. In the previous example, we examined satiation scenario 1. For comparison, we will now examine scenario 2, i.e., a situation where only some percentage of users, those that equal or exceed a probability threshold, experience satiation within the 7–14 day range. To achieve this, each user is randomly assigned a probability (from a uniform distribution) of between 1 and 100%. This modification of the simulation is set up so that only those users who equal or exceed a probability of 50% will experience satiation. To ensure a situation where users with low initial probabilities also eventually experience satiation, the probability of satiation is increased by a value of 0.5 each day for all users with an initial satiation probability <50% (once satiation occurs, the 50% threshold passed, these probabilities are reset to their default values). This seems realistic as, as more and more time goes by, it is probable that users will forget about the recommended message they received some time (up to 14 weeks) previously and revert back to a pre-existing behaviour. We show the results of this modification in Figure 2 (lower panel). From this figure, the solid line shows the results for recommended messages from NudgeAlong, while the dashed line shows the results for random messages (the control). Again, we find that, between the 1st November 2013 and the end of the simulation, the improvement in the average number of interactions, as a result of using recommended messages as opposed to random messages, is of the order of 10%.

Alternatively, if we look at a situation where strictly *only* those users who equal or exceed a satiation probability of 50% experience satiation, we get the results shown in Figure 2 (lower panel) by the dot-dash and dotted lines for recommended and random messages, respectively. Because approximately half of users do not now ever experience satiation the average number of interactions has increased. As all message types have a non-zero probability of changing user behaviour the case base can initially, following the initial "seeding", contain a majority of non-optimum message types for each segment. As users who do not experience satiation will not receive another message after they change behaviour (they will consistently behave thereafter), they will not contribute further to the development of the case base, and non-optimum messages can, as a result, remain a majority in the case base for each segment. The result of this is that message types recommended for a user may not be the optimum message type for any given segment. This potentially means that the difference in the level of interactions between using random messages and recommended messages is reduced, i.e., recommended messages may be no different or even worse than random messages in producing a positive behaviour change (a higher level of interactions). This is, in fact, what we observe, with the improvement between the results from the recommended and random messages being reduced to $\sim 3\%$ in this scenario. This result is perhaps not too surprising as, if users do not interact with the system, it cannot "learn" (expand the case base) and so improve its recommendations to users (select the optimum message type).

In Figure 3, we show the continuous variation over time of message types allocated to users in each of the 5 user segments. We note that the results in this figure are from our original set-up, scenario 1, i.e., a situation where *all* users are assumed to experience satiation after a period of between 7–14 days. As an example, we will concentrate on the first segment at the top of this figure, which shows the results for the Casual Programmers (CP) segment. In the first 5 weeks of the simulation, it can be seen that the system is preferentially allocating the Text1 message type to users and not the Email1 message type, which according to Table 1 has the highest probability of success. After 5 weeks, however, the system determines which message type is most appropriate for each user, i.e., Email1, and from then on this is the message type that is preferentially allocated to each user in this segment.

The relative success of the Text1 message type at the beginning of the simulation can be explained. At the beginning, it can be seen that 6 Text1 message

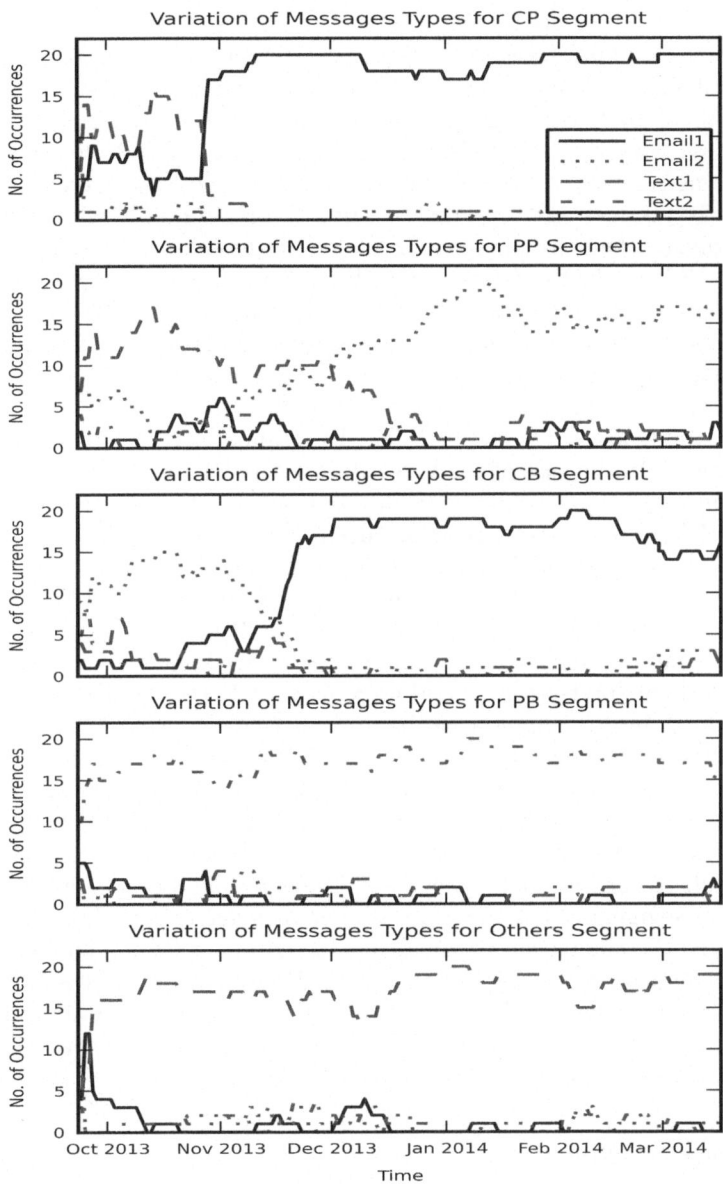

Fig. 3. Variation of the message types allocated for the different segments as a function of time. The different segments are, from top to bottom, Casual Programmer (CP), Power Programmer (PP), Casual Booster (CB), Power Booster (PB) and Others.

types (out of a possible 20) are randomly allocated to users, compared to 3 Email1 message types. After waiting the specified number of days, the system checks the results and finds that 2 out of the 6 Text1 message types are successful in causing an increase in the use of interactions in our simulated home heating systems, while only 1 out of the 3 Email1 message types is successful. In comparison, all 10 Text2 message types allocated at the beginning are found to fail. The case base, as a result, begins with a majority of cases corresponding to the Text1 message type (2 Text1 cases compared to 1 Email1 case) and this is consequently the message type that is initially preferentially allocated to users of this segment when their behaviour triggers a message to be sent. Over time, and by the 5th week, the number of failures from the Text1 message type has reached a point where the number of Email1 cases in the case base surpasses it. At this point, due to the higher probability of success of the Email1 message type, it soon dominates and, indeed, soon, at the end of April and the start of May, accounts for all possible message types allocated.

Similar results are found for the Power Programmer (PP) and Casual Booster (CB) segments. In the case of the Power Booster (PB) and Others segments, the "correct" message type is chosen essentially from the start of the simulation. By the start of May it can be seen that all segments are being allocated message types consistent with the "preferred" message type for their segment, i.e., the message types with the highest probabilities of success from Table 1. For example, Text2 for the PB segment, Email1 for the CB segment, etc. This remains the case for the remainder of the simulation, up to August. While users in the different segments preferentially and more frequently receive the particular message type that has the highest probability of success, they do not exclusively receive only those message types. As mentioned in Sect 2, a test is performed to see if a particular message type has recently failed to convince a user to change their behaviour. The test we use is that if a particular message type fails three times, out of the last five messages sent, then a new random message type is sent instead. As no message type has a 100% probability of success, users in the different segments, therefore, also receive message types other than the preferred one as the simulation progresses. This accounts for the small number of alternative message types, e.g. for the Power Booster segment these are Email1, Email2, Text1, present at all times up to the end of the simulation in August. It is clear, however, that the system is correctly finding the "correct" message types for each segment, as it was designed to do.

3.3 Real Deployment

Following the positive results of this simulation of the NudgeAlong system, we carried out an initial trial deployment of the system using actual users from Climote. The trial, using 9 users of the system, was carried out between the 15th February and the 28th March 2014. Messages to users were sent out using SMS, with 4 different message types (motivational, gamification (comparing user's usage to encourage competition), financial and environmental). With the proviso that this was a very small trial, and was not carried out as a blind experiment, it was found that within the first four weeks (a period within which messages were being sent out at least every 4 days) total system usage increased from on average approximately 7 daily interactions to approximately 17 interactions. While this was a very small and limited trial, carried out primarily to test the system "in the field", the initial results are promising and further trials are planned.

4   Related Work

The NudgeAlong system is at heart a recommender system. Recommender systems have been used successfully for many years in e-commerce applications to encourage users to buy more - more books, movies, trips, etc. Content-based recommenders [4] use the similarity between the items purchased or to be purchased whereas collaborative filtering recommenders [5] make recommendations based on similarities across users' opinions and preferences. Hybrid systems adopt elements of both approaches. Trust-based recommendations [6] use social network structures of their users to enhance and improve the quality of the recommendations made.

The NudgeAlong system differs from the more traditional types of recommenders as it uses user interaction data to drive the recommendations. The more typical use of user interactions in recommender systems has been in deriving ratings from implicit feedback for collaborative filtering systems, where the user preferences are modelled based on user interactions [7,8]. However there has been some work where user interactions are used directly to derive recommendations.

Trace-based reasoning [9] uses user traces, a sequence of events that occur as part of a user activity, as knowledge sources. Trace-based recommendation based on a trace-based reasoning approach has been used for contextual recommendation [10] in video recommendation [11] and in the acceleration of annotation and minimisation of user error in the user annotation process of digitised cultural heritage documents [12].

Esslimani et al. [13] proposed a variation to the standard collaborative filtering approach to recommendation which used usage traces, in this case users' interactions on a website including the pages they accessed and the times they were accessed. Their Behaviour Network based Collaborative Filtering approach to recommendation used these usage traces to determine behavioural similarities between users in order to recommend resources and other webpages on the site.

Related work by Liao et al. [14,15] predicts users' preferences by exploring their usage traces of apps on mobile devices. The usage trace in this work records the number of uses of a particular app.

The aim of the NudgeAlong system is to provide recommendations which will encourage change in user behaviour. There is little evidence of other recommender systems which attempt to do this but work by Nepal et al. [16] also focuses on recommendations to change user behaviour. They built a social recommender for an online community to deliver government services to citizens. Certain individuals in this community were moving from one payment to another which would potentially result in receiving less money and required them to go back into the workplace. The objective of the recommender system was to recommend people and activities to encourage behaviour change, i.e., to help those in the midst of such transitions. The recommender system captured the social behaviour of the individuals through their interactions with the online community, including viewing and commenting on forum posts and generating, accepting and declining friend invitations.

5   Conclusions

In this paper we presented the NudgeAlong system for encouraging behaviour change in users of digital consumer applications. The heart of this system is a case-based recommendation engine which uses similarities between user interaction behaviour to identify the appropriate types of messages to be sent to users to encourage behaviour change. The system incorporates a simulation framework to allow evaluation and we have presented a simulation evaluation of this system which is modelled closely on real user data from the Climote remote heating control system. Running a simulation, based on 100 users, and using four different message types (Email1, Email2, Text1, Text2), we show that the system can successfully find the most appropriate message type for each segment after a period of 1-2 months. We note that this system has also undergone preliminary trials using real Climote users and the initial results of this field testing have been found to be promising.

In the future we plan to perform significant evaluation deployments of the system with Climote and with other partner companies in different industries. There are significant improvements that could be made to the system. Firstly, the current case structure is relatively limited and we will work on integrating more extensive user profiles. This relates to an overall question regarding NudgeAlong: is there sufficient information in the user profiles to actually determine the likely success or failure of different message types? Another option would be to expand the case representation to include things like the desired behaviour change required for each user. The concept of a single message type being most successful for a particular user may be overly simplistic. Future work would involve an investigation into whether a mix of message types would be more successful. We also plan to modify the reuse aspect of our system to further take into account message failures, to build on the relatively simple test mentioned in Section 3, and fully exploit information regarding messages that did not result in behaviour change. Finally, we plan to extend the use of case-based approaches to the triggers as well as the selection of the message type.

References

1. Fogg, B.: A behavior model for persuasive design. In: Proceedings of the 4th International Conference on Persuasive Technology, p. 40. ACM (2009)
2. Fogg, B.: Captology: The study of computers as persuasive technologies. In: CHI 1997 Extended Abstracts on Human Factors in Computing Systems, CHI EA 1997, p. 129. ACM, New York (1997)

3. Jonsson, P., Wohlin, C.: An evaluation of k-nearest neighbour imputation using likert data. In: Proceedings of the 10th International Symposium on Software Metrics,
   METRICS 2004, pp. 108–118. IEEE Computer Society, Washington, DC (2004)
4. Pazzani, M.J., Billsus, D.: Content-based recommendation systems. In:
   Brusilovsky, P., Kobsa, A., Nejdl, W. (eds.) Adaptive Web 2007. LNCS, vol. 4321,
   pp. 325–341. Springer, Heidelberg (2007)
5. Su, X., Khoshgoftaar, T.M.: A survey of collaborative filtering techniques. Adv.
   Artificial Intellegence 2009 (2009)
6. Victor, P., Cornelis, C., De Cock, M., Teredesai, A.: Trust- and distrust-based
   recommendations for controversial reviews. IEEE Intelligent Systems 26(1), 48–55
   (2011)
7. Jawaheer, G., Szomszor, M., Kostkova, P.: Characterisation of explicit feedback in
   an online music recommendation service. In: Amatriain, X., Torrens, M., Resnick,
   P., Zanker, M. (eds.) RecSys, pp. 317–320. ACM (2010)
8. Palanivel, K., Sivakumar, R.: A study on collaborative recommender system using
   fuzzy-multicriteria approaches. IJBIS 7(4), 419–439 (2011)
9. Cordier, A., Lefevre, M., Champin, P.-A., Georgeon, O.L., Mille, A.: Trace-based
   reasoning - modeling interaction traces for reasoning on experiences. In: Boonthum-
   Denecke, C., Youngblood, G.M. (eds.) FLAIRS Conference. AAAI Press (2013)
10. Adomavicius, G., Mobasher, B., Ricci, F., Tuzhilin, A.: Context-aware recom-
    mender systems. AI Magazine 32(3), 67–80 (2011)
11. Zarka, R., Cordier, A., Egyed-Zsigmond, E., Mille, A.: Contextual trace-based
    video recommendations. In: Mille, A., Gandon, F.L., Misselis, J., Rabinovich, M.,
    Staab, S. (eds.) WWW (Companion Volume), pp. 751–754. ACM (2012)
12. Doumat, R., Egyed-Zsigmond, E., Pinon, J.-M.: User trace-based recommendation
    system for a digital archive. In: Bichindaritz, I., Montani, S. (eds.) ICCBR 2010.
    LNCS, vol. 6176, pp. 360–374. Springer, Heidelberg (2010)
13. Esslimani, I., Brun, A., Boyer, A.: From social networks to behavioral networks in
    recommender systems. In: Memon, N., Alhajj, R. (eds.) ASONAM, pp. 143–148.
    IEEE Computer Society (2009)
14. Liao, Z.-X., Pan, Y.-C., Peng, W.-C., Lei, P.-R.: On mining mobile apps usage
    behavior for predicting apps usage in smartphones. In: He, Q., Iyengar, A., Nejdl,
    W., Pei, J., Rastogi, R. (eds.) CIKM, pp. 609–618. ACM (2013)
15. Liao, Z.-X., Peng, W.-C., Yu, P.S.: Mining usage traces of mobile apps for dynamic
    preference prediction. In: Pei, J., Tseng, V.S., Cao, L., Motoda, H., Xu, G. (eds.)
    PAKDD 2013, Part I. LNCS, vol. 7818, pp. 339–353. Springer, Heidelberg (2013)
16. Nepal, S., Paris, C., Bista, S.K.: Srec: a social behaviour based recommender for on-
    line communities. In: Herder, E., Yacef, K., Chen, L., Weibelzahl, S. (eds.) UMAP
    Workshops. CEUR Workshop Proceedings, vol. 872. CEUR-WS.org (2012)

Explaining Probabilistic Fault Diagnosis and Classification Using Case-Based Reasoning

Tomas Olsson[1,2], Daniel Gillblad[2], Peter Funk[1], and Ning Xiong[1]

[1] School of Innovation, Design, and Engineering, Mälardalen University,
Västerås, Sweden
{tomas.olsson,peter.funk,ning.xiong}@mdh.se
[2] SICS Swedish ICT,
Isafjordsgatan 22, Box 1263, SE-164 29 Kista, Sweden
{tomas.olsson,daniel.gillblad}@sics.se

Abstract. This paper describes a generic framework for explaining the prediction of a probabilistic classifier using preceding cases. Within the framework, we derive similarity metrics that relate the similarity between two cases to a probability model and propose a novel case-based approach to justifying a classification using the local accuracy of the most similar cases as a confidence measure. As a basis for deriving similarity metrics, we define similarity in terms of *the principle of interchangeability* that two cases are considered similar or identical if two probability distributions, derived from excluding either one or the other case in the case base, are identical. Thereafter, we evaluate the proposed approach for explaining the probabilistic classification of faults by logistic regression. We show that with the proposed approach, it is possible to find cases for which the used classifier accuracy is very low and uncertain, even though the predicted class has high probability.

Keywords: Case-based Explanation, Machine Learning, Classification.

1 Introduction

Several papers from the last decades identify an intelligent system's ability to explain its predictions as a key factor for user acceptance [1–5]. Hence, a decision support system is less likely to be accepted if a user does not understand or trust its predictions or recommendations. For instance, in the medical domain, the physicians will not trust a system only because of good prediction performance but only if they understand the reasoning behind [6].

In a previous paper, we have proposed using case-based reasoning (CBR) as an intuitive approach for justifying (explaining) the predictions of a probabilistic model [7]. The idea is to support a non-expert user in assessing the system reliability by querying a CBR system for justifying explanations in form of a list of relevant, preceding cases, together with some sort of summary. While our previous work addressed the problem of explaining regression, that is, numerical predictions, this paper extends this approach to probabilistic classification, and specifically, for explaining fault diagnosis.

L. Lamontagne and E. Plaza (Eds.): ICCBR 2014, LNCS 8765, pp. 360–374, 2014.
© Springer International Publishing Switzerland 2014

Fault diagnosis is about detecting when a fault occurs, its location and thereafter identifying the fault type and severity [8, 9]. Both CBR and model-based machine learning approaches have been applied to fault diagnosis [10–12]. Yet, traditionally, CBR is not used when there is a sufficiently good model-based solution to a problem. Still, CBR is conceptually simpler and arguably more intuitive than many model-based approaches, and thus, a case-based explanation facility can make the classification of faults more understandable.

This work is inline with previous work in CBR that uses cases to explain model-based machine learning algorithms [13, 14]. The problem of explaining model-based algorithms with cases is twofold. First, in order to relate cases to the learned model, a similarity metric that measures the usefulness of a case relative to the model is needed. Since in many cases the model was not defined with this in mind, this is not a straightforward problem to solve. Second, a method for explaining the prediction based on the cases must be developed. Considering that the prediction is done with the learned model, it is also reasonable that the explanation is closely related to the model. This work makes therefore two contributions in order to solve this problem.

The first contribution of this paper is to use the generic, theoretically well-defined approach to define similarity metrics presented in [7] to probabilistic classification. As a basis for the definition of similarity, we have formulated *the principle of interchangeability* that two cases are similar or identical if they can replace each other with respect to the probability model and a statistical measure of similarity [7]. In the previous paper, we modeled cases using log-normal linear regression, while in this paper, we use logistic regression.

The second contribution is a novel approach for explaining classification predictions in form of the *local accuracy*. In [7], we used the local mean absolute error to explain regression. The local accuracy is the fraction of the most similar preceding cases that are correctly classified. We interpret the local accuracy as an estimation of how likely it is that the system's prediction is correct. The local accuracy together with a list of preceding cases is then used as a justification of the system performance.

The rest of the paper is organized as follows. Section 2 presents related work. In Sect. 3, we give some background to similarity metrics, statistical metrics and logistic regression. Section 4 presents the overall framework for explanation and four derived similarity metrics. Section 5 describes the application of the proposed approach to explain classification of faults. In Sect. 6, we make concluding remarks and describe future work.

2   Case-Based Explanation

This section presents related work that – similarly to the proposed approach – uses cases for explaining systems. This is a research field called case-based explanation (CBE) [4, 15–17]. CBE can, similarly to CBR, be divided into knowledge intensive and knowledge light CBE where the former makes use of explicit domain knowledge while the latter uses mainly knowledge already contained in the similarity metric and the case base [18]. The current work is an instance of knowledge light CBE with no explicit explanation model.

Furthermore, knowledge light CBE differs in how cases are explained. While our work uses CBR to make explanations of model-based machine learning algorithms, other work uses model-based methods to make explanations of CBR systems. The ProCon system described in [19, 20] uses a naive Bayes classifier trained on all cases to find which features of a case support or oppose a classification. The system presented in [21] by the same author generates rules from the nearest neighbors in order to explain the retrieved cases. Both of these systems investigate and present information to the user on what in the preceding cases support and oppose the classification. This is not considered in our approach.

A second type of research investigates which cases to present to a user as an explanation. The similarity metric that was used for classification might not be the best for explanation. In [22], the authors compare similarity metrics optimized for explanation with those optimized for classification, while in [23], the authors use the same similarity metric used for classification but explore different rules for selecting which case to use as an explanation. In [24], logistic regression is used to find cases, close to the classification border, that are assumed to better explain a classification. In comparison, the current work does not use CBR for classification, but for explanation, and we assume that the similarity metric that is best for explanation is also the best for estimating the local accuracy.

The third type of knowledge light CBE research addresses – similarly to our approach – the explanation of model-based machine learning methods using cases [13, 14, 25–27]. The first knowledge light CBE for model-based learning algorithms was presented in [13]. In this paper, the author sketches ideas on how to use the model of a neural network or a decision tree as a similarity metric. In case of neural networks the activation difference between two cases was proposed as a metric while the leaves in the decision tree naturally contain similar cases. The neural network activation metric resembles our approach in that model parts are compared, but in contrast to our approach, the metric is not theoretically well-defined. A generic CBE framework for black-box machine learning algorithms is presented in [14]. A neural network was locally approximated using a locally weighted linear model based on artificial cases generated from the neural network. Then, the coefficients of the linear model were used both as feature weights of a similarity metric and for identifying the important features of a prediction. In case of our approach, there is no need to approximate the machine learning model, since only probability distributions are compared. In addition, the similarity metrics in our work are theoretically well-founded [7], while there is no theoretical motivation in [14] to why the linear regression weights can be used in a similarity metric.

3   Preliminaries

In this section, we define the notion of a true metric that is important in order to index cases for fast retrieval and we discuss the relation between similarity and a distance. In addition, we present the J-divergence that is a statistical measure of similarity between probability distributions that we use in our definition of similarity between cases. Last, we describe multinomial logistic regression that we use as an example of a probabilistic classifier for classifying faults.

3.1   Similarity and True Metrics

In order to make fast retrieval of cases possible, similarity metrics should adhere to the axioms of a true metric. Given a true metric, the search space can be partitioned into smaller regions and organized so that there is no need to search through all regions. A true metric is a distance or a dissimilarity metric, while in CBR, we typically talk about the similarity between cases. However, similarity and distance are coupled concepts in that a distance can easily be transformed to a similarity or the other way around. So, we will not make a precise distinction between distance metrics and similarity metrics in this work.

In this paper, we use the term metric informally as any function that makes a comparison between two cases, while a true metric is a metric in a mathematical sense. This means that a true metric is a function d that satisfy the following three axioms where X denotes the case base with the set of all cases:

1. $d(\boldsymbol{x}, \boldsymbol{y}) \geq 0$ (non-negative and identity) with $d(\boldsymbol{x}, \boldsymbol{y}) = 0$ if and only if $\boldsymbol{x} = \boldsymbol{y}$, for all $\boldsymbol{x}, \boldsymbol{y} \in X$
2. $d(\boldsymbol{x}, \boldsymbol{y}) = d(\boldsymbol{y}, \boldsymbol{x})$ (symmetric) for all $x, y \in X$
3. $d(\boldsymbol{x}, \boldsymbol{z}) \leq d(\boldsymbol{x}, \boldsymbol{y}) + d(\boldsymbol{y}, \boldsymbol{z})$ (triangle inequality) for all $\boldsymbol{x}, \boldsymbol{y}, \boldsymbol{z} \in X$

There is a discussion in the CBR literature whether all of the above axioms are required for useful similarity and distance metrics [28, 29]. Common true metrics that we will use in this paper is the Manhattan distance (left) and the Euclidean distance (right) shown below:

$$d_m(\boldsymbol{x}, \boldsymbol{y}) = \sum_k |\boldsymbol{x}^k - \boldsymbol{y}^k| \qquad d_e(\boldsymbol{x}, \boldsymbol{y}) = \sqrt{\sum_k |\boldsymbol{x}^k - \boldsymbol{y}^k|^2}$$

where $|\dots|$ denotes the absolute value function and k denotes a case attribute.

3.2   Statistical Metrics

A commonly used statistical metric for comparing two probability distributions is the Kullback-Leibler divergence (KL) [30]. KL is also sometimes called the relative entropy or the information gain, since it is closely related to the entropy concept introduced by Shannon [31, 32].

The KL for the two probability distributions p_i, p_j, with parameter θ, that are two probability density functions for continuous variables and probability mass functions in case of discrete variables:

$$D(p_i \| p_j) = \int \log \left(\frac{p_i(\theta)}{p_j(\theta)} \right) p_i(\theta) d\theta \tag{1}$$

In case of discrete parameters, the integral is replaced with a sum.

KL is not symmetric but it can be made symmetric by computing the KL divergence in both directions and then add them together. This is an important characteristic if we desire a true metric as described in Sect. 3.1. The symmetric KL is often called Jeffreys divergence (J-divergence). The J-divergence will then be:

$$
\begin{aligned}
J(p_i, p_j) &= D(p_i \| p_j) + D(p_j \| p_i) \\
&= \int \log \left(\frac{p_i(\theta)}{p_j(\theta)} \right) p_i(\theta) d\theta + \int \log \left(\frac{p_j(\theta)}{p_i(\theta)} \right) p_j(\theta) d\theta \\
&= \int \log \left(\frac{p_i(\theta)}{p_j(\theta)} \right) \left(p_i(\theta) - p_j(\theta) \right) d\theta
\end{aligned}
\tag{2}
$$

In this paper, we use the J-divergence as basis for the similarity metrics, because it is a commonly used measure and it has a clear information theoretical interpretation. Other statistical metrics for comparing distributions are also available such as the total variation distance, the Euclidean distance and the Jensen-Shannon divergence [30, 33–36]. Later, we will see that there are connections between the J-divergence and other types of distances between probability distributions.

3.3 Logistic Regression

Logistic regression is a binary classifier that can be considered a discrete version of linear regression [37]. Thus, it is a linear classifier that can only separate between classes that are linearly separable. However, since no assumption is made of the distribution of the independent variables it is less restrictive than the related naive Bayes classifier [38]. Assuming a binary classification with $c \in \{0, 1\}$ and feature vector $\boldsymbol{x}$, for logistic regression, we have:

$$
\begin{aligned}
p(c = 1 | \boldsymbol{x}) &= \frac{1}{1 + \exp(\boldsymbol{\omega}^T \boldsymbol{x})} \\
p(c = 0 | \boldsymbol{x}) &= \frac{\exp(\boldsymbol{\omega}^T \boldsymbol{x})}{1 + \exp(\boldsymbol{\omega}^T \boldsymbol{x})}
\end{aligned}
\tag{3}
$$

where $\boldsymbol{\omega}$ is a weight vector with $K+1$ weights assuming that $\boldsymbol{x}$ has $K+1$ features including an extra feature that is 1 for all cases.

Logistic regression can be generalized to the multiclass situation, called multinomial logistic regression, by training one classifier for each class – using one class against all other classes - and then combine the classifiers' predictions. The probability of a class $z \in \{1, 2, \ldots, m\}$ is computed as follows:

$$
p(c = z | \boldsymbol{x}) = \frac{\exp(\boldsymbol{\omega}_z^T \boldsymbol{x})}{\sum_{z'=1}^m \exp(\boldsymbol{\omega}_{z'}^T \boldsymbol{x})}
\tag{4}
$$

where $\boldsymbol{\omega}_z$ is the fitted weight vector for each classifier and m is the number of classes. Then, a new case is classified with the most probable class.

4   The Case-Based Explanation Framework

This section applies the generic case-based explanation framework presented in [7] to classification. The framework justifies the predictions by estimating the system reliability case by case. By only considering probabilistic methods, we can give the framework a good theoretical foundation, while the explanation part can in principle be used for any classifier algorithm. The proposed approach for explaining classification is as follows:

1. *Classify a new case using the probabilistic model*
2. *Retrieve most similar previous cases using the defined similarity metric*
3. *For each previous case, classify the previous case using the probability model*
4. *Compute the local accuracy for the new case as the fraction of correctly classified previous cases*
5. *Present predicted class and the local accuracy together with most similar cases to the user*

Section 5 presents an application of this framework to a real example where we explain the predictions from a logistic regression model for diagnosing faults. However, before that, we will describe a generic approach to defining similarity metrics and derive metrics for comparing cases with respect to probabilistic classification.

4.1   Statistical Measures of Similarity for a Probabilistic Classifier

In this section, we present the principled approach to defining similarity metrics from probability distributions that was introduced in [7], and we apply it to probabilistic classification. We start by defining a basis for comparing two cases from a case base called *the principle of interchangeability*, and then, we derive four possible metrics. The principle of interchangeability is defined as follows:

Definition 1. *Two cases x_i, x_j in case base X are similar if they can be interchanged such that the two probability distributions P_i, P_j inferred from excluding x_i and x_j respectively from the case base – $X \setminus x_i$ and $X \setminus x_j$ – are identical with respect to some parameter(s) of interest.*

As starting point for deriving similarity metrics, we use the discrete version of J-divergence from Sect. 3.2. The J-divergence between two cases $(c_i, x_i), (c_j, x_j)$ in case base X, with respect to the class distributions, is then:

$$d(x_i, x_j) = J(p_i, p_j) = \sum_c \log \frac{p_i(c|x_i)}{p_j(c|x_j)} (p_i(c|x_i)) - p_j(c|x_j))) \tag{5}$$

where c is the class parameter and $p_i(c|x_i)$ and $p_j(c|x_j)$ are probability distributions of the class derived from the case base when excluding the cases x_i and x_j respectively.

The resulting measure between two cases can then be interpreted information theoretically as the sum of the information gain from including one over the other case and the information gain from including the other case over the first case in the case base. However, the resulting J-divergence distance is not a true metric, so an additional step might be needed that turns it into a final distance that fulfills the axioms of a true metric.

The Eq. 5 violates axiom 1 and axiom 3 of a true metric. However, by rewriting the Eq. 5, we can derive a lower and an upper limit that both are easily transformed into true metrics with respect to the class probability space and the class log-probability space respectively as follows

$$J(p_i, p_j) = \sum_c \left| \log(p_i(c|\boldsymbol{x}_i)) - \log(p_j(c|\boldsymbol{x}_j)) \right| \left| p_i(c|\boldsymbol{x}_i) - p_j(c|\boldsymbol{x}_j) \right|$$

Then, since $|x - y| \leq |\log(x) - \log(y)|$ and $\max(|x - y|) = 1$ for all $x, y \in (0, 1]$ we have:

$$\sum_c \left| p_i(c|\boldsymbol{x}_i) - p_j(c|\boldsymbol{x}_j) \right|^2 \leq J(P_i, P_j) \leq \sum_c \left| \log(p_i(c|\boldsymbol{x}_i)) - \log(p_j(c|\boldsymbol{x}_j)) \right| \quad (6)$$

So the J-divergence is greater than or equal to the square of the Euclidean distance in the probability space and lesser than or equal to the Manhattan distance in the log-probability space. Thus, the upper and lower limits results in two more possible distances that we can use.

Notice that the metric in Eq. 5 assumes that the true classes for both cases are known, but in this paper the goal is to predict the class of a new case. So, assuming that c_i is unknown for $\boldsymbol{x}_i$, we cannot estimate p_j since $(c_i, \boldsymbol{x}_i)$ cannot be included in the case base, and thereby, we cannot compute the J-divergence exactly. Yet, if we have a large case base, then p_i and p_j would anyway be approximately equal, and hence, this would not be a problem. So, with a large enough case base, we can approximate $p_j \approx p_i$.

If we do not want to approximate p_i and p_j as equal, we either have to compute the J-divergence analytically, which might not be easy, or estimate the probability distributions for each case in the case base, which might be computationally heavy. Another, more pragmatic approach for managing this problem is to consider that when the class is known for case $\boldsymbol{x}_j$, we can actually model that as $p_j(c_j|\boldsymbol{x}_j) = 1$ and $p_j(c|\boldsymbol{x}_j) = 0$ for all other classes c. But, for a new case $\boldsymbol{x}_i$ with an unknown class, we estimate $p_i(c_i|\boldsymbol{x}_i)$ using a probabilistic model from all known cases. Then, by assuming that $p_i(c|\boldsymbol{x}_i) > 0$ for all classes c, we have the following:

$$J(p_i, p_j) = \sum_c \log \left(\frac{p_i(c|\boldsymbol{x}_i)}{p_j(c|\boldsymbol{x}_j)} \right) (p_i(c|\boldsymbol{x}_i) - p_j(c|\boldsymbol{x}_j))$$

$$= \log \left(\frac{p_i(c_j|\boldsymbol{x}_i)}{1} \right) (p_i(c_j|\boldsymbol{x}_i) - 1) + \sum_{c \neq c_j} \log \left(\frac{p_i(c_j|\boldsymbol{x}_i)}{0} \right) (p_i(c|\boldsymbol{x}_i) - 0)$$

$$= \log \left(p_i(c_j|\boldsymbol{x}_i) \right) (p_i(c_j|\boldsymbol{x}_i) - 1) \ + \ \infty$$

Since we are only interested in comparing distances relatively each other to find the closest cases, we can choose to ignore the infinity part and only compare the first term. Thus, we will get yet an alternative distance between two cases as follows:

$$d'(\boldsymbol{x}_i, \boldsymbol{x}_j) = \log\left(p_i(c_j|\boldsymbol{x}_i)\right)\left(p_i(c_j|\boldsymbol{x}_i) - 1\right) \tag{7}$$

However, this leads to a different definition of similarity, since we include $(c_j, \boldsymbol{x}_j)$ in the case base for estimating both p_i, p_j, and that we use two different distributions conditioned on whether the class is known.

We have now derived four different distances for comparing the similarity between cases relating to our definition of similarity and the J-divergence. The four derived distances are listed in Table 1. From now, we assume a large case base so that the distributions p_i and p_j are approximately equal.

Table 1. The four derived distances and two standard distances

Name	Distance	From Equation
J-divergence	$\sum_c \log \frac{p_i(c|\boldsymbol{x}_i)}{p_j(c|\boldsymbol{x}_j)}\left(p_i(c|\boldsymbol{x}_i) - p_j(c|\boldsymbol{x}_j)\right)$	Eq. 5
Approximate Prob	$\sqrt{\sum_c \left|p_i(c|\boldsymbol{x}_i) - p_j(c|\boldsymbol{x}_j)\right|^2}$	Eq. 6
Approximate Log-Prob	$\sum_c \left|\log(p_i(c|\boldsymbol{x}_i)) - \log(p_j(c|\boldsymbol{x}_j))\right|$	Eq. 6
Pragmatic	$\log\left(p_i(c_j|\boldsymbol{x}_i)\right)\left(p_i(c_j|\boldsymbol{x}_i) - 1\right)$	Eq. 7

5   Explaining Fault Diagnosis

We will in this section apply the proposed approach for explaining fault diagnosis.

In fault diagnosis, faults are classified so that correct actions can be taken in order to minimize the cost of faults. Preferable, faults should be detected and classified before they harm the system. A measure of the confidence of a classification is also desirable so that no unnecessary actions are taken if the classification is wrong. Assuming that the probability model is correct, a high probability means a high confidence, while a low probability means a low confidence in the predicted value. However, if the probability model is locally error-prone, the probabilities cannot be trusted, as for instance, when the linearity assumption of logistic regression does not hold in the whole feature space. As one remedy, we propose to use the local accuracy as a complementing, and more intuitive, confidence measure. The local accuracy is computed locally, directly from the most similar cases, and thereby, it should be less affected by an error-prone probability model. Consequently, a high local accuracy should be able to justify a classification, while a low local accuracy should invalidate a classification. Thus, a user can use this approach to decide whether to trust a prediction case-by-case regardless of the global prediction accuracy of the algorithm.

In the following, given an error-prone logistic regression model (Sect. 5.1), we show that it is possible to train the k-nearest neighbor algorithm (kNN) to estimate the local accuracy (Sect. 5.2). Then, by looking at a case with inconsistent confidence measures – for instance, high probability and low local accuracy – we can detect bad prediction performance that is otherwise overlooked when only considering the class probability (Sect. 5.3). Last, we discuss how to interpret the local accuracy in other cases (Sect. 5.4).

For all experiments, we have used the implementation of logistic regression and kNN provided by the Scikit-learn Python module [39]. As example data set, we use the Steel plate faults data set [40] from the UC Irvine Machine Learning Repository [41]. The data set consists of more than 1900 cases with 7 types of steel plate faults and 27 different dependent variables listed in Table 2.

Table 2. Attributes: 27 independent variables, and the last rows shows the 7 fault classes

X_Minimum	X_Maximum	Y_Minimum	Y_Maximum
Pixels_Areas	X_Perimeter	Y_Perimeter	Sum_of_Luminosity
Min_of_Luminosity	Max_of_Luminosity	Length_of_Conveyer	TypeOfSteel_A300
TypeOfSteel_A400	Steel_Plate_Thickn.	Edges_Index	Empty_Index
Square_Index	Outside_X_Index	Edges_X_Index	Edges_Y_Index
Outside_Global_Index	LogOfAreas	Log_X_Index	Log_Y_Index
Orientation_Index	Luminosity_Index	SigmoidOfAreas	
1. Pastry	3. K_Scatch	5. Dirtiness	7. Other_Faults
2. Z_Scratch	4. Stains,	6. Bumps,	

5.1 Fitting Logistic Regression

In this section, we fit multinomial logistic regression to classify faults. As learning parameter, we use the $l1$-norm and the model regulation parameter is fine-tuned using grid search with 5-fold cross validation. Figure 1 shows the learning curve of the overall accuracy for classifying faults. The learning curve was computed by splitting the data set 10 times into 70% training set and a 30% testing, and the results were then averaged. As can be seen, the training curve and validation curve are converging, but at a low level just above 0.7. Thus, more features or a more complex learning algorithm would be needed to improve on this. Hence, this is an error-prone probability model that we will use as an example to illustrate the proposed approach.

5.2 Estimating Local Accuracy with kNN

After fitting the multinomial logistic regression, we train the kNN algorithm to estimate the local accuracy. So, in this case, instead of predicting the class of a case, we use kNN to estimate a confidence in whether it will be classified correctly. Therefore, the classification label is replaced with 1 if a case in the training set was correctly classified and 0 otherwise. Then, kNN estimates the local accuracy by averaging the ones and zeros from the k-nearest neighbors, and then, the mean squared error (MSE) is used for evaluation:

$$mse(X) = \sum_{x \in X} (C(\boldsymbol{x}) - A_k(\boldsymbol{x}))^2 \text{ with } A_k(\boldsymbol{x}) = \frac{\sum_{i=1}^{k} C(\boldsymbol{x}_i')}{k} \tag{8}$$

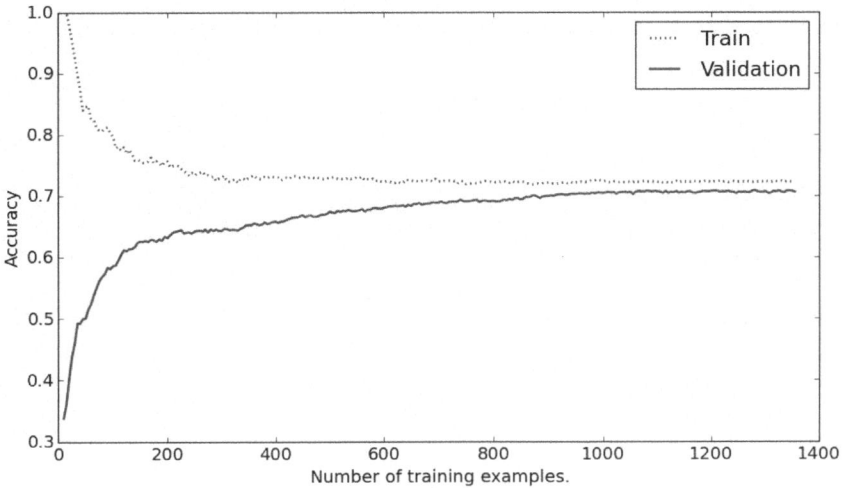

Fig. 1. The accuracy learning curve for classifying steel faults

where X is a set of cases, $A(x) \in [0, 1]$ is the local accuracy for the k most similar cases of x and $C(x) = 1$ and $C(x) = 0$ indicate correct and incorrect classification of x respectively. Then, if $A_k(x) > 0.5$, it is more likely than unlikely that the classification is correct. This also means that the local accuracy can be interpreted as the probability of a correct classification.

For fine-tuning kNN, we split the data set 10 times into 60% training set, 20% validation set, and 20% testing set. Thereafter, we compute the average MSE over the validation and test sets. The distances in Table 1 were used together with the Manhattan and Euclidean distances in combination with the derived distances to weigh in the similarity between the cases directly. Notice that the class probabilities are considered as additional features. For all distances but J-divergence, we normalize so that each feature has a mean of 0 and a standard deviation of 1. The results are shown in Table 3 where Approximate Log-Prob distance has the lowest validation MSE. Figure 2 plots the results for a varying number of k-neighbors.

In Fig 2, we notice that $k = 9$ is the best number of neighbors for Approximate Log-Prob, but we must also consider how convincing it is to support a claim with only 9 neighbors. Thus, since there seems to be no larger differences between the MSE of $k \in [9, 15]$, we can at least use $k = 10$ as a more convincing explanation.

5.3   Case-Based Explanation Examples

Given a fitted logistic regression classifier and a fine-tuned kNN algorithm, we will now demonstrate the approach using two example faults. Table 4 and Table 5 show two examples where the target case is the fault that is being diagnosed.

Table 3. The mean squared error (MSE) for different distances (best is in bold font)

Distance	Validation MSE	Test MSE	k neighbors
Manhattan	0.171	0.174	10
Euclidean	0.173	0.176	9
Pragmatic	0.214	0.215	60
Approximate Prob	0.175	0.173	17
Approximate Log-Prob	**0.166**	**0.166**	**9**
Approximate Log-Prob Euclidean	0.172	0.173	10
Approximate Log-Prob Manhattan	0.168	0.171	9
J-Divergence	0.176	0.173	19

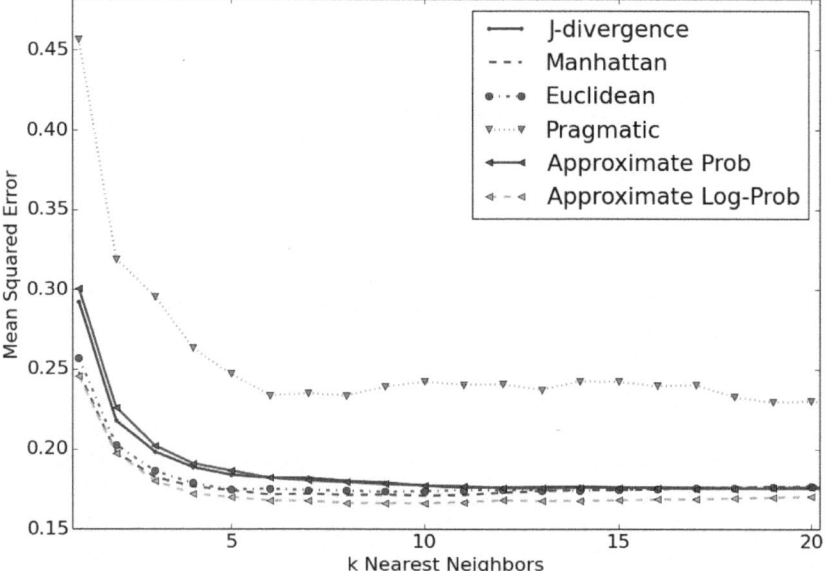

Fig. 2. MSE for the kNN algorithm using different distances and various k

Table 4. Fault 1: Low local accuracy 20% (2 of 10) and low class probability 33.4%

Attribute	Target	Case 1	Case 2	Case 3	Case 4	Case 5
X_Minimum	57.0	127.0	205.0	843.0	23.0	282.0
...						
SigmoidOfAreas	0.1753	0.2253	0.2359	0.215	0.2051	0.1954
True Class	(6)	7	6	7	7	6
Predicted Class	7	7	7	6	3	7
Probability of Class	0.334	0.503	0.5	0.476	0.444	0.364

For both faults, the estimated local accuracy is very low: only 1 or 2 out of 10 of the most similar cases are correctly classified. However, for the first fault, the probability of the predicted class is also low, and thus, it is consistent with the local accuracy. So, we have no reason to believe that the estimated probability is wrong. In contrast, for the second fault, the probability of the predicted class is quite high, despite that the local accuracy is very low. Then, if we also look at the probability of the predicted class for the five most similar cases shown in Table 5, it is quite high for all of them, although the wrong class is predicted in each case. So, we have good reason to doubt the prediction of the classifier, even though the classifier is quite confident in being right.

Table 5. Fault 2: Low local accuracy 10% (1 of 10) and high class probability 73.4%

Attribute	Target	Case 1	Case 2	Case 3	Case 4	Case 5
X_Minimum	57.0	61.0	623.0	10.0	1067.0	1338.0
. . .						
SigmoidOfAreas	0.1753	0.1659	0.2018	0.1753	0.215	0.2195
True Class	(6)	6	1	6	1	6
Predicted Class	7	7	7	7	7	7
Probability of Class	0.734	0.663	0.636	0.826	0.493	0.86

5.4 Analyzing the Local Accuracy

In the previous section, we saw two examples where in the first example the proposed approach justifies the probability model's prediction but invalidates the second example. Thus, we have shown that the proposed approach is able to detect when the probability model does not perform well. However, two more interesting situations to compare are when the prediction has high probability with a high local accuracy and when the prediction has low probability with a high local accuracy. In both cases, the local accuracy indicates that the prediction can be trusted, although the second case has low probability. In the latter case, this mean that the probability model does not make a good probability estimation but that the model anyway makes a good prediction of the most probable class. Clearly, the problem of deciding when to trust a prediction can be viewed as a decision theoretical problem and the solution will likely depend on the application domain, for instance, the cost of a misclassification. However, due to brevity, we will not further deal with the problem of how to decide whether the local accuracy is high or low, or what to do when it is neither high nor low, but leave that to future work.

6 Conclusions and Future Work

In this paper, we have extended the framework for knowledge light case-based explanation proposed in [7] to probabilistic classification, and we have applied it for explaining fault diagnosis. Thus, a major contribution of this work is a principled and theoretically well-founded approach to defining similarity metrics for retrieving cases relative to a probability model for classification.

A second contribution is a novel approach to justifying a prediction by computing the local accuracy as the fraction of the most similar cases that are classified correctly. Since the justification is based on real cases and not merely on the correctness of the probability model, we argue that this is a more intuitive justification of reliability than only considering the estimated probability as a measure of confidence. For instance, as noted in Sect. 5.3, with this approach, the users can easily detect when the prediction uncertainty of the probability model does not agree with the computed local accuracy, and therefore, judge for themselves whether to trust a prediction or not. Since the accuracy is computed locally, the proposed approach addresses the problem that a probability model might not perform consistently over the whole feature space.

An interesting future development – as already noted in previous section – is to further investigate the use of the local accuracy for deciding when to trust or not trust a prediction. In addition, this approach can also be used for selecting which classifier to use for a new case. Another future research direction is to develop CBR applications where the main task is a case-based prediction and not just as a complement to a probabilistic prediction.

Acknowledgments. The authors gratefully acknowledge the funding from the Swedish Knowledge Foundation (KK-stiftelsen) [42] through **ITS-EASY** Research School, the European **ARTEMIS** project ME3Gas (JU Grant Agreement number 100266), the ITEA 2 (ITEA 2 Call 5, 10020) and Swedish Governmental Agency for Innovation Systems (VINNOVA) grant no 10020 and JU grant no 100266 making this research possible.

References

1. Wick, M.R., Thompson, W.B.: Reconstructive expert system explanation. Artificial Intelligence 54(1), 33–70 (1992)
2. Ye, L.R., Johnson, P.E.: The impact of explanation facilities on user acceptance of expert systems advice. Mis Quarterly, 157–172 (1995)
3. Gregor, S., Benbasat, I.: Explanations from intelligent systems: Theoretical foundations and implications for practice. MIS Quarterly, 497–530 (1999)
4. Leake, D., McSherry, D.: Introduction to the special issue on explanation in case-based reasoning. Artificial Intelligence Review 24(2), 103–108 (2005)
5. Darlington, K.: Aspects of intelligent systems explanation. Universal Journal of Control and Automation 1, 40–51 (2013)
6. Langlotz, C.P., Shortliffe, E.H.: Adapting a consultation system to critique user plans. International Journal of Man-Machine Studies 19(5), 479–496 (1983)
7. Olsson, T., Gillblad, D., Funk, P., Xiong, N.: Case-based reasoning for explaining probabilistic machine learning. International Journal of Computer Science & Information Technology (IJCSIT) 6(2) (April 2014)
8. Isermann, R.: Supervision, fault-detection and fault-diagnosis methods–an introduction. Control Engineering Practice 5(5), 639–652 (1997)

9. Jayaswal, P., Wadhwani, A., Mulchandani, K.: Machine fault signature analysis. International Journal of Rotating Machinery (2008)
10. Olsson, E., Funk, P., Xiong, N.: Fault diagnosis in industry using sensor readings and case-based reasoning. Journal of Intelligent and Fuzzy Systems 15(1), 41–46 (2004)
11. Isermann, R.: Fault-diagnosis systems: an introduction from fault detection to fault tolerance. Springer (2006)
12. Olsson, T., Funk, P.: Case-based reasoning combined with statistics for diagnostics and prognosis. Journal of Physics: Conference Series 364(1), 012061 (2012)
13. Caruana, R., Kangarloo, H., Dionisio, J., Sinha, U., Johnson, D.: Case-based explanation of non-case-based learning methods. In: Proceedings of the AMIA Symposium, p. 212. American Medical Informatics Association (1999)
14. Nugent, C., Cunningham, P.: A case-based explanation system for black-box systems. Artificial Intelligence Review 24(2), 163–178 (2005)
15. Schank, R.C., Leake, D.B.: Creativity and learning in a case-based explainer. Artificial Intelligence 40(1), 353–385 (1989)
16. Aamodt, A.: Explanation-driven case-based reasoning. In: Wess, S., Richter, M., Althoff, K.-D. (eds.) EWCBR 1993. LNCS, vol. 837, pp. 274–288. Springer, Heidelberg (1994)
17. Doyle, D., Tsymbal, A., Cunningham, P.: A review of explanation and explanation in case-based reasoning, vol. 3. Dublin, Trinity college (2003), https://www.cs.tcd.ie/publications/tech-reports/reports
18. Cunningham, P., Doyle, D., Loughrey, J.: An evaluation of the usefulness of case-based explanation. In: Ashley, K.D., Bridge, D.G. (eds.) ICCBR 2003. LNCS, vol. 2689, pp. 122–130. Springer, Heidelberg (2003)
19. McSherry, D.: Explanation in case-based reasoning: an evidential approach. In: Proceedings of the 8th UK Workshop on Case-Based Reasoning, pp. 47–55 (2003)
20. McSherry, D.: Explaining the pros and cons of conclusions in CBR. In: Funk, P., González Calero, P.A. (eds.) ECCBR 2004. LNCS (LNAI), vol. 3155, pp. 317–330. Springer, Heidelberg (2004)
21. McSherry, D.: A lazy learning approach to explaining case-based reasoning solutions. In: Agudo, B.D., Watson, I. (eds.) ICCBR 2012. LNCS, vol. 7466, pp. 241–254. Springer, Heidelberg (2012)
22. Doyle, D., Cunningham, P., Bridge, D., Rahman, Y.: Explanation oriented retrieval. In: Funk, P., González Calero, P.A. (eds.) ECCBR 2004. LNCS (LNAI), vol. 3155, pp. 157–168. Springer, Heidelberg (2004)
23. Cummins, L., Bridge, D.: Kleor: A knowledge lite approach to explanation oriented retrieval. Computing and Informatics 25(2-3), 173–193 (2006)
24. Nugent, C.D., Cunningham, P., Doyle, D.: The best way to instil confidence is by being right. In: Muñoz-Ávila, H., Ricci, F. (eds.) ICCBR 2005. LNCS (LNAI), vol. 3620, pp. 368–381. Springer, Heidelberg (2005)
25. Wall, R., Cunningham, P., Walsh, P.: Explaining predictions from a neural network ensemble one at a time. In: Elomaa, T., Mannila, H., Toivonen, H. (eds.) PKDD 2002. LNCS (LNAI), vol. 2431, pp. 449–460. Springer, Heidelberg (2002)
26. Green, M., Ekelund, U., Edenbrandt, L., Björk, J., Hansen, J., Ohlsson, M.: Explaining artificial neural network ensembles: A case study with electrocardiograms from chest pain patients. In: Proceedings of the ICML/UAI/COLT 2008 Workshop on Machine Learning for Health-Care Applications (2008)
27. Green, M., Ekelund, U., Edenbrandt, L., Björk, J., Forberg, J.L., Ohlsson, M.: Exploring new possibilities for case-based explanation of artificial neural network ensembles. Neural Networks 22(1), 75–81 (2009)

28. Burkhard, H.D., Richter, M.M.: On the notion of similarity in case based reasoning and fuzzy theory. In: Soft Computing in Case Based Reasoning, pp. 29–45. Springer (2001)
29. Burkhard, H.D.: Similarity and distance in case based reasoning. Fundamenta Informaticae 47(3), 201–215 (2001)
30. Kullback, S., Leibler, R.A.: On information and sufficiency. The Annals of Mathematical Statistics 22(1), 79–86 (1951)
31. Ihara, S.: Information theory for continuous systems, vol. 2. World Scientific (1993)
32. Shannon, C.E.: A mathematical theory of communication. ACM SIGMOBILE Mobile Computing and Communications Review 5(1), 3–55 (2001)
33. Rachev, S.T., Stoyanov, S.V., Fabozzi, F.J., et al.: A probability metrics approach to financial risk measures. Wiley (2011)
34. Lin, J.: Divergence measures based on the shannon entropy. IEEE Transactions on Information Theory 37(1), 145–151 (1991)
35. Cha, S.H.: Comprehensive survey on distance/similarity measures between probability density functions. City 1(2), 1 (2007)
36. Dragomir, S.C.: Some properties for the exponential of the kullback-leibler divergence. Tamsui Oxford Journal of Mathematical Sciences 24(2), 141–151 (2008)
37. Murphy, K.P.: Machine learning: a probabilistic perspective. MIT Press (2012)
38. Ng, A.Y., Jordan, M.I.: On discriminative vs. generative classifiers: A comparison of logistic regression and naive bayes. In: Advances in Neural Information Processing Systems, vol. 2, pp. 841–848 (2002)
39. Pedregosa, F., Varoquaux, G., Gramfort, A., Michel, V., Thirion, B., Grisel, O., Blondel, M., Prettenhofer, P., Weiss, R., Dubourg, V., Vanderplas, J., Passos, A., Cournapeau, D., Brucher, M., Perrot, M., Duchesnay, E.: Scikit-learn: Machine learning in Python. Journal of Machine Learning Research 12, 2825–2830 (2011)
40. Steel Plates Faults Data Set. Source: Semeion, Research Center of Sciences of Communication, Via Sersale 117, 00128, Rome, Italy, www.semeion.it, https://archive.ics.uci.edu/ml/datasets/Steel+Plates+Faults (last accessed: May 2014)
41. Bache, K., Lichman, M.: UCI machine learning repository (2013)
42. KK-Stiftelse: Swedish Knowledge Foundation, http://www.kks.se (last accessed: September 2013)

Case-Based Prediction
of Teen Driver Behavior and Skill

Santiago Ontañón[1], Yi-Ching Lee[2], Sam Snodgrass[1], Dana Bonfiglio[2],
Flaura K. Winston[2], Catherine McDonald[3], and Avelino J. Gonzalez[4]

[1] Drexel University, Philadelphia, PA, USA
santi@cs.drexel.edu, sps74@drexel.edu
[2] Children's Hospital of Philadelphia (CHOP), Philadelphia, PA, USA
leey1@email.chop.edu, BonfiglioD@email.chop.edu, winston@email.chop.edu
[3] University of Pennsylvania, Philadelphia, PA, USA
mcdonalc@nursing.upenn.edu
[4] University of Central Florida, Orlando, FL, USA
gonzalez@ucf.edu

Abstract. Motor vehicle crashes are the leading cause of death for U.S.
teens, accounting for more than one in three deaths in this age group
and claiming the lives of about eight teenagers a day, according to the
2010 report by the *Center for Disease Control and Prevention*[1]. In order to inform new training methods and new technology to accelerate
learning and reduce teen crash risk, a more complete understanding of
this complex driving behavior was needed. In this application paper we
present our first step towards deploying case-based techniques to model
teenage driver behavior and skill level. Specifically, we present our results
in using case-based reasoning (CBR) to model both the vehicle control
behavior and the skill proficiency of teen drivers by using data collected
in a high-fidelity driving simulator. In particular, we present a new similarity measure to compare behavioral data based on feature selection
methods, which achieved good results in predicting behavior and skill.

Keywords: Driving behavior, similarity assessment, feature selection.

1 Introduction

Motor vehicle crashes are the leading cause of death for U.S. teens, accounting
for more than one in three deaths in this age group and claiming the lives of approximately eight teenagers per day, according to the 2010 report by the *Center
for Disease Control and Prevention*. In addition, teen drivers have four times
the fatal crash risk of adults [2]. Previous research has identified speed management as a major contributing factor for crashes involving newly licensed teen
drivers [14,2]. By comparison, experienced adult drivers are known to be better at adjusting and adapting driving behaviors and speed control when traffic

---

[1] http://www.iihs.org/iihs/topics/t/teenagers/fatalityfacts/teenagers/2010

L. Lamontagne and E. Plaza (Eds.): ICCBR 2014, LNCS 8765, pp. 375–389, 2014.
© Springer International Publishing Switzerland 2014

situations change, suggesting that learning and accumulation of driving experience contribute to declining crash risk with age and experience [7,14]. In order to develop and test new training methods and new technology to reduce teen crash risk, techniques are needed to better understand the complex behaviors associated with this increased risk.

This paper presents the results of a study on the feasibility of employing case-based reasoning for modeling and predicting teen driving behavior. Specifically, we examined two different problems: First, can CBR be used to predict the future behavior of a given driver? This problem addresses the need for in-vehicle monitoring that could predict potentially dangerous driving situations in sufficient time to alert the driver and provide feedback and assistance. Second, can CBR be used to automatically determine the skill level of a given driver purely based on her driving behavior. This second problem further enhances the design of automatic in-vehicle monitoring systems, as well as the design and evaluation of driver training programs, by providing automated ways to measure the effect of different training approaches and programs on teen driving skill.

Specifically, this applied paper presents a case-based reasoning approach to address both of these problems (driver behavior prediction and driver skill prediction) using a dataset collected from a high-fidelity driving simulator at the Children's Hospital of Philadelphia (CHOP) in which the performance of novice teen drivers within the first three months of licensure was recorded for common potential crash scenarios. This work presents two main contributions to case-based reasoning. First, a new potentially high-impact application of case-based reasoning is presented with the goal of improving our understanding of teen driving behavior, with preliminary results that suggest the utility of case-based reasoning for new in-vehicle personalized feedback technologies. Second, motivated by our application domain, we present a new set of similarity measures for behavioral data based on the idea of feature selection (see Section 5).

The remainder of this paper is organized as follows. Section 2 reviews the state of the art on driving behavior modeling. Section 3 presents the dataset used in our study. Sections 4 and 5 present our approaches to address behavior prediction and skill prediction respectively. Finally, we present the conclusions of our study and future lines of work.

2 Modeling Human Driving Behavior

Due to its many practical applications and its importance to road safety, driving behavior modeling has been approached from many perspectives. A significant body of work, illustrated by the work of Macadam [11], exists on the manual creation of control models that exhibit specific aspects of human driving behavior. For an in-depth description of existing techniques, see Markkula et al. [12].

A number of approaches employ machine learning methods to automatically acquire models of human driving behavior [4,19]. Concerning case-based approaches, the most closely related areas are that of case-based learning from demonstration [6,17,22,10] and CBR applied to sequential trace data [1]. These approaches learn behavior by observing an expert's performance, but except for the work of Rubin and Watson [22], focus on small-scale datasets, much smaller than the datasets required to analyze driving behavior.

Specific similarity measures have been proposed for the domain of driving behavior. For example, Nechyba and Xu [15] developed a similarity measure based on Hidden Markov Models (HMM) [20]. The results demonstrated the ability of the models to accurately differentiate driving traces generated from one driver from those generated from other drivers, however, the measures were based on relatively simple driving data (without other traffic). As elaborated below, our driving data are significantly more complex than that used by Nechyba and Xu (our dataset included other traffic).

Finally, previous work has utilized models to automatically predict driver states from driving data. Das et al. [3] showed that statistical measures over steering wheel data such as entropy and the Lyapunov exponent can determine whether a given driver is under the effect of alcohol or not. No previous work has attempted to automatically predict degree of driver skill.

The work presented on this paper closely builds upon work on case-based learning from demonstration, and applies it to the challenging domain of human driving behavior modeling and skill prediction.

3 Dataset

We used an existing dataset used in the development of a *Simulated Driving Assessment*. The three most common crash types and associated driving scenario configurations determined from the National Motor Vehicle Crash Causation Survey (NMVCCS)[2] were programmed for the Simulated Driving Assessment. One of the first objectives of that project was to validate the driving scenarios by differentiating behaviors between novice teen and experienced adult drivers. This existing dataset contains (1) simulator-derived behavioral and (2) eye-tracking data from four simulator drives (a practice drive plus three experimental drives), (3) a driver's education instructor's subjective ratings of driving behaviors (based on video review without simulator or eye tracker data), (4) images from three video cameras, and (5) self-report measures. Here, we briefly describe the apparatus, procedures, and the specific dataset used in this paper.

Apparatus. The simulator (see Figure 1) used was a high fidelity fixed-base simulator located within the Center for Injury Research Prevention at CHOP. The driving simulator consists of a Pontiac G6 driver seat, three-channel 46" LCD panels (160 degree field of view), with rearview, left side, and right side mirror images inlayed into the center, left, and right channels, respectively, active pedals and steering system, and a rich audio environment. The simulator is

---

[2] See: National Highway Traffic Safety Administration. NASS National Motor Vehicle Crash Causation Study: http://www.nhtsa.gov/Data/Special+Crash+Investigations+(SCI)/NASS+National+Motor+Vehicle+Crash+Causation+Study.

Fig. 1. The driving simulator at the Children's Hospital of Philadelphia, where the data was collected, and layout of *drive2*

powered by SimVista, a tile-based scenario authoring software, and the real-time simulation and modeling were controlled by SimCreator, a simulation tool, by Realtime Technologies, Inc.®.

Procedures. Based on the most frequent serious crash categories and configurations in NMVCCS for 16-18 year-old teen drivers driving alone or only with a peer passenger we included the following invents into the three experimental simulator drives (*drive2*, *drive3*, and *drive4*): 1) turning into opposite directions (turning left), 2) right roadside departure, and 3) rear-end events [13]. The locations of the scenarios within the drives were pre-determined and each participant experienced the three types of scenarios multiple times, for a total of 22 variations of scenarios with the potential for simulated crashes or run-off-the-road events. Weather conditions were also varied in that one of the drives contained continuous rain and fog and the other two had the appearance of a partly cloudy day. *Drive1*, a practice drive, allowed the participant to become familiar with driving the simulator and making turns and stopping. After *drive1*, the participants received a randomized order of the remaining three experimental drives. The total length of the study was approximately 35 minutes.

Variables. Raw variables[3] collected by the simulator included those related to participant vehicle's heading, speed, acceleration, deceleration, turning, distance to vehicle in front, location in the simulated route, and instructions provided to the driver (such as instructing her to "take the next left"). A number of derived variables have been computed to inform whether the participant vehicle collided with any other vehicle, if it was in the correct lane, if it came to a complete stop at stop-sign intersections, if pre-determined thresholds for hard braking, hard left turn, and rapid acceleration were met, and if participant was speeding:

---

[3] In this paper we will use the terms *variable* and *feature* interchangeably.

- *3 output variables* (*steer, throttle, brake*): these variables encode the current position of the steering wheel, and the amount the throttle and brake pedals were pressed. And are used to represent the driver's behavior.
- *3 previous output variables*: these variables encode the value of the output variables in the previous time instant.
- *33 input variables*: these variables include all the other raw and derived variables described above.

We used data from 17 novice teen drivers (licensed for less than 90 days). Therefore, for each of those 17 drivers, the dataset had data for the four drives described above. Each drive contained all the 39 variables described above sampled at 60Hz. Each of the drives had the following lengths (in seconds):

- *drive1*: 632.72s to 908.55s (avg. 719.44s, corresponding to 43166.65 samples).
- *drive2*: 538.22s to 859.00s (avg. 630.87s, corresponding to 37851.94 samples).
- *drive3*: 592.58s to 678.57s (avg. 638.29s, corresponding to 38297.12 samples).
- *drive4*: 530.40s to 672.95s (avg. 580.10s, corresponding to 34806.06 samples).

In total, the dataset has 12.13 hours of driving (corresponding to 2.6 million samples, where each sample contains the values for the 39 variables above). All the variables were normalized to be in the interval $[0, 1]$. We only used drives 2 and 3 for our study, since *drive1* was a practice drive, and *drive4* included rain.

In order to capture a rating of driver skill, a certified driver education instructor/evaluator reviewed video recordings of the participant's simulator drives without viewing any simulator or eye tracker data. He was instructed to rate on a scale of 1-10 each participant's crash likelihood and driving skill levels in comparison to all drivers as well as in comparison to teenage drivers. The instructor also evaluated each participant's driving in eight domains of behavior such as speed management, road positioning or gap selection.

Based upon this dataset, we addressed two separate problems: predicting driver behavior (Section 4) and predicting driver skill (Section 5).

4   Predicting Driving Behavior

The first task was to study with what level of accuracy can case-based learning predict simulated driving behavior regarding speed management. In this context, "predicting driver behavior" refers to the *regression task* of predicting the value of the output variables (*steer, throttle*, and *brake*) for a given situation (represented by the value of all the other variables). Moreover, because of its applications to the design of in-vehicle automatic monitoring systems, we also evaluated the accuracy with which *future behavior* can be predicted (i.e., the behavior of a driver a certain amount of time into the future of the current situation).

4.1   Methods

From a case-based reasoning point of view, we identified four main challenges to predicting driver behavior:

- *Scalability*: even when just trying to predict the behavior of a given driver by learning from samples collected from a single drive, the resulting case-base contained tens of thousands of cases (and more than two million if we attempted to create a model that learns from all the data that we had available). However, our data only contains 12.13 hours of driving. If behavior prediction systems are to be incorporated into actual vehicles, collecting data in real-time, techniques that can handle even larger sets of data are needed.
- *Dimensionality*: even if each of our samples contains a relatively low number of features (36, not including the output variables), many of those variables are not very relevant for predicting certain behaviors. Thus, in order to generate accurate predictions, feature selection is required.
- *Sequentiality*: driving behavior is inherently sequential; i.e., each data sample in our dataset is temporally linked to the previous and following sample. Thus, this should be taken into account. For example, if a sample positions the car at a stationary position in front of a green traffic light, the model would have to know how long the light had been green. If the light just turned green, the model should predict no immediate action (since humans take some time to react to a green light), but if the light has been green for 3 or 4 tenths of a second, then the model should predict a certain amount of throttle depression. Thus, to predict actions in a given sample, previous samples need to be considered.
- *Stochasticity*: even under the same circumstances, drivers sometimes behave differently, so there might be samples with identical or very similar input variable values, but different output variable values.

For the work presented in this paper, we only addressed the first two challenges, and minimized the third challenge (sequentiality) by introducing the 3 *previous output variables* as part of each sample, which provide some degree of sequentiality. However, as part of our future work, we would like to investigate the usefulness of sequential similarity measures such as *Dynamic Time Warping* [8], and of sequence-aware retrieval methods such as *temporal backtracking* [5,16]. We did not consider stochasticity in this project.

To experiment with driving behavior prediction, we set up a case-based prediction system in which each of the samples in our dataset was a case, and Euclidean distance was used for case retrieval. In addition, to address the scalability and dimensionality challenges, we incorporated two additional techniques:

- *kd*-trees [23] for scalability and to speed-up retrieval, which can achieve $O(log(n))$ retrieval times, rather than $O(n)$ as standard *k*-nn.
- To address the dimensionality problem, we employed a greedy additive wrapper feature selection method [24]. Given a training set C, where instances have a set of features F (initially F is the set of 36 input variables in our dataset), and a target output variable f, we used the following procedure to the select a subset of features $F^* \subseteq F$ that maximizes the performance of the system in predicting the value of f:

1. Divide the training set in two disjoint subsets: $C = A \cup B$.
2. Initialize $F' = \emptyset$
3. Measure the performance of the system (mean square error or MSE) when trained in A and tested in B when adding only one feature at a time to F' from the set of features F.
4. Select the feature f^* that achieved the best performance, and set $F' = F' \cup \{f^*\}$. Repeat this process until $F' = F$.
5. Select F^* as the set of features that achieved the maximum performance during the whole process.

Notice that this is basically a greedy search process over the space of subsets of features, trying to find the set of features that minimizes the . So, in this way, our system can select a subset that maximizes the performance of the system in predicting each of the output variables.

4.2   Experimental Validation

We evaluated the performance of the system by training the system in two ways. First, *drive2* behavior for a given driver was used to train the system with performance of the prediction evaluated by comparing predicted *drive3* behavior against actual *drive3* behavior. Second, in order to began to examine one of the main goals of this line of work, the feasibility of using case-based approaches for in-vehicle monitoring systems, we examined the utility of case-based learning to predict future driving based on current driving behavior in order to provide feedback, or intervene if a dangerous situation is forecasted. For this experiment, we evaluated the performance of the system in predicting the behavior of the driver both at the same instant as represented by a given sample, and also in predicting the behavior of the driver a certain amount of time Δ into the future (we evaluated $\Delta = 0.5s$, $\Delta = 1.0s$, $\Delta = 1.5s$, and $\Delta = 2.0s$).

Figures 2.b and 2.a illustrate the performance of the system in predicting *steer* in *drive3* (after learning from *drive2*) with and without feature selection respectively, and with $\Delta = 0s$ (i.e. without a time delay). The horizontal axis represents time, and the vertical axis is the value that the *steer* variable takes over time. We show both the actual and the predicted value of steer for one of the drivers in our dataset (predicted behavior is displayed in red; actual behavior is displayed in green). As we can see, the system is capable of predicting *steer* with high accuracy, but without feature selection, the system is less accurate. For example 2.a shows a large red spike, where our system predicted a left turn, but the driver actually did not perform such a turn. With feature selection, no such large prediction inaccuracies were observed in the output. The average MSE is also shown, illustrating, how much the performance of the system increases with feature selection (MSE reduction from 0.014 to 0.003 using feature selection).

The set of variables that were selected by feature selection to predict *steer* in Figure 2.b were: *previous_Steer*, *LatAcc*, *LateralVelocity*, *IntersectionRightIntersection*, *LeftTurn*, and *RightTurn*, meaning, respectively: the value of steer in the previous instant of time, the lateral acceleration, the lateral velocity, the distance to the next intersection where the driver has been instructed to make a

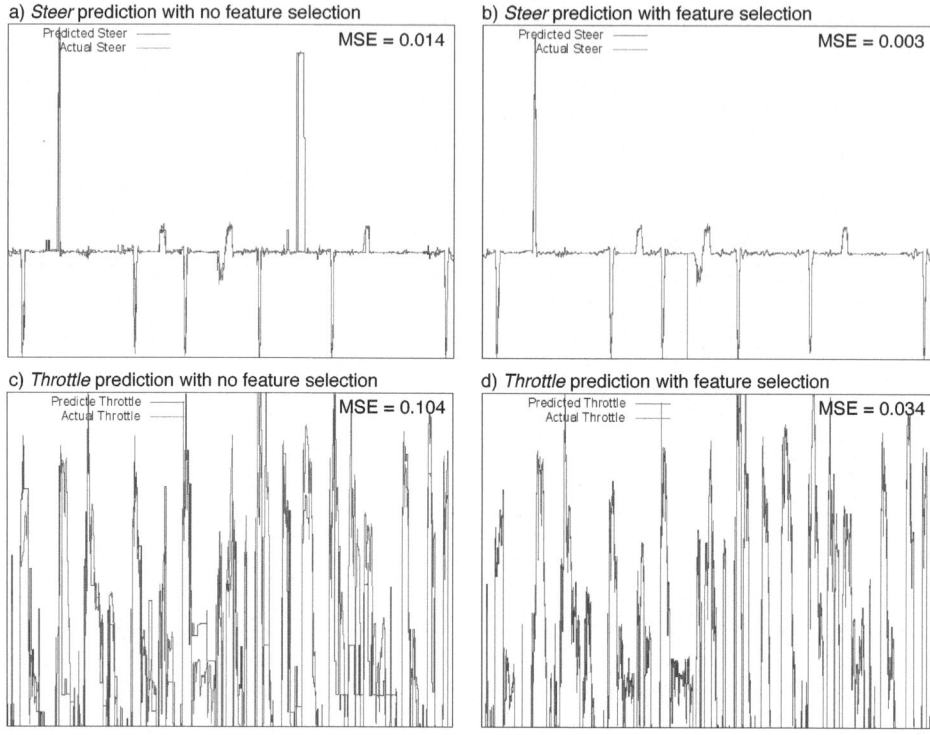

a) *Steer* prediction with no feature selection

Predicted Steer ——
Actual Steer ——   MSE = 0.014

b) *Steer* prediction with feature selection

Predicted Steer ——
Actual Steer ——   MSE = 0.003

c) *Throttle* prediction with no feature selection

Predicted Throttle ——
Actual Throttle ——   MSE = 0.104

d) *Throttle* prediction with feature selection

Predicted Throttle ——
Actual Throttle ——   MSE = 0.034

Fig. 2. Illustration of the performance of our case-based system in predicting *steer* and *throttle* for a given driver in our dataset with and without using feature selection

right turn (if any), whether the driver has been instructed to make a left turn, and whether the driver has been instructed to make a right turn. Notice that these are all variables that are indicative of steering behavior.

Figures 2.c and 2.d show the prediction of the *throttle* variable, with and without feature selection, respectively, and demonstrate worse prediction results than for steering. We observed that the *throttle* variable is subject to a high degree to non-determinism in our dataset, and thus, our system cannot predict its value as accurately. We believe that estimating a probability distribution over the possible values of the variable in a neighborhood instead of predicting a single value, the system could make more useful predictions. Also, we believe that incorporating retrieval techniques that account for the sequentiality of data, as mentioned before, can greatly increase the performance of the system.

Figure 3.a shows the performance of the case-based approach for different time delays with and without feature selection compared to a baseline prediction. The baseline prediction basically always predicts the average value for a given variable. Notice that for some variables (like *steer*), predicting the average value (representing the steering wheel being in the center position) is quite a good predictor when not turning, since in our dataset, about 92% of the time drivers keep the steering wheel no further than 5% away from the center position. As

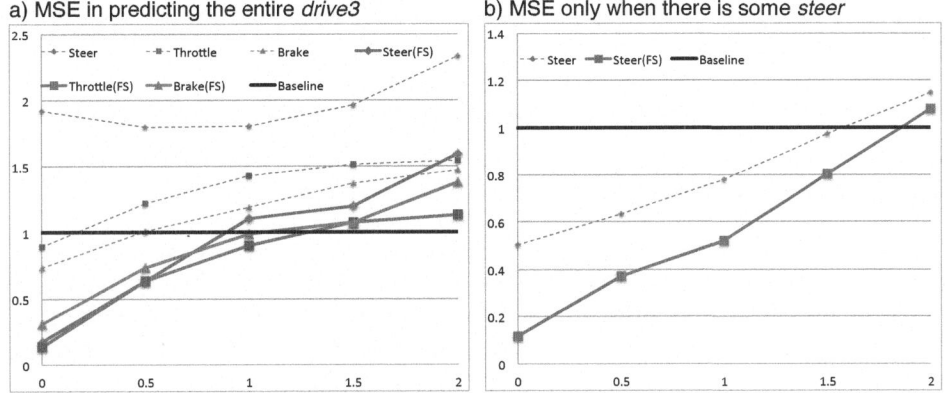

Fig. 3. Performance (in the vertical axis, lower is better) of our case-based system for predictions with different time delays (horizontal axis). Performance (vertical axis) is measured as the MSE, scaled to make the MSE of the baseline equal to 1.

expected, the predictive accuracy of the system decreases as the time delay increases. However, for time delays of $1.0s$ or less, our system was able to generate predictions that were better than the baseline (except for *steer*, for which this was the case only if $\Delta \leq 0.5s$). However, as can be seen in Figure 2.b, prediction for *steer* is very accurate, and thus there seems to be a contradiction between Figure 2.b and Figure 3.a. The issue is that in 92% of the instances the *steer* variable has a value very close to the average, and thus the baseline prediction is actually almost perfect in those 92% of the instances, biasing the results.

We thus performed a separate experiment only for those cases where *steer* did not have the average value (the remaining 8% of the instances, primarily involving instances when the driver was turning). Results, as shown in Figure 3.b demonstrate that in these situations (when the driver is actually turning), the predictions of our system were actually significantly better than the baseline. We did not perform similar experiments for *throttle* or *brake*, since those variables contain values different from the average in more than 50% of the instances, and thus the performance reported in Figure 3.a is accurate.

When predictions were performed with time delays $\Delta \geq 0.5$, variables that were considered relevant for $\Delta = 0$ (such as *previous_Steer*), were no longer relevant, and other variables (such as the offset position with respect to the lane) become more important for the model.

With respect to computational performance, our system is able to make predictions in 8.36 milliseconds when predicting with a case-base of an average of 37851.94 instances using all the features in the dataset (in a 2.8GHz intel Core i7), which is twice the speed required for real-time predictions. However, better scalability is required if models are to be learned from larger amounts of data.

In summary, our case-based approach using feature selection can be used to predict driver behavior with an acceptable degree of accuracy as long as the time delay is smaller than 1.0s. As part of our future work we want extend our system to take into account the sequential nature of the data, which we expect will significantly improve the accuracy of our system, with the goal of achieving accurate predictions with at least $\Delta = 2.0s$. However, this improved accuracy will have to be balanced with the potential for additional computational costs.

5   Predicting Driving Skill

The second task that we tried to address was significantly more challenging: predicting the skill level of a driver (likelihood of crashing relative to other drivers, and skill level relative to other drivers) based on available driving behavior data. Specifically, the dataset described above was used for prediction and a human expert (driving educator/evaluator) rated driving skill based on video review. The instructor rated the skill level of each driver along 18 dimensions such as *likelihood of crashing compared to all drivers and compared to teen drivers, speed management, vehicle control*, etc. Our goal in this task was to predict the ratings that the human instructor provided for a given driver solely based upon driving behavior data. These kind of predictions can be of great usefulness both for in-vehicle monitoring systems (to monitor the amount of assistance required by the driver), and also for assessment of driver skills during training. Moreover, in this study we focused on the first four dimensions, which are Boolean:

- Dimension 1: more likely to crash than the average driver?
- Dimension 2: more likely to crash than the average newly licensed teen driver?
- Dimension 3: more skilled than the average driver?
- Dimension 4: more skilled than the average newly licensed teen driver?

5.1   Methods

The main challenge that we faced in trying to predict driver skill using a case-based approach is that drivers in our dataset were not represented by a finite set of features among which we could assess similarity, but rather by long logs corresponding to their behavior in each of the four different drives. Therefore, we approached this problem via similarity measures between the driver's behavior.

The problem of measuring the similarity of behaviors has been addressed in the literature from different points of view. The most common approach is to use statistical models, such as Markov Chains or Hidden Markov Models (HMM) [20]. Given two behaviors, represented by two traces t_1 and t_2 (akin to the different drives we have for each driver), a statistical model is estimated for each of the two traces. This results in two models m_1 and m_2, which are defined by their probability distributions. Then, m_1 and m_2 can be compared using a similarity measure between probability distributions, such as the Kullback-Leibler Divergence [9]. This approach has been successfully deployed in driving domains as described above [15], and we explored it in our recent work, studying the circumstances under which different statistical models can be used for differentiating different types of behaviors [18]. However, the high dimensionality and large amount of data present in our domain make following that approach unfeasible. Thus, we followed an alternative approach, which we describe next.

Each case in our skill-prediction case-based system represented one driver. Each of these cases was composed of two parts: 1) The behavior of the user: represented by the four drives we have for each driver, and 2) The instructor rating: the rating of the instructor along the 18 different dimensions.

We defined three different similarity measures for this system:

- **Cross-Prediction Similarity** (S_{XP}): this similarity compares two drivers by measuring the average MSE with which the CBR system presented in Section 4 can predict the behavior of one driver when using all the drives from the other driver as the case-base. Since there were three output variables to predict (*steer, throttle, brake*), the similarity was defined as the average MSE among the three different output variables. The justification for this similarity measure was that if we can predict the behavior of one driver very accurately from cases obtained from another driver, then their behaviors were likely to be similar (e.g., they tend to drive at the same speed in the same situations).

- **Feature Selection Similarity** (S_{FS}): different drivers pay attention to different aspects of the road when driving. Many of those aspects were captured by the different input variables recorded for each of the drives. The justification for this similarity measure was that drivers that pay attention to similar aspects when driving should exhibit similar behaviors. The assumption made here is that variables selected by feature selection are a proxy for the variables to which drivers pay attention to. So, this similarity measure compares the variables that were selected by the feature selection process for each of the drivers in the following way:

  1. A feature selection process is run for each driver in the case-base, for each output feature (we used *drive2* for training and *drive3* for testing during feature selection).
  2. We re-represent each driver d in the case-base by a binary vector v^d of length 108 (36 input features times 3 output features), where each position v_i^d represents a pair of input variable/output variable. If one of the positions of this vector is set to 1, it means that the corresponding input feature was selected as important to predict the given output feature (the position is set to 0 otherwise).
  3. S_{FS} between two drivers d_1 and d_2 is defined as the Euclidean distance between the two vectors v^{d_1} and v^{d_2}.

- **Correlated-Feature Selection Similarity** (S_{CF}): following the same idea as the previous similarity measure, this measure uses the result of feature selection, but goes one step further, by computing which of those features are correlated or not with the target dimension.

  1. Given an instructor rating dimension r (e.g. "Is the driver more likely to crash than the average driver?") that we want to predict, we measured the average rating along dimension r for all drivers: $\mu(r)$ and its standard deviation $\sigma(r)$.

Table 1. Skill prediction accuracy for different similarity measures

		S_{XP}	S_{FS}	S_{CF}
More likely to crash	...driver?	23.53%	64.71%	82.35%
than the average...	...newly licensed teen driver?	17.64%	47.05%	82.35%
More skilled	...driver?	23.53%	64.71%	82.35%
than the average...	...newly licensed teen driver?	11.76%	47.05%	76.47%
	Average	19.12%	55.88%	80.88%

2. For each position i of the binary vector v^d, we computed the average score in rating r for those drivers for which $v_i^d = 1$: $\mu_i(r)$, and the average score for those drivers for which $v_i^d = 0$: $\mu_{\neg i}(r)$.
3. Then we measured the difference in scores (measured in standard deviations): $\Delta_{r,i} = \frac{\mu_i(r) - \mu_{\neg i}(r)}{\sigma(r)}$ (only for those values of i for which we had at least 3 instances where $v_i^d = 1$ and 3 where $v_i^d = 0$).
4. We then define two sets P_r and N_r, where $P_r = \{i | \Delta_{r,i} \geq 1.0\}$ are the positions of the vector for which the difference in average rating is more than 1 standard deviation higher than the average, and $N_r = \{i | \Delta_{r,i} \leq -1.0\}$ is defined analogously for 1 standard deviation below the average.
5. Finally, we re-represent each driver d by a pair of numbers (p_d, n_d), where $p_d = \sum_{i \in P_r} v_i^d$, and $n_d = \sum_{i \in N_r} v_i^d$.
6. S_{CF} between two drivers d_1 and d_2 is defined as the Euclidean distance between (p_{d_1}, n_{d_1}) and (p_{d_2}, n_{d_2})

The justification for this similarity was the assumption that only certain aspects were most related to achieving a high (or low) rating along a given rating dimension r. The sets P_r and N_r capture the set of aspects that are beneficial for such rating, and the sets of aspects that are detrimental, respectively. This similarity measures the difference in the number of beneficial and detrimental variables selected by feature selection.

The second and third similarity measures were developed given the surprising result that the first similarity measure (S_{XP}) obtained really low predictive accuracy. As reported in the next section, S_{CF} obtains a good predictive accuracy, meaning that in order to predict skill level, comparing which variables are important to predict the behavior of each driver is a promising approach.

5.2   Results

To evaluate the performance of our skill prediction case-based system, we employed the data from the 17 users in our dataset using all the three similarity measures described above. All the experimental evaluation was performed using a leave-one-out method. This means that the user being predicted was not used in any of the steps related to determining the sets of features to be used for the S_{CF} and S_{FS} measures, to ensure the statistical validity of the presented results. Table 1 presents the results obtained with each of the three similarity measures for the first four dimensions that the human instructor provided (see

Section 5). Trying to predict the other (more challenging) dimensions is part of our future work, and is out of the scope of this paper. The first observation is that neither S_{XP} nor S_{FS} obtained good results (S_{XP} achieves such a low accuracy in fact, that since the prediction task is binary, the reverse value of the predicted one would achieve a higher accuracy). This was surprising to us, since S_{XP} basically represents the standard measure to compare behavioral data, and usually obtains good results in these type of tasks [18]. The explanation to the fact that S_{XP} does not perform well is due to two main factors: First, S_{XP} basically measures how well the data coming from one driver can be used to predict other drivers' behavior. It happens that the data from some drivers covers a wider set of situations (some drivers never encountered a red traffic light, while others did), and thus those were retrieved most of the time. Thus, the differing driving situations and behaviors did not allow for the desired effect of retrieving "drivers with similar driving style". Second, S_{XP} is comparing the behavior of drivers during the *whole* drive, whereas we were unable to determine if certain events or behaviors were weighted differently in the review by the human instructor. Moreover, S_{CF} achieved very good results in predicting driver skill and likelihood to crash, indicating that feature selection data can provide useful information for driver behavior comparison.

Notice that we have very few cases in our case-base for this study (16 cases plus the one being predicted), and thus the performance of each of these similarity measures should increase with larger amounts of data. As future work, we want to extend our study to a larger dataset containing nearly one hundred drivers. Additionally, we would like to improve the skill prediction capabilities of our system. We are currently working on classifying the different drives in a series of *contexts* (such as "following a vehicle", "making a left turn", or "overtaking"). In this way, our system would be able to perform a more nuanced comparison of the behavior of different drivers under different circumstances and improve performance. Accurate skill prediction directly from driving data would allow a number of new applications, not currently possible today, such as automated driving performance comparisons for personalized training.

6   Conclusions and Future Work

This paper presented an application of case-based reasoning to two specific problems concerning human driving behavior modeling, for teen driving behavior: behavior prediction and skill prediction. Concerning behavior prediction, we presented the four main challenges (scalability, dimensionality, sequentiality and stochasticity), and an approach that addresses the first two of those challenges. Concerning skill prediction, we showed that standard behavior similarity measures do not achieve good results in this domain and presented a new family of similarity measures based on feature selection that obtain promising results.

As part of our future work, we would like to scale our evaluation to a larger dataset containing data from nearly one hundred drivers in order to further validate all the findings presented in this paper. Additionally, we would like to improve the performance of our behavior prediction system by incorporating both similarity measures that can take into account the sequential nature of the data, as well as stochasticity whenever necessary. Finally, we would like to study techniques to scale all the approaches presented here to big data scenarios. In our current experiments, each drive contained between 35000 and 43000 instances; however, if these techniques are to be deployed in real in-vehicle behavior monitoring systems that run for extended periods of time, the system needs to scale up to allow millions of instances. In order to achieve such scale, we plan to build upon recent work on sequential similarity measures that are capable of handling even larger datasets in an efficient way via the use of techniques such as early abandoning [21].

Acknowledgements. This project was partially funded by the National Science Foundation (NSF) Center for Child Injury Prevention Studies at the Children's Hospital of Philadelphia (CHOP). We would like to thank the Industry Advisory Board (IAB) members for their support, valuable input and advice. This project was also partly funded under a grant with the Pennsylvania Department of Health. The Department specifically disclaims responsibility for any analyses, interpretations or conclusions. Dr. Catherine C. McDonald is supported by the National Institutes of Health under Award Number K99NR013548. The content is solely the responsibility of the authors and does not necessarily represent the official views of the CHOP, NIH, NSF, or the IAB members.

References

1. Cordier, A., Mascret, B., Mille, A.: Extending case-based reasoning with traces. In: Grand Challenges for Reasoning from Experiences, Workshop at IJCAI, vol. 9, p. 31 (2009)
2. Curry, A.E., Hafetz, J., Kallan, M.J., Winston, F.K., Durbin, D.R.: Prevalence of teen driver errors leading to serious motor vehicle crashes. Accident Analysis & Prevention 43(4), 1285–1290 (2011)
3. Das, D., Zhou, S., Lee, J.D.: Differentiating alcohol-induced driving behavior using steering wheel signals. IEEE Transactions on Intelligent Transportation Systems 13(3), 1355–1368 (2012)
4. Fernlund, H.K.G., Gonzalez, A.J., Georgiopoulos, M., DeMara, R.F.: Learning tactical human behavior through observation of human performance. IEEE Transactions on Systems, Man, and Cybernetics, Part B 36(1), 128–140 (2006)
5. Floyd, M.W., Esfandiari, B.: Learning state-based behaviour using temporally related cases. In: Proceedings of the Sixteenth UK Workshop on Case-Based Reasoning (2011)
6. Floyd, M.W., Esfandiari, B., Lam, K.: A case-based reasoning approach to imitating robocup players. In: Proceedings of the Twenty-First International Florida Artificial Intelligence Research Society (FLAIRS), pp. 251–256 (2008)
7. Foss, R.D., Martell, C.A., Goodwin, A.H., O'Brien, N.P.: Measuring changes in teenage driver crash characteristics during the early months of driving. In: AAA Foundation for Traffic Safety (2011)

8. Keogh, E.J., Pazzani, M.J.: Scaling up dynamic time warping for datamining applications. In: Proceedings of the Sixth ACM SIGKDD International Conference on Knowledge Discovery and Data Mining, pp. 285–289. ACM (2000)
9. Kullback, S., Leibler, R.A.: On information and sufficiency. Ann. Math. Statist. 22(1), 79–86 (1951)
10. Lamontagne, L., Rugamba, F., Mineau, G.: Acquisition of cases in sequential games using conditional entropy. In: ICCBR 2012 Workshop on TRUE: Traces for Reusing Users' Experience (2012)
11. Macadam, C.C.: Understanding and modeling the human driver. Vehicle System Dynamics 40(1-3), 101–134 (2003)
12. Markkula, G., Benderius, O., Wolff, K., Wahde, M.: A review of near-collision driver behavior models. Human Factors: The Journal of the Human Factors and Ergonomics Society 54(6), 1117–1143 (2012)
13. McDonald, C.C., Tanenbaum, J.B., Lee, Y.-C., Fisher, D.L., Mayhew, D.R., Winston, F.K.: Using crash data to develop simulator scenarios for assessing novice driver performance. Transportation Research Record: Journal of the Transportation Research Board 2321(1), 73–78 (2012)
14. McKnight, A.J., McKnight, A.S.: Young novice drivers: careless or clueless? Accident Analysis & Prevention 35(6), 921–925 (2003)
15. Nechyba, M.C., Xu, Y.: Stochastic similarity for validating human control strategy models. IEEE Transactions on Robotics and Automation 14(3), 437–451 (1998)
16. Ontañón, S., Floyd, M.W.: A comparison of case acquisition strategies for learning from observations of state-based experts. In: Proceedings of FLAIRS 2013 (2013)
17. Ontañón, S., Mishra, K., Sugandh, N., Ram, A.: On-line case-based planning. Computational Intelligence Journal 26(1), 84–119 (2010)
18. Ontañón, S., Montaña, J.L., Gonzalez, A.J.: A dynamic-bayesian network framework for modeling and evaluating learning from observation. Expert Systems with Applications 41(11), 5212–5226 (2014)
19. Pomerleau, D.: Alvinn: An autonomous land vehicle in a neural network. In: Touretzky, D. (ed.) Advances in Neural Information Processing Systems 1. Morgan Kaufmann (1989)
20. Rabiner, L., Juang, B.-H.: An introduction to hidden markov models. IEEE ASSP Magazine 3(1), 4–16 (1986)
21. Rakthanmanon, T., Campana, B., Mueen, A., Batista, G., Westover, B., Zhu, Q., Zakaria, J., Keogh, E.: Searching and mining trillions of time series subsequences under dynamic time warping. In: Proceedings of the 18th ACM SIGKDD International Conference on Knowledge Discovery and Data Mining, pp. 262–270. ACM (2012)
22. Rubin, J., Watson, I.: On combining decisions from multiple expert imitators for performance. In: IJCAI, pp. 344–349 (2011)
23. Wess, S., Althoff, K.D., Derwand, G.: Using k-d trees to improve the retrieval step in case-based reasoning. In: Wess, S., Richter, M., Althoff, K.-D. (eds.) EWCBR 1993. LNCS, vol. 837, pp. 167–181. Springer, Heidelberg (1994)
24. Wettschereck, D., Aha, D.W., Mohri, T.: A review and empirical evaluation of feature weighting methods for a class of lazy learning algorithms. Artificial Intelligence Review 11(1-5), 273–314 (1997)

Adapting Propositional Cases Based on Tableaux Repairs Using Adaptation Knowledge*

Gabin Personeni[1,2,3], Alice Hermann[1,2,3], and Jean Lieber[1,2,3]

[1] Université de Lorraine, LORIA, UMR 7503 — 54506 Vandœuvre-lès-Nancy, France
{Gabin.Personni,Alice.Hermann,Jean.Lieber}@loria.fr
[2] CNRS — 54506 Vandœuvre-lès-Nancy, France
[3] Inria — 54602 Villers-lès-Nancy, France

Abstract. Adaptation is a step of case-based reasoning that aims at modifying a source case (representing a problem-solving episode) in order to solve a new problem, called the target case. An approach to adaptation consists in applying a belief revision operator that modifies minimally the source case so that it becomes consistent with the target case. Another approach consists in using domain-dependent adaptation rules. These two approaches can be combined: a revision operator parametrized by the adaptation rules is introduced and the corresponding revision-based adaptation uses the rules to modify the source case. This paper presents an algorithm for revision-based and rule-based adaptation based on tableaux repairs in propositional logic: when the conjunction of source and target cases is inconsistent, the tableaux method leads to a set of branches, each of them ending with clashes, and then, these clashes are repaired (thus modifying the source case), with the help of the adaptation rules. This algorithm has been implemented in the REVISOR/PLAK tool and some implementation issues are presented.

Keywords: Case-based reasoning, adaptation, tableaux repairs, propositional logic, belief revision, adaptation rules.

1 Introduction

Case-based reasoning (CBR [1]) is a reasoning paradigm based on the reuse of chunks of experience called cases. A *case* is a representation of a problem-solving episode, often separated in a problem part and a solution part: this separation is not formally necessary but is useful to the intuitive understanding of the notion of case. The input of a CBR system is a target case, which represents an underspecified case (intuitively, its problem part is well specified, whereas its solution part is not). A classical way to implement a CBR system consists in (1) selecting a case from a case base that is similar to the target case, (2) adapting this retrieved case in order to solve the target case, i.e., in order to add information to it (intuitively, this consists in specifying the solution part of the target case by reusing the solution part of the retrieved case). Variants of this approach to CBR and other steps related to it can be found in, e.g., [2].

---

* This research was partially funded by the project Kolflow of the French National Agency for Research (ANR), program ANR CONTINT (http://kolflow.univ-nantes.fr).

L. Lamontagne and E. Plaza (Eds.): ICCBR 2014, LNCS 8765, pp. 390–404, 2014.
© Springer International Publishing Switzerland 2014

This paper concentrates on step (2), *adaptation*. Despite its importance, this step of CBR has been a little bit neglected in the CBR literature, though it has recently received some attention (see, e.g., [3–6]). In particular, the paper concentrates on an approach to adaptation that we introduced some years ago [7] and studied according to various aspects (see [8] for a synthesis). This approach called (now) *revision-based adaptation* or $\dot{+}$-*adaptation* is based on a belief revision operator $\dot{+}$. According to the AGM postulates (called after the names of [9]), the *belief revision* of a belief base ψ by a belief base μ—$\psi \dot{+} \mu$—consists in minimally modifying ψ into ψ' such that the conjunction of ψ' and μ is consistent; then $\psi \dot{+} \mu$ is this conjunction. It must be noticed that, since there are many ways to "measure" modifications, there are also many ways to "minimally modify" a belief base, hence there are multiple revision operators satisfying the AGM postulates.

Roughly said, $\dot{+}$-adaptation consists in using $\dot{+}$ in order to modify the retrieved case so that it is consistent with the target case, and the adaptation process returns the result of this revision. So, implementing a revision-based adaptation amounts to implementing a revision operator. This implementation depends on the formalism used in the CBR system. In particular, we have studied how $\dot{+}$-adaptation can be implemented within the description logic $\mathcal{ALC}$ [10].[1] The approach to adaptation in $\mathcal{ALC}$ consisted in applying *tableaux repairs*. The same idea of tableaux repairs can be used for revision of ψ by μ according to the following principle: the tableaux method is applied separately on ψ and μ, then the consistent branches are combined, which leads to a set of inconsistent branches (unless the conjunction of ψ and μ is consistent). Then, tableaux repair consists in removing the parts of the clashes whose origin is ψ and the formulas of the branch from which these parts of clashes are deduced. This involves a weakening of ψ into ψ' such that ψ' is consistent with μ, and the result of the revision is the conjunction of ψ' and μ. This approach for belief revision algorithm has been found in parallel by Camilla Schwind [11], who has applied it to propositional logic with a finite number of variables and thus, this work can be used for $\dot{+}$-adaptation of propositional cases.

This paper proposes to go one step beyond Camilla Schwind's work, by integrating, in the tableaux repairs, some domain-specific adaptation knowledge in the form of *adaptation rules* (also called reformulations in [12]). Such rules represent the fact that, in a given context, a given part of a case can be substituted by something. Thus, the idea is to use such rules for tableaux repairs.

The paper is organized as follows. In Section 2, some preliminaries introduce the notions and the notations used throughout the paper. In Section 3, the adaptation process in CBR is presented, pointing out the notions of adaptation rules and of revision-based adaptation. Then, the algorithm for adaptation by tableaux repairs using adaptation rules is presented in Section 4. The approach has been implemented in an inference engine called REVISOR/PLAK: this system and some implementation issues are described, in Section 5, as well as a concrete example. Section 6 concludes the paper.

The research report [13] is a long version of this paper including the proofs.

---

[1] Technically, we have not defined a revision operator in $\mathcal{ALC}$, since the implemented operator violates some of the postulates of [9], but we have implemented an adaptation operator inspired from the ideas of revision-based adaptation.

2 Preliminaries

2.1 Propositional Logic

Let $V = \{a_1, \ldots, a_n\}$ be a set of n distinct symbols called propositional variables. A propositional formula built on V is either a variable a_i or of one of the forms $\varphi_1 \wedge \varphi_2$, $\varphi_1 \vee \varphi_2$, $\neg\varphi_1$, $\varphi_1 \Rightarrow \varphi_2$, and $\varphi_1 \Leftrightarrow \varphi_2$, where φ_1 and φ_2 are two propositional formulas. Let $\mathcal{L}$ be the set of the propositional formulas.

Let $\mathbb{B} = \{T, F\}$ be a set of two elements. Given $x = (x_1, \ldots, x_n) \in \mathbb{B}^n$ and $\varphi \in \mathcal{L}$, $\varphi^x \in \mathbb{B}$ is defined as follows: $a_i^x = x_i$, $(\varphi_1 \wedge \varphi_2)^x = T$ iff $\varphi_1^x = T$ and $\varphi_2^x = T$, $(\varphi_1 \vee \varphi_2)^x = T$ iff $\varphi_1^x = T$ or $\varphi_2^x = T$, $(\neg\varphi_1)^x = T$ iff $\varphi_1^x = F$, $(\varphi_1 \Rightarrow \varphi_2)^x = (\neg\varphi_1 \vee \varphi_2)^x$, and $(\varphi_1 \Leftrightarrow \varphi_2)^x = ((\varphi_1 \Rightarrow \varphi_2) \wedge (\varphi_2 \Rightarrow \varphi_1))^x$.

For $\varphi \in \mathcal{L}$, let $\mathcal{M}(\varphi) = \{x \in \mathbb{B}^n \mid \varphi^x = T\}$ called the set of models of φ. Given $\varphi_1, \varphi_2 \in \mathcal{L}$, $\varphi_1 \models \varphi_2$ if $\mathcal{M}(\varphi_1) \subseteq \mathcal{M}(\varphi_2)$, and $\varphi_1 \equiv \varphi_2$ if $\mathcal{M}(\varphi_1) = \mathcal{M}(\varphi_2)$. $(\mathcal{L}, \models)$ is called the propositional logic with n variables.

A literal ℓ is a propositional formula of the form a_i (positive literal) or $\neg a_i$ (negative literal), where $a_i \in V$. If $\ell = \neg a_i$ is a negative literal, then $\neg\ell$ denotes the positive literal a_i (instead of the equivalent formula $\neg\neg a_i$). A formula is in disjunctive normal form or DNF if it is a disjunction of conjunctions of literals. A formula is in negative normal form (NNF) if it contains only the connectives $\wedge$, $\vee$ and $\neg$, and if $\neg$ appears only in front of propositional variables. Every formula can be put in NNF by applying, as rewriting rules oriented from left to right, the following equivalences, until none of these equivalences is applicable (for $\varphi, \varphi_1, \varphi_2 \in \mathcal{L}$):

$$\varphi_1 \Leftrightarrow \varphi_2 \equiv (\varphi_1 \Rightarrow \varphi_2) \wedge (\varphi_2 \Rightarrow \varphi_1)$$
$$\varphi_1 \Rightarrow \varphi_2 \equiv \neg\varphi_1 \vee \varphi_2$$
$$\neg\neg\varphi \equiv \varphi$$
$$\neg(\varphi_1 \vee \varphi_2) \equiv \neg\varphi_1 \wedge \neg\varphi_2$$
$$\neg(\varphi_1 \wedge \varphi_2) \equiv \neg\varphi_1 \vee \neg\varphi_2$$

A formula in DNF is necessarily in NNF (the converse is false).

A set of literals L is often assimilated to the conjunction of its elements, for example $\{a, \neg b, \neg c\}$ is assimilated to $a \wedge \neg b \wedge \neg c$ and vice-versa. In particular, L is satisfiable iff there is no literal $\ell \in L$ such that $\neg\ell \in L$.

An implicant of φ, I is a conjunction or a disjunction of literals, such that $I \models \varphi$ and I is satisfiable. However, as there exists a duality between conjunctive and disjunctive implicants, only conjunctive implicants will be considered in this paper. An implicant I of φ is prime if for any conjunction of literals C such that $C \subset I$, $C \not\models \varphi$. That is, a prime implicant I is minimal. Let $\mathrm{PI}(\varphi)$ be the set of prime implicants of φ. Then:

$$\varphi \equiv \bigvee_{I \in \mathrm{PI}(\varphi)} I$$

An algorithm to find efficiently prime implicants of a formula is detailed in [14].

2.2 Distances

Let $\mathcal{U}$ be a set. In this paper, a distance on $\mathcal{U}$ is a function $d : \mathcal{U} \times \mathcal{U} \rightarrow [0; +\infty]$ such that $d(x, y) = 0$ iff $x = y$ (the other properties of a distance function are not required in this paper). Given $A, B \in 2^{\mathcal{U}}$ and $y \in \mathcal{U}$, the following shortcuts are used:

$$d(A, y) = \inf_{x \in A} d(x, y) \qquad\qquad d(A, B) = \inf_{x \in A, y \in B} d(x, y)$$

where $\inf X$ denotes the infimum of the set X, with the convention $\inf \varnothing = +\infty$ (e.g., $d(A, \varnothing) = +\infty$).

The Hamming distance on propositional logic interpretations, d_H, is defined by $d_H(x, y) = |\{i \mid i \in \{1, 2, \ldots, n\}, x_i \neq y_i\}|$, for $x, y \in \mathbb{B}^n$. In other words, $d_H(x, y)$ is the number of variable flips to go from x to y.

2.3 Belief Revision

Let ψ be the beliefs of an agent about the world, expressed in a logic $(\mathcal{L}, \models)$. This agent is confronted to new beliefs expressed by $\mu \in \mathcal{L}$. These new beliefs are assumed to be non revisable, whereas the old beliefs ψ can be changed. If μ does not contradict ψ (i.e. if $\psi \wedge \mu$ is consistent), then the new beliefs are simply added to the old beliefs. Otherwise, according to the *minimal change principle* [9], ψ has to be modified minimally into $\psi' \in \mathcal{L}$ such that $\psi' \wedge \mu$ is consistent, and then the revision of ψ by μ, denoted by $\psi \dotplus \mu$, is this conjunction.

There are multiple ways of measuring modifications of beliefs, hence multiple revision operators $\dotplus$. In [9], a set of postulates has been defined that a revision operator is supposed to verify. In [15], these postulates have been reformulated in propositional logic and a family of revision operators based on distances has been defined as follows. Let d be a distance on $\mathcal{U} = \mathbb{B}^n$. Let ψ and μ be two formulas. The revision of ψ by μ according to $\dotplus^d$ ($\psi \dotplus^d \mu$) is a formula whose models are the models of μ that are the closest ones to the models of ψ according to d (the change from an interpretation x to an interpretation y is measured by $d(x, y)$). Formally, $\psi \dotplus^d \mu$ is such that:

$$\mathcal{M}(\psi \dotplus^d \mu) = \{y \in \mathcal{M}(\mu) \mid d(\mathcal{M}(\psi), y) = d^*\}$$
$$\text{with } d^* = d(\mathcal{M}(\psi), \mathcal{M}(\mu))$$

Note that this definition specifies $\psi \dotplus^d \mu$ up to logical equivalence, but this is sufficient since a revision operator has to satisfy the irrelevance of syntax principle (this is one of the [15]'s postulates: if $\psi_1 \equiv \psi_2$ and $\mu_1 \equiv \mu_2$ then $\psi_1 \dotplus \mu_1 \equiv \psi_2 \dotplus \mu_2$).

2.4 A* Search

A* is a heuristic-based best-first search algorithm [16]. It is suited for searching state spaces where there are a finite number of transitions from a given state to its successor states and for which the cost of a path is additive (the cost of a path is the sum of the cost of the transitions it contains). A search problem is given by a finite set of initial states and by a goal giving a condition for a state to be final. Given a state S, let $\mathcal{F}^*(S)$

be the minimum of the costs of the paths from an initial state to a final state that contain S. $\mathcal{F}^*(\text{S}) = \mathcal{G}^*(\text{S}) + \mathcal{H}^*(\text{S})$ where $\mathcal{G}^*(\text{S})$ (resp., $\mathcal{H}^*(\text{S})$) is the minimal of the costs of the paths from an initial state to S (resp., from S to a final state). In general, $\mathcal{F}^*$ is unknown and is approximated by a function $\mathcal{F} = \mathcal{G} + \mathcal{H}$. $\mathcal{F}$ is said to be *admissible* if $\mathcal{G} \geq \mathcal{G}^*$ and $\mathcal{H} \leq \mathcal{H}^*$. If $\mathcal{F}$ is admissible, then the A* procedure is optimal: if there is a solution (a path from an initial to a final state), then this solution has a minimal cost. The A* procedure consists in searching the state space, starting from the initial states, by increasing $\mathcal{F}(\text{S})$: among two successors of the current state, the one that is minimum for $\mathcal{F}$ is preferred. Usually, $\mathcal{G}(\text{S})$ is the cost of the path that has already been generated for reaching S and thus, $\mathcal{G} \geq \mathcal{G}^*$. Then, the main difficulty is to find an admissible $\mathcal{H}$ (the constant function 0 is an admissible $\mathcal{H}$—and using it corresponds to dynamic programming—but the closer an admissible $\mathcal{H}$ is to $\mathcal{H}^*$, the faster the search is).

3   Adaptation in Case-Based Reasoning

Let $(\mathcal{L}, \models)$ be the logic in which the knowledge containers of the CBR application are defined. A source case Source $\in \mathcal{L}$ is a case of the case base. Often, such a case represents a specific experience: $\mathcal{M}(\text{Source})$ is a singleton. However, this assumption is not formally necessary (though it has an impact on the complexity of the algorithms). A target case Target $\in \mathcal{L}$ represents a problem to be solved. This means that there is some missing information about Target and solving Target leads to adding information to it. So, adaptation of Source to solve Target consists in building a formula ComplTarget $\in \mathcal{L}$ that makes Target precise in the sense that it adds information and, therefore, reduces the set of models: ComplTarget $\models$ Target. To perform adaptation, the domain knowledge DK $\in \mathcal{L}$ can be used. Therefore, the adaptation process has the following signature:

$$\text{Adaptation} : (\text{DK}, \text{Source}, \text{Target}) \mapsto \text{ComplTarget}$$

(Source, Target) is called the adaptation problem (DK is supposed to be fixed).

Of course, this does not completely specify the adaptation process. Several approaches are introduced in the CBR literature. Two of them are presented below, followed by a combination of them.

Revision-Based Adaptation. Let $\dotplus$ be a revision operator in the logic $(\mathcal{L}, \models)$ used for a given CBR system. The $\dotplus$-adaptation is defined as follows:

$$\text{ComplTarget} = (\text{DK} \wedge \text{Source}) \dotplus (\text{DK} \wedge \text{Target})$$

Intuitively, the source case is modified minimally (according to $\dotplus$) so that it satisfies the target case. Both cases are considered w.r.t. the domain knowledge.

Rule-Based Adaptation. is a general approach to adaptation relying on domain-specific adaptation knowledge in the form of a set AK of adaptation rules (see, e.g., [12], where adaptation rules are called reformulations). An adaptation rule R $\in$ AK, when applicable on the adaptation problem (Source, Target), maps Source into ComplTarget which makes Target precise. Adaptation rules can be composed (or chained): if $R_1, R_2, \ldots, R_q \in AK$ are such that there exist $q + 1$ cases $C_0, C_1, \ldots, C_q$ verifying:

- $C_0 = $ Source,
- C_q makes Target precise ($C_q \models$ Target),
- for each $i \in \{1, \ldots, q\}$, R_i is applicable on (C_{i-1}, C_i) and maps C_{i-1} into C_i,

then $C_q = $ ComplTarget is the result of the adaptation of Source to solve Target. The sequence $R_1; R_2; \ldots; R_q$ is called an *adaptation path*.

Given an adaptation problem (Source, Target), there may be several adaptation paths to solve it. In order to make a choice among them, a cost function is introduced: $\text{cost} : R \in AK \mapsto \text{cost}(R) > 0$. The cost of an adaptation path is the sum of the costs of its adaptation rules.

In this paper, an adaptation rule R is defined by two sets of literals, left and right, and is denoted by $R = $ left $\rightsquigarrow$ right. Let (Source, Target) be an adaptation problem. Several cases have to be considered:

- If Source is a conjunction of literals represented by a set of literals L, then R is applicable on Source if left $\subseteq$ L. In this situation:

$$R(\text{Source}) = R(L) = (L \setminus \text{left}) \cup \text{right}$$

- If Source is in DNF, such that Source $= \bigvee_i L_i$, where the L_i's are conjunctions of literals, R is applicable on Source if it is applicable on at least one L_i. Then:

$$R(\text{Source}) = \bigvee_i \begin{cases} R(L_i) & \text{if R is applicable to } L_i, \\ L_i & \text{otherwise.} \end{cases}$$

- If Source is not in DNF, it is replaced by an equivalent formula in DNF.

Given an adaptation rule $R = $ left $\rightsquigarrow$ right, repairs$(R) = $ left $\setminus$ right. left is the set of literals whose presence is necessary to apply R, and is then removed by application of R. Every adaptation rule must be such that repairs$(R) \neq \varnothing$.

Combining Rule-Based and Revision-Based Adaptation. Let us consider the following distance on $\mathcal{U} = \mathbb{B}^n$ (for $x, y \in \mathcal{U}$):

$$\delta_{AK}(x, y) = \inf\{\text{cost}(p) \mid p: \text{adaptation path from } x \text{ to } y \text{ based on rules from AK}\}$$

(x and y are interpretations that are assimilated to conjunctions of literals). By convention, $\inf \varnothing = +\infty$ and thus, if there is no adaptation path relating x to y, then $\delta_{AK}(x, y) = +\infty$ and vice-versa. Otherwise, it can be shown that the infimum is always reached and so, $\delta_{AK}(x, y) = 0$ iff $x = y$ (corresponding to the empty adaptation path).

It has been shown [8] that rule-based adaptation can be simulated by revision-based adaptation with no domain knowledge (i.e., DK is a tautology) and with the $\dotplus^{\delta_{AK}}$ revision operator. When rule-based adaptation fails (no adaptation path from Source to Target), $\dotplus^{\delta_{AK}}$-adaptation gives ComplTarget equivalent to Target (no added information).

A failure of rule-based adaptation is due to the fact that no adaptation rules can be composed in order to solve the adaptation problem. In order to have an adaptation that always provides a result, the idea is to add $2n$ adaptation rules, one for each literal: given a literal ℓ, the flip of the literal ℓ is the adaptation rule $F_\ell = \ell \rightsquigarrow \top$, where $\top$ is the empty conjunction of literals.[2] It can be noticed that a cost is associated to each literal flip and that it may be the case that $\text{cost}(F_{\neg\ell}) \neq \text{cost}(F_\ell)$.

Now, let d_{AK} be the distance on $\mathcal{U} = \mathbb{B}^n$ defined, for $x, y \in \mathcal{U}$, by

$$d_{\text{AK}}(x, y) = \inf \left\{ \text{cost}(p) \;\middle|\; \begin{array}{l} p\text{: adaptation path from } x \text{ to } y \text{ based on} \\ \text{rules from AK and on flips of literals} \end{array} \right\}$$

Let $\dotplus_{\text{AK}} = \dotplus^{d_{\text{AK}}}$. $\dotplus_{\text{AK}}$-adaptation is a revision-based adaptation using the adaptation rules, thus combining rule-based adaptation and revision-based adaptation, and that can take into account domain knowledge.

It can be noticed that if $\text{AK} = \varnothing$ and $\text{cost}(F_\ell) = 1$ for every literal ℓ then $d_{\text{AK}} = d_H$. Moreover the following assumption is made:

$$\text{For any R} \in \text{AK}, \text{cost(R)} \leq \sum_{\ell \in \text{repairs(R)}} \text{cost}(F_\ell) \tag{1}$$

In fact, this assumption does not involve any loss of generality: if an adaptation rule violates (1), then it does not appear in any optimal adaptation path, since applying the flips of all the literals of its repair part will be less costly then applying the rule itself.

4   Algorithm of Adaptation Based on Tableaux Repairs

The adaptation algorithm is based on the revision of the source case by the target case, both w.r.t. the domain knowledge. As such, the algorithm performs the revision of a formula ψ by a formula μ, with:

$$\psi = \text{DK} \wedge \text{Source} \qquad \mu = \text{DK} \wedge \text{Target}$$
$$\text{and so:} \qquad \text{ComplTarget} = \psi \dotplus_{\text{AK}} \mu$$

This section presents the algorithm, which uses a heuristic function $\mathcal{H}$, that is defined and proven to be admissible. Then, an example is detailed. Finally, the termination and the complexity of the algorithm are studied.

4.1   Algorithm

Let φ be a formula and let $\text{branches}(\varphi)$ be the set of branches of the tableaux of φ, or implicants of φ, as a set of sets of literals, that is:

– For any literal ℓ: $\text{branches}(\ell) = \{\{\ell\}\}$;

---

[2] In the literature, a flip is more frequently a rule of the form $\ell \rightsquigarrow \neg\ell$ but it can be shown that at the definition level, this amounts to the same kind of adaptation (using either $\ell \rightsquigarrow \top$ or $\ell \rightsquigarrow \neg\ell$) but the first form makes explanations easier for the paper.

– For any formulas φ_1 and φ_2:
  $\texttt{branches}(\varphi_1 \vee \varphi_2) = \texttt{branches}(\varphi_1) \cup \texttt{branches}(\varphi_2)$
  $\texttt{branches}(\varphi_1 \wedge \varphi_2) = \{B_1 \cup B_2 \mid B_1 \in \texttt{branches}(\varphi_1), B_2 \in \texttt{branches}(\varphi_2)\}.$

Furthermore, in order to preserve the independence to the syntax of the algorithm, we introduce the function $\texttt{min-branches}$, that converts the output of the $\texttt{branches}$ function into the set of prime implicants of the formula, represented as sets of literals.

For any set of literals L, let $\text{L}^{\neg} = \{\neg\ell \mid \ell \in \text{L}\}$. For example, if $\text{L} = \{a, \neg b, \neg c\}$ then $\text{L}^{\neg} = \{\neg a, b, c\}$. L is said to be consistent if and only if $\text{L} \cap \text{L}^{\neg} = \varnothing$.

The algorithm is an A^* search, where a state is an ordered pair (L, M) of consistent sets of literals. Given the propositional formulas ψ and μ, the set of the initial states is:

$$\texttt{min-branches}(\psi) \times \texttt{min-branches}(\mu)$$

A final state is a state (L, M) such that $\text{L} \cap \text{M}^{\neg} = \varnothing$ (which is equivalent to the fact that $\text{L} \cup \text{M}$ is consistent). If (L, M) is a non final state, then there exits $\ell \in \text{L}$ such that $\ell \in \text{L} \cap \text{M}^{\neg}$: there is a clash on ℓ.

The transitions from a state to another state are defined as follows. There is a transition σ from the state $x = (\text{L}_x, \text{M}_x)$ to the state $y = (\text{L}_y, \text{M}_y)$ if $\text{M}_x = \text{M}_y$ and:

– there exists a literal $\ell \in \text{L}_x$ such that $\text{L}_y = \text{F}_\ell(\text{L}_x)$ or
– there exists an adaptation rule R such that R is applicable on L_x and $\text{L}_y = \text{R}(\text{L}_x)$.

The cost of a transition σ is the cost of the rule (literal flip or adaptation rule) it uses.

Using a slight variant of the A^* algorithm, it is possible to determine the set of least costly transition sequences leading to final states. That is, the algorithm does not stop after finding the first optimal solution, but after finding every other optimal solutions (i.e., the solutions with the same minimal cost). The detailed algorithm is presented in Algorithm 1 and is explained hereafter.

The algorithm first creates the initial states (line 3). For every initial state $\text{S}_0, \mathcal{G}(\text{S}_0) = 0$, that is the cost of reaching S_0 is 0. In the case no value of $\mathcal{G}$ is set for a state S, $\mathcal{G}(\text{S})$ evaluates to $+\infty$, meaning there is no known path from an initial state to S. The algorithm then finds a state S_c minimizing $\mathcal{F}(\text{S}_c) = \mathcal{G}(\text{S}_c) + \mathcal{H}(\text{S}_c)$, that is the cost of reaching S_c from an initial state plus the heuristic cost of reaching a final state from S_c (line 10). For each rule or flip σ that can be applied on S_c, it creates the state S_d such that $\text{S}_c \xrightarrow{\sigma} \text{S}_d$. If $\mathcal{G}(\text{S}_d) > \mathcal{G}(\text{S}_c) + \texttt{cost}(\sigma)$, that is there is no known less or equally costly path to S_d, then $\mathcal{G}(\text{S}_d)$ is set to $\mathcal{G}(\text{S}_c) + \texttt{cost}(\sigma)$. Each generated S_d is added to the set of states to explore, while S_c is removed from it (lines 15 to 20). The algorithm repeats the previous step until it finds a state S_f minimizing $\mathcal{F}(\text{S}_f)$ and that is a final state (lines 11 to 13). From this point, the algorithm carries on but ignores any state S_c such that $\mathcal{F}(\text{S}_c) > \mathcal{F}(\text{S}_f)$, which cannot lead to a less costly solution. It stops when there is no more states that can lead to a less costly solution (line 9). Finally, all final states S_f minimizing $\mathcal{F}(\text{S}_f)$ are returned in the form of tableaux branches (line 22), that is for each state S_f defined by (L_f, M_f), the branch containing the literals $\text{L}_f \cup \text{M}_f$ is returned.

```
1 revise(ψ, μ, AK, cost)
  Input:
     – ψ and μ, two propositional formulas in NNF. ψ has to be revised by μ.
     – AK, the set of adaptations rules, used for repairing clashes.
     – cost, a function that associates to every literal flip and adaptation rule a positive real
       number.

  Output: ψ +ₐₖ μ in DNF
2 begin
      // open-states is initiated by initial states with cost 0.
3       open-states ← min-branches(ψ) × min-branches(μ)
4       𝒢(S) evaluates to +∞ by default
5       𝒢(S) ← 0 for each S ∈ open-states
6       ℱ(S) evaluates to 𝒢(S) + ℋ(S)
7       Solutions ← ∅
8       solutionCost ← +∞
9       while {S | S ∈ open-states and ℱ(S) ≤ solutionCost} ≠ ∅ do
          // (Lᶜ, Mᶜ) is the current state.
10         (Lᶜ, Mᶜ) ← one of the S ∈ open-states that minimizes ℱ(S)
11         if (Lᶜ ∩ Mᶜ⁻) = ∅ then
12             Solutions ← Solutions ∪ {Lᶜ ∪ Mᶜ}
13             solutionCost ← ℱ((Lᶜ, Mᶜ))
14         else
15             for each σ ∈ AK ∪ {Fℓ | ℓ: literal} such that σ is applicable on Lᶜ do
16                 open-states ← open-states ∪ {(σ(Lᶜ), Mᶜ)}
17                 𝒢((σ(Lᶜ), Mᶜ)) ← min(𝒢((σ(Lᶜ), Mᶜ)), 𝒢((Lᶜ, Mᶜ)) + cost(σ))
18             end
19         end
20         open-states ← open-states \ {(Lᶜ, Mᶜ)}
21     end
22     return Solutions
23 end
```

Algorithm 1: Algorithm of revision based on tableaux repairs using adaptation knowledge.

4.2 Heuristics

The function $\mathcal{H}$ used in our algorithm is defined by: for $S = (L, M), \mathcal{H}(S) = \text{ERC}(L \cap M^-)$ where ERC is defined as follows (ERC stands for Estimated Repair Cost):

$$\text{ERC}(\{\ell\}) = \min\{\text{cost}(F_\ell)\} \cup \left\{ \frac{\text{cost}(R)}{|\text{repairs}(R)|} \;\middle|\; \begin{array}{l} R \in AK, \text{ such that} \\ \ell \in \text{repairs}(R) \end{array} \right\} \quad \text{for any literal } \ell$$

$$\text{ERC}(L) = \sum_{\ell \in L} \text{ERC}(\{\ell\}) \qquad\qquad\qquad \text{for any set of literals } L$$

Proposition 1. $\mathcal{H}$ *is admissible.*

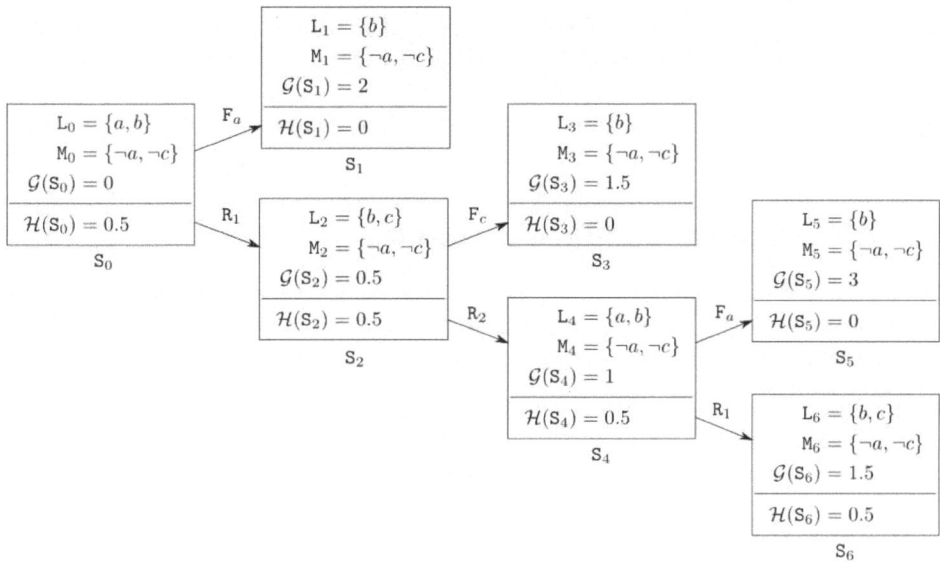

Fig. 1. Application of the algorithm on the running example

4.3   Example

Let us consider the following inputs:

$$\psi = a \wedge b \qquad\qquad \mu = \neg a \wedge \neg c$$
$$\mathtt{AK} = \{R_1, R_2\}$$
$$R_1 = a \wedge b \rightsquigarrow b \wedge c \qquad \mathtt{cost}(R_1) = 0.5$$
$$R_2 = b \wedge c \rightsquigarrow a \wedge b \qquad \mathtt{cost}(R_2) = 0.5$$
$$\mathtt{cost}(F_\ell) = 1 \text{ for } \ell \neq a$$
$$\mathtt{cost}(F_a) = 2$$

Here, $\mathtt{branches}(\psi) = \{\{a, b\}\}$ and $\mathtt{branches}(\mu) = \{\{\neg a, \neg c\}\}$. Figure 1 shows the different branches developed by the algorithm. At the initial state S_0, the cost is 0. The algorithm repairs the clash on the variable a, for a minimal cost of 0.5. Two branches are developed, the first using a flip on a and the second using the rule R_1. The state S_1 is a final state with a cost of 2. The variable $\mathtt{solutionCost}$ is set to that cost of 2. That cost is defined as the best final cost. The state S_2 has a clash on the variable c. To repair that clash, the minimal cost is 0.5. Adding the cost of that state, the cost is lower than $\mathtt{solutionCost}$. Thus, the clash is repaired to possibly find the solution with a lower final cost. To repair the clash on the variable c, the rule R_2 and the flip F_c are used. The use of F_c leads to a final state S_3 with a final cost of 1.5. That cost becomes the best final cost. The use of R_2 gets back to a state S_4 equivalent to the initial state S_0 but with a cost of 1. To repair the clash on the variable a, the minimal cost is 0.5. As the cost to reach that state plus the cost to repair that clash is equal to $\mathtt{solutionCost}$, the clash on the variable a is repaired again. The state S_5 is reached using F_a. That final state is not a best solution because the cost of that state is higher than solutionCost. The state S_6 is not final but that branch is not developed because the cost to reach that state plus the cost to repair the clash on the variable c is higher than solutionCost. The algorithm returns the solution $L_3 \cup M_3$, i.e. $\{b, \neg a, \neg c\}$.

4.4   Termination and Complexity of the Algorithm

In this section, the termination and complexity of the algorithm are studied when no heuristics is used ($\mathcal{H} = 0$). Since the heuristics introduced above is admissible, the complexity without heuristics is an upper bound for the complexity with heuristics.

Proposition 2. *The algorithm described in this paper always terminates. Its complexity in the worst case is in $O(4^n \times t^{\varrho+1} \times (n + \varrho \log(t)))$, where n is the number of variables, $t = |AK| + |\mathcal{V}|$, and ϱ is the sum of each flip cost divided by the minimum transition cost.*

According to [17], the complexity of a revision operator is $\mathsf{P}^{\mathsf{NP}[O(\log n)]}$-hard[3] hence this high worst-case complexity of the algorithm is not a surprise.

4.5   Optimization

The algorithm can be improved by considering the pairs of rules that commute. That optimization is an application of symmetry breaking constraints [18]. The adaptation rules R_1 and R_2 commute if the two sequences of rules R_1-R_2 and R_2-R_1 are applicable on the same set of formulas and give equivalent results. Formally, R_1 and R_2 commute if, for every set of literals L

- The two following assertions are equivalent:
    - (i) R_1 is applicable on L and R_2 is applicable on $R_1(L)$.
    - (ii) R_2 is applicable on L and R_1 is applicable on $R_2(L)$.
- and, when (i) (or (ii)) holds, $R_2(R_1(L)) = R_1(R_2(L))$.

So, the algorithm is changed as follows. First, an arbitrary total order $\leq$ is defined on the set of rules (including the F_ℓ's). Then, the algorithm only enables the application of a rule $\sigma \in AK \cup \{F_\ell \mid \ell: \text{literal}\}$ on a state such that there is no rule σ' commuting with σ and such that $\sigma' < \sigma$, that has been applied at the previous step to build the current state (this can be changed on line 15).

For example, if R_1 and R_2 commute and $R_1 < R_2$, then with the previous version of the algorithm there may be two branches developed in the search space that contain respectively the sequences R_1-R_2 and R_2-R_1 while, without loss of results, the new version of the algorithm generates only the former sequence.

In order to determine whether two adaptation rules commute, the following proposition can be used.

---

[3] More precisely, the complexity of the Dalal revision operator $\dot{+}_{\text{Dalal}}$ is $\mathsf{P}^{\mathsf{NP}[O(\log n)]}$-complete and the algorithm presented in this paper can be used for computing $\dot{+}_{\text{Dalal}}$: $\dot{+}_{\text{Dalal}} = \dot{+}_{\text{AK}}$ with $AK = \varnothing$ and $\text{cost}(F_\ell) = 1$ for each literal ℓ.

Table 1. Average time and standard deviation under REVISOR/PLAK according to d^*, with and without optimization

d^*	REVISOR/PLAK without optimization		REVISOR/PLAK with optimization	
	Average time (ms)	Standard deviation	Average time (ms)	Standard deviation
1	0.521	0.941	0.466	0.415
2	1.965	1.978	1.309	1.146
3	127.993	708.208	10.366	24.77
4	1349.659	2347.341	43.331	54.685

Proposition 3. *Let* $R_1 = \mathtt{left}_1 \rightsquigarrow \mathtt{right}_1$ *and* $R_2 = \mathtt{left}_2 \rightsquigarrow \mathtt{right}_2$ *be two adaptation rules.* R_1 *and* R_2 *commute iff the two following conditions hold:*

$$\mathtt{repairs}(R_1) \cap \mathtt{repairs}(R_2) = \varnothing$$
$$(\mathtt{right}_1 \cup \mathtt{right}_2) \cap (\mathtt{repairs}(R_1) \cup \mathtt{repairs}(R_2)) = \varnothing$$

5 Implementation Issues

5.1 REVISOR/PLAK

REVISOR/PLAK implements the algorithm for revision-based and rule-based adaptation based on tableaux repairs in propositional logic. REVISOR/PLAK has been implemented in Java and is available for download on the site `http://revisor.loria.fr`. Java version 7 is required to launch the software.

So far, empirical data shows that the computing time is largely dependent upon ϱ. However, a million-fold increase of ϱ may result in only a tenfold increase of computing time. The current implementation is vulnerable to sets of rules $\{(\mathtt{left}_i, \mathtt{right}_i) \mid i \in \{0, 1, \ldots, n\}\}$ such that $\mathtt{left}_i \subseteq \mathtt{right}_{i-1}$ for $i \in \{1, \ldots, n\}$ and $\mathtt{left}_0 \subseteq \mathtt{right}_n$, that is, rules that could potentially create infinite paths. While the algorithm always terminates, the algorithm could generate very long finite paths using such rules, if their cost is very small compared to the cost of the solution. The most important factor for computing time is the distance between ψ and μ, d^*. Table 1 and Figure 2 present the average time of computation of REVISOR/PLAK according to d^*, with and without optimization. The tested adaptation problems are for $n = 50$ variables and 40 adaptation rules. Each rule and flip has a cost of 1. The average time are computed on series of 1000 tests, for each value of d^*, on a computer with a 2.60 Ghz processor and 10 GB of available memory. For example, for $d^* = 3$, without optimization REVISOR/PLAK solves the problems in 127.993 ms in average (standard deviation of 708.208) and with optimization in 10.366 ms in average (standard deviation of 24.77). Figure 2 shows that REVISOR/PLAK with optimization is much more efficient.

5.2 A Concrete Example

Bob searches for a pie recipe without cinnamon, egg or pear. No recipe of the case base exactly matches Bob request. The only pie recipe is a recipe of pear pie containing eggs:

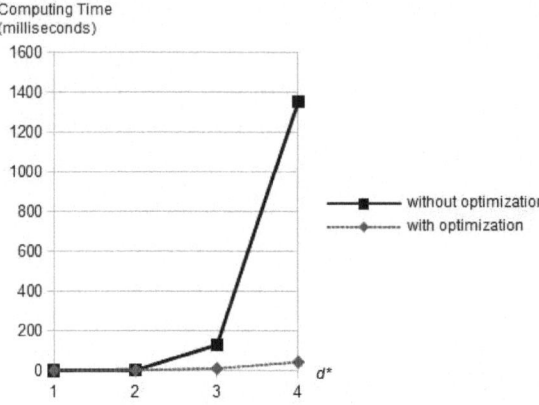

Fig. 2. Average time under REVISOR/PLAK according to d^*, with and without optimization

$$\texttt{Source} = \text{pie} \wedge \text{pie\_shell} \wedge \text{pear} \wedge \text{sugar} \wedge \text{egg}$$
$$\texttt{Target} = \text{pie} \wedge \neg\text{pear} \wedge \neg\text{cinnamon} \wedge \neg\text{egg}$$

REVISOR/PLAK adapts Source to solve Target, using the following knowledge:

$$\texttt{DK} = (\text{apple} \vee \text{peach} \vee \text{pear}) \Leftrightarrow \text{fruit}$$

(apples, peaches and pears are fruits and, conversely, the only available fruits are apples, peaches and pears)

$$\texttt{AK} = \{R_1, R_2, R_3, R_4, R_5, R_6\}$$
$$R_1 = \text{cake} \wedge \text{egg} \rightsquigarrow \text{cake} \wedge \text{banana}$$
$$R_2 = \text{pie} \wedge \text{egg} \rightsquigarrow \text{pie} \wedge \text{flour} \wedge \text{cider\_vinegar}$$
$$R_3 = \text{pear} \rightsquigarrow \text{peach}$$
$$R_4 = \text{pear} \rightsquigarrow \text{apple} \wedge \text{cinnamon}$$
$$R_5 = \text{cinnamon} \rightsquigarrow \text{orange\_blossom}$$
$$R_6 = \text{cinnamon} \rightsquigarrow \text{vanilla\_sugar}$$

Each rule has a cost of 0.3 except R_3 which has a cost of 0.7. Indeed, pears are considered to be more similar to apples than peaches. Each flip has a cost of 1.

REVISOR/PLAK gives the following result (in 1.4 ms and in 1.2 ms with the optimization after the computation of the function min-branches in 9.0 ms):

$$\text{pie} \wedge \text{pie\_shell} \wedge \text{sugar} \wedge \text{fruit} \wedge \neg\text{egg} \wedge \text{flour} \wedge \text{cider\_vinegar} \wedge$$
$$\neg\text{pear} \wedge \text{apple} \wedge \neg\text{cinnamon} \wedge (\text{vanilla\_sugar} \vee \text{orange\_blossom})$$

Two recipes are proposed. In both, pears have been replaced by apples, and eggs by flour and cider vinegar. For the cinnamon, two choices were available: either vanilla sugar or orange blossom could replace it: since $\text{cost}(R_5) = \text{cost}(R_6)$, the disjunction of there possibilities is presented; if $\text{cost}(R_5) < \text{cost}(R_6)$, only the orange blossom alternative would have been given.

6   Conclusion and Future Work

This paper proposes an original algorithm for revision-based and rule-based adaptation based on tableaux repairs in propositional logic. This algorithm modifies minimally the source case so that it becomes consistent with the target case. The tableaux method is applied separately on the source and target cases. Then the consistent branches are combined, which leads to a set of branches, each of them ending with a clash (unless the source case is consistent with the target case and need not to be adapted). Those clashes are repaired with adaptation rules which modify the source case. Adaptation rules allow to substitute a given part of a source case by something. If no rule is available, a flip on the literal leading to the clash deletes it from the source case.

This algorithm has been implemented in the REVISOR/PLAK tool. For each literal ℓ, a heuristic function computes the minimal cost to repair a clash on ℓ. Thus, at each step, the branch developed is the one for which the cost plus the cost to repair the clash is the lowest. All branches whose cost is lower or equal to the best final cost are developed. The algorithm is optimized by considering adaptation rules that commute which experimentally proves to improve the computational efficiency.

Belief revision is one of the operations of belief change. There are other ones (see [19], for a synthesis) such as contraction ($\psi \dot{-} \mu$ is a belief base obtained by minimally modifying ψ so that it does not entail μ) or integrity constraint belief merging [20] ($\Delta_\mu(\{\psi_1, \ldots, \psi_n\})$ is a belief base obtained by minimally modifying the belief bases ψ_i into ψ_i' such that $\mu \wedge \bigwedge_i \psi_i'$ is consistent). A possible line of future work consists of studying how the tableaux repair approach presented in this paper can be modified for such belief change operations. This would have an impact on CBR; in particular, integrity constraint belief merging can be used for multiple case adaptation (i.e., combining several source cases to solve the target case), see [21] for details.

References

1. Riesbeck, C.K., Schank, R.C.: Inside Case-Based Reasoning. Lawrence Erlbaum Associates, Inc., Hillsdale (1989)
2. Aamodt, A., Plaza, E.: Case-based Reasoning: Foundational Issues, Methodological Variations, and System Approaches. AI Communications 7(1), 39–59 (1994)
3. Minor, M., Bergmann, R., Görg, S., Walter, K.: Towards Case-Based Adaptation of Workflows. In: Bichindaritz, I., Montani, S. (eds.) ICCBR 2010. LNCS, vol. 6176, pp. 421–435. Springer, Heidelberg (2010)
4. Manzano, S., Ontañón, S., Plaza, E.: Amalgam-Based Reuse for Multiagent Case-Based Reasoning. In: Ram, A., Wiratunga, N. (eds.) ICCBR 2011. LNCS, vol. 6880, pp. 122–136. Springer, Heidelberg (2011)
5. Coman, A., Muñoz-Avila, H.: Diverse plan generation by plan adaptation and by first-principles planning: A comparative study. In: Díaz-Agudo, B., Watson, I. (eds.) ICCBR 2012. LNCS, vol. 7466, pp. 32–46. Springer, Heidelberg (2012)

6. Rubin, J., Watson, I.: Opponent type adaptation for case-based strategies in adversarial games. In: Díaz-Agudo, B., Watson, I. (eds.) ICCBR 2012. LNCS, vol. 7466, pp. 357–368. Springer, Heidelberg (2012)
7. Lieber, J.: Application of the Revision Theory to Adaptation in Case-Based Reasoning: The Conservative Adaptation. In: Weber, R.O., Richter, M.M. (eds.) ICCBR 2007. LNCS (LNAI), vol. 4626, pp. 239–253. Springer, Heidelberg (2007)
8. Cojan, J., Lieber, J.: Applying belief revision to case-based reasoning. In: Prade, H., Richard, G. (eds.) Computational Approaches to Analogical Reasoning: Current Trends. SCI, vol. 548, pp. 133–162. Springer, Heidelberg (2014)
9. Alchourrón, C.E., Gärdenfors, P., Makinson, D.: On the Logic of Theory Change: partial meet functions for contraction and revision. Journal of Symbolic Logic 50, 510–530 (1985)
10. Cojan, J., Lieber, J.: An Algorithm for Adapting Cases Represented in an Expressive Description Logic. In: Bichindaritz, I., Montani, S. (eds.) ICCBR 2010. LNCS, vol. 6176, pp. 51–65. Springer, Heidelberg (2010)
11. Schwind, C.: From Inconsistency to Consistency: Knowledge Base Revision by Tableaux Opening. In: Kuri-Morales, A., Simari, G.R. (eds.) IBERAMIA 2010. LNCS, vol. 6433, pp. 120–132. Springer, Heidelberg (2010)
12. Melis, E., Lieber, J., Napoli, A.: Reformulation in Case-Based Reasoning. In: Smyth, B., Cunningham, P. (eds.) EWCBR 1998. LNCS (LNAI), vol. 1488, pp. 172–183. Springer, Heidelberg (1998)
13. Personeni, G., Hermann, A., Lieber, J.: Adapting propositional cases based on tableaux repairs using adaptation knowledge – extended report (2014), http://hal.archives-ouvertes.fr/docs/01/01/17/51/PDF/report_on_revisor_plak.pdf
14. Marquis, P., Sadaoui, S.: A new algorithm for computing theory prime implicates compilations. In: AAAI/IAAI, vol. 1, pp. 504–509 (1996)
15. Katsuno, H., Mendelzon, A.: Propositional knowledge base revision and minimal change. Artificial Intelligence 52(3), 263–294 (1991)
16. Pearl, J.: Heuristics – Intelligent Search Strategies for Computer Problem Solving. Addison-Wesley Publishing Co., Reading (1984)
17. Eiter, T., Gottlob, G.: On the Complexity of Propositional Knowledge Base Revision, Updates, and Counterfactuals. Artificial Intelligence 57, 227–270 (1992)
18. Gent, I., Petrie, K., Puget, J.-F.: Symmetry in constraint programming. In: Rossi, F., van Beek, P., Walsh, T. (eds.) Handbook for Constraint Programming, ch. 10, pp. 329–376. Elsevier (2006)
19. Peppas, P.: Belief Revision. In: van Harmelen, F., Lifschitz, V., Porter, B. (eds.) Handbook of Knowledge Representation, ch. 8, pp. 317–359. Elsevier (2008)
20. Konieczny, S., Pino Pérez, R.: Merging information under constraints: a logical framework. Journal of Logic and Computation 12(5), 773–808 (2002)
21. Cojan, J., Lieber, J.: Belief Merging-Based Case Combination. In: McGinty, L., Wilson, D.C. (eds.) ICCBR 2009. LNCS, vol. 5650, pp. 105–119. Springer, Heidelberg (2009)

Least Common Subsumer Trees
for Plan Retrieval

Antonio A. Sánchez-Ruiz[1] and Santiago Ontañón[2]

[1] Dep. Ingeniería del Software e Inteligencia Artificial,
Universidad Complutense de Madrid, Spain
antsanch@fdi.ucm.es
[2] Computer Science Department,
Drexel University,
Philadelphia, PA, USA 19104
santi@cs.drexel.edu

Abstract. This paper presents a new hierarchical case retrieval method called *Least Common Subsumer Trees* (LCS trees). LCS trees perform a hierarchical clustering of the cases in the case base by iteratively computing the least-common subsumer of pairs of cases. We show that LCS trees offer two main advantages: First, they can enhance the accuracy of the CBR system by capturing regularities in the case base that are not captured by the similarity measure. Second, they can reduce retrieval time by filtering the set of cases that need to be considered for retrieval. We present and evaluate LCS trees in the context of plan retrieval for plan recognition, and present procedures for both assessing similarity and computing the least common subsumer of plans using refinement operators.

Keywords: Case-based reasoning, case retrieval, plan similarity.

1 Introduction

Similarity assessment, case retrieval and case base organization are some of the most studied topics in Case Based Reasoning (CBR). In this paper we extend our previous work regarding similarity assessment in structured representations [17,15,18] and particularly focus on retrieval of plans. We present both a similarity measure for plans using refinement operators and a new hierarchical organization of the case base, *LCS trees*, based on computing the least-common subsumer (LCS) of pairs of plans.

An LCS tree is a tree where the leaves are the cases in the case base, and each node is the LCS of its children. The intuitive idea of LCS trees is that they partition the case base into hierarchical clusters, based on whether cases subsume or not each of the nodes in the tree. Thus, LCS trees offer two main advantages with respect to standard linear case-retrieval: First, they can enhance the accuracy of the CBR system by capturing regularities in the case base that are not captured by a similarity measure; and second, they can reduce retrieval time by

L. Lamontagne and E. Plaza (Eds.): ICCBR 2014, LNCS 8765, pp. 405–419, 2014.
© Springer International Publishing Switzerland 2014

filtering the set of cases that need to be considered for retrieval. Compared to other tree-based case-retrieval approaches, LCS trees thus do not only focus on efficiency, but also on accuracy.

Although in this paper we focus on similarity between plans, LCS trees can be used with any structured representation as long as there exists a mechanism to compute the least-common subsumer of two domain entities. In this sense, the similarity measures presented in our previous work, based around computing the LCS [17,15,18], and which we extend in this paper with a similarity measure between plans, can be used build LCS trees for other representation formalisms, such as description logics or feature terms.

The rest of the paper is organized as follows. Section 2 introduces our plan representation. Section 3 describes a refinement operator for partial plans, which is then used in Section 4 to introduce a similarity measure between plans that also computes their least-common subsumer. Section 5 introduces the concept of LCS tree and explains how to use it to hierarchically organize the case base. In Section 6, we perform a empirical evaluation of both the similarity measure based on refinements and the LCS tree structure. The paper closes with related work, conclusions and directions for future research.

2   Plans and Partial Plans

In the context of this paper, we will consider a plan to be a sequence of actions. For example, consider a maze navigation domain where the only available domain actions are *move-forward*, *turn-right*, and *turn-left*. An example plan with 3 actions is:

$$\langle move\text{-}forward, turn\text{-}right, move\text{-}forward \rangle$$

Actions are usually formalized using preconditions and effects that describe, respectively, the conditions required to execute them and the changes that they produce. In this paper, however, we will compare plans only from a structural point of view (the actions involved, the order in which they are executed, and their parameters) without considering the preconditions or effects of the actions.

A partial plan is a compact way to describe sets of plans used in partial order planning [10]. A partial plan $\pi = (A_\pi, <_\pi)$ consists of a set of actions $A_\pi = \{a_1, a_2, \ldots, a_n\}$ and a set of ordering constraints of the form $a_i < a_j$, denoting that action a_i is executed before action a_j.

For example, consider the following partial plan consisting of 3 actions:

$$\pi_1 = (\{a_1, a_2, a_3\}, \{a_1 < a_3\})$$

According to this partial plan, a_1 has to be executed before a_3, but nothing is said about the relative execution order among the other actions. The usual interpretation of π_1 is that it implicitly describes the set of plans consisting of those 3 actions in which a_1 appears before a_3, i.e., $\langle a_1, a_2, a_3 \rangle$, $\langle a_1, a_3, a_2 \rangle$ and $\langle a_2, a_1, a_3 \rangle$. We say that a partial plan is *consistent* when the order induced over the actions by $<$ does not contain cycles.

In this work we will extend the usual interpretation of a partial plan π to include all the plans consisting *at least* of the actions in π as long as the actions appear in an order compatible with the ordering constraints in π. For example, using the new interpretation the previous partial plan π_1 implicitly describes $\langle a_1, a_2, a_3 \rangle$, $\langle a_1, a_3, a_2 \rangle$ and $\langle a_2, a_1, a_3 \rangle$, but also plans like $\langle a_1, a_2, a_3, a_4 \rangle$ or $\langle a_5, a_1, a_3, a_2, a_4 \rangle$. The set of plans implicitly represented by a partial plan is the *coverage* of the partial plan. Note that, if we do not limit the length of the plans, the coverage of a consistent partial plan is always an infinite set of plans.

Definition 1. *Given two partial plans π_1 and π_2, we say that π_1 subsumes π_2, denoted as $\pi_2 \sqsubseteq \pi_1$, and meaning that π_1 is more general than π_2 iff coverage(π_2) $\subseteq$ coverage(π_1).*

This subsumption relation between partial plans induces a semi-lattice where the most general plan $\top = (\varnothing, \varnothing)$ (a plan with no actions) is the root, and partial plans get more specific as we move away from the root. We have described two ways to make a plan more *specific*: adding a new action, or adding a new ordering constraint. We call *plan refinement* to this operation of transforming a plan into one that is more specific.

Until now we have considered actions in plans as atomic entities, but actions are in fact ground instances of planning *operators*. Planning operators use typed variables to describe sets of related actions. For example, consider a transportation domain that defines the following operator to move between different locations:

$$move(?from : location, ?to : location)$$

This operator implicitly describes all the actions that result from replacing each of those parameters with a constant compatible with the type *location*. If we extend the definition of partial plan to allow partially instantiated operators instead of ground actions, we can define additional plan refinements based on specializing the operator parameters. In particular, we can specialize types ($move(?from : store, ?to : location)$), we can unify variables ($move(?x : location, ?x : location)$) and we can replace typed variables with compatible constants ($move(bakery1, airport1)$).

The ideas of plan refinement, described in detail in the following section, and plan subsumption, form the basis of both the similarity measure (Section 4) and LCS trees (Section 5) presented in this paper.

3   Refinement Operators for Partial Plans

This section briefly summarizes the notion of *refinement operators* and the concepts relevant for this paper (see [12] for a more in-depth analysis of refinement operators). Refinement operators are defined over *quasi-ordered sets*.

Definition 2. *A quasi-ordered set is a pair $(S, \leq)$, where S is a set, and $\leq$ is a binary relation among elements of S that is reflexive ($a \leq a$) and transitive (if $a \leq b$ and $b \leq c$ then $a \leq c$).*

If $a \leqslant b$ and $b \leqslant a$, we say that $a \approx b$, or that they are *equivalent*. Concerning partial plans, the set of partial plans together with the *plan subsumption* operation (Definition 1) form a quasi-ordered set. A downward refinement operator is defined as follows:

Definition 3. *A downward refinement operator over a quasi-ordered set* $(S, \leqslant)$ *is a function* ρ *such that* $\forall a \in S : \rho(a) \subseteq \{b \in S | b \leqslant a\}$.

In the context of this paper, a downward refinement operator generates elements of S which are "more specific" (the complementary notion of upward refinement operator, corresponds to functions that generate elements of S which are "more general", but are not used in this paper). A common use of refinement operators is for navigating sets in an orderly way, given a starting element.

- A refinement operator ρ is *locally finite* if $\forall a \in S : \rho(a)$ is finite.
- A downward refinement operator ρ is *complete* if $\forall a, b \in S | a \leqslant b : a \in \rho^*(b)$.
- A refinement operator ρ is *proper* if $\forall a, b \in S \; b \in \rho(a) \Rightarrow a \not\approx b$.

where ρ^* means the *transitive closure* of a refinement operator. Intuitively, *locally finiteness* means that the refinement operator is computable, *completeness* means we can generate, by refinement of a, any element of S related to a given element a by the order relation $\leqslant$ (except maybe those which are equivalent to a), and *properness* means that a refinement operator does not generate elements which are equivalent to a given element a. When a refinement operator is locally finite, complete and proper, we say that it is *ideal*.

3.1 A Downward Refinement Operator for Partial Plans

We will define refinement operators as a set of *rewriting rules*. A rewriting rule is composed of three parts: the applicability conditions of the rewriting rule (shown between square brackets), the original partial plan (above the line), and the refined partial plan (below the line). Given a partial plan $\pi = (A_\pi, <_\pi)$ the following rewriting rules define a downward refinement operator ρ_c:

(R1) Add operator:

$$[a \text{ is a domain operator with fresh variables}] \quad \frac{\pi = (A_\pi, <_\pi)}{\pi' = (A_\pi \cup \{a\}, <_\pi)}$$

fresh variables means that a contains only variables, and none of them appears in π.

(R2) Add ordering constraint:

$$\begin{bmatrix} a_i, a_j \in A_\pi \; \wedge \\ a_i \not\ll a_j \; \wedge \\ a_j \not\ll a_i \end{bmatrix} \quad \frac{\pi = (A_\pi, <_\pi)}{\pi' = (A_\pi, <_\pi \cup \{a_i < a_j\})}$$

where $a_i \ll a_j = (a_i < a_j) \in <_\pi \vee (a_i < a') \in <_\pi \wedge a' \ll a_j$.

(**R3**) Variable unification:

$$\left[\begin{array}{c} x_1 : t, x_2 : t \text{ appear in } A_\pi \ \wedge \\ \theta = \{x_1 : t \mapsto x_2 : t\} \end{array}\right] \quad \frac{\pi = (A_\pi, <_\pi)}{\pi = (A_\pi\theta, <_\pi \theta)}$$

(**R4**) Type specialization:

$$\left[\begin{array}{c} x : t \text{ appears in } A_\pi \ \wedge \\ t' \text{ is a direct subtype of } t \ \wedge \\ \theta = \{x : t \mapsto x : t'\} \end{array}\right] \quad \frac{\pi = (A_\pi, <_\pi)}{\pi' = (A_\pi\theta, <_\pi \theta)}$$

(**R5**) Constant introduction:

$$\left[\begin{array}{c} x : t \text{ appears in } A_\pi \wedge \\ c \text{ has type } t \wedge \\ \theta = \{x : t \mapsto c : t\} \end{array}\right] \quad \frac{\pi = (A_\pi, <_\pi)}{\pi = (A_\pi\theta, <_\pi \theta)}$$

The above rules correspond to the 5 different ways to specialize a partial plan that we introduced in Section 2: adding a new planning operator, adding a new ordering constraint, unifying variables, replacing the type of a variable with a more specific type, and replacing a variable with constants of the same type (rules R3-5 use a substitution θ to specialize the parameters). ρ_c is locally finite and complete for the set of consistent partial plans (two partial plans are equivalent if one can be obtained by just renaming variables in the other).

Although ρ_c is not ideal, in practice it behaves quite well because the number of times it can be used without specializing a partial plan is limited (and usually quite small). Note that rules R1 and R2 always specialize the plan, and rules R3-5 do not produce a specialization only in very particular scenarios (when a type has only a subtype in the domain model or when there is only one constant compatible with a type).

4   Refinement-Based Similarity between Plans

Given the refinement operator ρ_c, we can measure similarity between plans following the recent idea of refinement-based similarity measures [15,18,17], illustrated in in Figure 1, and based on the following intuitions:

- First, given two partial plans π and π' such that $\pi' \sqsubseteq \pi$, it is possible to reach π' from π by applying a complete downward refinement operator ρ to π a finite number of times, i.e. $\pi' \in \rho^*(\pi)$.
- Second, the number of times a refinement operator needs to be applied to reach π' from π is an indication of how much more specific π' is than π. The length of the chain of refinements to reach π' from π, which will be noted as $\lambda(\pi \xrightarrow{\rho} \pi')$, is an indicator of how much information π' contains that was not contained in π. It is also an indication of their similarity: the smaller the length, the higher their similarity.

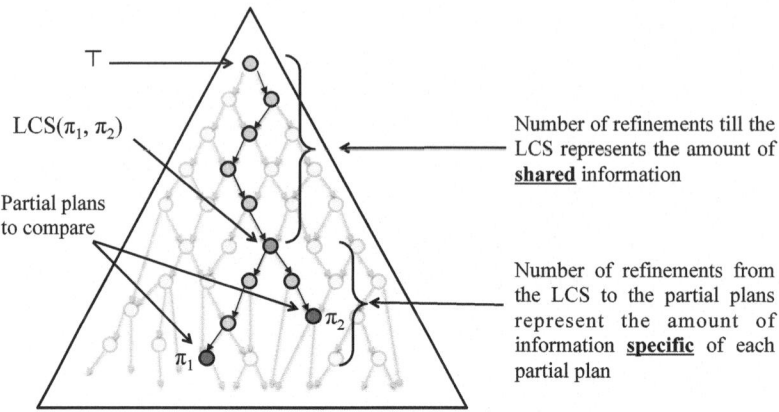

Fig. 1. Partial plan distance based on refinements

- Third, given any two partial plans, their *least common subsumer* (LCS) is the most specific partial plan which subsumes both. The larger the LCS (more actions and more ordering constraints), the more the two partial plans share. Given two partial plans π_1 and π_2, the LCS can be computed by starting with the most general plan $\top$ and refining it with the refinement operator until no more refinements can be applied that result in a partial plan that still subsumes π_1 and π_2.

The LCS of two partial plans contains all that is shared between them, and the more they share the more similar they are. $\lambda(\top \overset{\rho}{\to} LCS(\pi_1, \pi_2))$ measures the distance from the most general partial plan, $\top$, to the LCS and, thus, it is a measure of the amount of information shared by π_1 and π_2. Similarly, $\lambda(LCS(\pi_1, \pi_2) \overset{\rho}{\to} \pi_1)$ and $\lambda(LCS(\pi_1, \pi_2) \overset{\rho}{\to} \pi_2)$ measure the amount of information specific to each partial plan and that is not shared.

Using these ideas, the similarity between two partial plans π_1 and π_2 can be measured as the ratio between the amount of information contained in their LCS and the total amount of information contained in π_1 and π_2. These ideas are collected in the following formula:

$$S_\rho(\pi_1, \pi_2) = \frac{\lambda_1}{\lambda_1 + \lambda_2 + \lambda_3}$$

where:

$$\lambda_1 = \lambda(\top \overset{\rho}{\to} LCS(\pi_1, \pi_2))$$
$$\lambda_2 = \lambda(LCS(\pi_1, \pi_2) \overset{\rho}{\to} \pi_1)$$
$$\lambda_3 = \lambda(LCS(\pi_1, \pi_2) \overset{\rho}{\to} \pi_2)$$

Intuitively, if two plans are identical, then λ_2 and λ_3 are 0, and thus the similarity is 1. On the other hand, if two plans share nothing, then λ_1 is 0, and thus the similarity is 0.

Constants
$p1 : location, p2 : location, b1 : box, b2 : box$

Plans to partial plans
$\langle take(b1), move(p1, p2) \rangle \rightarrow \pi_1 : (\{a_1 : take(b1), a_2 : move(p1, p2)\}, \{a_1 < a_2\})$
$\langle move(p1, p2), drop(b2) \rangle \rightarrow \pi_2 : (\{b_1 : move(p1, p2), b_2 : take(b2)\}, \{b_1 < b_2\})$

Refinements from $\top$ to the LCS
$0 : (\{\}, \{\})$
$1 : (\{c_1 : move(?x1 : location, ?x2 : location)\}, \{\})$
$2 : (\{c_1 : move(\underline{p1}, ?x2 : location)\}, \{\})$
$3 : (\{c_1 : move(\underline{p1}, \underline{p2})\}, \{\})$

Refinements from the LCS to π_1
$1 : (\{c_1 : move(p1, p2), c_2 : take(?x3 : object)\}, \{\})$
$2 : (\{c_1 : move(p1, p2), \overline{c_2 : take(?x3 : object)}\}, \{\underline{c_2 < c_1}\})$
$3 : (\{c_1 : move(p1, p2), c_2 : take(?x3 : \underline{box})\}, \{c_2 < c_1\})$
$4 : (\{c_1 : move(p1, p2), c_2 : take(\underline{b1})\}, \{c_2 < c_1\})$

Refinements from the LCS to π_2
$1 : (\{c_1 : move(p1, p2), c_2 : drop(?x4 : object)\}, \{\})$
$2 : (\{c_1 : move(p1, p2), c_2 : take(?x4 : object)\}, \{\underline{c_1 < c_2}\})$
$3 : (\{c_1 : move(p1, p2), c_2 : take(?x4 : \underline{box})\}, \{c_1 < c_2\})$
$4 : (\{c_1 : move(p1, p2), c_2 : take(\underline{b2})\}, \{c_1 < c_2\})$

Similarity $\quad S_\rho(\pi_1, \pi_2) = \frac{3}{3+4+4} = 0.27$

Fig. 2. Similarity assessment example (refinements are underlined)

In order to compute the similarity between (non-partial) plans, we first transform them into the most specific partial plans that cover them and then use the previous similarity measure. This transformation is quite straightforward since the most specific partial plan that covers a plan $p = \langle a_1, \ldots, a_k \rangle$, is $\pi = (\{a_1, \ldots, a_k\}, \{(a_1 < a_2), \ldots, (a_{k-1} < a_k)\})$, i.e., a partial plan with the same actions that defines ordering constraints between consecutive actions.

Figure 2 shows a detailed example of the similarity assessment between 2 plans using the procedure described in this section.

5   Least Common Subsumer Trees

The retrieval task in a CBR system involves finding the most relevant cases in the case base to solve a given query. In our system the case base is a repository of plans annotated with the type of problem they try to solve. Given a plan query, a standard way to perform retrieval is to find the *nearest neighbors* using a similarity measure. The standard approach to find the nearest neighbor is to perform linear search, but efficient approaches such as kd-trees [21] or cover trees [1] exist that can find the nearest neighbor in near logarithmic time.

In this paper, we present a new hierarchical structure that we call an *LCS tree* (Least Common Subsumer tree) to perform retrieval. The LCS tree exploits the fact that our refinement-based similarity measures not only provides a numerical value of similarity but an explicit description of the structure shared between two plans (the LCS). In this section, we describe how to use this additional information to hierarchically organize the plan base. Thanks to this additional information, our hierarchical structure not only helps in decreasing retrieval time, but also exploits the LCS information to increase performance by retrieving better cases. An LCS tree is defined in the following way: each node in the tree is a partial plan; each node that is not a leaf is the LCS of its children. Thus, a given node is always subsumed by all of its ancestors, and a given node always subsumes all of its descendants.

Algorithm 1. Construction of the LCS tree.

1: **function** BUILDLCSTREE(*PlanBase*)
2:      $T = \emptyset$
3:      **for all** *Plan* $\in$ *PlanBase* **do**
4:          $T = T \cup \{$ TREE(PARTIALPLAN(*Plan*)) $\}$
5:      **end for**
6:      **while** $|T| \neq 1$ **do**
7:          $MSP =$ PAIRSOFMOSTSIMILARTREES(T)
8:          **for all** $(t_1, t_2) \in MSP$ **do**
9:              $T = T \setminus \{t_1, t_2\} \cup \{$TREE($t_1$, $LCS(t_1, t_2)$, t_2) $\}$
10:          **end for**
11:      **end while**
12:      **return** only tree in T
13: **end function**

Algorithm 1 shows how to build a LCS tree from a collection of plans. The algorithm builds the tree from the leaves to the root and consists of two different stages. The first stage (lines 3-5) transforms each plan from the input into a partial plan using the approach described in Section 4, and then generates a set T with one tree leaf containing each of the partial plans. The second part of the algorithm (lines 6-11) matches the trees than contain the most similar partial plans in pairs (according to our similarity measure) and adds them as children of a new tree whose root contains the LCS of both plans. That way, each pass through the loop generates a new level of the tree with approximately half of the previous level nodes. At the end of the process we obtain a binary tree in which each node contains a partial plan which subsumes (in more general than) all the partial plans in the nodes below. And the leaves of the tree contain the partial plans corresponding to the original plans in the plan base.

Figure 3 shows an example LCS tree generated with Algorithm 1 from a set of three plans. As can be seen, in the LCS tree, each node subsumes all of its descendants, and, in particular, the root subsumes all the plans in the case base. An interesting aspect of LCS trees is that if there is a set of features that a

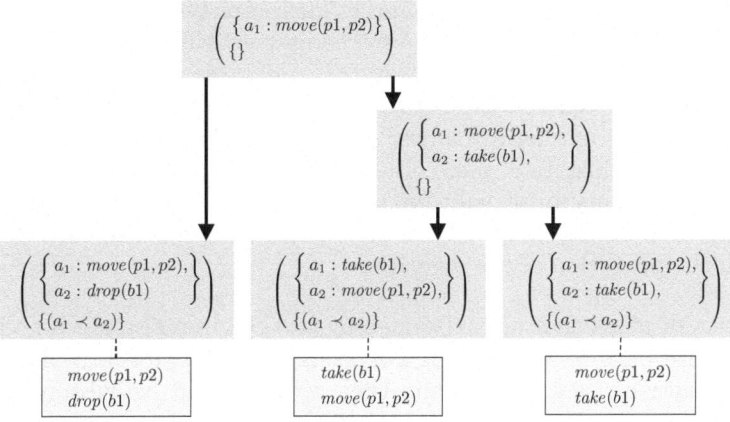

Fig. 3. Example LCS tree constructed from three plans in the transportation domain. Each node subsumes all its descendants, and the leaves are the plans in the case base.

large number of plans in the case base share, then those would most likely be captured by some node in the tree as part of the LCS. This is interesting, since if there is some feature that is shared amongst a large set of plans, probably means that that feature is important. Thus, in a sense, the LCS tree can be likened to a hierarchical clustering technique [9] that computes a description (the LCS) of each cluster, or to a subset of the concept lattice that would be built by Formal Concept Analysis [7]. Given a new problem, we can determine whether the new problem belongs to one of these clusters or not via subsumption by the appropriate LCS. This idea defines our retrieval algorithm.

Algorithm 2. Retrieve partial plan candidates.

```
 1: function RETRIEVE(LCSTree, Query, MaxDepth)
 2:     if ISLEAF(LCSTree) or DEPTH(LCSTree) ≥ MaxDepth then
 3:         C = PARTIALPLANSINLEAVES(LCSTree)
 4:     else
 5:         C = ∅
 6:         for all SubTree of LCSTree do                    ▷ left or right subtree
 7:             if ISLEAF(SubTree) or PARTIALPLAN(SubTree) subsumes Query then
 8:                 C = C∪ RETRIEVE(SubTree, Query, MaxDepth)
 9:             end if
10:         end for
11:         if C = ∅ then
12:             C = C∪ PARTIALPLANSINLEAVES(LCSTree)
13:         end if
14:     end if
15:     return C                                            ▷ Candidates
16: end function
```

The retrieval is shown in Algorithm 2. There are two intuitions behind this algorithm. First, similar plans will be stored in nearby leaves of the tree because the more similar the plans are the closer they will be to their LCS node. Second, the differences between the LCSs that are in the higher levels of the tree are likely to be differences that are of key importance, since they are the ones that distinguish larger clusters of plans. Basically, given a plan query the algorithm looks for similar plans in the subtrees containing a partial plan which subsumes (more general than) the query. We need to make several considerations in order to explain the operation of the algorithm:

- The sets of plans that are subsumed by two children of a given node do not necessarily have to be disjoint. For example, in Figure 3, a plan that has a move action, then a drop, and then a take would be subsumed by both children of the root node. Therefore, during retrieval we might have to visit both of them. In practice, the tree is useful to filter an important number of plans (the leaves of the tree), but in the worst scenario the search remains linear (line 6).
- The algorithm does not only return a plan but a sequence of candidates. During the search process it is possible to reach a node such that none of its children subsume the query. In this situation the retrieval algorithm returns all the plans in the leaves of the current tree (lines 11-13) as candidates. The list of candidates is processed afterwards linearly to find the most similar plan among them to the query.
- Although both the similarity measure and the tree construction are based on the idea of subsumption, there is no guarantee that the resulting list of candidates will contain the nearest neighbor to the query. The list of candidates will contain, however, some of the most similar plans in the tree and, as we discuss in Section 6, it outperforms nearest neighbor retrieval in our experiments.
- The parameter $MaxDepth$ limits the maximum depth of the search and it is useful to configure the behavior of the algorithm (line 2). Smaller values of $MaxDepth$ produce larger sets of candidates (the probability of missing the most similar case decreases) and increases the retrieval time, while larger values of $MaxDepth$ produce smaller sets of candidates and improve the retrieval times. So we can use this parameter to balance the desired level of accuracy and performance.

6   Evaluation

In order to evaluate the approach presented in this paper, we used two datasets, commonly used in the plan recognition literature: *Linux* and *Monroe*. The Linux dataset [4] consists of 457 plans, classified into 18 different classes, where each class represents one of 18 different goals that a Linux user can be trying to achieve (create a file, compress a folder, etc.). Each plan contains between 1 to 60 actions (with an average of 6.1), and the actions are Linux command-line commands including their parameters. The Monroe dataset [2] contains plans for a disaster management domain. It consists of 5000 plans classified into 10 different classes, and each plan contains between 1 to 29 actions (with an average of 9.6). Plans in the Linux domain correspond to recorded traces of human users (real data) while plans in the Monroe domain were generated using a modified version of the SHOP2 planner [14] (synthetic data).

We performed two sets of experiments. In the first one, we evaluated the performance of LCS trees in the context of a nearest neighbor classifier, when trying to predict the correct class for each of the plans in each of the datasets. In a second experiment, we evaluated the performance of our LCS trees in determining the class of a plan when only part of the plan can be observed (i.e., when we can only observe the first few actions of the plan). This second experiment is relevant for many tasks, such as intrusion detection or plan recognition (where we want to identify a plan before it is actually completed). All of our experiments were performed using a *leave-one-out* method, where to classify one plan, we compared it against all the other plans in our training set.

In both experiments we compared LCS trees using our refinement-based similarity measure S_ρ, against linear standard retrieval using both S_ρ and a collection of other similarity measures:

- *Random*: just returns a random number between 0 and 1.
- *Jaccard*: given two plans, their similarity is determined using the Jaccard index, i.e., the number of common actions divided by the total number of different actions in both plans.
- *Edit-distance (action-level)*: assuming that a plan is a sequence of symbols, and where each action is a different symbol, we compute the distance between two plans p and q as the edit distance [13] between these sequences.
- *Edit-distance (symbol-level)*: this similarity is the same as the previous one, but where consider that each action name, and each action parameter is a different symbol. So, this distance is more fine-grained.
- *Edit-distance (character-level)*: finally, going one step further, we convert every plan to a string of characters, used to compute the edit distance (at a character level). The rationale behind this distance is that similar parameters or actions tend to have similar string representations, and thus, without having to add additional domain knowledge to our planning domains about the relation between their actions or parameter types, we can compute similarity at an even finer granularity level.

Table 1 shows the performance in both domains. The Linux domain is well known to be a very hard domain (the original paper presenting the Linux domain reports precision and recall values of 0.351 and 0.236 respectively, when detecting the plan class) because the dataset contains real data. Nevertheless our refinement-based similarity measure S_ρ outperforms all the other similarity measures significantly. Additionally, when we compare the linear and the hierarchical plan base using LCS trees, we realize that the LCS trees not only substantially improve the retrieval time (reduced to a 29.5% of the original time

Table 1. Classification accuracy and computation time using several similarity measures in a nearest neighbor classifier

	Linux		Monroe	
	Accuracy (%)	Time (ms)	Accuracy (%)	Time (ms)
Random	5.69	0	12.6	0
Jaccard	52.74	1	77.6	2
Edit(Char)	52.30	38	97.8	220
Edit(Symbol)	51.86	5	96.6	10
Edit(Action)	49.23	0	81.0	1
S_ρ linear	56.67	549	99.8	288
LCS tree (S_ρ) depth 1	57.55	467	99.8	242
LCS tree (S_ρ) depth 3	57.55	412	99.8	233
LCS tree (S_ρ) depth 5	57.77	162	99.4	103

with $MaxDepth = 5$), but also improve the accuracy. The explanation is that, in this domain, the filtering performed by the LCS tree, equivalent to hierarchical clustering, captures additional information that the similarity measure does not.

The Monroe domain is an easier domain than the Linux one, and the literature on plan recognition reports prediction accuracies between 95% to 99% for different methods (for example, [3] report a precision of 95.6% in this dataset). Again, we observe that our refinement-based similarity outperforms all other similarity measures, reaching an accuracy of 99.8%. Edit distance at the character level is the similarity measure that came closer, reaching an accuracy of 97.8%. Another interesting result is that using LCS trees and $MaxDepth = 5$ we are able to reduce the retrieval time to 35.7% of the original time and we only loose a few tenths of accuracy.

Figure 4 shows the performance of the different best similarity measures in our second set of experiments. For different plan lengths k starting from 1 and up to the length of the longest plan in each dataset, we classify each plan but only using the first k actions of the plan. The intuition is that when only using the first few actions, classifying the plan should be harder, and the more actions being considered, the higher the expected performance. The idea is to model the situation (common in the plan recognition literature) of trying to recognize the first actions of a plan p against a library of complete plans. As expected, the classification accuracy improves quite fast as we consider more actions at the beginning and stabilizes around 10-13 actions for both S_ρ and the edition (char) distance. In this experiments using an LCS tree not only improves the performance but, in most of the lengths, classifies better the plans than the linear plan base, illustrating again, that the LCS-based clustering used in LCS trees can capture aspects of the problem domain that the similarity measure does not.

In summary, our results point out that the refinement-based approach to plan similarity achieves very good accuracy, higher than all the other similarity measures we used in our evaluation. Additionally, this accuracy can be improved in some domains by the structure imposed by LCS trees, achieving a side effect of significantly improving retrieval times.

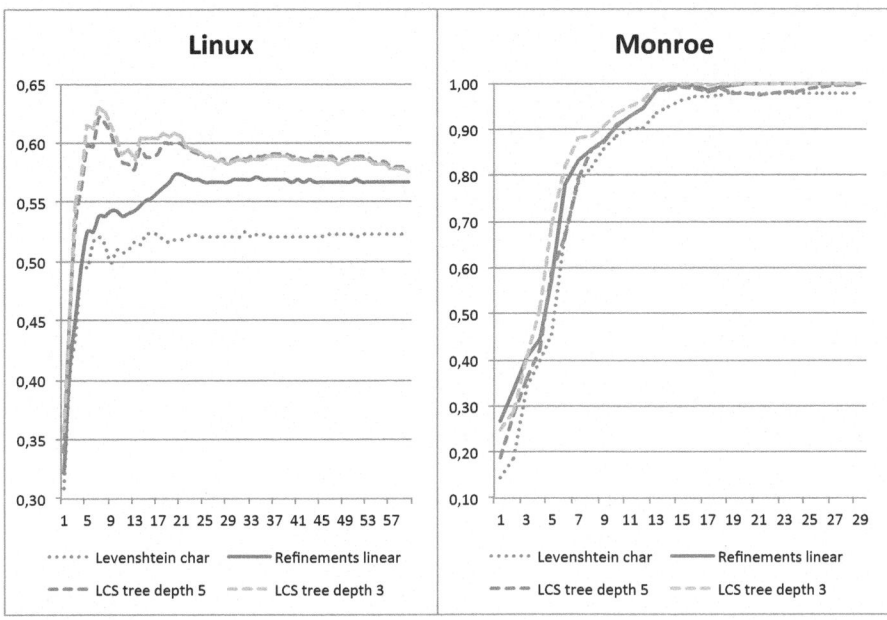

Fig. 4. Classification accuracy of NN for increasing plan length in the *Linux* and *Monroe* datasets

7 Related Work

Our *LCS trees* resemble other approaches such as *kd*-trees [21] or cover trees [1]. However, LCS trees are not just an organization to improve the retrieval time, they serve as an additional filter to retrieve better cases. In this sense, LCS trees can be related to hierarchical clustering [9].

Concerning our similarity between plans, similarity measures for plans have been studied mostly in the field of case-based planning [5]. Most of these similarities, like the *foot-print* similarity used in the PRODIGY system [20] or the Action-Distance Guided similarity [19] are based on comparing the initial and end states when executing plans. Our similarity measure focuses on the plan structure instead, and can be used for tasks like plan recognition in which the goal of one of the plans is unknown.

The refinement operator presented in this paper resembles the search process carried out by partial-order planners like UCPOP [16]. However, the space of partial plans is traversed with a different perspective since we are not solving a planning problem and we do not need to keep *casual links* between actions or an *agenda* of open goals.

Our work is related to the general idea of assessing similarity between structured representations. In our previous work we presented similarity measures for instances represented using Description Logics [17] and Feature Terms [15]. Other relevant work on structured similarity measures is that of RIBL

(Relational Instance-Based Learning), an approach to apply lazy learning techniques while using Horn clauses [6]. For a comprehensive overview on this topic, the reader is referred to [15].

Plan similarity is also related to the area of *plan recognition* [11] [8], where the goal is to identify plans, or their goals or intents. One of the main differences with respect to our work is that plan recognition concerns with the problem of identifying plans from an already existing plan model (e.g. a plan grammar, or a given set of goals), whereas in the work presented in this paper we are interesting on measuring similarity between two concrete plans.

8   Conclusions

In this paper we have introduced the *Least Common Subsumer Trees* (LCS trees), a new hierarchical case retrieval method that clusters the cases according to their least-common subsumers. LCS trees are able to capture implicit patterns in the case base and can be used in conjunction with another similarity measures to improve the accuracy of a CBR system. LCS trees also reduce the retrieval time by filtering the set of cases that need to be considered for retrieval.

Since LCS trees are based on subsumption, they can be used in domains formalized using structured representations. We have empirically shown their advantages in the context of plan recognition, using a nearest neighbor classifier and a repository of annotated plans. We have also described a similarity measure for partial plans based on refinements that can be used to compute the LCS and to build the tree. The combination of the LCS tree to retrieve a set of candidate plans and our similarity measure based on refinements to select the most similar among them, achieves very good accuracy results in our experiments.

As part of our future work, we would like to study the types of patterns that LCS trees are able to capture, as well as their relation with other hierarchical clustering techniques, to better understand the interplay between the unsupervised induction performed by the LCS tree and similarity-based case-retrieval.

References

1. Beygelzimer, A., Kakade, S., Langford, J.: Cover trees for nearest neighbor. In: Proceedings of the 23rd International Conference on Machine Learning, pp. 97–104. ACM (2006)
2. Blaylock, N., Allen, J.: Generating artificial corpora for plan recognition. In: Ardissono, L., Brna, P., Mitrović, A. (eds.) UM 2005. LNCS (LNAI), vol. 3538, pp. 179–188. Springer, Heidelberg (2005)
3. Blaylock, N., Allen, J.: Fast hierarchical goal schema recognition. In: Proceedings of the National Conference on Artificial Intelligence, vol. 21, p. 796. AAAI Press, MIT Press, Menlo Park, Cambridge (2006)
4. Blaylock, N., Allen, J.F.: Statistical goal parameter recognition. In: ICAPS, vol. 4, pp. 297–304 (2004)
5. Cox, M.T., Munoz-Avila, H., Bergmann, R.: Case-based planning. The Knowledge Engineering Review 20(03), 283–287 (2005)

6. Emde, W., Wettschereck, D.: Relational instance based learning. In: Saitta, L. (ed.) Machine Learning - Proceedings 13th International Conference on Machine Learning, pp. 122–130. Morgan Kaufmann Publishers (1996)
7. Ganter, B., Wille, R.: Formal concept analysis, vol. 284. Springer, Berlin (1999)
8. Goldman, R.P., Geib, C.W., Miller, C.A.: A new model of plan recognition. In: Proceedings of the Fifteenth Conference on Uncertainty in Artificial Intelligence, pp. 245–254. Kaufmann Publishers Inc. (1999)
9. Johnson, S.C.: Hierarchical clustering schemes. Psychometrika 32(3), 241–254 (1967)
10. Kambhampati, S., Knoblock, C.A., Yang, Q.: Planning as refinement search: A unified framework for evaluating design tradeoffs in partial-order planning. Artificial Intelligence 76(1), 167–238 (1995)
11. Kautz, H.A., Allen, J.F.: Generalized plan recognition. In: AAAI, vol. 86, pp. 32–37 (1986)
12. van der Laag, P.R.J., Nienhuys-Cheng, S.H.: Completeness and properness of refinement operators in inductive logic programming. Journal of Logic Programming 34(3), 201–225 (1998)
13. Levenshtein, V.I.: Binary codes capable of correcting deletions, insertions and reversals. Soviet Physics Doklady 10, 707 (1966)
14. Nau, D.S., Au, T.C., Ilghami, O., Kuter, U., Murdock, J.W., Wu, D., Yaman, F.: Shop2: An htn planning system. J. Artif. Intell. Res. (JAIR) 20, 379–404 (2003)
15. Ontañón, S., Plaza, E.: Similarity measures over refinement graphs. Machine Learning 87(1), 57–92 (2012)
16. Penberthy, J.S., Weld, D.S., et al.: Ucpop: A sound, complete, partial order planner for adl. KR 92, 103–114 (1992)
17. Sánchez-Ruiz, A.A., Ontañón, S., González-Calero, P.A., Plaza, E.: Refinement-based similarity measure over DL conjunctive queries. In: Delany, S.J., Ontañón, S. (eds.) ICCBR 2013. LNCS, vol. 7969, pp. 270–284. Springer, Heidelberg (2013)
18. Sánchez-Ruiz, A.A., Ontañón, S., González-Calero, P.A., Plaza, E.: Measuring similarity in description logics using refinement operators. In: Ram, A., Wiratunga, N. (eds.) ICCBR 2011. LNCS, vol. 6880, pp. 289–303. Springer, Heidelberg (2011)
19. Tonidandel, F., Rillo, M.: An accurate adaptation-guided similarity metric for case-based planning. In: Aha, D.W., Watson, I. (eds.) ICCBR 2001. LNCS (LNAI), vol. 2080, pp. 531–545. Springer, Heidelberg (2001)
20. Veloso, M.M., Carbonell, J.G.: Derivational analogy in prodigy: Automating case acquisition, storage, and utilization. In: Case-Based Learning, pp. 55–84. Springer (1993)
21. Wess, S., Althoff, K.D., Derwand, G.: Using kd trees to improve the retrieval step in case-based reasoning. In: Wess, S., Richter, M., Althoff, K.-D. (eds.) EWCBR 1993. LNCS, vol. 837, pp. 167–181. Springer, Heidelberg (1994)

Supervised Semantic Indexing Using Sub-spacing

Sadiq Sani, Nirmalie Wiratunga, Stewart Massie, and Robert Lothian

IDEAS Research Institute,
Robert Gordon University,
Aberdeen AB10 7GJ, Scotland, UK
{s.a.sani,n.wiratunga,s.massie,r.m.lothian}@rgu.ac.uk

Abstract. Indexing of textual cases is commonly affected by the problem of variation in vocabulary. Semantic indexing is commonly used to address this problem by discovering semantic or conceptual relatedness between individual terms and using this to improve textual case representation. However, representations produced using this approach are not optimal for supervised tasks because standard semantic indexing approaches do not take into account class membership of these textual cases. Supervised semantic indexing approaches e.g. sprinkled Latent Semantic Indexing (SPLSI) and supervised Latent Dirichlet Allocation (SLDA) have been proposed for addressing this limitation. However, both SPLSI and SLDA are computationally expensive and require parameter tuning. In this work, we present an approach called Supervised Sub-Spacing ($S3$) for supervised semantic indexing of documents. $S3$ works by creating a separate sub-space for each class within which class-specific term relations and term weights are extracted. The power of $S3$ lies in its ability to modify document representations such that documents that belong to the same class are made more similar to one another while, at the same time, reducing their similarity to documents of other classes. In addition, $S3$ is flexible enough to work with a variety of semantic relatedness metrics and yet, powerful enough that it leads to significant improvements in text classification accuracy. We evaluate our approach on a number of supervised datasets and results show classification performance on $S3$-based representations to significantly outperform both a supervised version of Latent Semantic Indexing (LSI) called Sprinkled LSI, and supervised LDA.

Keywords: Textual case-based reasoning, textual case representation, semantic indexing, supervised semantic indexing.

1 Introduction

Textual case-based reasoning (TCBR) has many important applications including sentiment classification, spam filtering and text classification. A common feature of these applications is that each document (textual case) is associated with a class label (sentiment class, spam/not spam label, topic label etc.). Thus, modelling each application in a TCBR system involves modelling the content of each document as a case description, and the class label as the case solution. Accordingly, the task of the TCBR system is to predict the label of a new document, given the class labels of its k nearest neighbours from the casebase of textual cases.

L. Lamontagne and E. Plaza (Eds.): ICCBR 2014, LNCS 8765, pp. 420–434, 2014.
© Springer International Publishing Switzerland 2014

An important step in the TCBR process is textual case indexing where documents are converted from raw text to a representation that is more suitable for computational algorithms. The bag-of-words (BOW) model is typically employed to represent textual cases as un-ordered collections of their constituent terms. However, the BOW model is not able to cope with variation in natural language vocabulary (e.g. synonymy and polysemy) which often requires semantic indexing approaches [12].

The general idea of semantic indexing is to discover terms that are semantically related and use this knowledge to identify conceptual similarity even in the presence of vocabulary variation. The result is the generalisation of textual case representations away from low-level expressions to high-level semantic concepts. Many different approaches have been proposed for semantic indexing e.g. Generalised Vector Space Model (GVSM) [16], Latent Semantic Indexing (LSI) [8] and Latent Dirichlet Allocation (LDA) [3]. Some of these approaches e.g. GVSM rely on explicit computation of pairwise semantic relatedness between terms using one of several available metrics e.g. document co-occurence [6]. Although unsupervised semantic indexing techniques have generally proven quite beneficial, further improvements can still be achieved by utilising information on class membership of documents [5]. This intuition is guided by the knowledge that supervised techniques generally outperform unsupervised ones.

Supervised semantic indexing approaches e.g. sprinkled Latent Semantic Indexing (SpLSI) and supervised Latent Dirichlet Allocation (sLDA) have been proposed to address the limitations with traditional semantic indexing. However, both SpLSI and sLDA are computationally expensive and require parameter tuning. Accordingly, our contribution in this paper is a novel supervised semantic indexing approach called Supervised Sub-Spacing ($S3$). $S3$ is a powerful technique that modifies document representations such that documents that belong to the same class are made more similar to one another. $S3$ is flexible enough to work with any semantic relatedness metric and yet, powerful enough that it leads to consistent improvements in text classification accuracy. Our evaluation on 30 datasets shows supervised semantic indexing using $S3$ to significantly outperform both a baseline BOW representation, as well as unsupervised semantic indexing. Significant improvements are also achieved using sLDA. However, sLDA seems to suffer on small datasets.

The rest of this paper is organised as follows, related work is presented in Section 2. In Section 3, we present $S3$ and describe how supervision is introduced into semantic indexing of documents. Section 4 describes the datasets we use for evaluation. In Section 5 we present our evaluation of $S3$ against several baseline supervised and unsupervised techniques, followed by conclusions in Section 6.

2   Related Work

A common representation approach employed for TCBR is the BOW model where documents are represented using individual terms as features [15]. However, this representation approach is unable to cope with variation in natural language vocabulary. Several approaches have been proposed for transforming document representations from the space of individual terms to that of latent concepts or topics that better capture the underlying semantics of the documents. The work we present here is related to this general class of feature transformation techniques.

A typical example of semantic document transformation approaches is a generative probabilistic model called latent dirichlet allocation (LDA) where each document in a collection is modeled as an infinite mixture over an underlying set of topics [3]. The set of LDA topics is taken to represent the underlying latent semantic structure of the document collection. Another approach is latent semantic indexing (LSI) which uses singular value decomposition (SVD) to transform an initial term-document matrix into a latent semantic space [8]. LSI reduces sparseness in representations through the discovery of higher-order relationships from term co-occurrences.

The main limitation of LDA and LSI for text classification is that these techniques are agnostic to class knowledge. Thus, the concept/topic spaces extracted by these approaches are not necessarily optimised for the class distribution of the document collection[1]. A supervised version of LDA called sLDA is presented in [2]. Here, a response variable (class label, real value, cardinal or ordinal integer value) associated with each document is added to the LDA model. Thus the topic model is learnt jointly for the documents and responses such that the resultant topics are good predictors of the response variables. The predictive performance of sLDA on two regression tasks compared with LDA and lasso (L_1-regularized linear regression) suggests moderate improvements on both tasks. However, choosing the optimal number of topics for sLDA is non-trivial. In addition, sLDA is a very computationally expensive process to run, taking several hours to complete even on small datasets [17]. This calls for further research into alternative semantic indexing approaches e.g. $S3$.

An extension of LSI called supervised LSI (SLSI) that iteratively computes SVD on term similarity matrices of separate class is presented in [11]. A separate term-doc matrix is constructed for each class and in each iteration, SVD is performed on each class-specific term-doc matrix. The most discriminative eigen vector accross all categories is selected as the basis vector in the current iteration. The effect of the selected eigen vector is then subtracted from the original term-document matrix. The iteration continues until the dimension of the resultant space reaches a predefined threshold. The evaluation compared three types of representations: standard BOW without transformation, unsupervised LSI and SLSI using kNN and SVM classifiers. Results show SLSI performs better than LSI. However, SLSI only achieved marginal gains over BOW using kNN while both SLSI and LSI failed to perform better than SVM. A more promising supervised extension to LSI uses an approach called sprinkling where class-specific artificial terms are appended to representations of documents of the corresponding class [5]. LSI is then applied on the sprinkled term-document space resulting in a concept space that better reflects the underlying class distribution of documents. Sprinkled LSI was compared with unsupervised LSI and SVM on a number of classification tasks. Results showed sprinkled LSI to significantly out perform both unsupervised LSI and SVM. However, sprinkling is only applicable to term relatedness techniques that exploit higher order associations between terms.

An important consideration for sprinkling is the number of artificial terms to sprinkle. In [4], the authors found sprinkling sixteen terms per-class to give optimal performance. In [6], the authors present an approach called adaptive sprinkling which optimises the number of sprinkled terms for each individual dataset based on dataset complexity. Adaptive sprinkling exploits the confusion matrix of each dataset produced by a classifier. A confusion matrix records the performance of a classifier such that the columns of the matrix represent the instances predicted by the classifier and the rows represent the actual instances that belong to the class. The non-diagonal entries of the confusion matrix therefore represent the instances that are misclassified by the classifier. The larger the entry in a non-diagonal cell, the harder that class is to the classifier. Thus, adaptive sprinkling allocates more artificial terms to the harder classes.

A major limitation of sprinkling and adaptive sprinkling is that both techniques are only applicable to higher order term relations. This is because the 'sprinkled' term-document space has no effect on first-order term relations. Therefore, there is a need for a more general approach for introducing class knowledge into semantic relatedness. Particularly, we need a method that is independent of the type and order of term relations. Furthermore, adaptive sprinkling requires the number of artificial terms used for sprinkling to be optimised for each individual dataset. In the next section, we present our approach which is not limited to any specific term-relation technique, nor any specific order of term relations.

3 Supervised Sub-spacing

The success of semantic indexing for TCBR depends on how it addresses the discovery of the following: semantic relatedness between terms and the importance of terms which are usually captured as term weights. The intuition behind $S3$ is to discover these knowledge separately in class-partitioned term-document sub-spaces before uniting into a common space. Isolating this discovery allows the relation between any two terms t_i and t_j in class c_k to reflect how semantically close the two terms are in class c_k. Likewise, the weight of any term t_i in class c_k should also indicate how important t_i is with respect to class c_k. To achieve this, we assume that the entire term-document space is composed of N term-document sub-spaces, one for each of N classes in the training corpus. We then apply a transformation function, which consists of assigning term relatedness and term weighting, to each sub-space such that documents that belong to the same class are processed together and separate from documents of other classes. Computing semantic relatedness and term weights in class-partitioned subspaces has the desired effect of making documents that belong to the same class more similar to one another while making them more dissimilar to documents of other classes.

More formally, a standard term-documents matrix D is initially created from the training corpus where

$$D = \bigcup_{k=1}^{N} D_k = D_1 \cup D_2 \cup ...D_N \tag{1}$$

D is an $m \times n$ matrix where m is the total number of documents in the corpus, n is the number of terms in the indexing vocabulary and N is the total number of classes. Each sub matrix D_i has dimensions $p \times n$ where $p \leq m$ i.e. sub-space D_i contains at most the same number of documents as D (in the case where there is just a single class) and has the same row dimension as D. We define a linear transformation function:

$$H : D_k \to D'_k \tag{2}$$

which transforms each document vector $v \in D_k$ into its semantic representation equivalent $v' \in D'_k$. The details of the linear function H are as follows.

$$H(D_k) = D_k \times T_k \times W_k \tag{3}$$

Where T_k is an $n \times n$ matrix such that each entry $t_{ij} \in T_k$ represents the strength of the semantic relatedness between vocabulary terms t_i and t_j. Each entry in T_k is normalised between 0 and 1 with all entries along the leading diagonal (t_{ij} where $i = j$) equal to 1 i.e. the relation between any term and itself is 1 (maximum similarity). W_k is a diagonal matrix of class specific term weights where each entry $w_{i,i} \in W_k$ represents the relevance of terms t_i to class c_k. Note that working with just the subspace D_k implicitly biases the discovery of semantic relatedness and term importance by class.

The transformed semantic term-document space (D') can be constructed from the individual semantic sub-spaces as follows:
$D' = \bigcup_{k=1}^{N} D'_k = D'_1 \cup D'_2 \cup ...D'_N$

In the following sub-sections, we focus on the remaining components of our linear transformation function in equation 3. In particular, we describe how to generate the T matrix which contains the semantic relatedness knowledge and the W matrix which holds the term weights.

3.1   Semantic Relatedness Extraction

Statistical approaches to semantic term relatedness extraction is based on the premise that co-occurrence patterns of terms in a corpus can be used to infer semantic associations [13]. Thus, the more two terms co-occur in a specified context, the stronger their semantic relatedness. In the following sub-sections, we describe two different approaches for estimating term relatedness from corpus co-occurrence. Note that for $S3$, semantic relatedness is calculated exclusively for each class. Thus, the frequency counts presented in the following subsections are computed exclusively from documents that belong to the same class.

Document Co-occurrence. Documents are considered similar in the vector space model (VSM) if they contain a similar set of terms. In the same way, terms can also be considered similar if they appear in a similar set of documents. Given a term-document matrix D_k where column vectors represent documents and the row vectors represent terms, the similarity between two terms can be determined by finding the distance between their vector representations. The relatedness between two terms, t_1 and t_2 using the cosine similarity metric is given in equation 4.

$$Sim_{DocCooc}(t_1, t_2) = \frac{\sum_{i=0}^{m} t_{1,i} t_{2,i}}{|t_1||t_2|} \tag{4}$$

Normalised Positive Pointwise Mutual Information (NPMI). The use of mutual information to model term associations is demonstrated in [7]. Given two terms t_1 and t_2, mutual information compares the probability of observing t_1 and t_2 together with the probability of observing them independently as shown in equation 5. Thus, unlike document co-occurrence, PMI is able to disregard co-occurrence that could be attributed to chance.

$$PMI(t_1, t_2) = log_2 \frac{P(t_1, t_2)}{P(t_1)P(t_2)} \tag{5}$$

The probability of a term t in any class can be estimated by the frequency of occurrence of t normalised by the frequency of occurrence of all term pairs in the corpus as shown in Equation 6.

$$P(t) = \frac{df(t)}{\sum_{i=1}^{m} \sum_{j=1}^{m} df(t_i, t_j)} \tag{6}$$

Where $df(t)$ is the document frequency of t, $df(t_i, t_j)$ is a count of the documents that contain both t_i and t_j, and m is the number of terms in the vocabulary. If a significant association exists between t_1 and t_2, then the joint probability $P(t_1, t_2)$ will be much larger than the independent probabilities $P(t_1)$ and $P(t_2)$ and thus, $PMI(t_1, t_2)$ will be greater than 0. Positive PMI (PPMI) is obtained by setting all negative PMI values to 0. PMI values do not lie within the range 0 to 1 as is the case with typical term relatedness metrics [10]. Thus we need to introduce a normalisation operation. We compute Normalised PMI (NPMI) as shown in equation 7.

$$Sim_{Npmi}(t_1, t_2) = \frac{PPMI(t_1, t_2)}{-log_2 P(t_1, t_2)} \tag{7}$$

3.2   Class Relevance Term Weighting

The assignment of class-specific relevance term weights (CRW) for each class c_k, is key to the $S3$ semantic indexing approach. It is intuitive to assume that given a term $t_j \in T$ and candidate class $c_k \in C$, the higher the probability that a document belonging to class c_k contains t_j, the more t_j is considered to be predictive of c_k. This means that the class specific weighting for any term t_j with respect to class c_k can be derived as a function of the probability of observing t_j in a document belonging to class c_k. Accordingly, we can define a simple class relevance weighting (CRW) function as the conditional probability that a document belonging to the class c_k contains the term t_j as shown in equation 8.

$$\text{CRW}(t_j, c_k) = p(c_k | t_j) \tag{8}$$

The conditional probability $p(c_k | t_j)$ can be decomposed using Bayes' theorem. Recall that a document is simply a set of terms $d_i = \{t_j\}$. Therefore, according to Bayes' theorem, the conditional probability $p(c_k | t_j)$ can be re-written as shown in equation 9.

$$\text{CRW}(t_j, c_k) = p(c_k |_j) = \frac{p(t_j | c_k) p(c_k)}{p(t_j)} \tag{9}$$

Where $p(t_j | c_k)$ is the conditional probability that a document contains the term t_j given that the document belongs to class c_k and $p(t_j)$ is the probability that any document in the collection contains the term t_j, regardless of the class membership of that document, and $p(c_k)$ is the probability of the class c_k. Both probabilities $p(t_j | c_k)$

and $p(t_j)$ can be estimated from observed frequency counts in the corpus as shown in equation 11.

$$p(t_j|c_k) = \frac{df(t_j, c_k)}{m_{c_k}} \tag{10}$$

$$p(t_j) = \frac{df(t_j)}{m} \tag{11}$$

Where $df(t_j, c_k)$ is the number of documents that belong to class c_k that contain term t_j, $df(t_j)$ is the number of documents in the entire collection that contain t_j, m_{c_k} is the number of documents that belong to class c_k and m is the number of documents in the entire collection.

One can argue that other functions can equally be applied to learn class-predictive term weights. The first proposal might be to use just the probability that a document contains the term given the class i.e. $p(t_j|c_k)$ as calculated from equation 10. Surely, the higher the conditional probability $p(t_j|c_k)$, the more likely it is that t_j is relevant to c_k. However, one major fault with this argument is that we are assuming higher relevance of the term t_j to the class c_k on the basis of higher document frequency of t_j in c_k. In other words, terms will only have a high weight if they appear in most documents in the class. Given that it is unlikely to have more than a handful of terms appearing most documents in any given class, using $p(t_j|c_k)$ for term weights will not produce ideal class-predictive term weights.

A more common weighting scheme that can be adopted from information theory is Mutual Information (MI) which measures the mutual dependence between any two given variables. Accordingly, we can derive class-specific weights for any term t_j as the mutual information of t_j with the class c_k as shown in equation 12.

$$MI(t_j, c_k) = \frac{p(t_j, c_k)}{p(t_j)p(c_k)} \tag{12}$$

Indeed, equation 12 has been widely used as a measure of term-goodness for feature selection. However, note that mutual information is affected by marginal probabilities of terms. This means that MI tends to assign higher weights to rare terms [18]. MI is also aggressive at assigning 0 weight to terms that are not considered to be mutually dependent with the target class. However, this aggressive strategy is not likely to be beneficial for the purpose of assigning class-specific term weights as many of the terms will then be eliminated from indexing.

We illustrate the difference between the three weighting schemes with the aid of a trivial example. Let t_1, t_2 and t_3 be terms and c_k the class for which we wish to calculate class-predictive term weights. This sample corpus contains four hundred documents, one hundred of which belong to class c_k. Let the distribution of terms t_1, t_2 and t_3 in the corpus be as shown in table 1. Term t_1 occurs in seven documents in class c_k and once in a document that does not belong to c_k. Thus, the numbers shown under the columns c_k and $\bar{c}_k$ are document frequencies of the corresponding terms within and outside of class c_k respectively. Accordingly, the CRW, Prob and MI weighting for class c_k and its complement, $\bar{c}_k$, is also shown. Note that we have normalised the values of MI to between 0 and 1 to make it comparable with the other two weighting metrics.

Table 1. Trivial example of term weighting schemes for t_1, t_2 and t_3 with document frequency (df) distributions over classes c_k and $\bar{c}_k$

Term	c_k				$\bar{c}_k$			
	df	CRW	Prob	MI	df	CRW	Prob	MI
t_1	7	0.875	0.070	0.310	1	0.125	0.003	0.000
t_2	7	0.539	0.070	0.190	6	0.461	0.020	0.000
t_3	30	0.857	0.300	0.476	5	0.143	0.017	0.000

Note that terms t_1 and t_3 have a much higher occurrence in documents of class c_k and thus are good predictors of this class. However, this fact is only recognised by the CRW function which assigns a correspondingly high weight to both t_1 and t_3. Prob assigns the same weight to t_1 and t_2 despite the fact that t_2 is not a good predictor of class. This is because Prob. does not utilise information on the occurrence of a term outside of the class of interest. Also, note that none of the terms is assigned a high weight by Prob which illustrates the likelihood of Prob to assign low weight to predictive terms. These reasons obviously make Prob unsuitable for class-predictive term weighting.

MI on the other hand is very sensitive to the occurrence of terms outside of the target class c_k. Note that term t_1 only manages to achieve a weighting of 0.310 despite the fact that t_1 occurs in only a single document outside of c_k. This sensitivity is further highlighted in the case of t_3 which occurs just 5 times outside of the c_k, yet MI assigns this a weight of 0.476. This shows the tendency of MI to downplay the importance of terms that can be considered to be highly predictive of class. Thus, MI is not ideal for learning class-predictive term weights. In contrast, the properties of CRW make it very suitable for class-predictive term weighting.

4   Datasets

For evaluation, we used a number of different supervised corpora covering various different domains including online incident reports, news stories, online reviews, and medical journal abstracts. Such diverse collections allow for a more robust evaluation of our approach. Also, these corpora are designed for a variety of different supervised tasks e.g. sentiment classification (Movie Reviews), topic classification (Reuters Volume 1, Ohsumed, 20 Newsgroups) and semantic classification (Incident Reports). We describe these corpora in detail in the following sub-sections.

20 Newsgroups: 20,000 documents collected from Newsnet newsgroups messages. We created one multi-class dataset (Science) from this corpus comprising of the 4 science categories and four binary class datasets. All datasets contain 500 documents per class.

Ohsumed: 348,566 medical references from medical journals from the MEDLINE database. The Ohsumed collection is unequally divided into 23 classes according to different disease types e.g. Virus Diseases. We created 4 multi-class and 8 binary class datasets from this corpus. All the datasets contain 500 documents per class.

Movie Reviews: a sentiment classification corpus comprising movie reviews from the Internet Movie Database (IMDB) [9]. We used version 1 of this corpus which contains 1400 reviews, equally distributed on class.

Incident Reports: created using incident reports crawled from the Government of Western Australia's Department of Mines and Petroleum website[1]. These are binary-class datasets where each document is assigned to the class 'Injury' and 'NoInjury', depending on whether or not injuries were sustained in the incident. All datasets contains 100 documents in each class

Reuters Volume 1: comprises news stories produced by Reuters journalists where the class label of a document represents the subject area of the news story. We create 4 binary-class datasets from this corpus (NewProdRes, ProdNewProd, MarketAdvert, and OilGas), containing 500 documents per class.

5   Evaluation

The aim of our evaluation is two-fold. Firstly, we wish to determine how standard term relatedness metrics are affected by the introduction of supervision using our $S3$ approach. To achieve this we compare classification performance on document representations obtained using the following strategies.

- BASE: Basic BOW representation without semantic indexing
- COOC: Unsupervised semantic indexing using with document co-occurrence (COOC) for semantic relatedness (see Section 3.1);
- NPMI: Unsupervised semantic indexing using with NPMI for semantic relatedness (see Section 3.1)
- S3COOC: Supervised semantic indexing using our $S3$ approach with COOC for semantic relatedness
- S3NPMI: Supervised semantic indexing using our $S3$ approach with NPMI for semantic relatedness

Our expectation is that in comparison with COOC and NPMI, S3COOC and S3NPMI should lead to better text classification performance. The results for BASE serve as a baseline to measure the improvement achieved using semantic indexing.

Secondly, we compare the performance of the two $S3$-based techniques, S3COOC and S3NPMI, to state-of-the-art text classification algorithms. Thus we include a comparison with the following approaches:

- SVM: Basic BOW representation with a Support Vector Machine classifier.
- SPLSI: Supervised semantic indexing using Sprinkled Latent Semantic Indexing approach (see [5]) with kNN classifier.
- sLDA: Supervised semantic indexing using supervised Latent Dirichlet Allocation (see [2]).

---

[1] http://dmp.wa.gov.au

Standard preprocessing operations i.e. lemmatisation and stopwords removal are applied to all datasets. For all experiments (except SVM and sLDA), we use a similarity weighted kNN classifier (with k=3) and using the cosine similarity metric to identify the neighbourhood. For sLDA, we use the c++ implementation by [14][2] with default parameter settings, while for SPLSI, we use the Java Matrix (JAMA) package for SVD and 16 artificial terms for sprinkling based on the findings in [4]. We report classification accuracy averaged over 5 runs of 10-fold cross validation. Statistical significance is reported at 95% using the paired t-test.

5.1   Results

Results of comparison between BASE, COOC, NPMI, S3COOC and S3NPMI are presented in Table 2. Datasets are grouped according to the source corpus they were created from. Values with $^+$ represent a significant improvement over BASE while values with $^-$ represent a significant drop in classification accuracy compared to BASE. Values in the S3COOC and S3NPMI columns that are presented with * represent a significant improvement over their unsupervised counterparts i.e. COOC and NPMI respectively. Best results in each row are presented in bold.

Overall results indicate $S3$-based representations to be significantly superior to their non-supervised counterparts. Comparing COOC and S3COOC, our $S3$ approach produced statistically significant improvements in accuracy on 26 out of 30 datasets. On the other hand, S3NPMI produced significantly better results on 25 datasets compared to NPMI. Note also that no significant depreciation in performance compared to BASE was observed with any of the $S3$-based representations. Compare this with significant drop in accuracy observed on 4 datasets with COOC and on 8 datasets with NPMI. This indicates that $S3$ successfully addresses the problem of noisy term relatedness that could harm classification performance.

The bottom 5 rows of Table 2 compare BASE, COOC, NPMI, S3COOC and S3NPMI on multi-class datasets. Note that the results are consistent with that of binary classification. Both S3COOC and S3NPMI significantly outperform the unsupervised approaches, NPMI and S3COOC. Also note the Science and Ohsumed02 datasets where the performance of NPMI and COOC is worse that BASE. Again, the use of supervision by S3COOC and S3NPMI produces significant improvements compared to BASE which further supports that supervision addresses the problem of noise associated with unsupervised semantic relatedness.

5.2   Comparison with State-of-the-Art

Table 3 compares the results our two $S3$ approaches with those of SVM, Sprinkled LSI (SPLSI) and supervised LDA (sLDA). Values in bold represent the best results in each row. The overall significant improvement over SVM indicates a clear advantage from $S3$-based representations for text classification. For instance S3COOC is significantly better than SVM on 16 out of 25 datasets, while S3NPMI is significantly better than SVM on 15 datasets. In comparison with SPLSI, S3COOC performs better on 18 datasets

---

[2] Available at http://www.cs.cmu.edu/~chongw/slda/

Table 2. Comparison of supervised ($S3$) and unsupervised semantic indexing

Dataset	BASE	COOC	NPMI	S3COOC	S3NPMI
BactV	85.1	88.6$^+$	90.0$^+$	90.3^{+*}	**90.6^{+*}**
CardR	90.0	92.2$^+$	93.8$^+$	**94.3^{+*}**	94.0$^+$
NervI	91.4	91.0	92.9$^+$	**94.0^{+*}**	93.1^{+*}
MouthJ	89.9	92.2$^+$	92.9$^+$	94.0^{+*}	**94.1^{+*}**
NeopE	91.6	93.8$^+$	94.2$^+$	**95.4^{+*}**	**95.4^{+*}**
PregN	89.7	90.4	90.9$^+$	**92.8^{+*}**	92.2^{+*}
ImmunoV	78.7	82.5$^+$	84.8$^+$	**85.5^{+*}**	**85.5^{+*}**
RespENT	87.2	88.1	91.0$^+$	92.0^{+*}	**93.1**
Hardw	90.1	90.9$^+$	91.3$^+$	92.5^{+*}	**92.64^{+*}**
MeastM	95.6	95.3	94.9	**95.8**	94.7
GunsM	93.7	94.0	94.0	**94.1**	93.9
AutoC	93.7	95.1	96.2$^+$	**95.8^{+*}**	96.2$^+$
EqtyB	95.5	95.5	94.8$^-$	**95.9^{+*}**	**95.9^{+*}**
FundA	89.4	92.0$^+$	89.9	**92.6^{+*}**	91.5^{+*}
InRelD	92.3	94.1$^+$	91.7	**94.2$^+$**	93.9^{+*}
NProdRes	85.5	86.9	80.4$^-$	**89.6^{+*}**	86.5^{+*}
ProdNP	87.7	89.3$^+$	88.4	**90.2^{+*}**	89.9^{+*}
OilGas	87.3	86.3$^-$	85.7$^-$	**88.1***	87.7*
ElectG	88.7	84.6$^-$	84.0$^-$	87.1*	**88.3***
MovieRev	70.7	78.6$^+$	81.8$^+$	83.4^{+*}	**85.0^{+*}**
Fire	87.3	93.4$^+$	92.3$^+$	92.7$^+$	**94.1^{+*}**
Collision	88.6	91.2$^+$	93.3$^+$	93.9^{+*}	**95.7^{+*}**
Rollover	86.1	89.5$^+$	90.7$^+$	**92.2^{+*}**	**92.2^{+*}**
CollRoll	90.6	93.9$^+$	93.4$^+$	**96.1^{+*}**	95.5^{+*}
MiscInc	81.5	84.4$^+$	89.8$^+$	88.7^{+*}	**90.4^{+*}**
Science	80.8	77.9	73.2	82.6^{+*}	**83.0^{+*}**
Ohsumed01	52.0	51.7	52.5	56.7^{+*}	**58.0^{+*}**
Ohsumed02	45.2	44.3	43.0	55.0^{+*}	**55.6^{+*}**
Ohsumed03	47.8	50.2	50.0	**58.4^{+*}**	56.1^{+*}
Ohsumed04	31.9	33.5	32.8	**40.6^{+*}**	39.2^{+*}

(significantly on 10). On the other hand, S3NPMI outperforms SPLSI on 20 of the datasets with significant improvements also on 10 of these datasets.

Comparing S3COOC with sLDA, S3COOC is better on 19 datasets (significantly on 16). S3NPMI is better than sLDA on 19 datasets with significant improvements on 13. In contrast, sLDA is significantly better than S3COOC on only 4 datasets. Observe that sLDA performs particularly poorly on the incident report datasets, Fire, Collision, Rollover, CollRoll, MiscInc and ShovFP. These datasets have a total of only 200 documents (100 documents per class). This indicates that perhaps the number of documents in these datasets is too small for sLDA to learn accurate supervised topic models. Note that accuracy is about 50% for these datasets (about 46% for Fire and Collision). This indicates that sLDA is particularly unsuitable for situations where there is labelled

Table 3. Comparison of $S3$ techniques with SVM, SPLSI and sLDA

Dataset	SVM	SPLSI	sLDA	S3COOC	S3NPMI
BactV	90.2	88.6	89.3	90.3	**90.6**
CardR	93.7	93.7	92.76	**94.3**	94.0
NervI	92.2	90.3	91.9	**94.0**	93.1
MouthJ	91.8	93.4	92.3	**94.0**	**94.1**
NeopE	93.5	94.5	94.8	**95.4**	**95.4**
PregN	89.6	91.9	89.6	**92.8**	91.4
ImmunoV	82.4	83.3	81.0	**85.5**	83.6
RespENT	90.5	92.0	90.2	92.0	**93.1**
Hardw	92.4	92.9	91.3	92.5	**92.64**
MeastM	95.7	93.2	95.0	**95.8**	94.7
GunsM	92.2	93.5	92.68	**94.1**	93.9
AutoC	95.9	95.6	**97.0**	95.8	96.2
EqtyB	**96.1**	96.0	95.2	95.9	95.9
FundA	90.9	91.3	**93.1**	92.6	91.5
InRelD	92.0	93.4	**94.9**	94.2	93.9
NProdRes	85.2	85.8	87.7	**89.6**	86.5
ProdNP	86.4	89.3	87.8	**90.2**	89.9
OilGas	**88.8**	86.6	88.6	88.1	87.7
ElectG	90.6	89.2	**93.02**	87.1	88.3
MovieRev	82.3	82.6	83.28	83.4	**85.0**
Fire	91.9	92.3	46.4	92.7	**94.1**
Collision	89.7	95.5	46.2	93.9	**95.7**
Rollover	91.5	89.8	50.2	**92.2**	**92.2**
CollRoll	93.8	96.2	50.8	**96.1**	95.5
MiscInc	**92.5**	89.1	50.6	88.7	90.4

training data. However, S3COOC and S3NPMI produce the best accuracies on these datasets except on MiscInc and ShovFP where SVM performs best. This shows that semantic indexing with $S3$ is effective even when labelled training data is limited.

5.3 Term Space Visualisation

To further illustrate how our $S3$ approach works, we present visualisations of a typical binary-class term-document space before and after semantic indexing with $S3$. Figure 1 shows a default term-document space of a single, binary-class dataset. The column dimensions of the space represent documents and the row dimensions represent terms. Each light coloured point in the space represents a non-zero value indicating the presence of a term in a document. The dark points are zero-valued indicating the absence of the corresponding term (row) in the corresponding document (column). The space has been organised such that the left half contains documents that belong to the first class and the second half contains documents that belong to the second class. Figure 2 shows the same term-document space after $S3$ semantic indexing. Note the difference between the left and right sides of the space is now clearly visible. This indicates how document

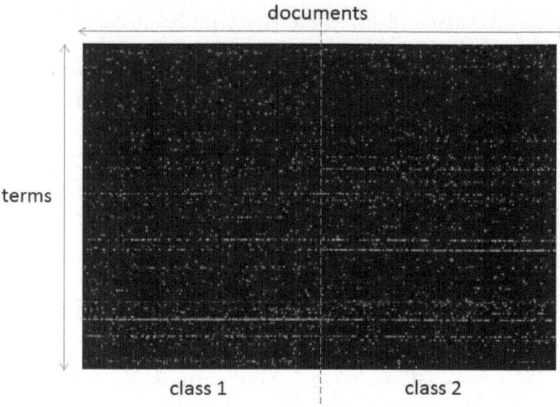

Fig. 1. Original term-document space

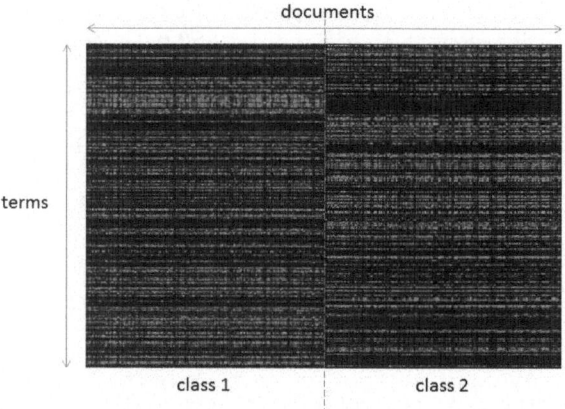

Fig. 2. Term-document space after $S3$ transformation

vectors belonging to the same class have been transformed to be similar to one another whilst empahising their difference to documents of the other class. Contrast this with Figure 1 where in the absence of semantic indexing, the term-document space amounts to a uniform image over classes.

6   Conclusion

In this paper, we introduced a novel technique called Supervised Sub-Spacing ($S3$) for supervised semantic indexing of text documents. We presented a detailed evaluation of this approach on 30 datasets from a variety of different domains including news stories, medical abstracts and online reviews. We investigated applying $S3$ with two semantic relatedness metrics: document co-occurrence (COOC) and Normalised Point-wise

Mutual Information (NPMI). Results show $S3$ with COOC semantic relatedness leads to best classification performance, out-performing all other representations including sprinkled LSI (SPLSI) and supervised LDA (sLDA) on 8 datasets.

The assignment of term weights is key to the performance of $S3$. Accordingly, we also introduced probability-based class relevance weights (CRW) which uses Bayesian probability to learn the relevance of a term, given a class.

The effectiveness of $S3$ lies in its ability to transform document representations such that documents that belong to the same class are made more similar to one another. We presented visualisations of a typical term-document space before and after $S3$ transformation in order to demonstrate the effect of $S3$ on document representations.

The $S3$ technique we presented here has a number of advantages compared to other supervised semantic modelling approaches. Firstly, unlike typical feature transformation approaches e.g. sLDA and SPLSI, our approach is not tied to any particular model of term associations. We demonstrated this by using $S3$ with both COOC and NPMI. Secondly, unlike sprinkling, $S3$ does not require higher order term relations. This means $S3$ is flexible enough to be used with a wider variety of term semantic relatedness metrics. A second advantage is that $S3$ does not require any parameter tuning whereas sprinkling requires a predetermined number of k artificial terms to be injected into the vocabulary. Thirdly, $S3$ requires less computing time and memory to execute compared to both SPLSI and sLDA. Also, because the term-document space of each individual class gets processed separately, $S3$ is convenient for distributed and parallel processing.

References

1. Aggarwal, C.C., Zhai, C. (eds.): Mining Text Data. Springer (2012)
2. Blei, D., McAuliffe, J.: Supervised topic models. In: Platt, J., Koller, D., Singer, Y., Roweis, S. (eds.) Advances in Neural Information Processing Systems 20, pp. 121–128. MIT Press, Cambridge (2008)
3. Blei, D.M., Ng, A.Y., Jordan, M.I.: Latent dirichlet allocation. J. Mach. Learn. Res. 3, 993–1022 (2003)
4. Chakraborti, S., Lothian, R., Wiratunga, N., Watt, S.: Sprinkling: Supervised latent semantic indexing. In: Lalmas, M., MacFarlane, A., Rüger, S.M., Tombros, A., Tsikrika, T., Yavlinsky, A. (eds.) ECIR 2006. LNCS, vol. 3936, pp. 510–514. Springer, Heidelberg (2006)
5. Chakraborti, S., Mukras, R., Lothian, R., Wiratunga, N., Watt, S., Harper, D.: Supervised latent semantic indexing using adaptive sprinkling. In: Proceedings of the 20th International Joint Conference on Artifical Intelligence, IJCAI 2007, pp. 1582–1587 (2007)
6. Chakraborti, S., Wiratunga, N., Lothian, R., Watt, S.: Acquiring word similarities with higher order association mining. In: Weber, R.O., Richter, M.M. (eds.) ICCBR 2007. LNCS (LNAI), vol. 4626, pp. 61–76. Springer, Heidelberg (2007)
7. Church, K.W., Hanks, P.: Word association norms, mutual information, and lexicography. Computational Linguistics 16(1), 22–29 (1990)
8. Deerwester, S.C., Dumais, S.T., Landauer, T.K., Furnas, G.W., Harshman, R.A.: Indexing by latent semantic analysis. Journal of the American Society of Information Science 41(6), 391–407 (1990)
9. Pang, B., Lee, L., Vaithyanathan, S.: Thumbs up?: sentiment classification using machine learning techniques. In: Proceedings of the ACL 2002 Conference on Empirical Methods in Natural Language Processing, EMNLP 2002, vol. 10, pp. 79–86. Association for Computational Linguistics, Stroudsburg (2002)

10. Rohde, D.L.T., Gonnerman, L.M., Plaut, D.C.: An improved model of semantic similarity based on lexical co-occurence. Communications of the ACM 8, 627–633 (2006)
11. Sun, J.T., Chen, Z., Zeng, H.J., Lu, Y.C., Shi, C.Y., Ma, W.Y.: Supervised latent semantic indexing for document categorization. In: IEEE International Conference on Data Mining, pp. 535–538 (2004)
12. Tsatsaronis, G., Panagiotopoulou, V.: A generalized vector space model for text retrieval based on semantic relatedness. In: Proceedings of the Student Research Workshop at EACL 2009, pp. 70–78 (2009)
13. Turney, P.D., Pantel, P.: From frequency to meaning: vector space models of semantics. J. Artif. Int. Res. 37, 141–188 (2010)
14. Wang, C., Blei, D., Fei-fei, L.: Simultaneous image classification and annotation. In: Proceedings of Computer Vision and Pattern Recognition (2009)
15. Weber, R.O., Ashley, K.D., Bruninghaus, S.: Textual case-based reasoning. Knowledge Engineering Review 20(3), 255–260 (2005)
16. Wong, S.K., Ziarko, W., Raghavan, V.V., Wong, P.C.: On modeling of information retrieval concepts in vector spaces. ACM Trans. Database Syst. 12(2), 299–321 (1987)
17. Xu, Z.E., Chen, M., Weinberger, K.Q., Sha, F.: An alternative text representation to tf-idf and bag-of-words. In: Proceedings of the 21st ACM Conferece of Information and Knowledge Management, CIKM (2012)
18. Yang, Y., Pedersen, J.O.: A comparative study on feature selection in text categorization. In: Proceedings of the Fourteenth International Conference on Machine Learning, ICML 1997, pp. 412–420 (1997)

Estimation of Machine Settings for Spinning of Yarns – New Algorithms for Comparing Complex Structures

Beatriz Sevilla-Villanueva[1], Miquel Sànchez-Marrè[1], and Thomas V. Fischer[2]

[1] Knowledge Engineering and Machine Learning Group,
Universitat Politècnica de Catalunya-BarcelonaTech. Jordi Girona 1-3,
08034 Barcelona, Catalonia
{bsevilla,miquel}@lsi.upc.edu
[2] DITF Denkendorf, Center for Management Research, Koerschtalstr. 26,
73770 Denkendorf, Germany
thomasvfischer@ditf-mr-denkendorf.de

Abstract. The textile industry in Europe is facing a new challenge in order to stay competitive into the textile market. They need to be flexible, cost efficient and produce with high quality. The setting of the machinery parameters is therefore an important aspect that combines implicit knowledge of workers and engineers with explicit knowledge. This makes it an ideal domain for CBR. It is used for an automatic parameter setting but the data cannot be reduced to a flat representation, as yarns and fabrics are multicomponent artefacts. Therefore we propose a combination of 4 algorithms to evaluate the similarity of the yarns. The application was successfully applied for spinning and it can be applied in the following steps of the textile processes like weaving.

Keywords: Similarity Assessment, Complex Case Structure, Retrieval, CBR, Textile industry, Yarns, IDSS.

1 Introduction

The textile industry in Europe is facing an important challenge in order to survive to the strong competition of other countries with lower cost of labor. Therefore, this industry has a challenge in the production of new textiles.

Textile manufacturing is a complex and a distributed process. This complexity depends on the processes that are involved and on the complexity of the textile product. The most common sub-processes integrated into the production are spinning, weaving, knitting, non-woven and finishing.

In Europe, the textile manufacturing tends to produce more complex textile products such as technical or medical textiles. But also the production of textiles for fashion and clothing is facing challenges, as a lot of raw materials are natural products such as cotton, silk, and wool. These raw materials vary slightly in terms of physical properties, e.g. elongation or resistance. The variation may be small, but optimal process settings are sensitive to such changes.

L. Lamontagne and E. Plaza (Eds.): ICCBR 2014, LNCS 8765, pp. 435–449, 2014.
© Springer International Publishing Switzerland 2014

Currently the production of a new textile product requires a high cost in terms of time, effort and money. Thus, one of the main objectives of the global textile industry is the reduction of this high cost of production of new textiles. To do this, many companies / industries are investing to include technological tools that can help in achieving this goal. For instance, spinning companies are focusing on recycled materials and therefore, they have more than 300 components to consider for the production of a yarn.

In a factory about 50% of orders usually refer to existing products or to products which are very similar. Therefore, the remaining 50% require a new configuration of the parameters of the machines involved in the process. This leads to adjust 100 or more parameters for each machine involved in the manufacturing process. Therefore, the incidence of reconfiguration of several textile machines is a problem of considerable volume. Thus, the developed system helps to configure the machine parameters in the following situations:

- New products (new combination of materials) for a customer request
- Existing products with one component substituted by a similar one due to limited availability of raw materials in the warehouse
- Change of settings during production due to low efficiency or quality

In addition, due to the critical economic situation, the textile industries tend to work on demand with small amounts of products (small lot size). This implies a constant variation of textile products to be manufactured, and thus a continuous reconfiguration of the parameters of the different machines in the manufacturing process, further increasing the production costs and the response times to orders.

The most common practice is to produce small samples of textile and the parameters are adjusted until the desired product with all desired features is obtained through a process of trial and error – based on existing standard settings. Therefore, this process, as mentioned, is very expensive and slow, resulting in loss of money and time on the part of the textile industries.

In general, the knowledge in the textile industry is documented only partially. Today companies have records of what is produced, but few companies store how it is produced. Very often only standard recipes and settings are stored, but the settings for real production is not documented. This situation leads to a huge loss of textile knowledge. Knowledge that could be exploited by the companies, especially to produce similar items or to reproduce the same products faster and better

All textile machines have a user manual that explains each of the parameters that can be set on the machine. However, the setting of the parameters is done by trial and error, since the values defined in the manual do not match reality either by the type of material used (with different compositions) or any other external factor such as the thermal conditions. Furthermore, it is not possible to define all the characteristics and parameters of the textile structure because of the difficulty of measuring them. This makes it very difficult to set up the machines involved in the manufacture of textiles.

For this reason, the setting of parameters is done on the basis of expert knowledge. This expertise has been generated over time long processes of trial and error. So, this knowledge is being lost as the experts are being retired and carrying this knowledge with them without being registered. Therefore, this setting is made by experts, considering past experiences and knowledge of the similarity between two products, which depends largely on the similarity of the yarns that form them, since similar products require a similar setting of the machine parameters.

This configuration process can be simulated computationally using an Intelligent Decision Support System (IDSS) based on Case-Based Reasoning. This system needs to define how similar are two processes. Because the configuration, by the experts, is limited to their own experience and memory, an automated system can take into account much more information and therefore be very valuable in this sector[1].

The degree of similarity of two textile processes depends on the similarity of the different parts of the process including the textile products and thus, on the comparison of the material that compose these products. In spinning, the end products are yarns and the raw material fibers and in the rest of processes the yarns are the raw material. The calculation of the degree of similarity between two yarns is extremely difficult as the yarns can be composed of different fiber types with different percentages. In addition, the calculation of the similarity of two yarns is influenced by other properties such as thickness, twist, target sector, etc. The latter properties can be modeled by numerical or categorical values and their respective degrees of similarity can be easily calculated. It could be easy to compare two yarns composed of the 100% same cotton type. In reality the yarns are composed of many different fibers, for example, a yarn composed of 80% regenerated cotton and 20% viscose has to be compared with a yarn composed of a 40% cotton, 20% polyester, 20% wool and 20% elastane.

In next section the related work about the comparison of complex structures in CBR, Then, the methodology that we propose is explained in Section 4. Section 4 presents the results and in Section 5, the conclusions are discussed.

2   Related Work

Advanced systems in the textile industry are able to simulate a textile product, but it is a visual representation of the product and therefore, these systems cannot provide an evaluation of the mechanics or physical properties. Thus, these systems are not suitable for configuring the machine settings. Neither these systems are suitable for comparing two products or two yarns.

However, in literature we found works which deal with yarns of the same composition and some different physical characteristics that can be numerically modelled, such as tenacity [2,3,4]. Also, the optimization of the textile process is focused on maximizing the quality of the final product and minimizing the cost and possible defects. In [5] there is an optimization of the spinning process for obtaining a higher quality yarn using neural networks and genetic algorithms. However, the attributes used in the data are measurable physical characteristics, and the composition of the yarn is not taken into account. In [6] is presented an application for detection and classification of the spliced yarn join using quantization and dynamic time warping. This is made by a visual inspection of the yarn joints. In [7], the fabric defects are recognized using CBR. The cases are images represented by a gray level co-occurrence matrix.

Previously we had already applied CBR to the textile industry [1]. In that work, the case base is automatically built for each case with processes where the yarns have the same composition and only physical properties were taken into account.

Besides, since the Retrieve task requires the comparison of cases, the (dis)similarity metric is an important part of this task. There are many distance metrics and most of them deal with two vectors of values which are numerical or categorical. The most used distances are the Euclidean distance and Manhattan. Other distances such as Canberra [8] or Clark [9] do not need to have scaled data and are centered into the origin. Also, there are a wide variety of distances that simultaneously work with heterogeneous data [10,11,12,13,14]. This wide variety of metrics is narrowed as the complexity of the objects grows. Some fuzzy measures have been proposed in the literature to deal with fuzzy sets, such as in [15]. Some of them are based on union and intersection operations [16]. The Hausdorf [17] metric that deals with sets and it is widely used in image recognition. Also, there are different complex structures for representing a case in CBR and their corresponding similarity function [18,19,20] such as hierarchies, episodes or objects. However, authors are not aware of any previous works that fulfill the complexity of the yarn composition.

3   Methodology

We propose a new procedure to face one of the major problems encountered in the manufacture of new textile products: the configuration of the machinery parameters. Under the assumption that similar products have similar configurations, methods that compare and evaluate the degree of similarity of these products are needed. Then, the comparison of the materials used to create the product is also needed. The yarn is present in most of the processes: in spinning as an end product and in the rest of processes as a raw material. Hence, the comparison and evaluation of the degree of the similarity between two yarns is fundamental to this methodology. Then, the aim of this work is to design a procedure to compare two processes which includes calculating the similarity or dissimilarity between two yarns based on their physical characteristics and including the composition of these yarns. Two yarns of different composition can have similar behavior from the textile point of view and, therefore, one may be a substitute for the other and the machinery settings can be reused.

Given two yarns, their physical properties and the composition of their fibers are compared. The physical characteristics of the yarn that are measurable can be compared using their numeric value with existing distance metrics such as the Euclidean. Typically, these characteristics refer to different physical aspects of the yarn such as thickness, torsion, elongation or resistance. These characteristics depend on how the yarn is produced and the materials it is composed of. Other features like the sector are qualitatively modelled because they cannot be modeled numerically and we only know if they are equal or different. The composition of the yarn is a combination of different fibers types with a percentage of presence.

Fibers can be classified into different families depending on the material that are composed: cotton, viscose, silk, wool, etc. At the same time, each family has different types of fibers. The differences between fiber types from the same family are based on certain physical characteristics of the fibers such as the length and/or fineness. However, in general, materials from the same family with different physical characteristics are more similar than those from different families with but similar physical characteristics according to the experts' knowledge. In Fig. 1, a hierarchy of the families and subtypes of the fibers is shown.

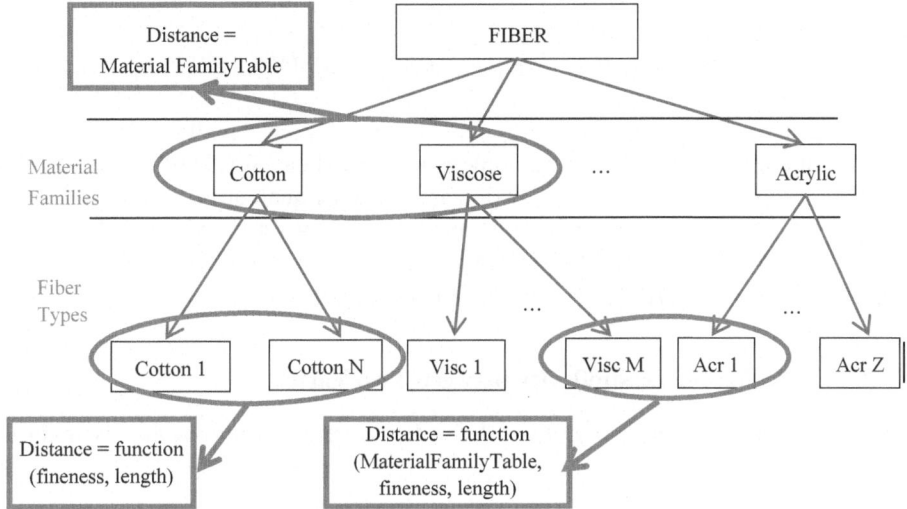

Fig. 1. Hierarchy of Fibers

Therefore, given two yarns, its composition and the physical characteristics of their fibers, a new procedure is proposed to calculate how similar these two yarns are. The complexity of this procedure lies in comparing different compositions since both the number of components and their percentage are variable.

3.1   Definition

A yarn can be understood as a composition of n components and each component has a certain percentage of presence. A component can be a composition itself (yarn) or be composed of fibers of the same type. Usually, the main component (higher presence) defines the behavior of the yarn and is therefore more important than the other components.

The different fiber types are classified into different material families (see Fig.1). Fiber types of the same family can have different physical characteristics. Typically these characteristics include the *fineness* and the *length* of the fibers.

Given two yarns (H_1, H_2) the degree of (dis)similarity is calculated taking into account:

- The characteristics of the yarn: count, sector, etc.
- The composition
  - How to compare the different components and its percentages
  - How to compare two components

A yarn H_i can be modelled as follows:

$$H_i = < COUNT(H_i), SECTOR(H_i), LC(H_i) >$$
$$LC(H_i) = < COMP_1(H_i), ..., COMP_N(H_i) >$$
$$COMP_j(H_i) = < PER_j(H_i), MATFAM_j(H_i), FINENESS_j(H_i), LENGTH_j(H_i) >$$

COUNT (Nm) is the number of meters of yarn per kg (smaller values indicate higher yarn diameter) and it can be numerically modelled. SECTOR is a qualitative label that designates the area of production and it can be qualitatively modelled. LC is a list of components, where each component $COMP_j$ has a percentage of presence PER_j, a material family $MATFAM_j$ and some physical characteristics of the fibers such as the fineness $FINENESS_j$ and length $LENGTH_j$.

3.2   Calculating the Dissimilarity between Two Yarns

The dissimilarity between two yarns is defined as a weighted sum of the dissimilarity of their features:

$$Disim(H_1, H_2) = W_{COUNT} * Disim_{COUNT}(H_1, H_2) + W_{SECTOR} * Disim_{SECTOR}(H_1, H_2)$$
$$+ W_{LC} * Disim_{LC}(H_1, H_2)$$

$$\sum W_i = W_{COUNT} + W_{SECTOR} + W_{LC} = 1 \tag{1}$$

Calculation of Dissimilarity According to the COUNT Attribute

According to the expert opinion, the dissimilarity between two small values is higher than among larger values. Thus, the dissimilarity does not follow a linear growth and, therefore, a relative measure that takes into account this effect is proposed to be used:

$$Disim_{COUNT}(H_1, H_2) = \frac{|COUNT(H_1) - COUNT(H_2)|}{\max(COUNT(H_1), COUNT(H_2))} \tag{2}$$

Calculation of Dissimilarity According to the Attribute SECTOR

SECTOR is a qualitative feature since it cannot be measured numerically. Therefore, only whether both yarns belong to the same sector or not can be assessed according to the following formula:

$$Disim_{SECTOR}(H_1, H_2) = \begin{cases} 0 \ if \ SECTOR(H_1) = SECTOR(H_2) \\ 1 \ if \ SECTOR(H_1) \neq SECTOR(H_2) \end{cases} \tag{3}$$

Calculation of Dissimilarity According to the List of Components (LC)

For the composition of the yarn, the combination of four algorithms is proposed. These algorithms are weighted sums of different combinations of pairs of components, where the weights depend on the presence of these components in the yarn. Each algorithm has a different strategy for choosing the pairs to be compared and its weights.

$$Disim_{LC}(LC(H_1), LC(H_2)) = \frac{A_1(H_1, H_2) + A_2(H_1, H_2) + A_3(H_1, H_2) + A_4(H_1, H_2)}{4} \tag{4}$$

$A_1(H_1, H_2) \equiv \text{MinAlg}((LC(H_1), LC(H_2))$
$A_2(H_1, H_2) \equiv \text{MainMinAlg}((LC(H_1), LC(H_2))$
$A_3(H_1, H_2) \equiv \text{CrossAlg}((LC(H_1), LC(H_2))$
$A_4(H_1, H_2) \equiv \text{MainHigherAlg}((LC(H_1), LC(H_2))$

For clarity, the following definition $LC(H_1)$ is abbreviated as Y_1 and $LC(H_2)$ as Y_2.

Algorithm A1 (MinAlg).
  The main idea of the first algorithm *MinAlg* is to compare pairs of components with the same percentage. The selection of the pairs to be compared depends on how similar are, so those pairs of components having a lower dissimilarity are selected and the weight is the smallest percentage of the two components. Thus, components are compared with the same percentage. That is, for example, if we have a yarn of 60% cotton and 40 % wool and other yarn of 80% viscose and 20% cotton, first cotton is compared and weighted by 20%, then compare the 40% cotton and 40% wool 80% viscose and then the remaining pairs will be chosen with lower dissimilarity between (cotton, viscose) and (wool, viscose) .

```
MinAlg (Y1, Y2)
   minAlg = 0
   while(not empty Y₁ and Y₂)
        < C₁, C₂ >= arg max    (Disim_COMP(C, C'))
                     C∈Y₁,C'∈Y₂
        percentage = min (PER(C₁), PER(C₂))
        minAlg += percentage * Disim_COMP(C₁, C₂))
        Y₁ = Y₁ - < percentage, C₁ >
        Y₂ = Y₂ - < percentage, C₂ >
   endWhile
   Return(minAlg)
endAlgorithm
```

Algorithm A2 (MainMinAlg)
  The second algorithm *MainMinAlg* is a variant of the first algorithm that takes into account the main component (higher presence). In this algorithm, first, the components with higher presence (main) and their common percentage to weight are used, i.e. if there is a yarn with 60% cotton and other with 80% viscose, cotton and viscose

are selected and weighted with 60%. Then, pairs of components with higher similarity are selected and the common percentage is used in the same way as in the first algorithm (*MinAlg*).

$$MainMinAlg(Y_1, Y_2) =$$
$$\min\left(PERC(Main(Y_1)), PERC(Main(Y_2))\right) * Disim_{COMP}\left(Main(Y_1), Main(Y_2)\right) +$$
$$MinAlg(Y_1', Y_2') \tag{5}$$

That is, if for example, a yarn of 60% cotton and 40% wool and other yarn of 80% viscose and 20% cotton, the main components would be cotton and wool respectively and would be weighted with 60%. Therefore Y_1' would be 40% wool and Y_2' 20% viscose and 20% cotton. Then, Y_1' and Y_2' are assessed with the first algorithm.

Algorithm A3 (CrossAlg)

The third algorithm, *CrossAlg* proposes removing the common part from both yarns. The common part is defined as the set of pairs of components with the same percentage and dissimilarity equal to 0. If the percentages are not equal, then only the lower percentage should be extracted. And the remaining components are all compared against all. So being Y_1' and Y_2' the remaining components of the yarns Y_1 and Y_2 after removing the common part respectively, the algorithm is defined by the following formula:

$$CrossAlg(Y_1', Y_2') =$$
$$\sum_{i=1}^{N} \sum_{j=1}^{M} \frac{PERC_i(Y_1') * PERC_j(Y_2')}{remain\ percentage} * Disim_{COMP}(COMP_i(Y_1),\ COMP_j(Y_2)), \tag{6}$$

where N is the number of remaining components from yarn 1, M is the number of remaining components of the yarn 2, $PERC_i(Y_1')$ is the percentage of the component i of Y_1', $PERC_j(Y_2')$ is the percentage of the component j of Y_2', *remain percentage* is $1 - common\ percentage$. $Disim_{COMP}\left(COMP_i(Y_1),\ COMP_j(Y_2)\right)$ is the dissimilarity of the components i and j from yarns 1 and 2 respectively.

Algorithm A4 (MainHigherAlg)

The fourth algorithm, *MainHigherAlg*, compares pairs of components by percentage. So, for each yarn their components are ordered by their percentage from higher to lower and then, the first is compared with the first, the second with the second and so on. In case that the number of components are not the same, the component without pair is added with the maximum distance value. Given a yarn Y_1 with equal or more components than yarn Y_2, the algorithm is defined by the following formula:

$$MainHigherAlg(Y_1, Y_2) = \sum_{i=0}^{M} mean\left(PERC(COMP_i(Y_1)), PERC(COMP_i(Y_2))\right) *$$
$$Disim_{COMP}\left((COMP_i(Y_1), (COMP_i(Y_2))\right) + \sum_{i=M+1}^{N} \frac{PERC(COMP_i(Y_1))}{2} * Dis_{MAX} \tag{7}$$

where M is the number of components of Y_2, N is the number of components of Y_1 and $N \geq M$.

Calculation of the Dissimilarity between Two Components (COMP)

The four algorithms have in common the comparison of two components independently of whether they are from the same or different family. The general dissimilarity between families is defined by the experts. The expert can quantify how different or similar the material families of fibers are or rather can quantify how compatible it is to replace a family with other. This quantification is defined for fibers with characteristics within normal values. Hence the value of the expert will be considered for those fibers having these features within the normal values while for those fibers which are not within normal values an increase of the dissimilarity based on other physical characteristics is calculated. For this calculation, the following information is available for each pair of families (A, B):

- Dissimilarity of replacing the fibers of family B by fibers of family A $\in [0,1]$
- $RANGE_{FINENESS} =$
  $\left[\text{Minimum Ratio} \left(\frac{FINENESS(A)}{FINENESS(B)}\right), \text{Maximum Ratio} \left(\frac{FINENESS(A)}{FINENESS(B)}\right)\right] =$
  $\left[RATIO_{FINENESS_{MIN}}, RATIO_{FINENESS_{MAX}}\right]$
- $RANGE_{LENGTH} =$
  $\left[\text{Minimum Ratio} \left(\frac{LENGTH(A)}{LENGTH(B)}\right), \text{Maximum Ratio} \left(\frac{LENGTH(A)}{LENGTH(B)}\right)\right] =$
  $\left[RATIO_{LENGTH_{MIN}}, RATIO_{LENGTH_{MAX}}\right]$

Therefore, for comparing two components the following dissimilarities are calculated:

- Dissimilarity between the families of the components 1 and 2
- Dissimilarity of the *fineness* of components 1 and 2
- Dissimilarity of the *length* of components 1 and 2

The dissimilarity is a weighted sum of these dissimilarities and according to the experts the fineness is more important than length, so this weight is higher. The general formula between two components is the following:

$$Disim_{COMP} \left(COMP_i(Y_1), COMP_j(Y_2)\right) =$$
$$W_{MATFAM} * Disim_{MATFAM} \left(MATFAM_i(Y_1), MATFAM_j(Y_2)\right) + W_{FINENESS} *$$
$$Disim_{FINENESS} \left(FINENESS_i(Y_1), FINENESS_j(Y_2)\right) + W_{LENGTH} *$$
$$Disim_{LENGTH} \left(LENGTH_i(Y_1), LENGTH_j(Y_2)\right) \tag{8}$$

where

- all dissimilarities $\in [0,1]$
- $w_{MATFAM} + w_{FINENESS} + w_{LENGTH} = 1$
- $w_{FINENESS} > w_{LENGTH}$

Calculation of the dissimilarity between Family Materials (MATFAM).

As mentioned, the dissimilarity between families of materials is given by the experts. There is a table containing for each pair of families the associated dissimilarity.

$$Disim_{MATFAM}\left(MATFAM_i(Y_1), MATFAM_j(Y_2)\right) =$$
$$MATERIAL\ TABLE[MATFAM_i(Y_1), MATFAM_j(Y_2)] \tag{9}$$

Calculation of dissimilarity according to fineness attribute.

However, the fineness and length depend on the range of ratios that are defined as ideal/normal. This means that if, for example, given the range of diameter of the fibers [0.9, 1.2], the ratio between the fibers to be compared is expected to be between these values. But if it is not, the difference between the actual value of the fineness and the expected value is calculated.

$$Disim_{FINENESS}\left(FINENESS_i(Y_1), FINENESS_j(Y_2)\right) =$$
$$\begin{cases} 0\ if\ ratio_{real}\left(\frac{FINENESS_i(Y_1)}{FINENESS_j(Y_2)}\right) \in ratio_{exp} \\ dist(FINENESS_j(Y_2), expected(FINENESS_i(Y_2)) \end{cases} \tag{10}$$

Similarly to the COUNT characteristics, comparing the diameter of the fibers is not linear; therefore, a relative measure is used:

$$dist\left(FINENESS_j(Y_2), expected(FINENESS_i(Y_2))\right) =$$
$$\frac{|FINENESS_j(Y_2), -expected(FINENESS_i(Y_2))|}{max\ (FINENESS_j(Y_2), expected(FINENESS_i(Y_2)))} \tag{11}$$

where the $expected(FINENESS_i(Y_2))$ depends on whether the real ratio is smaller or larger than expected ratio:

- Actual ratio < $RATIO_{FINENESS_{MIN}}$, $\left(expected(FINENESS_i(Y_2)) = \frac{FINENESS(Y_2)}{RATIO_{FINENESS_{MIN}}}\right)$

- Actual ratio > $RATIO_{FINENESS_{MAX}}$, $\left(expected(FINENESS_i(Y_2)) = \frac{FINENESS(Y_2)}{RATIO_{FINENESS_{MAX}}}\right)$

Calculation of dissimilarity according to LENGTH attribute.

The *length* dissimilarity is very similar to the *fineness* dissimilarity:

$$Disim_{LENTGH}\left(LENTGH_i(Y_1), LENTGH_j(Y_2)\right) =$$
$$\begin{cases} 0\ if\ ratio_{real}\left(\frac{LENGTH_i(Y_1)}{LENGTH_j(Y_2)}\right) \in ratio_{exp} \\ dist(LENGTH_j(Y_2), expected(LENGTH_i(Y_2)) \end{cases} \tag{12}$$

In this case *length* is linear and then, the standardized difference is used to ensure the values to be between 0 and 1. To standardize the maximum length value ($LENGTH_{MAX}$) and the lowest ($LENGTH_{MIN}$) which have been registered in the database are used.

$$dist(LENTGH_j(Y_2), expected(LENTGH_i(Y_2))) = \frac{|LENGTH_j(Y_2) - expected(LENGTH_i(Y_2))|}{LENGTH_{MAX} - LENGTH_{MIN}}$$

(13)

The $expected(LENGTH_i(Y_2))$ is calculated as analogously to the expected value of the fineness.

4   Application

The experiments presented in this work are the real validation performed in one of the spinning industries in the TexWIN project[1]. First, an expert evaluation is presented. In this evaluation, the algorithms for comparing the composition of the yarns are evaluated. Second, in Section 3.2 the whole system is shown.

4.1   Expert Evaluation

The evaluation of the dissimilarity of two yarns is focused on the comparison of the composition of the yarns because this is an important part and also, it is the most complicated and novel aspect.

This expert evaluation consists of comparing the given distances by the expert criteria and the distances obtained with the proposed algorithms. In this experimentation, 5 different yarns were compared with 9 or 10 yarns giving a set of 46 pairs of yarns. Two textile experts evaluated how similar are these 46 pairs. Finally, the experts suggest using the mean of the both evaluations because they differ in some cases. Then the 46 pairs were compared using the four proposed algorithms and the mean of them.

The results of the algorithms have been compared with the experts' estimations using the mean absolute error (MAE). Table 1 displays the results: the first row includes all cases and the second one excludes the six cases that have a higher error.

Table 1. MAE of the 4 algorithms and their mean

	MainMinAlg	MinAlg	CrossAlg	MainHigherAlg	Mean of Algs
46 cases	0.166	0.162	0.153	0.169	0.142
40 cases	0.128	0.125	0.116	0.147	0.106

Both the results of the experts and the results of the algorithms are shown in Fig.2. In this graphic, the horizontal axis represents the pairs of yarns; the vertical axis is the distance between the pairs. The pairs of yarns are ordered by the MAE between the experts' estimations and the mean of the 4 algorithms (*MeanAlg*).

---

[1] TexWIN (CP-FP 246193-2) is European Project from the framework FP7-NMP-2009-SMALL-3 Adaptive control systems for responsive factories.

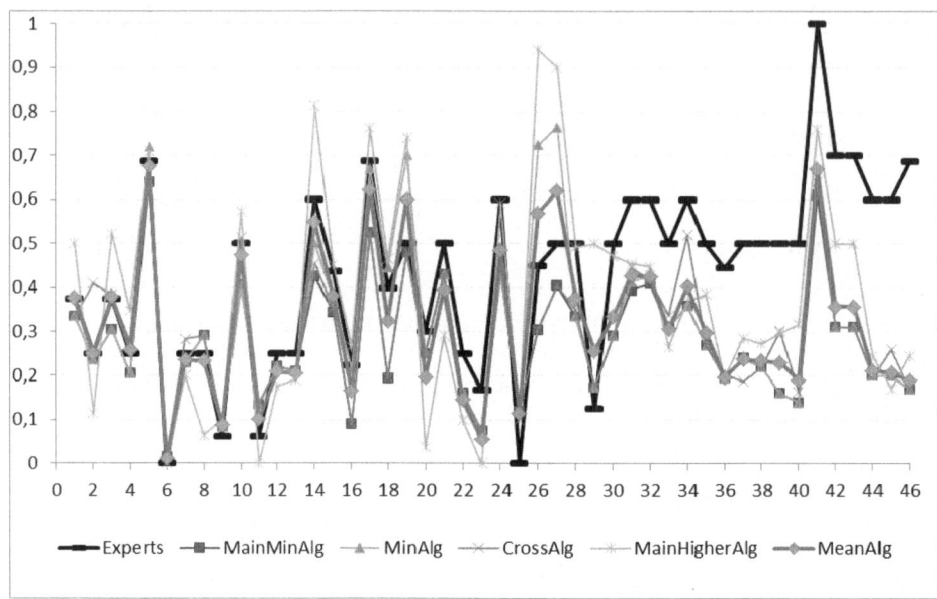

Fig. 2. Distances between 46 pairs of yarns according to the experts, the 4 algorithms and their mean

Observing the results in Table 1, the MAE of the MeanAlg is lower than the others because, as it can be observed in the Fig.2, the values are compensated by themselves. Moreover, analyzing all the results, we realized that those yarns having worse results also have a higher different evaluation from the two experts in the algorithms. In addition, we observe that those cases that compare acrylic and viscose material families are evaluated by the experts as more different, but in the material table have a small dissimilarity. In Fig.2, the last six cases (41-46) with higher MAE are the cases which compare a higher percentage of acrylic and viscose. For this reason, these 6 cases with higher error and comparing acrylic with viscose are deleted. Then, the total percentage of error decreases to 10% (see second row of Table 1). Therefore, this means that the family table needs to be reviewed to fix some values.

4.2 Experimental Evaluation

In this section, a whole use case from spinning is presented. The specific use case was to set the machine settings for a new yarn.

Given the description of the desirable yarn and the sector, the settings of the machines are estimated. The description of this yarn is the composition of the material and count. In Fig.3, an example of this use case is shown. Also, the CBR estimation of the results are shown in the right part of the Fig.3 (adapted case) and compared with the real case (left part).

Then, following a process of leave-one-out, 46 cases were evaluated. Given a case base of 46 correct cases, 46 CBR executions were run. For each execution, one case from the case base was used as a new case, and the rest of cases were the case base. Then, we run our CBR system with this new case and the new case base. Finally, the estimations of the settings were compared with the real settings from the cases. The MAE of the estimations and the real values were in between [2% - 15%] depending on the parameters.

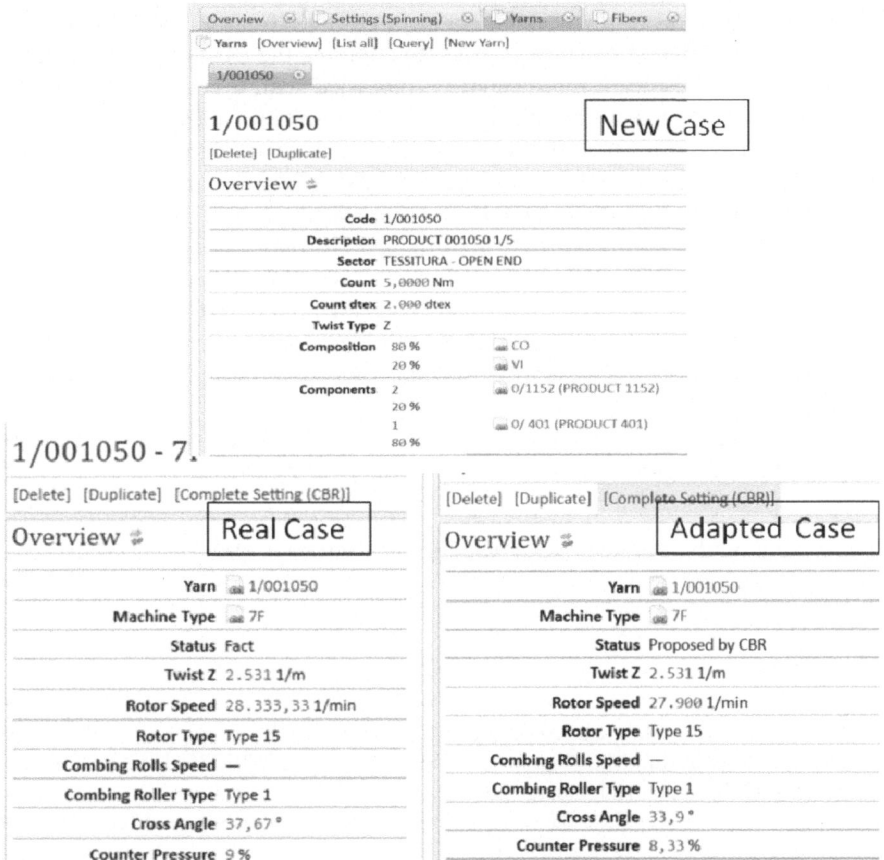

Fig. 3. Predicting machinery settings with CBR

Analyzing all the results, those cases which obtain worse estimations are those that the new case contains acrylic or viscose. As it was shown in the expert evaluation (subsection 4.1) results are sensible to the domain knowledge included in the system.

5   Conclusions and Future Work

In this work we have seen the importance of creating new tools for the textile industry that nowadays needs to be more efficient for competing with other countries. Therefore, tools that reduce the cost and improve the quality of the textile production are really appreciable. As a solution, the framework of the TexWIN project, an Intelligent Decision Support System has been built using CBR for predicting the machine settings which is one of the main problems for reducing the costs of the production.

The instance model was not fitting the data obtained from these industries. For that reason, we develop a specific CBR to handle this kind of data and so, handling the requirements of this application.

One of the principal differences is that each attribute from the description can be compared with a different distance metrics. In addition, the most different attribute is the composition of the yarn that cannot be modelled with a simple value. For that reason, in this work, the mean of four new algorithms to evaluate this structure is proposed. As it was shown in the application section, these algorithms are accurate although they are sensible to the expert knowledge that they use.

In addition, these algorithms could be extrapolated to other objects that can be modelled as the yarns are modelled. This structure is a list of components where each component has an associated percentage. In fact, it has already been adapted to weaving sector. In weaving, similar concepts can be applied but for a fabric that consists of different yarns. This can be used in other domains such as comparing paints, colors or recipes. Thus, the proposed similarity computation can be generalized to other complex structures in other domains.

The system has been successfully implanted in the industries involved in the Tex-WIN project and this procedure is under a process of being patented.

Acknowledgements. Authors acknowledge the European project TexWIN for the funding and support.

References

1. Sevilla Villanueva, B., Sànchez-Marrè, M.: Case-Based Reasoning Applied to Textile Industry Processes. In: Dáaz-Agudo, B., Watson, I. (eds.) ICCBR 2012. LNCS, vol. 7466, pp. 428–442. Springer, Heidelberg (2012)
2. Cheng, Y., Cheng, K.: Case-based reasoning system for predicting yarn tenacity. Textile Research Journal 74, 718–722 (2005)
3. Lü, Z., Yang, J., Xiang, Q., Wang, X.: Support vector machines for predicting worsted yam properties. Indian Journal of Fibre and Textile Research 32, 173–178 (2007)
4. Ruzhong, J., Zhijun, L., Jianguo, Y.: Estimating a product quality by support vector machines method. In: Mechatronics and Automation, ICMA 2007, pp. 3907–3912 (2007)
5. Sette, S., Boullart, L., Van Langenhove, L., Kiekens, P.: Optimizing the fiber-to-yarn production process with a combined neural network/genetic algorithm approach. Textile Research Journal 67(2), 84–92 (1997)
6. Issa, K., Nagahashi, H.: New Approach for unsupervised detection and classification of the spliced yarn joint. Autex Research Journal 99(4), 347–358 (2008)
7. Lin, J.J.: Pattern Recognition of Fabric Defects Using Case Based Reasoning. Textile Research Journal 80(9), 794–802 (2010)
8. Lance, G.N., Williams, W.T.: A general theory of classificatory sorting strategies II. Clustering systems. The Computer Journal 10(3), 271–277 (1967)

9. Eidenberger, H.: Distance measures for mpeg-7-based retrieval. In: Proceedings of the 5th ACM SIGMM Int. Workshop on Multimedia Information Retrieval (2003)
10. Gower, J.C.: A general coefficient of similarity and some of its properties. Biometrics, 857–871 (1971)
11. Chidananda Gowda, K., Diday, E.: Symbolic clustering using a new dissimilarity measure. Pattern Recognition 24(6), 567–578 (1991)
12. Gibert, K., Nonell, R., Velarde, J.M., et al.: Knowledge discovery with clustering: impact of metrics and reporting phase by using Klass. Neural Network World 4, 319–326 (2005)
13. Ichino, M., Yaguchi, H.: Generalized Minkowski metrics for mixed feature-type data analysis. IEEE Transactions on Systems, Man and Cybernetics 24(4), 698–708 (1994)
14. Sànchez-Marrè, M., Cortés, U., Roda, I., et al.: L'Eixample Distance: a New Similarity Measure for Case Retrieval. In: 1st Catalan Conf. on Artificial Intelligence, ACIA Bulletin; vol.14-15, pp. 246–253 (1998)
15. Liao, T.W., Zhang, Z.: Similarity measures for retrieval in case-based reasoning systems. Applied Artificial Intelligence 12, 267–288 (1998)
16. Núñez, H., Sànchez-Marrè, M., et al.: A comparative study on the use of similarity measures in case-based reasoning to improve the classification of environmental system situations. Environmental Modelling & Software 19(9), 809–819 (2004)
17. Huttenlocher, D.P., Klanderman, G.A., Rucklidge, W.J.: Comparing images using the Hausdorff distance. IEEE Transactions on Pattern Analysis and Machine Intelligence 15(9), 850–863 (1993)
18. Aamodt, A., Plaza, E.: Case-based reasoning: Foundational issues, methodological variations, and system approaches. AI Communications 7(1), 39–59 (1994)
19. Bergmann, R., Kolodner, J., Plaza, E.: Representation in case-based reasoning. The Knowledge Engineering Review 20(03), 209–213 (2005)
20. Pal, S.K., Shiu, S.C.: Foundations of soft case-based reasoning, vol. 8. John Wiley & Sons (2004)

Linking Cases Up: An Extension
to the Case Retrieval Network

Shubhranshu Shekhar, Sutanu Chakraborti, and Deepak Khemani

Department of Computer Science and Engineering,
Indian Institute of Technology Madras, Chennai, India - 600036
{shekhars,sutanuc,khemani}@cse.iitm.ac.in

Abstract. In many domains, cases are associated with each other though
this is not easily explained by the set of features they share. It is hard,
for example to explicitly enumerate features that make a movie romantic. We present an extension to the Case Retrieval Network architecture,
a spreading activation model initially proposed by Burkhard and Lenz,
by allowing cases to influence each other independently of the features.
We show that the architecture holds promise in improving effectiveness
of retrieval in two distinct experimental domains.

Keywords: Case Retrieval Network, Spreading Activation, Human
Cognition, Utility.

1   Introduction

Case Based Reasoning (CBR) is centered around the idea of reusing successful
problem solving episodes to solve new problems. The effectiveness of CBR is crit-
ically dependent on how closely the case representation and similarity measures
reflect on the intended function of the case(s) in problem solving.

This paper is centered around one single theme, that we call structure-function
correspondence. If a person is suffering from kidney malfunction and is looking
for a transplant, the transplant can be done because a kidney from a donor
will function similar to the patients own kidneys. However, a brain transplant
is harder, since the correspondence between the brain structure and its function
is weaker than that of kidney. In the context of CBR systems, if the representation of a case is strongly indicative of its function in solving problems, the
structure-function correspondence is strong; it is weak otherwise. The choice of
representation is closely tied to the domain. Features of a flat such as locality,
type and number of bedrooms that describe the problem side of a case may be
fairly predictive of its price, which is the solution component. In such a case,
the structure-function correspondence is strong. In contrast, it is hard to define features that make a movie romantic. Pixels in a frame of a movie interact
non-linearly with each other, those frames interact with each other in complex
ways as well, and finally the totality of frames interacts with the storehouse of
experiences of the person watching the movie to finally declare it to be romantic or otherwise. In Machine Learning, we do not yet know how to address this

L. Lamontagne and E. Plaza (Eds.): ICCBR 2014, LNCS 8765, pp. 450–464, 2014.
© Springer International Publishing Switzerland 2014

massive juggernaut. Such models of emergent concepts (e.g. romantic), even if devised, would have far too many parameters and interactions between them, to be trainable in practice. The easy way out, and a clever one at that, is collaborative filtering [11]. It does not target the holy grail, but given enough data of how users rate movies, it estimates relatedness of movies by relatedness of users who rate them, rather than a model of contents of the movies. The knowledge of users enriches the model and bypasses, to an extent, the need to build sophisticated models of movie content.

The paper is motivated by the realization that CBR systems would benefit from an architecture that splits concerns between two components, the first that models case similarity based on features they share, and the second that brings together cases without committing explicitly on the features. Thus two books could be similar because they share similar authors or belong to the same genre, but they could also be regarded as similar because they are often bought together (though the features present do not seem to indicate why this so). We present an extension to the Case Retrieval Network (CRN) architecture, a spreading activation model initially proposed by Lenz [13], by allowing cases to influence each other independently of the features. We show that the architecture holds promise in improving effectiveness of retrieval in two distinct experimental domains. The essential contribution, however, is that the idea is fairly general to be adapted to a large class of practical CBR systems, which can use the architecture to elegantly model a neat divide between two distinct sources of knowledge. With social networks playing an increasingly important role in large classes of search and recommender systems, we envisage that the need to break free from a rigid feature theoretic model will be severely felt and there will be several interesting ways in which direct case-case associations can be exploited.

The structure of the paper is as follows. Section 2 describes the background of our work for the choice of formalism. A detailed description of proposed framework is presented in Section 3. Evaluation methodology and experiments are presented in Section 4, and discussions on the obtained results and future work is described in Section 5. We summarize our contribution in Section 6.

2   Background

Lenz and Burkhard [13] proposed CRN as a representation formalism for CBR. CRN provides a suitable architecture for CBR as it can facilitate flexible and efficient retrieval in CBR. It facilitates spreading activation over a representation of cases in terms of feature value and similarity knowledge. CRN has been deployed successfully for real world CBR systems [2, 8, 14]. We revisit the basic CRN for retrieval in CBR in the following section. The central hypothesis in CBR is that *"similar problems have similar solutions"*. A typical CBR system models case similarity based on the case features. It is important to measure the effect of the similarity on the CBR hypothesis. This can be measured by looking at the alignment of problem side with that of solution side of the case-base which can be characterized by complexity. Complexity measures the extent to which CBR hypothesis is applicable. We further discuss the complexity in this section later.

2.1   Case Retrieval Net

To illustrate the basic idea underlying a CRN, let us consider documents (cases) in the Deerwester collection [7] to be our case-base shown in Fig. 1(a). The case-base has nine documents which come from two concepts - human interaction and mathematical graphs. Terms appearing in more than one document are identified as keywords which are referred to as Information Entities (IEs) in CRN parlance. Each case is represented as a column of binary values. The first five cases (columns $c1$ through $c5$) come from the theme of human-computer interaction and the rest of the cases (columns $m1$ through $m4$) relate to topic of mathematical graphs.

Fig. 1(b) depicts the representation of the case-base mapped onto a CRN. The IEs (keywords) are denoted by rectangles and the cases (documents) by ovals. IE nodes are related to each other by weighted symmetric *similarity* arcs denoting the semantic similarity between two keywords (IEs). For example, "human" is semantically more similar to "user" than "interface". The semantic similarity can be estimated from thesauri like Wordnet [15] or an encyclopedia like Wikipedia, however acquiring domain-specific similarity from an expert may be more involved. IEs are linked to case nodes by weighted relevance arcs which capture the strength of relevance of the term to a document. For our documents in the example, the relevance of a keyword is 1 if the keyword is in the document, 0 otherwise. The relevance value can be a real number. Here the relevance values are simply the corresponding values in the term-document matrix given in Fig. 1(a).

Let us formalize the retrieval using CRN as descibed above. A CRN is defined over a finite set of s IE nodes $\mathbb{E}$ and a finite set of m case nodes $\mathbb{C}$. Using the conventions from [5, 13], we define a similarity function σ over IE nodes:

$$\sigma : \mathbb{E} \times \mathbb{E} \to \mathbb{R}$$

which specifies similarity $\sigma(e_i, e_j)$ between IE nodes e_i and e_j and a relevance function:

$$\rho : \mathbb{E} \times \mathbb{C} \to \mathbb{R}$$

which depicts the relevance $\rho(e, c)$ of IE node e to case node c. We also associate propagation function with each node n in $\mathbb{E} \cup \mathbb{C}$ defined as

$$\pi_n : \mathbb{R}^{|\mathbb{E}|} \to \mathbb{R}.$$

The propagation functions are annotation to the nodes whose role is to aggregate all the incoming activations to the node. Case retrieval is done in three steps by propagating activations from IE nodes in the given CRN as described below:

Step 1: Given a query, initial activation α_0 is determined

Step 2: Similarity Propagation: The activation α_0 is propagated to all IE nodes $e \in \mathbb{E}$:

$$\alpha_1(e) = \pi_e(\sigma(e_1, e).\alpha_0(e_1), \dots, \sigma(s, e).\alpha_0(e_s))$$

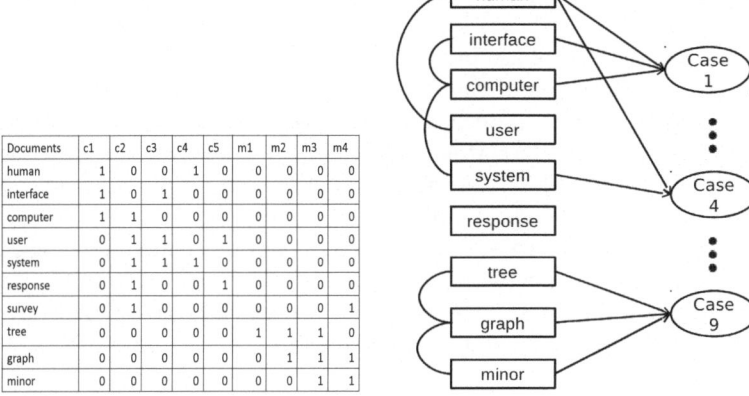

Documents	c1	c2	c3	c4	c5	m1	m2	m3	m4
human	1	0	0	1	0	0	0	0	0
interface	1	0	1	0	0	0	0	0	0
computer	1	1	0	0	0	0	0	0	0
user	0	1	1	0	1	0	0	0	0
system	0	1	1	1	0	0	0	0	0
response	0	1	0	0	1	0	0	0	0
survey	0	1	0	0	0	0	0	0	1
tree	0	0	0	0	0	1	1	1	0
graph	0	0	0	0	0	0	1	1	1
minor	0	0	0	0	0	0	0	1	1

(a) Adapted term-document matrix (b) Case-base mapped onto CRN
from Deerwester collection

Fig. 1. An example CRN

For simplicity, we can assume that propagation function π_n used is summation and we will continue our analysis with current choice of π_n. Therefore, the α_1 can be re-written as follows:

$$\alpha_1(e) = \sum_{i=1}^{s} \sigma(e_i, e).\alpha_0(e_i) \qquad (1)$$

Step 3: Relevance Propagation: The IE node activations obtained from above step are propagated to all the case nodes using the summation as propagation function.

$$\alpha_2(c) = \sum_{i=1}^{s} \rho(e_i, c).\alpha_1(e_i) \qquad (2)$$

The cases are ranked according to α_2 scores and top k cases are retrieved in case of the CRN.

Once the case-base is mapped onto the CRN, it can be used for retrieval of cases whose problem component is similar to query case. The query case is parsed and keywords are extracted. The IEs corresponding to these keywords are then activated and a similarity propagation through similarity arcs is initiated to find out the relevant IEs which in turn are also activated. All the activated IE nodes take part in relevance propagation from IEs to case nodes through relevance arcs. At each of the case node, activation score is computed based on the incoming activations to that node. The cases are then ranked according to the obtained scores and the top k cases are retrieved.

2.2   Complexity

Complexity [3] measures the extent to which *similar problems have similar solutions*. The complexity measure helps in assessing the suitability of CBR for a particular task. It can be estimated from the case-base alignment given by the similarity function. This can be adapted to classification tasks as well, where solution component is replaced by the label denoting the class knowledge. To evaluate the complexity of a classification dataset $GAME_{class}$ [4] is defined. $GAME_{class}$ is an extension of the Global Alignment MEasure (GAME) for the classification domain. The underlying idea is to see whether the nearest neighbors belong to the same class when the cases are sorted in the order given by problem side similarities. This determines whether the problems similar to query can be used to predict the class label. Cases are ordered in the case-base based on the problem side similarities. Solutions of the cases are arranged based on the ordering imposed by the problem side. In an ideal scenario where the case-base is perfectly aligned, all the same class labels are stacked together. Compression ratio of the solution side stacking is used to estimate the complexity.

3   Extended Case Retrieval Network

In this section we present Extended CRN (ECRN). We introduce spreading activation from case nodes as well, in addition to those from IEs. To facilitate spreading activation from case nodes, we introduce case-to-case links. To motivate the introduction of case-links, consider the process of recommendation in a shop. A person looking for a camera may ask a shopkeeper for a product specifying certain features of the camera. The shopkeeper uses this information along with the knowledge of interaction with the past users to recommend a product. This knowledge of past interactions plays a critical part in the recommendation process. We propose a framework that could enable us to encode this tacit knowledge in the form of case-case links. We formalize the ECRN framework in this section assuming the presence of case-to-case links.

In our structure, we introduce directed links between cases that indicate how activation propagates. The semantics of the directed arc has been discussed in the following subsection along with an algorithm to construct the arcs. For analytical purposes, the ECRN can be thought of as a CRN with three layers: layer one with all the IE nodes and all the case nodes in the remaining two layers each as illustrated in Fig. 2. Cases from layer two will connect to themselves in layer three. The dashed arrow in Fig. 2 shows the case-link that has been introduced based on case interaction.

The spreading activation starts at IE nodes and follows the same steps as explained above for the simple CRN. However, in this case one more spreading happens from case node to case nodes. We define a function over the case nodes for case-case interactions:

$$\varphi : \mathbb{C} \times \mathbb{C} \to \mathbb{R}$$

which describes the strength of association $\varphi(c_i, c_j)$ from a case c_i to c_j. The retrieval using the new structure will have an additional step.

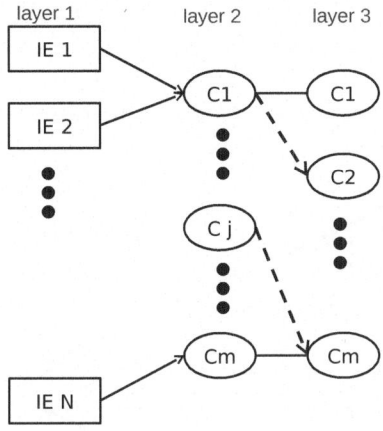

Fig. 2. Extended CRN with case links

Case activation propagation: The activation α_2 (Eq. 2) is, then, propagated via *case-links* to m case nodes.

$$\alpha_3(c) = \sum_{i=1}^{m} \varphi(c_i, c).\alpha_2(c_i) \qquad (3)$$

The cases are sorted in descending order of α_3 and the top k cases are retrieved. We observe that Step 2 and Step 3 account for most of the computation for retrieval at run time. Chakraborti et al. [5] describe a method for speeding up computation by introducing *effective relevance*. We extend the idea further to be applicable to our structure.

We substitute the α_1 from Eq. 1 into Eq. 2 to obtain case activation.

$$\alpha_2(c) = \sum_{j=1}^{s} \rho(e_j, c).\sum_{i=1}^{s} \sigma(e_i, e_j).\alpha_0(i) \qquad (4)$$

Effective relevance of IE e_i to case node c is given as

$$\Lambda(e_i, c) = \sum_{j=1}^{s} \rho(e_j, c).\sigma(e_i, e_j)$$

Therefore, by using Λ into Eq. 5 we get final case activation as

$$\alpha_2(c) = \sum_{i=1}^{s} \Lambda(e_i, c).\alpha_0(i) \qquad (5)$$

We now give a formulation for α_3 similar to α_2 as presented in Eq. 5. To arrive at that, the influence of IE e_i over case c in the presence of *case-links* is formalized. To find the influence we need to consider all the paths via which

e_i can influence c. Spreading activation from e_i to other nodes is followed by propagation of relevance to case nodes. These case nodes in turn propagate case activation to other case nodes. Thus, total aggregated activation at a case node c can be computed given σ, ρ and φ. Fig. 3 illustrates various paths from which e_i can influence a case node c.

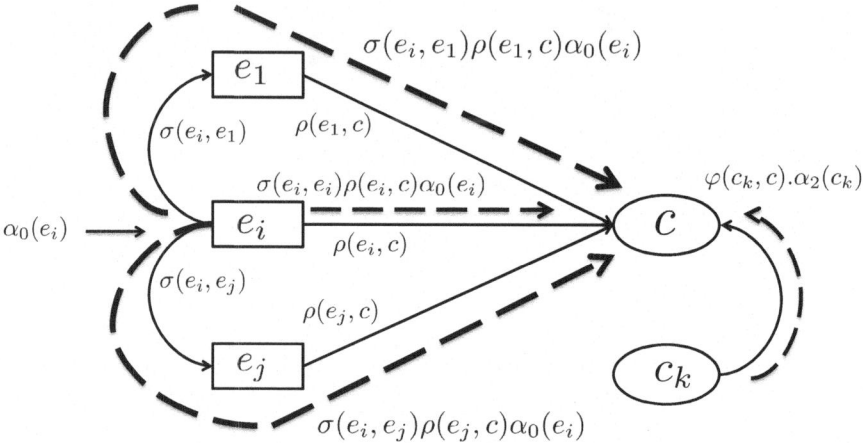

Fig. 3. Activation paths for c from an IE e_i and a case c_k

Consequently, we can formulate effective relevance in presence of *case-links* as follows:

$$\Phi(e_i, c) = \sum_{k=1}^{m} \sum_{j=1}^{s} \sigma(e_i, e_j).\rho(e_j, c_k).\varphi(c_k, c)$$

We can write it in terms of effective relevance Λ:

$$\Phi(e_i, c) = \sum_{k=1}^{m} \Lambda(e_i, c_k).\varphi(c_k, c)$$

We observe that Φ can be pre-computed. We now substitute α_2 from Eq. 5 into Eq. 3. That gives:

$$\alpha_3(c) = \sum_{j=1}^{m} \varphi(c_j, c). \sum_{i=1}^{s} \Lambda(e_i, c_j).\alpha_0(i)$$

Expanding the Λ, rearranging the terms and using definition of the Φ in above expression, we get:

$$\alpha_3(c) = \sum_{i=1}^{s} \Phi(e_i, c).\alpha_0(i) \tag{6}$$

Here, Φ can be precomputed that facilitates single propagation from the IEs to the cases. The above result is significant because it compacts the extended CRN representation and reduces relevance sparsity.

3.1   Forming Case Links

We present an algorithm to form the case-links assuming a textual domain since we compute case similarities using Wikipedia. The algorithm is designed for classification tasks over extended CRN. A classification problem can be viewed as identifying the class label for a new instance, based on the instances in training set whose class labels are known a priori. A training instance is represented as a flat attribute-value vector. For example, an instance $x = (v_1^b, ..., v_n^b)$ is a vector with n values where v_i^b is the value of feature f_i. The attribute values form the IE nodes and cases form the case nodes. Top k retrieved cases determine the class label for input test query. If the predicted class label is incorrect, we may say that the feature representation is not adequate enough to capture class knowledge for the given instance. We can say that the structure-function correspondence is weak for the instance in terms of feature representation. Therefore, we hypothesize that interaction among instances can be exploited to account for the misclassification. The interaction can be characterized by introducing case-to-case arcs with weights that signify the strength of influence of one instance over the other.

Algorithm 1 explains the steps to identify cases with weak structure-function correspondence. Each instance of the training data is used as query case over simple CRN for classification following the leave-one-out strategy. In doing so, cases are associated with a strength factor as described in the algorithm. Strength factor is the confidence associated with a case that the case is assigned actual class label. If top k retrieved cases belong to same class for the input query, then the strength factor associated with the query will be 1. We use the strength factors for weighting the case-links as described in Algorithm 2.

Algorithm 1. Algorithm to find cases with weak structure-function correspondence

1. Initialize, weakCases $\leftarrow$ {}
2. For each case d_i in trainingData, do: ▷ *Finding Weak Documents*
    (a) Remove d_i from case nodes in CRN
    (b) Activate IEs that appear in d_i, and do similarity and relevance activation on CRN
    (c) Aggregate activation score on each case node.
    (d) Assign class label l_i to d_i, based on top k cases retrieved
    (e) Associate strength with d_i

$$strengthFactor(d_i) \leftarrow \frac{average\ aggregation\ score\ for\ actualClassLabel(d_i)}{\sum\limits_{l \in classLabels} average\ aggregation\ score\ for\ class\ l}$$

    (f) If actualClassLabel(d_i) $\neq l_i$; weakCases $\leftarrow d_i$
    (g) Add d_i again to case nodes in CRN

While reading a book, words appearing in the book trigger the world knowledge. The words are the actual content of the book, while world knowledge is acquired through experiences. The reader may find the book similar to some other book by forming associations among experiences from the two books. These experiences relate the two books, not the words present in these books. Analogously, to quantify the interaction between two cases, Wikipedia is leveraged as a source for the world knowledge. Gabrilovich et al. [9] proposed a method of computing semantic relatedness of texts using Wikipedia-based explicit semantic analysis (ESA). ESA creates a weighted inverted index of words in terms of Wikipedia concepts. A document is then represented as weighted vector of these concepts. Once we have the vector representation of documents in terms of Wikipedia concepts, similarity between documents can be computed using standard vector comparison. Using the ESA representation we characterize the link strength between two cases. For ESA concepts, discriminating Wikipedia articles with respect to the classes were considered as explained in [17]. It is to be noted that the external knowledge is encoded only in the case links formed. For spreading activation on the extended CRN, original feature-value representation is retained. The external knowledge encoded in case links complements the spreading activations from low-level features through the propagation of case activations.

Algorithm to form case links is described for the textual domain since Wikipedia is readily available to serve as external knowledge to compute semantic relatedness of the cases. The cases exhibiting weak structure-function correspondence are obtained from Algorithm 1. Each of these cases is then linked up to the near neighbors of the case with a link strength equal to the similarity score for this case for the neighbor. The links are directed from neighbors to weak case and are weighted by strength factor of the case as explained in Algorithm 1. The weights of the case links are further revised to arrive at a better case-case link representation. A case-link is penalized in terms of down votes; if a document is misclassified after the introduction of the link, the down vote count is increased. A link with more down votes will be assigned relatively lower weight, the weight can become 0.0 if all the misclassification is due to the introduction of the link. The algorithm to form case-to-case links is described in Algorithm 2.

The algorithm to form case links can be formalized better as an optimization problem. However, we present a greedy approach because the goal of this paper is to convey the idea of case links. Although, the algorithm to learn case links is presented with respect to a classification problem, this can be extended for traditional case-based reasoners. Definition of strength factors as introduced in Algorithm 1 would be required to be adapted to the problem domain.

4   Evaluation

In our experiments, we evaluate the effectiveness of introducing *case-links* in the proposed extension of CRN framework in the context of document classification and book recommendation.

Algorithm 2. Algorithm to create case-case links

1. For each case wd_i in weakCases, do: ▷ *Obtained using Algorithm 1*
   (a) Compute distance $dist_{i,j}$ of wd_i from each case d_j in trainingData
   (b) For wd_i, retrieve top k nearest cases nnk
   (c) Link wd_i to each of the case c_j in nnk by case-arcs as per following rule:
      i. If actualClassLabel(wd_i) == actualClassLabel(c_j) then, assign weight
         wikiSim(wd_i, c_j) * strengthFactor(wd_i) to the link,
         where wikiSim(c_i, c_j) is the similarity of two cases based on Wikipedia
         concept representation
      ii. Else, assign weight 0 to the link
2. For each case d_i in trainingData, do: ▷ *Improving case-link representation*
   1. Classify d_i using extended network with initial case-arcs formed in Step 1.
   2. If d_i is misclassified and $d_i \notin weakCases$, then, For link l_j in case-arcs:
      (a) remove link l_j from case-arcs
      (b) classify d_i
      (c) If d_i classified correctly: DownVote(l_j) $\leftarrow$ + 1
3. For link l_j in case-arcs: ▷ *Compute revised arc weights*
   1. $weight(l_j) \leftarrow weight(l_j) * \frac{numVotes - DownVote(l_j)}{numVotes}$, where numVotes = number
      of misclassified cases in Step 2.

4.1 Datasets

To demonstrate the effectiveness of using the proposed framework, we apply the framework over two different tasks - classification and recommendation. For the task of recommendation, we created a dataset for book recommendation containing 407 books. These books are taken from philosophical fiction and fantasy genres. For simplicity, we have limited ourselves to only two genres to show the effectiveness of our framework. We have obtained recommendations for 30 books from Shelfari.com, a community based social cataloging platform. We treat these suggestions as ground truth because these suggestions have been edited by actual users collaboratively. 20 of these books are used as validation set and 10 of them as test set. We have carefully chosen validation set such that it has some relation with test set. We also obtain recommendations for books in validation set from Amazon.com from the "Customers Who Bought This Item Also Bought" section for each of these books. These recommendations obtained from Amazon.com capture the collaborative behavior of users which exhibit some relationship between the book and its recommendations. Content knowledge about all the 407 books are extracted from Wikipedia article of these books. The content of each book is preprocessed and keywords are extracted. We then project the books onto a CRN with 20502 IE nodes and 407 case nodes.

We have tried classification task on four standard datasets obtained from 20 Newsgroup [12]. The 20 Newsgroups data set is a collection of approximately 20,000 newsgroup documents. Each collection forms a newsgroup about a particular topic. Dataset is partitioned evenly across 20 different newsgroups. The four datasets are procured from this collection: 1. HARDWARE 2. RELPOL

3. SCIENCE 4. RECREATION. The topics under HARDWARE newsgroup is closely related which are *ibm* and *mac*, while the topics *religion* and *politics* are moderately related to each other under RELPOL dataset. SCIENCE and RECREATION are multi-class (4 classes) datasets formed from science and recreation related groups respectively. Each dataset is divided into train and test sets that are equal in size. An instance of train and test partition contains 20% of documents randomly selected from the original corpus. Each train-test split is stratified such that it preserves the class distribution of the original corpus. For repeated trials, fifteen such train-test partitions [4] are formed for each of the mentioned datasets. Documents are pre-processed by removing stop-words and functional words which are ineffective in discriminating between classes. ESA is used on the processed documents to compute document-document similarities which uses Wikipedia as knowledge source.

Performance Measure. We measure the performance in terms of classification accuracy for the task of classification. For a test set T, classification accuracy:

$$accuracy = \frac{|N_c|}{|T|}$$

where N_c is number of correctly classified instances.

Suppose for an item i the predicted recommendation set P_i and actual recommendation set A_i is given. Then we measure the performance in terms of set similarity between two sets given by *Jaccard* similarity coefficient.

$$similarity = \frac{|P_i \cap A_i|}{|P_i \cup A_i|}$$

Higher the value for similarity, the better is the result. However, similarity doesn't take order of recommendation into account. For larger datasets, it would be interesting to study the trade-offs for precision@k and recall measures.

4.2   Experimental Results and Observations

In the recommendation task, recommendation using spreading activation of the CRN is generated for books in validation set. The recommendation list for each book is compared with that of recommendation obtained from Shelfari.com for that book. If the recommendation from CRN deviates significantly (similarity less than 0.15), then we identify the book as exhibiting weak structure-function correspondence. For each such book, we introduce case arcs to CRN using the collaborative knowledge obtained from Amazon.com. The weights to the links are set to 1 when books appear in Amazon's recommendations, 0 otherwise. We then predict the recommendation for the books in test set using CRN and extended CRN. Fig. 4 shows the similarity between recommended set and actual recommendation set for each of the test books on the two architectures.

We see in the Fig. 4 that proposed architecture outperforms CRN for most of the books. Top k cases are used as recommendation where k is specified for each

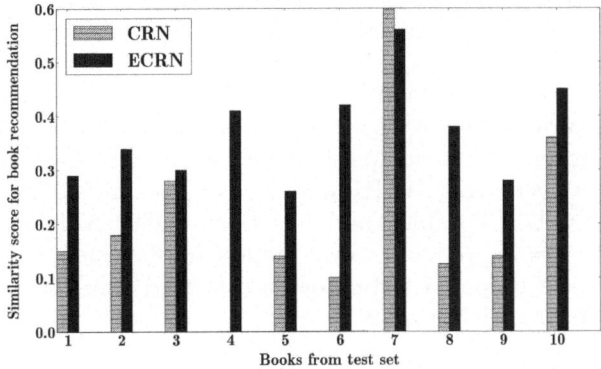

Fig. 4. Comparison of CRN and ECRN for recommendation task

book. Books 3 and 10 do not show any changes in the recommendation, this is because these books do not activate cases which incorporate *case links*. For Book 4, CRN is unable to recommend any relevant book as per suggestions from ground truth. On the other hand, ECRN is able to retrieve few relevant recommendations. However, Book 7 is showing slight decrease in similarity because one of the irrelevant books is recommended through the case-link activated by book 7. The above results give evidence that the proposed architecture is promising at unifying content representation of case with that of collaborative knowledge, which is external to the system and beyond content characterization.

Table 1. Classification accuracy of datasets

Dataset	CRN	ECRN
RELPOL	91.28 %	91.76 %
HARDWARE	70.02 %	72.86 %
SCIENCE	72.71 %	73.91 %
RECREATION	77.92 %	79.09 %

Table 2. $GAME_{class}$ score for datasets

Dataset	$GAME_{class}$
RELPOL	2.0358
HARDWARE	1.0028
SCIENCE	1.0492
RECREATION	1.1629

For classification task, the performance of the extended CRN is reported against simple CRN in Table 1. We also report the complexity of the datasets in terms of $GAME_{class}$ in Table 2. We observe that HARDWARE dataset is comparatively more complex. It is evident from the results that introduction of *case-links* is more effective in case of HARDWARE dataset. This observation falls in line with our intuition. The introduction of case links change the representation of cases in feature space by means of these arcs. Although, a weak case may be far apart from cases of same class, the case links alter the feature space to bring the weak case closer to its neighbors from same class. The more complex dataset provides more opportunity for such perturbations, thus providing more improvements as compared to a less complex dataset. We obtain $\approx 4\%$ improvement over HARDWARE datset as opposed to $\approx 0.5\%$ on RELPOL dataset, $\approx 2\%$ on SCIENCE dataset and $\approx 1.5\%$ on RECREATION. Experiments are performed for $k = 3$. We have shown modest improvement in the results as the case links in the proposed framework are formed using simplistic update rules as outlined in the Algorithm 2.

5   Discussion and Future Work

In studies on human cognition [19,20], the difference between two kinds of knowledge, rational and nonconscious, is often emphasized. Rational knowledge can be logically explained, for example the IE-case links exist because the IE is a property of the case. Non-conscious knowledge is hard to explicitize, for example we recognize a person's face where we do not know how we do it, that means we are unable to elicit the steps to do it. The case-case links aim to capture such knowledge. The proposed architecture elegantly models the non-conscious knowledge in addition to the explicit rational knowledge.

Spreading activation models have been effectively used in systems such as case-based systems [5, 14], information retrieval [6], understanding human cognition [1]. Thagard et al. [22] proposed an architecture for retrieval using multi-constraint satisfaction networks. These models exploit the features, on the other hand the the proposed ECRN architecture is able to model the non-conscious knowledge as well. ECRN can suitably applied for hybrid recommenders. The hybrid recommenders [16,18,21] combine the two types of knowledge to arrive at a better recommendation. Unlike existing methods, ECRN provides a principled approach to use collaborative knowledge with the content knowledge i.e. when the content is *weak*, case-links complement the content activation.

Case-links formation can be posed as an optimization problem. However, we do not form optimal case to case links, instead simplistic rules have been used to form the links. It is interesting to observe that although background knowledge is captured in terms of the case-links, original feature representation of the cases remain preserved. The knowledge contained in case links are invoked only when a case with weak structure representation is activated as a result of spreading activation. The *case-links* have been introduced selectively when the attribute-value representation is inadequate to solve the problem at hand. The *Case-links* are introduced independent of the case characterization, which allows for the flexibility for characterization of the case.

Maintenance in a CBR system is mostly about addition and deletion of cases independently of other cases. However, in current architecture, addition or deletion of a case will have implications on other cases as well, much similar to how human memory works.

In the proposed extended CRN, we incorporate background knowledge in terms of case interactions. When a query comes, spreading activations utilize these case links to arrive at better neighborhood of the query. This spreading action happens only in forward direction. In future we wish to explore backward propagation from case nodes to IE nodes, similar to the ideas proposed in [10], which talk about how we complement lower level information with our expectations learned over time. Since the case nodes capture domain knowledge, spreading activation from cases to IE may help us uncover relevant combination of features activated because of the *case-links*. If sufficient number of times a certain combination of the features are getting activated because of a particular case side arc, then we may remove that case-link and instead introduce similarity relation among the features that were getting triggered. This forward-backward propagation may enable us to enrich the case representation and to facilitate more efficient retrieval.

6   Conclusion

The main contribution of this work has been to extend the CRN framework to realize a separation of concerns between case similarity based on features, and models that bring together cases without committing explicitly on the features. We have presented an algorithm to learn case associations using the background domain knowledge. It is evident from experimental results that case-links are complementing the basic CRN to facilitate better retrieval. The reported experiments have been conducted at a small scale. We wish to carry out experiments on a larger scale and diverse domains.

References

1. Anderson, J.R.: A spreading activation theory of memory. Journal of Verbal Learning and Verbal Behavior 22(3), 261–295 (1983)
2. Balaraman, V., Chakraborti, S.: Satisfying varying retrieval requirements in case based intelligent directory assistance. In: FLAIRS Conference, pp. 160–165 (2004)
3. Chakraborti, S., Cerviño Beresi, U., Wiratunga, N., Massie, S., Lothian, R., Khemani, D.: Visualizing and evaluating complexity of textual case bases. In: Althoff, K.-D., Bergmann, R., Minor, M., Hanft, A. (eds.) ECCBR 2008. LNCS (LNAI), vol. 5239, pp. 104–119. Springer, Heidelberg (2008)
4. Chakraborti, S., Cerviño Beresi, U., Wiratunga, N., Massie, S., Lothian, R., Khemani, D.: Visualizing and evaluating complexity of textual case bases. In: Althoff, K.-D., Bergmann, R., Minor, M., Hanft, A. (eds.) ECCBR 2008. LNCS (LNAI), vol. 5239, pp. 104–119. Springer, Heidelberg (2008)
5. Chakraborti, S., Lothian, R., Wiratunga, N., Orecchioni, A., Watt, S.: Fast case retrieval nets for textual data. In: Roth-Berghofer, T.R., Göker, M.H., Güvenir, H.A. (eds.) ECCBR 2006. LNCS (LNAI), vol. 4106, pp. 400–414. Springer, Heidelberg (2006)
6. Crestani, F.: Application of spreading activation techniques in information retrieval. Artificial Intelligence Review 11(6), 453–482 (1997)

7. Deerwester, S., Dumais, S.T., Furnas, G.W., Landauer, T.K., Harshman, R.: Indexing by latent semantic analysis. Journal of the American Society for Information Science 41(6), 391–407 (1990)
8. Fdez-Riverola, F., Iglesias, E.L., Díaz, F., Méndez, J.R., Corchado, J.M.: Applying lazy learning algorithms to tackle concept drift in spam filtering. Expert Systems with Applications 33(1), 36–48 (2007)
9. Gabrilovich, E., Markovitch, S.: Computing semantic relatedness using wikipedia-based explicit semantic analysis. In: Proceedings of the 20th International Joint Conference on Artifical Intelligence, IJCAI 2007, pp. 1606–1611. Morgan Kaufmann Publishers Inc., San Francisco (2007)
10. Hawkins, J., Blakeslee, S.: On Intelligence. Times Books (2004)
11. Jannach, D., Zanker, M., Felfernig, A., Friedrich, G.: Recommender Systems: An Introduction, ch. 2. Cambridge University Press (2010)
12. Lang, K.: Newsweeder: Learning to filter netnews. In: Proceedings of the 12th International Machine Learning Conference, ML 1995 (1995)
13. Lenz, M., Burkhard, H.D.: Case retrieval nets: Basic ideas and extensions. In: Görz, G., Hölldobler, S. (eds.) KI 1996. LNCS, vol. 1137, pp. 227–239. Springer, Heidelberg (1996)
14. Lenz, M., Burkhard, H.D.: CBR for document retrieval: The fallq project. In: Leake, D.B., Plaza, E. (eds.) ICCBR 1997. LNCS, vol. 1266, pp. 84–93. Springer, Heidelberg (1997)
15. Miller, G.A.: Wordnet: A lexical database for English. Communications of the ACM 38, 39–41 (1995)
16. Miranda, T., Claypool, M., Gokhale, A., Mir, T., Murnikov, P., Netes, D., Sartin, M.: Combining content-based and collaborative filters in an online newspaper. In: Proceedings of ACM SIGIR Workshop on Recommender Systems. Citeseer (1999)
17. Patelia, A., Chakraborti, S., Wiratunga, N.: Selective integration of background knowledge in TCBR systems. In: Ram, A., Wiratunga, N. (eds.) ICCBR 2011. LNCS, vol. 6880, pp. 196–210. Springer, Heidelberg (2011)
18. Pazzani, M.J.: A framework for collaborative, content-based and demographic filtering. Artificial Intelligence Review 13(5-6), 393–408 (1999)
19. Schank, R.C.: Where's the ai? AI Magazine 12(4), 38 (1991)
20. Schank, R.C.: Dynamic Memory Revisited. Cambridge University Press, New York (1999)
21. Soboroff, I., Nicholas, C.: Combining content and collaboration in text filtering. In: Proceedings of the IJCAI, vol. 99, pp. 86–91 (1999)
22. Thagard, P., Holyoak, K.J., Nelson, G., Gochfeld, D.: Analog retrieval by constraint satisfaction. Artif. Intell. 46(3), 259–310 (1990)

Acquisition and Reuse of Reasoning Knowledge from Textual Cases for Automated Analysis

Gleb Sizov, Pinar Öztürk, and Jozef Štyrák

Department of Computer Science,
Norwegian University of Science and Technology,
Trondheim, Norway
{sizov,pinar}@idi.ntnu.no, jozef.styrak@gmail.com

Abstract. Analysis is essential for solving complex problems such as diagnosing a patient, investigating an accident or predicting the outcome of a legal case. It is a non-trivial process even for human experts. To assist experts in this process we propose a CBR-based approach for automated problem analysis. In this approach a new problem is analysed by reusing reasoning knowledge from the analysis of a similar problem. To avoid the laborious process of manual case acquisition, the reasoning knowledge is extracted automatically from text and captured in a graph-based representation, which we dubbed *Text Reasoning Graph* (TRG), that consists of causal, entailment and paraphrase relations. The reuse procedure involves adaptation of a similar past analysis to a new problem by finding paths in TRG that connect the evidence in the new problem to conclusions of the past analysis. The objective is to generate the best explanation of how the new evidence connects to the conclusion. For evaluation, we built a system for analysing aircraft accidents based on the collection of aviation investigation reports. The evaluation results show that our reuse method increases the precision of the retrieved conclusions.

Keywords: Practical reasoning knowledge, causal relation extraction, knowledge acquisition, case reuse, textual CBR, automated analysis.

1 Introduction

Which is more exciting about Sherlock Holmes stories: learning who the murderer was, or following Sherlock's reasoning on the road from evidence to conclusion leading him to the murderer?

Our overarching goal is to facilitate the reuse of reasoning knowledge residing in documents written by problem solvers. Such reasoning knowledge can be found embedded in, for example, accident reports. CBR research recognized accident reports as reusable experiences, and a special workshop challenge about analysing air investigation reports was organized as a part of the 4th Workshop on Textual case-based Reasoning in 2007 [6]. Accident reports were used by CBR researchers also before this workshop [11,10,22,9]. Much of these work

L. Lamontagne and E. Plaza (Eds.): ICCBR 2014, LNCS 8765, pp. 465–479, 2014.
© Springer International Publishing Switzerland 2014

combines CBR with information retrieval and focuses on the retrieval stage of the CBR cycle. It may be considered close to the "weakly textual" end of the weakly-strongly textual scale defined in [24]. Weakly textual means either that the concerned documents are sufficiently structured which reduces the need for sophisticated natural language processing (NLP), or that the type of task under consideration does not need deep NLP. Weakly textual CBR is appropriate for tasks that focus on retrieval and classification. However, when moved beyond retrieval or simple reuse of classes without adaptation, and toward the reuse of reasoning knowledge, strongly textual CBR becomes more appropriate because reasoning knowledge is often buried in text making it difficult to discern or process without deep NLP techniques. Bruninghaus and Ashley [8] are pioneers in strongly textual CBR who took in use information extraction and sophisticated NLP techniques. The work presented in this paper also uses deep NLP techniques for case acquisition (i.e. mining the reasoning knowledge that in turn comprises the problem solution part of a case) and reuse purpose.

Causality is recognized as the most fundamental type of reasoning knowledge [18]. In textual CBR research, Orecchioni et al. [17] attempted to extract causal knowledge by classifying sentences in accident reports as causal or factual. In our work we also aim to extract and reuse causal knowledge, not as a set of isolated causal sentences but rather as a chain of causal relations reflecting the problem solver's line of reasoning from the evidence to the conclusion. The rationale behind the proposed CBR-based analysis approach is that computers may support human analysts in understanding a problem through reuse of previously constructed reasoning paths that show how the conclusion *explains* the evidences. Such a CBR system essentially involves adaptation of past reasoning paths. Two challenges pertinent to the textual CBR task we are targeting and the type of data we are concerned with are: (i) representation and extraction of reasoning knowledge contained in text (ii) automated adaptation of this explanatory/reasoning knowledge to a new problem.

The rest of this paper is organized as follows. Section 2 provides an overview of our approach to CBR-based problem analysis. Section 3 describes Text Reasoning Graph (TRG), a graph-based representation for capturing reasoning knowledge extracted from text. In section 4, we present the procedure for automatic adaptation of reasoning knowledge to a new problem. Section 5 explains details for extraction of TRG from text. Empirical evaluation of the approach is described in section 6 with the results and error analysis in section 7. In section 8 we look at the related work. The conclusion and the future work directions are presented in section 9.

2   Automated Problem Analysis

In this section we introduce our approach to the automated problem analysis through reuse of reasoning knowledge extracted from textual cases. Under this approach, the analysis for a new problem is generated by retrieving a similar problem and adapting its analysis to fit the new problem. For this to work,

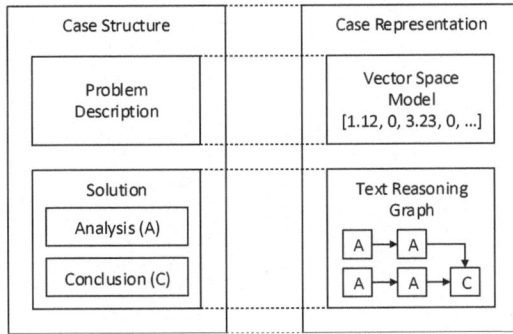

Fig. 1. Case structure and representation

cases need to contain an analysis as part of their solution, as shown in Figure 1. The *analysis* should describe the reasoning of an expert connecting *evidences* to *conclusions* where the evidence is part of the problem description that serves as the starting point for the analysis while a conclusion refers to a decision or judgement that solves or explains the problem, e.g. the cause of an accident, a diagnosis for a patient or an outcome of a legal case.

Manual case acquisition is a laborious task. It becomes even more challenging when a case includes also an analysis part. To overcome the manual knowledge engineering problem and to take the advantages of abundant free-text analysis reports existing either in organizations or on the web, we target automated extraction of cases from such reports. First, we propose a case structure with a hybrid representation, as shown in Figure 1. In this structure, problem description is represented by a vector space model (VSM) with TFIDF weights, which is a well known representation for free text documents, often used in information retrieval (IR) and for document classification. We introduce a different representation for representation of the solution part which we dubbed *Text Reasoning Graph* (TRG), a graph-based representation with expressive power to represent the chain of reasoning underlying the analysis as well as to facilitate the adaptation of a past analysis to a new problem.

An overview of our approach is shown in Figure 2. In the *case acquisition* stage, the case base is populated by converting each free-text document to a case of which the problem description part is represented with VSM and the solution part as a TRG. Given the description of a new problem, the *retrieval process* finds a case with the most similar problem description. This is implemented by converting the new problem description to a VSM representation and then computing its *cosine* similarity with problem descriptions of the cases in the case base. Solution of the retrieved best case in the TRG form, further referred to as *CaseGraph*, is then adapted to the new problem description in the *reuse step*. The result of the reuse step is the *ReuseGraph* representing the analysis of the new problem in the TRG form - the details are described in section 4. Finally, the ReuseGraph is visualized for manual interpretation by a user.

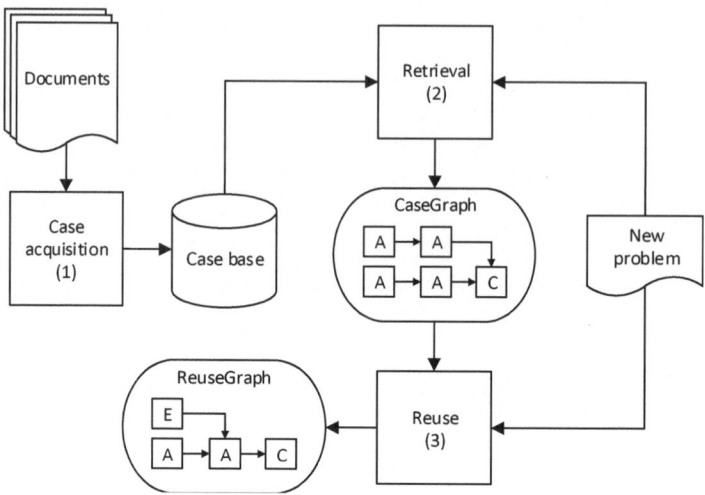

Fig. 2. System diagram

For our empirical work, we used a collection of aviation investigation reports where each report documents an investigation of an aircraft accident including a brief description of the accident, an analysis, and conclusions reached by a human expert. The reports have a consistent structure, with different sections corresponding to different parts in the case structure (Figure 1). For instance, the "Summary"/"Synopsis" section in a report is considered as the problem description, the "Analysis" section as the analysis, and several sections with titles similar to "Finding as to causes and contributing factors", "Findings as to risk" and "Other findings" as the conclusion. Hence, the conclusion consists of one or more sentences highlighting the possible causes and findings.

3   Representation of Reasoning Knowledge

For our approach, we needed a representation that is able to capture the line of reasoning embedded in text. Consider this excerpt from the analysis section of the aviation investigation report a06q0091:

> The oil that burned away did not return to the tank and, after a short time, the oil level became very low, causing the engine oil pump to cavitate and the engine oil pressure to fluctuate. Furthermore, since the oil did not return to the tank, the oil temperature did not change, or at least not significantly, and the pilot falsely deduced that the engine oil pressure gauge was displaying an incorrect indication.

The type of knowledge contained in this passage is important for understanding the accident because it describes how the human expert reasoned about the situation. The reasoning (e.g. causal) knowledge is mostly of relational nature.

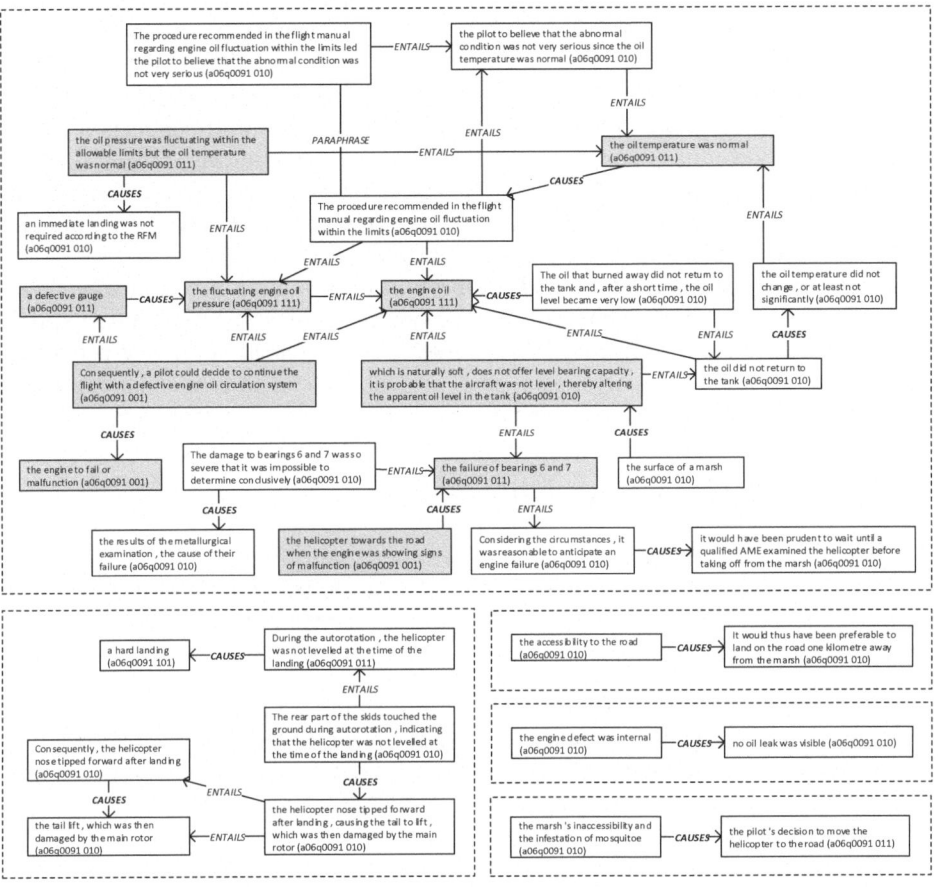

Fig. 3. Structure of the CaseGraph generated from report a06q0091. The content of the darker nodes can be seen in Figure 4.

A graph-based representation is therefore more appropriate than a frame-based representation, which is commonplace in classical CBR, because the chains/links of relations can explicitly be represented through edges in a graph. As noted in [19] causal graphs extracted from domain-specific documents provide a powerful representation of the expert's knowledge. The TRG representation comprises mainly causal relations, complemented with *textual entailment and paraphrase relations* that make the graph more connected. Automatic extraction of TRG from text relies on various NLP techniques, as described in section 5.

Figure 3 provides an overview of the CaseGraph structure generated from the report a06q0091. The graph consists of five connected components (surrounded by dashed lines) with one of them containing most nodes and relations in the graph. The content of the darker nodes is shown in Figure 4. As it can be seen from these figures, nodes in a TRG contain phrases (or sentences) that are arguments for causal relations extracted from text. Contents of the arguments are determined through causal patterns described in section 5.1. Each node is also labelled with two codes: the report id and the code that indicates part(s) of the report the the node was extracted from, e.g. a node labelled a06q0091 010 is extracted from the analysis part (010) of the report a06q0091. Other binary codes correspond to the problem description (100) and conclusion (001). A combination of binary codes means that the concerned phrase is contained in several parts of the same report, e.g. 101 stands for the problem description and the conclusion parts together. Further in this paper, nodes extracted from the conclusion part will be referred to as *conclusion nodes*.

One of the advantages of the TRG representation is that it can be visualized and interpreted easily. It represents the reasoning knowledge in a more explicit way than the original text does. This makes TRG a visual summary providing a quick overview of the analysis of a case, which is useful in particular when the user needs to study several similar past cases to solve a new problem.

4   Reuse of Reasoning Knowledge

The flexibility of the TRG representation allows automatic adaptation of the retrieved solution to a new problem. The adaptation process generates a Reuse-Graph from the retrieved CaseGraph and the description of a new problem. This process consists of the following steps:

1. Find *evidence nodes* (labelled 100 in Figure 4), i.e. nodes in the CaseGraph that are entailed by sentences in the description of the new problem.
2. Activate *reasoning paths*, i.e. one or several shortest paths (type and direction are ignored) that connect the evidence nodes (found in step 1) to conclusion nodes in the CaseGraph.
3. Construct a ReuseGraph by combining activated reasoning paths and then by adding sentences from the new problem description that entail the evidence nodes from step 1.

The first step corresponds to identifying the important pieces of evidence in the new problem description. These evidences are then used as starting points for reasoning in the second step where they are connected to the conclusions. The result of this procedure is the ReuseGraph that represents the analysis of a new problem based on the analysis of a previously solved problem. Notice that ReuseGraph may contain fewer conclusion nodes than the CaseGraph if some evidences in the new case do not link to the conclusions in the past case.

Figure 4 shows the reuse graph for the accident in report a08a0095 (new case) generated from the CaseGraph for report a06q0091 (past case) shown in Figure 3. Two nodes with bold frames on the left are sentences from report a08a0095 providing evidence that there is an engine failure or malfunction. This evidence can be explained by the defective engine oil circulation system as pointed out by the node to the right. There are several reasoning paths pointing to the failure of bearings, the oil level, the fluctuating oil pressure and, the defective gauge. As it turns out, one of the correct conclusions for report a08a0095 is inadequate

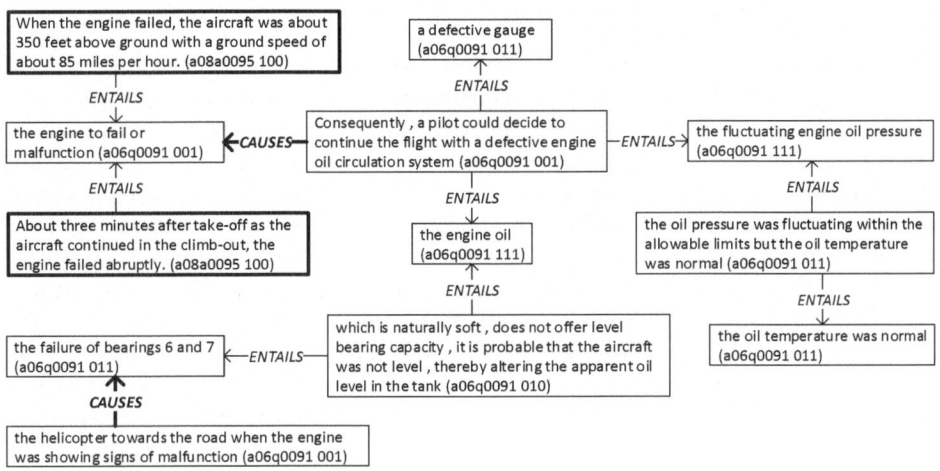

Fig. 4. ReuseGraph representing analysis of the accident a08a0095

lubrication of linkpin bushings (type of bearing), which is close to failure of bearings and the oil level.

One major advantage of a ReuseGraph as the solution compared to a short answer, label or category, is that it provides a justification for the reached conclusions. At the same time, compared to the whole retrieved case as the output, a ReuseGraph is specifically adjusted to the new problem and eliminates all unnecessary information. A ReuseGraph tends to be much smaller than the corresponding CaseGraph, e.g. the ReuseGraph in Figure 4 contains only 12 nodes compared to 34 nodes in the corresponding CaseGraph in Figure 3.

5   Acquisition of Reasoning Knowledge

TRG contains three types of relations: causal, entailment and paraphrase relations. These relations are extracted automatically from text and combined in one graph. The process of generating a TRG is as follows:

1. Preprocessing text documents.
2. Extracting the causal relations.
3. Determining the informativeness of arguments in causal relations.
4. Combining relations where both arguments are informative into one graph.
5. Adding entailment and paraphrase relations between arguments.

Steps involved in the preprocessing stage are HTML scraping, tokenization, POS tagging, syntactic parsing, lemmatization and stop word removal. HTML scraping is implemented using custom component specifically designed to obtain the text and the structure of an investigation report from Transportation Board of Canada website. The rest of the preprocessing is accomplished using Stanford CoreNLP pipeline. Further in this section we describe steps 2, 3, 5 in more detail.

5.1   Causal Relation Extraction

The causal relation extraction component implements the approach described by Khoo [12] which uses manually constructed lexico-syntactic patterns. The extraction algorithm applies the set of patterns to a given sentence or a pair of neighbour sentences. If matching succeeds, the cause and the effect phrases are extracted according to the applied pattern. The patterns contain elements such as part-of-speech tags, phrase types and specific tokens. For example, given a pattern "[effect] because [cause]" and a sentence "The rotor blade failed because its structural integrity was compromised by a manufacturing defect" as the input, the system will extract "The rotor blade failed" as the effect and "its structural integrity was compromised by a manufacturing defect" as the cause. The patterns are built around causal triggers, words that indicate a causal relation between phrases. Based on Altenberg's [3] *typology of causal linkage*. Khoo defines a list of 651 causal patterns, 382 sub-patterns and 2115 causal verbs in his PhD thesis [12] where he describes also a pattern matching algorithm. We were unable to obtain the original implementation of the system and reimplemented the system using CoreNLP pipeline. We have updated Khoo's list of patterns, sub-patterns and causal triggers to improve the performance for our task.

5.2   Node Informativeness

The quality of a TRG depends on the quality of the constituent nodes and relations. The quality of a node is determined by its informativeness. Nodes that do not carry a concrete piece of information are considered uninformative. Nodes in a TRG are the arguments of causal relations extracted from text, which are phrases of different size. Smaller phrases tend to be less informative than longer ones. Even if a word is relevant to the domain, e.g. "pilot", "flight", "procedure", "altitude", it often does not carry specific enough information to be used in a reasoning process. For phrases that contain more than one word we need to measure the informativeness of each word first. For this purpose we use inverse document frequency (IDF), a statistical measure commonly used in IR. Words with IDF values above 1.0 are considered informative. This threshold was manually determined by finding an IDF value that would filter out general terms for our dataset. For a phrase to be considered informative it should contain at least two non-stop words where at least one is informative.

5.3   Textual Entailment and Paraphrase Relations

Causal relations are typically scattered in text and rarely form connected graphs with more than two or three relations in one graph. In order to link these graphs to form larger and more connected ones, we rely on entailment and paraphrase relations between the arguments of the causal relations. For this purpose, we employ methods for textual entailment and paraphrase detection.

Paraphrase detection is the task of recognizing text fragments with approximately the same meaning. For example, "the engine failure during take off" and "in the climb-out, the engine failed abruptly" are paraphrases. In textual entailment the task is to determine whether one text fragment, called text (T), infers another text fragment, called hypothesis (H), i.e. if T is true H is also true. Unlike paraphrase, entailment is not symmetrical, e.g. "the engine failure during take off" entails "the engine failed" but "the engine failed" does not entail "the engine failure during take off" because part of the information, i.e. "during take off", is missing. A nice overview of the existing methods for detection of paraphrases and entailments can be found in the survey paper by [4].

In our system, paraphrase detection and textual entailment are based on the same text similarity measure. This measure assigns words in one text fragment to words in another fragment through a word similarity measure. LCH [14] is used for word similarity measure, which is based on a shortest path between the corresponding senses in WordNet[16]. The text similarity is computed by solving the assignment problem, where each word in one text fragment is assigned to a word in another text fragment so that no two words are assigned to the same word and the sum of similarities between assigned words is maximized. The final value is obtained by normalizing this sum by the average number of words in both fragments.

For paraphrase detection we compute the text similarity between two arguments. If the resulting value is equal to or above a threshold value, a paraphrase relation between these arguments is added to the graph. For entailment, if H is smaller than T we extract all substrings of T with the same number of words as in H + 1. Then the text similarity is computed between T and each substring of H. If the maximum of the obtained similarities is equal to or above the threshold value, the entailment relation from T to H is added to the graph. The similarity threshold value is set to 0.6, which was manually determined by considering the number of correctly identified entailment and paraphrase relations in TRG graphs constructed with different thresholds.

6 Experimental Evaluation

For evaluation, we implemented a system that, given a short textual description of an accident and a collection of investigation reports for previous accidents, automatically generates an analysis of the new accident following the approach described in section 2. As the dataset, we use a collection of 494 aviation investigation reports from Transportation Board of Canada for years 1994-2008.

An analysis generated by our system is represented by a ReuseGraph. Evaluation of a ReuseGraph directly is problematic because of the lack of standard evaluation measures or baseline systems. To overcome this problem, we evaluate only conclusions in a ReuseGraph. Our assumption is that since reasoning paths in a ReuseGraph link the evidences in the new problem with conclusions in a similar past accident analysis, validity of the conclusions will reflect the quality of the reasoning paths and thus of the whole ReuseGraph. Validity of conclusions is determined by the similarity between the actual conclusions in the test case report and the ones generated by the system. For the baseline we use conclusions in the retrieved report, i.e. immediately after retrieval, without adaptation.

6.1   Evaluation Procedure

Every investigation report in our dataset contains one or two sections that correspond to what we refer to as conclusions. These sections contain sentences enumerating findings, causes and contributing factors for an accident. The following is from report a06q0091:

1. The area adjacent to bearings 6 and 7 had exceeded a temperature of 900 °C. The bearings were destroyed for undetermined reasons, causing an engine failure.
2. Moving the helicopter towards the road when the engine was showing signs of malfunction contributed to the failure of bearings 6 and 7.
3. During the auto-rotation, the helicopter was not levelled at the time of the landing, which resulted in a hard landing.

Conclusion nodes in a ReuseGraph usually do not contain full sentences from the conclusion part but rather phrases extracted from these sentences or entailed by them. Evaluation procedure outlined in the algorithm 1 contains separate functions for evaluation of the retrieval (i.e., baseline) and reuse steps. In both of them conclusion sentences are compared with the actual conclusions in the test case. The difference between them is that the first one takes all the conclusion sentences in the retrieved case while the latter makes use of the ReuseGraph to select the subset of the conclusion sentences.

Algorithm 1. Evaluation procedure

```
 1: function EVALUATERETRIEVAL(TestCase, CaseBase)
 2:     RetrievedCase = Retrieve(TestCase.Problem, CaseBase)
 3:     RetrievedCS = RetrievedCase.Conclusion.Sentences
 4:     ReferenceCS = TestCase.Conclusion.Sentences
 5:     RetrievalScore = COMPARE(RetrievedCS, ReferenceCS)
 6:     return RetrievalScore
 7: end function
 8: function EVALUATEREUSE(TestCase, CaseBase)
 9:     RetrievedCase = Retrieve(TestCase.Problem, CaseBase)
10:     ReuseGraph = Reuse(TestCase.Problem, RetrievedCase.CaseGraph)
11:     ReusedCS = ReuseGraph.Conclusion.Nodes.Sentences
12:     ReferenceCS = TestCase.Conclusion.Sentences
13:     ReuseScore = COMPARE(ReusedCS, ReferenceCS)
14:     return ReuseScore
15: end function
16: function COMPARE(CandidateCS, ReferenceCS)
17:     Cost = Assignment(CandidateCS, ReferenceCS, TextSimilarity)
18:     Precision = Cost / |CandidateCS|
19:     Recall = Cost / |ReferenceCS|
20:     F-score = 2 · Precision · Recall / (Precision + Recall)
21:     return (Precision, Recall, F-score)
22: end function
```

In the retrieval evaluation, a case most similar to the problem description of the test (i.e, new) case is retrieved from the case base (line 2). Conclusion sentences are obtained (line 3) from this case and compared with reference conclusion sentences, i.e. the conclusion in the test case (line 4 and 5). The result of this comparison is the retrieval score which is based on the similarity between the retrieved and the reference conclusion sentences.

Reuse is evaluated in a similar way. The difference is that after the retrieval step (line 9), the reuse procedure is applied to generate a ReuseGraph based on the problem description of the test case and the CaseGraph of the retrieved case. A subset of conclusion sentences from the retrieved case is obtained through the conclusion nodes in the ReuseGraph (line 11). Unlike evaluation of the retrieval step where all the conclusion sentences are taken, only sentences that entail or are the source of one or more conclusion nodes in the ReuseGraph are retained. This results in fewer conclusion sentences than in the retrieval evaluation because the ReuseGraph contains only the conclusion nodes that are connected to evidence in the problem description of the test case, discarding all other nodes. The retained conclusion sentences are then compared with reference conclusion sentences in the same way as for the retrieval evaluation (line 13).

For both reuse and retrieval evaluation we use the same evaluation measure outlined in lines 16-22. This measure is based on the similarity between two sets of sentences, candidate conclusion sentences (CandidateCS) and the reference conclusion sentences (ReferenceCS). CandidateCS correspond to conclusions under evaluation, i.e. either RetrievedCS or ReusedCS. ReferenceCS are the correct conclusion sentences, i.e., the ones in the test case. Higher similarity between a CandidateCS and the ReferenceCS results in a higher evaluation score. The similarity is computed by using the notion of the *assignment problem* where each sentence in CandidateCS is assigned to a sentence in ReferenceCS so that no two sentences are assigned to the same sentence and the sum of text similarities between assigned sentences is maximized. This sum is referred to as Cost (line 17) and is used to compute precision, recall and F-score in lines 18-20. For the text similarity we use the same measure as described in section 5.3.

Our evaluation procedure follows leave-one-out cross-validation strategy, where one case is selected as the test case while the rest are considered as the case base. The evaluation scores are obtained for each test case to compute mean and standard deviation. Paired difference tests are carried out to compare the scores for retrieval and reuse steps.

7   Results and Error Analysis

Our system was able to generate ReuseGraphs for 118 of 494 reports that were used as the test cases. For the remaining cases, no connection was found between the evidence in the problem description of the test case and the conclusions of the most similar case from the case base. This is mostly because the "Summary" sections used as the problem descriptions did not provide enough evidence for reuse.

Table 1. Evaluation results, mean ± standard deviation in % values

Step	Precision	Recall	F-score
retrieval	19.71 ± 7.11	20.97 ±7.40	18.70 ± 5.18
reuse	25.03 ± 8.90	11.35 ± 7.99	13.74 ± 6.99
reuse - retrieval	5.32	-9.62	-4.96

Table 1 summarizes the results. Retrieval scores indicate the correctness of conclusions in a retrieved case before the reuse procedure, and the reuse results those of after the reuse procedure. Since the reuse can't add any new conclusions but merely removes conclusions that can't be connected to the evidence in the new/test case, the recall score after reuse is expected to be equal to or lower than that of before reuse. At the same time, reuse is expected to increase the precision score because conclusions are selected based on the reasoning paths originating from the new evidence. In contrast, a randomly selected subset of retrieved conclusions should not change the precision.

Evaluation results confirm our hypothesis about an increase in the precision and a decrease in the recall scores after reuse. The F-score decreased as well. Paired t-test and Wilcoxon signed-rank test show that the difference in scores is statistically significant with p-value < 0.0001. Increased precision indicates that the analysis generated by the system is able to connect the evidences in the test case to the correct conclusions, at least to some degree. It can also be argued that for our task precision is more important than recall because it reduces information overload for the user of the system by eliminating conclusions that the system is not able link to the new evidences. If necessary, the user can always see the solution before the reuse.

Manual inspection of the generated CaseGraphs and ReuseGraph indicates that although a lot of relations and paths through these graphs are coherent, there are many errors as well. These errors are mostly attributed to automatic generation of TRG from text. The used dataset is quite complex having long sentences and domain-specific terminology. Components in our NLP pipeline that rely on supervised machine learning are mostly trained on news articles and other corpora unrelated to the aviation domain which results in suboptimal performance. These errors propagate further in the system and decrease the accuracy of causal extraction, entailment and paraphrase identification components, which introduce errors of their own. We believe it is possible to improve each of these components by combining multiple approaches together and adjusting them specifically to the target domain.

8   Related Work

Adeyanju's PhD work [1] extends Case Retrieval Network (CRN) [15] to Case Retrieval Reuse Network (CR2N) to enable case reuse. Nodes in CR2N represent cases and keywords from solutions of these cases, and the edges connect cases to keywords or keywords to each other. While both CR2N and our approach use graph-based representations and have a focus on reuse, the content of the representation and the reuse process in the two approaches are substantially different. First of all, CR2N is a more general approach, not aiming to capture the reasoning knowledge. In the reuse process CR2N identifies reusable keywords in the retrieved solution,while our system translates the retrieved solution to a CaseGraph and, in turn, modifies it to a ReuseGraph. This modification involves addition and deletion of solution elements, which can be considered a structural form of reuse in terms of adaptation models described by Wilke et al. [23]. Although CR2N is also considered as a transformational form of reuse, our approach does not only identify what can be reused but generates the actual solution to be reused.

There are several other textual CBR systems that do structural reuse but without transforming a textual case to a more structured representation similar to TRG. Most of these are aimed at assisting users in text authoring. For example Lamontagne et al. [13] system adapts email responses by determining a subset of sentences relevant to the new request and replacing some specific information in these sentences such as individuals, locations and addresses. A somewhat similar approach has been applied by Swanson et al. [21]. The authors developed "Say Anything" CBR system for interactive storytelling. Given the story so far, the system proposes next sentences. The adaptation involves modification of the retrieved sentences by replacing proper nouns and pronouns with corresponding frequent substitutes from the already written part of the story. GhostWriter-2.0 developed by Bridge et al. [7] assists users in writing product reviews. The system suggests phrases extracted from previous reviews similar to what the user has already written. In the work by Recio-Garcíia et al. [20], the authors experimented with the same aviation accident report dataset as we do. The user is supposed to rewrite the solution of a retrieved similar report using text spans retrieved from other reports. The system assists the user by finding and clustering sections from the past aviation investigation reports that are relevant to the user's query. One interesting aspect of these text authoring systems is that they are focusing on providing assistance to the user rather than doing adaptation automatically. This assistance involves a graphical user interface to support human interaction, which often makes systems more practical for real-life applications. A branch of CBR research called conversational CBR[2] studies such systems. Currently, we don't have a conversational component in our system but it is highly relevant for future work.

The importance of reasoning knowledge, such as contained in the analysis section of aviation accident reports, is underrated in textual CBR research. Our work captures this knowledge in TRG predominantly through causal relations. A representation containing causal relations similar to TRG has been previously proposed by Ashley et al. [5]. For their LARGO system, the authors developed a diagrammatic representation capturing arguments in a legal case. Their diagrams very much resembles TRG but with stronger semantics, which allows more sophisticated types of domain-specific adaptation. However, the diagrams were constructed manually by human experts, which is a very laborious process compared to automatic generation of TRG from text. In general, reasoning knowledge is crucial for legal cases and TRG representation seems like a good fit for this domain.

9   Conclusion and Future Work

We have presented a novel CBR approach that opens up for reuse of reasoning knowledge in textual cases. Our approach relies on a textual graph-based representation TRG, which captures reasoning/explanatory knowledge through causal, entailment and paraphrase relations automatically extracted from text. We also developed a reuse procedure to adapt the explanatory path used for a previous problem to a new problem. The approach was evaluated on a collection of aviation investigation reports, demonstrating the ability to capture and adapt the analysis of a previous accident to a new accident.

The novelty of the approach suggests many directions for future work. First, automatic extraction of TRG from text is a challenging tasks, with subtasks such as causal relation extraction, entailment and paraphrase identification leaving a lot of room for improvement. Second, conversational elements can be introduced to allow users to guide the analysis. Third, the reuse procedure can be enhanced by upgrading the shortest path with a more sophisticated heuristic that takes type and directionality of relations into consideration. Forth, structured domain knowledge can be integrated in the approach to improve TRG graphs. Fifth, it makes sense to use TRG in the retrieval step as well as in the reuse. Finally, extrinsic evaluation by human experts is required to confirm the validity of the approach in practical settings.

References

1. Adeyanju, I.: Case reuse in textual case-based reasoning. Ph.D. thesis, Robert Gordon University (2011)
2. Aha, D.W., Breslow, L.A., Muñoz-Avila, H.: Conversational case-based reasoning. Applied Intelligence 14(1), 9–32 (2001)
3. Altenberg, B.: Causal linking in spoken and written English. Studia Linguistica 38(1), 20–69 (1984)
4. Androutsopoulos, I., Malakasiotis, P.: A survey of paraphrasing and textual entailment methods. Journal of Artificial Intelligence Research 38, 135–187 (2010)
5. Ashley, K., Lynch, C., Pinkwart, N., Aleven, V.: Toward modeling and teaching legal case-based adaptation with expert examples. In: McGinty, L., Wilson, D.C. (eds.) ICCBR 2009. LNCS, vol. 5650, pp. 45–59. Springer, Heidelberg (2009)
6. Bridge, D., Gomes, P., Seco, N.: Analysing air incident reports: workshop challenge. In: Proc. of the 4th Workshop on Textual Case-Based Reasoning (2007)
7. Bridge, D., Healy, P.: Ghostwriter-2.0: Product reviews with case-based support. In: Research and Development in Intelligent Systems XXVII, pp. 467–480. Springer (2011)

8. Brüninghaus, S., Ashley, K.D.: Progress in textual case-based reasoning: predicting the outcome of legal cases from text. In: Proceedings of the National Conference on Artificial Intelligence, vol. 21, p. 1577. AAAI Press, MIT Press, Menlo Park, Cambridge (2006)

9. Carthy, J., Wilson, D.C., Wang, R., Dunnion, J., Drummond, A.: Using T-ret system to improve incident report retrieval. In: Gelbukh, A. (ed.) CICLing 2004. LNCS, vol. 2945, pp. 468–471. Springer, Heidelberg (2004)

10. Cassidy, D., Carthy, J., Drummond, A., Dunnion, J., Sheppard, J.: The use of data mining in the design and implementation of an incident report retrieval system. In: 2003 IEEE Systems and Information Engineering Design Symposium, pp. 13–18. IEEE (2003)

11. Johnson, C.: Using case-based reasoning to support the indexing and retrieval of incident reports. In: Proceeding of European Safety and Reliability Conference (ESREL 2000): Foresight and Precaution, Balkema, Rotterdam, the Netherlands. pp. 1387–1394. Citeseer (2000)

12. Khoo, C.S.G.: Automatic identification of causal relations in text and their use for improving precision in information retrieval. Ph.D. thesis, The University of Arizona (1995)

13. Lamontagne, L., Lee, H.-H.: Textual reuse for email response. In: Funk, P., González Calero, P.A. (eds.) ECCBR 2004. LNCS (LNAI), vol. 3155, pp. 242–256. Springer, Heidelberg (2004)

14. Leacock, C., Miller, G.A., Chodorow, M.: Using corpus statistics and wordnet relations for sense identification. Computational Linguistics 24(1), 147–165 (1998)

15. Lenz, M., Burkhard, H.D.: Case retrieval nets: Basic ideas and extensions. In: Görz, G., Hölldobler, S. (eds.) KI 1996. LNCS, vol. 1137, pp. 227–239. Springer, Heidelberg (1996)

16. Miller, G.A.: Wordnet: a lexical database for English. Communications of the ACM 38(11), 39–41 (1995)

17. Orecchioni, A., Wiratunga, N., Massie, S., Chakraborti, S., Mukras, R.: Learning incident causes. In: Proc. of the 4th Workshop on Textual Case-Based Reasoning (2007)

18. Pearl, J.: Causality: models, reasoning and inference, vol. 29. Cambridge Univ. Press (2000)

19. Pechsiri, C., Piriyakul, R.: Explanation knowledge graph construction through causality extraction from texts. Journal of Computer Science and Technology 25(5), 1055–1070 (2010)

20. Recio-Garcia, J.A., Diaz-Agudo, B., González-Calero, P.A.: Textual cbr in jcolibri: From retrieval to reuse. In: Proceedings of the ICCBR 2007 Workshop on Textual Case-Based Reasoning: Beyond Retrieval, pp. 217–226. Citeseer (2007)

21. Swanson, R., Gordon, A.S.: Say anything: Using textual case-based reasoning to enable open-domain interactive storytelling. ACM Transactions on Interactive Intelligent Systems (TiiS) 2(3), 16 (2012)

22. Tsatsoulis, C., Amthauer, H.A.: Finding clusters of similar events within clinical incident reports: a novel methodology combining case based reasoning and information retrieval. Quality and Safety in Health Care 12(suppl. 2), ii24–ii32 (2003)

23. Wilke, W., Bergmann, R.: Techniques and knowledge used for adaptation during case-based problem solving. In: Mira, J., Moonis, A., de Pobil, A.P. (eds.) IEA/AIE 1998. LNCS, vol. 1416, pp. 497–506. Springer, Heidelberg (1998)

24. Wilson, D.C., Bradshaw, S.: Cbr textuality. In: Proceedings of the Fourth UK Case-Based Reasoning Workshop, pp. 67–80. Citeseer (1999)

Enriching Case Descriptions Using Trails in Conversational Recommenders

Skanda Raj Vasudevan and Sutanu Chakraborti

Department of Computer Science and Engineering
IIT Madras, Chennai 600036, India
{skandavs, sutanuc}@cse.iitm.ac.in

Abstract. Case based recommenders often use similarity as a surrogate for utility. For a given user query, the most similar products are given as recommendations. Similarities are designed in such a way that they closely approximate utilities. In this paper, we propose ways of estimating robust utility estimates based on user trails. In conversational recommenders, as the users interact with the system trails are left behind. We propose ways of leveraging these trails to induce preference models of items which can be used to estimate the relative feature specific utilities of the products. We explain how case descriptions can be enriched based on these utilities. We demonstrate the effectiveness of PageRank style algorithms to induce preference models which can in turn be used in re-ranking the recommendations.

Keywords: Recommender Systems, Critiquing, Utility, PageRank.

1 Introduction

Conversational recommenders facilitate an interaction with the users allowing them to iteratively refine the requirements by exploring the product space. Items are often represented as attribute-value pairs and each of these can be a potential solution to the requirements given by the customer. For a given user query, the system recommends most similar items and interaction with the user is facilitated through means such as natural language dialogues or feedback on recommendations thereby learning user preferences which is in turn used to improve the recommendations.

The holy grail in designing recommender systems is to arrive at a robust estimate of the utility of a product with respect to a user and a set of expressed preferences. While similarity is often used as a surrogate for utility, it is at best an a priori approximation closely tied to choices made in representation and similarity measures [2]. One challenge in arriving at a robust utility estimate is that the features used to estimate the utility of a product may not be from within the features of the product itself. For example while purchasing a mobile, *ease of use* can be factor of utility however it may not be present in product features. Besides this, in some scenarios calibrating case utility for a given user requirement is difficult. If a user asks for *romantic* movies, the recommender

L. Lamontagne and E. Plaza (Eds.): ICCBR 2014, LNCS 8765, pp. 480–494, 2014.
© Springer International Publishing Switzerland 2014

may not have a precise measure of the *extent of romance* in a movie and hence recommendation becomes difficult.

In this paper we try to address these issues by making use of user trails. In any conversational system as the users interact with the system, trails are left behind. Each trail is a sequence of actions such as feedbacks or preference elicitations which ultimately lead the user to a desired product. These trails are rich knowledge sources of user behavior and interests and therefore can be used to arrive at case representations in such a way that they closely reflect user interests. The essential contribution of our work is to present ways of estimating the relative feature-specific utilities based on preferences mined from the user trails. We construct feature-specific preference graphs based on pairwise preferences obtained from the trails. Based on these graphs, we present a class of PageRank style algorithms and show their effectiveness to induce an ordering compared to majority voting techniques in scenarios where there are a few pairwise preferences available [7]. Also by estimating the feature specific utilities, we essentially enrich case descriptions that can in turn be used to enhance recommendations. To evaluate this idea, we propose a novel methodology centered around the actual user trails.

We focus on critiquing based recommenders which are commonly used to guide the user through the product space by facilitating interaction in the form of critiques. Here the user views a product and may give a critique like *show a similar product but a cheaper one*, to which the system responds by recommending the products that satisfy the critique and are most similar to the current product. The motivation behind critiquing in recommender systems is that it is easier for users to critique a product than construct formal queries. Our experiments are based on *Entree*[5], a restaurant recommender where the features over which users express desired functionality(*i.e, critiques*) are different from the features used to describe the cases. Hence we propose ways of estimating relative utilities for these critiquing features based on preferences mined from user trails.

In Section 2, the required background and related work is discussed. In Section 3.1, we give an overview of *Entree* and in Sections 3.2, 3.3 we discuss our approach to mine the preferences and use them to define various algorithms that can be used to rank the recommendations. In Section 4, the evaluation framework and results are presented.

2   Background and Related Work

Case based recommenders are a category of knowledge based recommenders where the cases are retrieved based on similarity to the query [18]. By modality of interaction, they can be classified into single-shot and conversational recommenders. In single-shot recommenders, a single list of recommendations are given and no further interaction is facilitated. In conversational recommenders on the other hand, an artificial sales agent is simulated which interacts with the user in a variety of ways thereby learning preferences of the user and updating the recommendations. They also help users who have a vague idea about their needs to explore the product space effectively and in turn form the preferences during the process.

Critiquing based recommenders are a particular category of conversational systems where the users request for change in feature values in the form of critiques like *show a similar product but with cheaper price*. FindMe systems developed by Burke et al.[5] are some the earliest critiquing based recommenders. Following this, critiquing based recommenders have been studied from various perspectives like ease of interaction and ways to improve the effectiveness of search thereby reducing the user effort to reach a target[9]. In compound critiquing, change request can be over multiple features. McCarthy et al.[12] and Zhang et al.[21] elaborate techniques to dynamically generate these compound critiques. McSherry et al.[14] discuss the essential problems and factors affecting critiquing and Chen et al.[8] explain evaluation of these systems.

Bergmann et al.[2] discuss the inadequacy of similarity as a surrogate for utility. This motivated research into novel methods to improve the similarity measures based on utilities inferred from user feedback over recommendations. Stahl et al.[19] and Xiong et al.[20] used machine learning techniques to learn similarity measures. Here, the implicit or explicit feedback given by users lead to a partial ordering of cases reflecting their utilities which are inturn used to learn similarities. However, our work contrasts with the above in the way we try to arrive at a robust estimates of relative utilities.

Some of the earlier work in conversational recommenders focussed on analyzing user logs to enhance recommendations. Burke proposed a hybrid recommender adding collaborative flavor to the critiquing[4]. Here user trails in critiquing based recommenders are used to construct the user-item rating matrix which is essential in a collaborative setting to give recommendations. Dong et al. [10], extract sentiment features by mining user opinions from product reviews. Based on the reviews given to a product the sentiment scores for these features are computed. Products represented based on these feature values are used as the basis for a form of recommendation that emphasises similarity and sentiment. Since these feature values are derived from product reviews given by users, it essentially helps in harnessing past experiences into recommender systems. In contrast to extracting feature-values from trails, McCarthy et al.[13] treat each successful trail in the past (i.e, where user reached a target) as a problem-solution pair which forms a case. The problem part of the case is the path user followed to reach the target while the solution is the target product of the session. For a given new session, the similar sessions are identified based on the overlap of critiquing patterns and corresponding solutions are ranked and presented to the user. However besides critique patterns, the types of items that the user has been recommended so far are not considered which may lead to unexpected recommendations. To overcome this, Salem et al. [16] added a new componentwhich takes into account item similarity when selecting relevant sessions. Modeling preferences using graphs and estimating utilities was studied by Gupta et al.[11] in the context of preference based recommenders. In this case, the domain specific dominance knowledge is combined with SimRank based similarity to give recommendations. The utilities are estimated using PageRank on the dominance graphs formed during each cycle of user interaction. Besides recommender systems, trails are extensively studied in the context of web search and retrieval. Cao et al.[6] and Singla[17] discuss mining clickthrough logs of users to understand user behavior, improve search results and also help users in navigating through the web. Chandar et al.[7] compare majority voting and PageRank procedures to rank documents based on the pairwise relevance judgements. In particular here the effectiveness of PageRank in case of sparse preferences is discussed. Our work contrasts with the above in the way we try to enrich case representations based on the relative utilities estimated. Also, we present ranking algorithms which are essentially decoupled from the internals of systems and solely dependant on the trails.

3 Our Approach

In E-commerce, there is an essential gap between the vocabulary of the customer and the salesperson[2]. In conversational recommenders like critiquing, the users give desired changes in the form of critiques which may not be from within the product features. *Entree* is an example where the critique space is different from product space. Besides this, implementing critiquing might require calibrating certain features of the cases. Similar to the movie example given in Section 1, while purchasing a dress the customer may look at a dress and ask for a *more attractive* one. Now the system should recommend a dress that is *more attractive* than the present dress. However, quite often all we have from the case base is whether a dress is *attractive* or not. Moreover even if there exists a precise measure, there can be gap between calibrating ways of the customer and the system. To address these issues, in this section we propose ways of estimating relative critique specific utilities based on user trails and show their effectiveness in re-ranking the recommendations.

In the context of critiquing based recommenders, a trail can be defined as the path the user followed starting from a product, critiquing and choosing the preferred product in each cycle until she reaches the product she is satisfied with, which is generally termed as the target. In a restaurant domain where $r'_i s$ are restaurants, an example trail is:

$$r_1 \xrightarrow{cheaper} r_3 \xrightarrow{nicer} r_6 \xrightarrow{creative} r_2 \xrightarrow{cheaper} r_5$$

When users critique a product and make choices, they implicitly assess utilities and hence by mining these user preferences relative utilities can be estimated. an present

3.1   Entree: A Brief Overview

Entree is a restaurant recommender for the city of Chicago[5,4]. A user starts the interaction by submitting an entry point, either a known restaurant in some other city to which she seeks a local counter-part or a set of criteria expressed in terms of case features to which the system responds with "similar" restaurants. In [5] a sketch of how similarities are estimated is given. In Fig. 1, the user is looking for a restaurant in Chicago that is similar to Los Angeles restaurant "Chinois On Main". *Entree* recommends "Yoshi's Cafe"(Fig. 1) preserving Asian and French influences[1]. Now, say the user is interested in a similar restaurant but with a slightly lesser price and hence gives the critique *"Less $$"*. This results in "Lulu's"(Fig. 1), a creative Asian restaurant as the next recommendation. However, now the French influence is lost - one consequence of moving to a lower price basket([3]).

In this way, by interactively redirecting the search using critiques the user finds an acceptable option. Critiques used in *Entree* are *Cheaper, Creative, Lively, Nicer, Quieter, Traditional.* Our experiments are based on the *Entree* dataset [1].It consists of trails where each trail is a sequence of user interactions while reaching the target. In *Entree* once a critique is given, the system responds with maximally similar restaurants satisfying the critique based on its own sim-

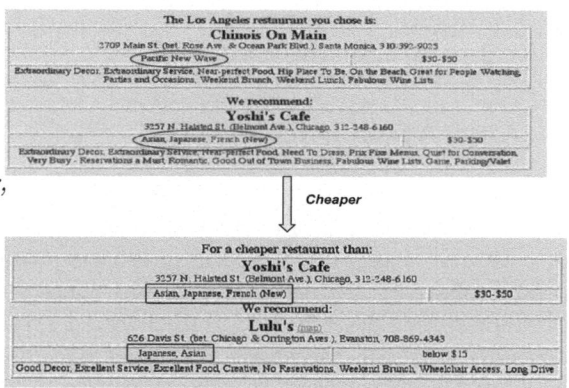

Fig. 1. "Yoshi's Cafe" given as the best possible match to *"Chinois On Main"* and *Lulu's* is the best possible match that is cheaper to *Yoshi's Cafe*

ilarity model. However, this model which is constructed based on the domain knowledge is not available in public. Only the trails of users along with restaurant descriptions are available from the dataset. In each critiquing cycle of these trails, the user can browse through the list of recommendations and select one for the next critiquing cycle or stop the trail indicating if she is satisfied with it. An example of such a trail is shown in Table 1.

Table 1. Structure of a trail in *Entree*

$$r_1 \xrightarrow{cheaper} r_3 \xrightarrow{browse} r_{10} \xrightarrow{browse} r_2 \xrightarrow{cheaper} r_5 \xrightarrow{cheaper} r_9 \xrightarrow{nicer} r_7 \xrightarrow{browse}$$
$$r_8 \xrightarrow{browse} r_{23} \xrightarrow{livelier} r_{12}$$

Initially user critiques r_1 with *cheaper* and prefers r_2 after viewing r_3 and r_{10}. This sub-trail from r_1 through r_2 constitutes one critiquing cycle. The number of sub-trails en route r_1 through r_{12} is five; note that this is the path length discounting the edges labeled *browse*. After a few cycles of critiquing the session comes to an end with the user reaching a target or giving up.

---

[1] Note that the connection between "Pacific New Wave" cuisine and its Asian and French culinary components is part of the system's knowledge base of cuisines([3]).

3.2   Preference Graphs

In the trail shown in Table 1, the user preferred restaurant r_2 as a *cheaper* alternative to restaurant r_1 among the given recommendations, similarly she rated r_{23} *nicer* than r_9 and r_4 *livelier* than r_{23}. We use preference graphs to represent these preferences. A preference graph is constructed for each critique by aggregating all trails as shown in Algorithm 1. Examples of such graphs are shown in Fig 2 and these are formed based on the four trails shown in Table 2. For example, in the graph for *cheaper* the weight of edge (r_1, r_2) is 2 since r_2 is preferred over r_1 in T_1 and T_4, similarly for (r_5, r_9) since r_9 is preferred over r_5 in T_1 and T_2.

Algorithm 1: Critique Specific Preference Graph

Input: Trails from the training data
Output: Preference graphs for each critique

1 **begin** *Procedure*
2   *Graphs* : a container of preference graphs
    /* In this loop each graph is initialized with all the restaurants as vertices */
3   **for** *critique in CritiqueList* **do**
4   *Graphs[critique]* ← WeightedDiGraph()
5   **for** *each restaurant r_i in CaseBase* **do**
6   *Graphs[critique].add_vertex(r_i)*

7   **for** *each trail in the training set* **do**
8   **for** *each critiquing cycle in the trail* **do**
9   *Critique* ← critique given by user in this cycle
10   r_{src} ← critiqued case /* restaurant over which critique is given   */
11   r_{des} ← preferred case
12   **if** *Graphs[Critique].has_edge(r_{src}, r_{des})* **then**
13   w_{new} ← *Graphs[Critique].edge$_{weight}$(r_{src}, r_{des})* + 1
14   *Graphs[Critique].set_edgeWeight($r_{src}, r_{des}, weight = w_{new}$)*
15   **else**
16   *Graphs[Critique].add_edge($r_{src}, r_{des}, weight=1$)*

17   **return** *Graphs*

Table 2. Examples indicating preferences(B-*browse*, C-*cheaper*, L-*livelier*, N-*nicer*)

T_1	$r_1 \xrightarrow{C} r_3 \xrightarrow{B} r_{10} \xrightarrow{B} r_2 \xrightarrow{C} r_5 \xrightarrow{C} r_9 \xrightarrow{N} r_7 \xrightarrow{B} r_8 \xrightarrow{B} r_{23} \xrightarrow{N} r_{14} \xrightarrow{L} r_{12} \xrightarrow{B} r_4$
T_2	$r_5 \xrightarrow{C} r_3 \xrightarrow{B} r_8 \xrightarrow{B} r_6 \xrightarrow{B} r_9 \xrightarrow{N} r_{10} \xrightarrow{B} r_{23} \xrightarrow{N} r_{14} \xrightarrow{L} r_2 \xrightarrow{L} r_{15} \xrightarrow{B} r_4$
T_3	$r_2 \xrightarrow{C} r_3 \xrightarrow{B} r_5 \xrightarrow{C} r_9 \xrightarrow{B} r_6 \xrightarrow{B} r_{14} \xrightarrow{L} r_4 \xrightarrow{L} r_9 \xrightarrow{N} r_5 \xrightarrow{N} r_7 \xrightarrow{B} r_1 \xrightarrow{N} r_{23}$
T_4	$r_9 \xrightarrow{L} r_1 \xrightarrow{L} r_5 \xrightarrow{C} r_6 \xrightarrow{B} r_3 \xrightarrow{B} r_{14} \xrightarrow{C} r_1 \xrightarrow{C} r_2 \xrightarrow{N} r_4 \xrightarrow{N} r_1 \xrightarrow{N} r_5$

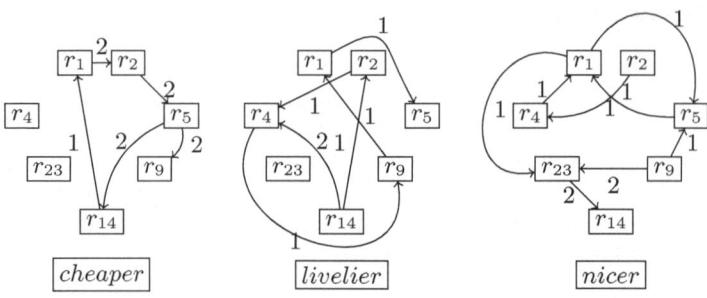

Fig. 2. Preference graphs for *cheaper, livelier* and *nicer*. Here, the weight of an edge is the number of times r_{des} is preferred after critiquing r_{src} (some isolated nodes are not shown here).

3.3 Ranking Algorithms

In this section we propose five methods for ranking *Entree* recommendations based on preference graphs. The trail log is partitioned into training and test data; the goal is to use preference graphs to induce a preference model over training data and use the same to rank recommendations for each test sub-trail (refer Section 3.1), once users provide critiques.

3.3.1 Critique Specific Utilities

Preference graphs constructed in Section 3.2 can be used to define algorithms which essentially induce an ordering among the cases for each critique. In this section we present two algorithms which ranks the nodes in each graph by estimating utilities.

Critique Specific Majority Voted Ranking(CMR): In this scheme, the relative utilities correspond to relative number of times a case is preferred over any other case. The utility of a case r_i for a particular critique based on the preference graph of the corresponding critique can be formulated as:

$$Util(r_i) = \frac{degree_{in}(r_i)}{\sum_{r_j \in R} degree_{in}(r_j)} \tag{1}$$

where R is the set of all restaurants. Intuitively, this is the number of votes to r_i over total number of votes. Given a restaurant and its critique in a test sub-trail, CMR ranks *Entree* recommendations based on majority of votes they get across that critique. Given the training data in Table 2, if $< r_1, r_2, r_4 >$ is the sub trail in test corpus and the critique applied is *cheaper*, the order induced by CMR is $< r_2, r_1, r_4 >$. This comes from the graph corresponding to *cheaper* in Fig. 2 where r_2 has 2 votes, r_1 has 1 and r_4 has none.

Critique Specific PageRank(CPR): Consider the simple preference sub-graph in Fig. 3, the order induced by CMR on these five cases is $< r_4 r_5 r_2 r_9 r_{14} >$.
Despite the fact that r_9 has fewer votes than r_4, it seems intuitive to promote the ranking of r_9, since it is preferred twice over r_4 which in turn has maximum votes. This observation is of central to Critique specific PageRank (CPR), where the hypothesis is: *'The utility of a restaurant is high across a certain critique, if it is preferred over high utility restaurants over the same critique'*

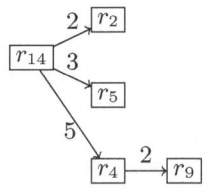

Fig. 3. An example sub-graph

This formulation has an inherent circularity and has its close analogue in that of PageRank[15]. The PageRank algorithm assigns a real number to each node in a graph with an intent that the higher the PageRank score of a node, the more "important" it is. PageRank is generally used in web scenario to measure the importance of web pages based on a similar hypothesis: *'A page is important if it is pointed to by important pages'*. With a small modification of the standard PageRank approach, it can be adapted for weighted graphs as well. The utility score is computed as shown below:

$$U(r_i) = \sum_j U(r_j) \times \frac{w_{ji}}{W_j} \ , where \ W_j = \sum_k w_{jk} \tag{2}$$

where $U(r_i)$ denotes utility of r_i, w_{ji} is the weight of edge from r_j to r_i. It is important here to note that these utilities are local to the critiquing dimension i.e, if we consider the graph for *nicer* and do a PageRank computation on it, the scores we get for each restaurant is the utility of that restaurant in the *nicer* dimension. Given a restaurant and its critique in a test sub-trail, CPR ranks *Entree* recommendations based on their PageRank utility scores across that critique. Note that unlike CMR, CPR takes into account higher order associations between restaurants in the preference graph i.e, in Fig. 3, r_9 is indirectly preferred over r_{14} through r_4.

3.3.2   Item and Critique Specific Utilities
Recommendation rankings generated by the algorithms in previous section are agnostic to the specific restaurant critiqued. They just give a global ranking of cases specific to each critique. In this section, we present similar algorithms but sensitive to restaurant critiqued.

Item and Critique Specific Majority Voted Ranking(ICMR): This is a baseline approach similar to CMR but sensitive to restaurant critiqued. Recommendations are ranked based on the frequency with which they are preferred over the current restaurant when the same critique is applied over the latter in the trail corpus. Given the training data in Table 2, consider a test case where a restaurant r_{14} is shown and the user applies the critique *livelier*. If *Entree* recommends r_2, r_4 and r_9, then a frequency based ranking induces an order $[r_4, r_2, r_9]$.

This is because the graph for *livelier* (Fig. 2) suggests that r_4 is preferred two times over r_{14} and r_2 is preferred once, but r_9 is never preferred over r_{14}.

Item and Critique specific PageRank(ICPR): In this section, we propose a method of making PageRank sensitive to each restaurant. Here preference graphs of critiques are formed specifically for each restaurant. There is an important difference in the way edges are added to the graph. Consider a critiquing cycle of a trail from r_1 through r_6,

$$r_1 \xrightarrow{cheaper} r_2 \xrightarrow{browse} r_4 \xrightarrow{browse} r_3 \xrightarrow{browse} r_6 \xrightarrow{livelier} ....$$

here for critique *cheaper* against r_1, r_6 is preferred over r_2, r_4, r_3. Intuitively, this can be considered as pairwise preference judgements given by users between rejected recommendations(r_2, r_4, r_3) and target(r_6) after a critique is given over a specific restaurant. Hence we add the corresponding edges (r_2, r_6), (r_3, r_6), (r_4, r_6) to the graph of critique *cheaper* corresponding to r_1. In this manner the graphs are formed by aggregating preferences from all the trails. Some examples of restaurant and critique specific preference graphs are shown in Fig. 4 based on the trails in Table 2.

Once these graphs are formed, the PageRank procedure is used to induce an ordering among the restaurants which can in turn be used in ranking the recommendations. For a new sub-trail in the test data, the relative critique specific utilities personalised to each restaurant are used to order the candidates.

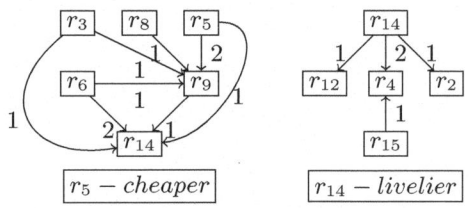

Fig. 4. Preference graphs of r_5 for *cheaper* and r_{14} for *livelier*(some isolated nodes are not shown here)

3.3.3   Fusion

The algorithms proposed in Section 3.3.2 essentially require considerable training data. However, quite often preferences obtained for specific restaurants are sparse and hence ICMR and ICPR are less demanding since they may not induce an ordering. On the other hand, the CPR and CMR approaches do not consider the restaurant over which critique is given and the utilities remain the same irrespective of the context. Hence we propose fusion approach which is a linear combination of CPR and ICPR,

$$Fusion_{score} = \alpha * ICPR_{score} + (1 - \alpha) * CPR_{score} \tag{3}$$

where the value of α is determined using cross-validation. The fusion approach attempts to make the best of both worlds, by using CPR for smoothing the preference models and compensate for ICPR sparsity.

3.4 Enriching Case Descriptions

Each of the algorithms described above implicitly generates a representation of cases that is reflective of relative utilities across the critiquing dimensions and this enriches the existing case descriptions. For example, based on the CPR scores each case can be positioned in the 6-dimensional critique space where the feature values indicate the relative utilities. Now since the user requests come in the form of critiques, the case descriptions derived here can be used to rank the solutions. This essentially bridges the gap between the representation on which users provide feedback and those internally used by the recommender. This also helps in effective filtering of cases since now each critique has a crisp value for every case. For example in movies domain, initially *romantic* might be boolean feature(*yes* or *no*), however now an order is induced for these features amongst the cases based on utilities.

4 Evaluation

Dataset: The dataset consists of user interaction sessions with the *Entree* system.[2] It has 50672 trails collected over a period of 3 years. For all our experiments we split the dataset into train and test partitions. Utilities are estimated over the train data using approaches in Section 3.3 and their effectiveness over the test trails is evaluated. The results are reported after averaging across 10 different train and test partitions. In this section, we propose two evaluation schemes centered around the actual user trails.

4.1 Re-ranking Effectiveness

From the *Entree* dataset used for our experiments we do not have access to the information about the complete list of recommended restaurants in each critiquing cycle. All we know from the trails are the restaurants that users browsed before picking one. One way of empirically testing the effectiveness of our approach would be to use the standard offline approach as in [12] which is a leave one out strategy where each case is removed from the case base and used in two ways. First, to generate a query based on a subset of features and second, to decide a target that is most similar to removed case. Now, an optimal user behavior is simulated and reduction in number of cycles to reach target is reported. However, doing this kind of evaluation would require us to have access to *Entree*'s similarity measures and not evaluate the effectiveness of incorporating the actual user interactions that we have access to. We thus came up with an alternate evaluation scheme tailored to critique trails.

---

[2] https://archive.ics.uci.edu/ml/datasets/Entree+Chicago+
Recommendation+Data

4.1.1   Evaluation Criteria

In each critiquing cycle, we see if the order induced by a ranking algorithm helps in reducing the number of steps needed to reach the preferred restaurant. Supposing that the restaurants are ranked by the algorithm and presented to the user, the rank of the one finally picked by the user is indicative of the effectiveness of the approach. If the user viewed k restaurants before reaching the preferred one, the rank r of the preferred restaurant among these k restaurants as induced by the algorithm is observed. If the rank r is less than k then the number of steps that can be reduced is k - r. For example, in the first cycle from r_1 through r_2 of the trail shown in Table 1, if the rank(r) of the target r_2 is 2 and the total number of cases browsed(k) is 3, then the number of steps reduced is 1.

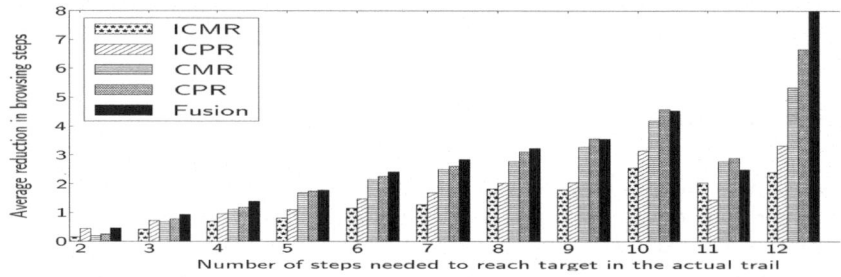

Fig. 5. The results are divided into bins based on the number of recommendations(k) browsed before reaching target

4.1.2   Results and Observations

The performance of different methods are discussed here. In Fig. 5, results are separately shown based on the number of recommendations user browsed. In Fig. 6 the overall reduction in number of steps by various methods is compared. The reduction in percentage of steps is computed by the formula,

$$\% \ reduction = \frac{k-r}{k} \times 100 \qquad (4)$$

where k is size of recommendation list, r is number of steps needed to reach preferred restaurant.

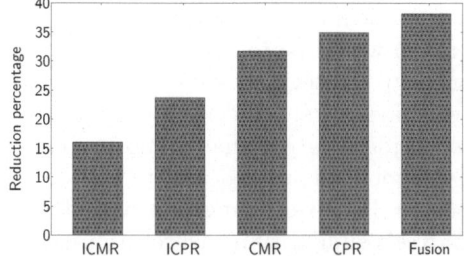

Fig. 6. Overall percentage reduction in number steps needed to reach preferred restaurant

PageRank vs. Majority Voting: From the results it can be observed that the PageRank style algorithms clearly out perform the majority voting algorithms i.e, CPR better than CMR and ICPR better than ICMR. In particular, the number of preferences available for item specific algorithms are low and the effectiveness of PageRank to induce an order can be observed from relative performance gain of

ICPR over ICMR. This can be also be attributed to the higher order associations that are implicitly taken care by PageRank.

Analysis: The results show that ICPR is doing better than ICMR but not as good as CPR. This is due to the sparseness associated with the preferences available to each restaurant i.e, the restaurant and critique specific graphs are sparse. Also, it is important to note that though CPR does not explicitly take into account the restaurant being critiqued, this limitation is, to some extent, being compensated by the fact that in all our experiments ranking is only restricted to products reckoned by Entree as similar to the critiqued restaurant. We note the fusion of CPR and ICPR is performing the best. From our experiments the best value of α in Fusion method is observed to be 0.9 which is in turn determined by cross validation.

4.2 Sanity Check

In the earlier section, effectiveness of PageRank to infer preferences is explained. Now we present a scheme to test the sanity of the utilities estimated using PageRank. Since ICPR is observed to be affected by sparsity, we limit this evaluation to CPR approach. For a given critique, *Entree* recommends cases that satisfy the critique and similar to the current case. By this hypothesis, the utilities estimated based on training data should have higher utility scores for the recommended restaurants than the critiqued ones. Hence, a measure of sanity is fraction of times recommended restaurant has a higher utility score than the critiqued restaurant over all the trails. This is formulated as:

$$Sanity\ success\ rate = \frac{\#Successful\ comparisons}{total\ \#comparisons} \times 100 \qquad (5)$$

a Successful comparison is the one where a recommendation has higher score than critiqued restaurant. Consider an example where critique *cheaper* is applied over r_4 with CPR score 0.25 and the recommendations are $< r_2, r_5, r_6, r_{12}, r_{14} >$ with CPR scores [0.21 0.24 0.3 0.35 0.38] respectively. Here three recommendations have higher scores and the sanity success rate is 60%. The results are shown in Fig. 7a and these are determined by aggregating all the comparisons from all the trails in test data. It can be observed that as the browsing length increases the sanity success rate decreases. This can be attributed to the noise from the user behavior while browsing through the recommendations.

4.2.1 Denoising the Preference Models

As mentioned above, the noise in user trails may affect the preference models inferred. In *Entree*, once a critique is given not all the recommendations presented satisfy the critique. This is due to the fact that sometimes there may not be enough cases satisfying the critique and hence the similar restaurants that may not satisfy the critique are also recommended. In such situations when the user prefers a restaurant that do not satisfy the critique, it results in a noisy preference modeling and this noise is to be removed before looking at the sanity of the scores.

Among the critiques used in *Entree*, the *cheaper* critique has a corresponding *price* attribute used in the case description. For each restaurant the *price* attribute has one of the following four categories : *below $50, $15-$30, $30-$50, over $50*. In order to remove noise while training, an edge is added to the preference graph only if the preferred target in the cycle belongs to either same or lower level price category than the critiqued restaurant. Similarly in the test sub-trail, among the recommended restaurants only the ones which are in lower or same level price category are considered and sanity of the scores are computed. The results after removing the noise are shown in Fig. 7b, however these are the results based on the critiquing cycles for *cheaper* alone. It can be observed that after denoising, the sanity of the scores are much better which indicate the soundness of the algorithm to estimate the relative utilities.

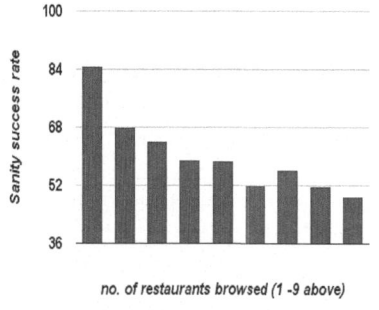
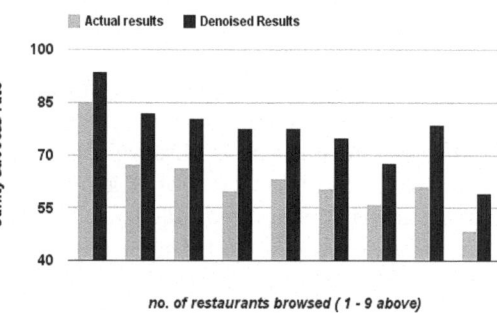

(a)   Sanity results of the utility scores computed.

(b)   Comparison of the Sanity results for the critique *cheaper* before and after denoising

Fig. 7. Sanity results

5   Discussion

5.1   Comparison with Related Work

Our contributions in this paper are threefold. First, we proposed a novel way of enriching case descriptions based on the relative utilities mined from user trails. Second, we demonstrated the effectiveness of PageRank to infer user preferences mined from trails. Third, we proposed a novel way of evaluation centered around the actual user trails there by overcoming the need for similarity measures in evaluation.

Earlier work on utility based recommendations mostly focused on learning more robust similarity measures that approximate the utilities as close as possible based on the user interactions with the system. These are generally inferred on the fly when the user is interacting. Our work contrasts in the way we try to arrive at robust utility estimates and there by use them to rank the recommendations. In [10], a similar attempt has been made by considering the product reviews. These reviews are used to mine the sentiment features, represent products based on these features and there by indirectly estimating some form of utility. However these would require the reviews for the products which are not generally not available for case based recommenders and hence estimating the utilities from implicit user feedback is effective in these situations. Reusing past critiquing experiences in the form of trails to find a desired case is also studied in [13], however this requires a lot of computation online and estimating similarities between trails is a real challenge. Hence our approach of enriching the case descriptions based on utilities can be used in re-ranking the recommendations.

5.2 Outlook

In this paper, we proposed ways of enriching case descriptions based on trails in the context of critiquing based recommenders. However, the same can be adapted to other classes of conversational systems though the kind of explicit preferences available in critiquing systems may no longer be available. The ranking algorithms presented in this paper can be used as a tool or plug-in which is decoupled from similarity measures and other internals and hence can be adapted to a large class of systems that engage with end users to progressively narrow down on target. Also in systems like *Entree*, the utilities estimated can be combined with the feature specific similarities handcrafted by system designers thereby enhancing the recommendations.

6 Conclusion

Trails left behind by the users are a very important knowledge containers for user preferences. In CBR, these preferences can be used to estimate utilities in such a way they complement the similarities. Similarly, in conversational recommenders these preferences can be used to enrich case descriptions so that match the vocabulary of user requirements and essentially this can be useful in bridging the gap between the vocabulary of the user and the system. We have proposed a class of PageRank style algorithms to estimate the relative utilities of the cases and use them in re-ranking the recommendations. Here the relative utilities complement the existing feature specific similarities thereby enhancing the recommendations. We proposed a simple evaluation scheme centered around actual user behavior recorded in logs and empirical results demonstrate the effectiveness of the proposed approaches in improving the rank of the target recommendation. Since we are inferring preference models from user trails, the holy grail of CBR to reuse the past experiences is also added to conversational recommenders.

References

1. Bache, K., Lichman, M.: UCI machine learning repository (2013)
2. Bergmann, R., Richter, M.M., Schmitt, S., Stahl, A., Vollrath, I.: Utility-oriented matching: A new research direction for case-based reasoning. In: Professionelles Wissensmanagement: Erfahrungen und Visionen. Proceedings of the 1st Conference on Professional Knowledge Management. Shaker. Citeseer (2001)

3. Burke, R.: A case-based reasoning approach to collaborative filtering. In: Blanzieri, E., Portinale, L. (eds.) EWCBR 2000. LNCS (LNAI), vol. 1898, pp. 370–379. Springer, Heidelberg (2000)
4. Burke, R.D.: Hybrid recommender systems: Survey and experiments. User Model. User-Adapt. Interact. 12(4), 331–370 (2002)
5. Burke, R.D., Hammond, K.J., Young, B.C.: The findme approach to assisted browsing. IEEE Expert 12(4), 32–40 (1997)
6. Cao, B., Shen, D., Wang, K., Yang, Q.: Clickthrough log analysis by collaborative ranking. In: AAAI (2010)
7. Chandar, P., Carterette, B.: Using pagerank to infer user preferences. In: Proceedings of the 35th International ACM SIGIR Conference on Research and Development in Information Retrieval, SIGIR 2012, pp. 1167–1168. ACM, New York (2012)
8. Chen, L., Pu, P.: Evaluating critiquing-based recommender agents. In: Proceedings of the National Conference on Artificial Intelligence, vol. 21, p. 157. AAAI Press, MIT Press, Menlo Park, Cambridge (2006)
9. Chen, L., Pu, P.: Critiquing-based recommenders: survey and emerging trends. User Model. User-Adapt. Interact. 22(1-2), 125–150 (2012)
10. Dong, R., Schaal, M., O'Mahony, M.P., McCarthy, K., Smyth, B.: Opinionated product recommendation. In: Delany, S.J., Ontañón, S. (eds.) ICCBR 2013. LNCS, vol. 7969, pp. 44–58. Springer, Heidelberg (2013)
11. Gupta, S., Chakraborti, S.: UtilSim: Iteratively helping users discover their preferences. In: Huemer, C., Lops, P. (eds.) EC-Web 2013. LNBIP, vol. 152, pp. 113–124. Springer, Heidelberg (2013)
12. McCarthy, K., Reilly, J., McGinty, L., Smyth, B.: On the dynamic generation of compound critiques in conversational recommender systems. In: De Bra, P.M.E., Nejdl, W. (eds.) AH 2004. LNCS, vol. 3137, pp. 176–184. Springer, Heidelberg (2004)
13. McCarthy, K., Salem, Y., Smyth, B.: Experience-based critiquing: Reusing critiquing experiences to improve conversational recommendation. In: Bichindaritz, I., Montani, S. (eds.) ICCBR 2010. LNCS, vol. 6176, pp. 480–494. Springer, Heidelberg (2010)
14. McSherry, D., Aha, D.W.: The ins and outs of critiquing. In: IJCAI, pp. 962–967 (2007)
15. Page, L., Brin, S., Motwani, R., Winograd, T.: The pagerank citation ranking: Bringing order to the web. Technical report, Stanford University (1998)
16. Salem, Y., Hong, J.: History-aware critiquing-based conversational recommendation. In: Proceedings of the 22nd International Conference on World Wide Web Companion, WWW 2013 Companion, Republic and Canton of Geneva, Switzerland, pp. 63–64. International World Wide Web Conferences Steering Committee (2013)
17. Singla, A., White, R., Huang, J.: Studying trailfinding algorithms for enhanced web search. In: Proceedings of the 33rd International ACM SIGIR Conference on Research and Development in Information Retrieval, pp. 443–450. ACM (2010)
18. Smyth, B.: Case-based recommendation. In: Brusilovsky, P., Kobsa, A., Nejdl, W. (eds.) Adaptive Web 2007. LNCS, vol. 4321, pp. 342–376. Springer, Heidelberg (2007)
19. Stahl, A.: Approximation of utility functions by learning similarity measures. In: Lenski, W. (ed.) Logic versus Approximation. LNCS, vol. 3075, pp. 150–172. Springer, Heidelberg (2004)
20. Xiong, N., Funk, P.: Building similarity metrics reflecting utility in case-based reasoning. Journal of Intelligent and Fuzzy Systems 17(4), 407–416 (2006)
21. Zhang, J., Pu, P.: A comparative study of compound critique generation in conversational recommender systems. In: Wade, V.P., Ashman, H., Smyth, B. (eds.) AH 2006. LNCS, vol. 4018, pp. 234–243. Springer, Heidelberg (2006)

Case-Based Plan Recognition
Using Action Sequence Graphs

Swaroop S. Vattam[1], David W. Aha[2], and Michael Floyd[3]

[1] NRC Postdoctoral Fellow, Naval Research Laboratory (Code 5514), Washington, DC
[2] Navy Center for Applied Research in Artificial Intelligence,
Naval Research Laboratory (Code 5514), Washington, DC, USA
[3] Knexus Research Corporation, Springfield, VA, USA
{swaroop.vattam.ctr.in,david.aha}@nrl.navy.mil,
michael.floyd@knexusresearch.com

Abstract. We present SET-PR, a novel case-based plan recognition algorithm
that is tolerant to missing and misclassified actions in its input action sequences.
SET-PR uses a novel representation called action sequence graphs to represent
stored plans in its plan library and a similarity metric that uses a combination of
graph degree sequences and object similarity to retrieve relevant plans from its
library. We evaluated SET-PR by measuring plan recognition convergence and
precision with increasing levels of missing and misclassified actions in its input.
In our experiments, SET-PR tolerated 20%-30% of input errors without
compromising plan recognition performance.

Keywords: Case-based reasoning, plan recognition, error tolerance, graph
representation of plans, approximate graph matching.

1 Introduction

Plan recognition is considered the inverse problem of plan synthesis. It involves an
observed and an observing agent. Given an input sequence of actions executed by the
observed agent, the observing agent attempts to map this observed sequence to a plan
such that the observed sequence is a subsequence of actions in the recognized plan.

One of the fundamental assumptions of classical plan recognition is that observed
actions are reliable. This assumption is unrealistic for agents acting in the real world
who may frequently fail to notice the actions of others (because they have to attend to
several actors and events in their environment) or misclassify the observed actions
(due to uncertainty in the real world and incomplete or inaccurate agent models). We
relax this assumption, and present a single-agent keyhole plan recognition algorithm
that is tolerant to two kinds of input errors: missing and misclassified actions.

Our plan recognition algorithm, called *Single-agent Error-Tolerant Plan
Recognizer* (SET-PR), assumes the existence of a plan library consisting of a set of
cases. Each case includes a specification of a planning problem and a fully-grounded
plan that is a solution to this problem. Inputs to SET-PR are subsequences of plans,
which are matched to plans in the plan library to retrieve candidate plans. Currently,

L. Lamontagne and E. Plaza (Eds.): ICCBR 2014, LNCS 8765, pp. 495–510, 2014.
© Springer International Publishing Switzerland 2014

the top-ranked plan is selected as the solution (the recognized plan). For online or dynamic plan recognition, SET-PR is executed each time a new set of observations arrive, obtaining an any-time hypothesized plan in each observation cycle.

Although case-based plan recognition (CBPR) is not novel, we explore a new representation for stored plans that impacts the similarity function we propose for case retrieval. More specifically, we use *action sequence graph*s to represent plans in a plan library (case base); they encode a detailed topology of a plan trace (a sequence of action-state pairs). Our similarity function uses a combination of graph degree sequences and object similarity for computing the similarity between input action sequences and stored plans.

Our paper is organized as follows. Section 2 describes related work on plan recognition. Section 3 then introduces essential notation that formalizes the CBPR problem. Section 4 introduces action sequence graphs and their use in SET-PR. Section 5 details our similarity function. Finally, Section 6 presents an initial empirical study that evaluates the robustness of SET-PR's plan recognition algorithm in the presence of input errors. We found that SET-PR is highly tolerant to increasing levels of input error until a yield point is reached, after which its performance degrades sharply.

2   Related Work

The ability to recognize the plans and goals of other agents is a fundamental aspect of intelligence that allows one to reason about what other agents are doing, why they are doing it, and what they will do next. Many AI researchers have focused on plan recognition approaches which can be broadly classified into *keyhole* or *intended* plan recognition. In keyhole recognition, the observing agent monitors the actions of an ambivalent observed agent. In contrast, in intended recognition, the observed and observing agents cooperate to convey the intentions of the observed agent. Another dimension of classification relates to the presence of single or multiple observed agents. We restrict ourselves to the single-agent keyhole plan recognition problem.

Several approaches has been proposed to address the problem of plan recognition (Sukthankar et al., 2014), including *consistency-based* approaches (e.g., Hong, 2001; Kautz & Allen, 1986; Lesh & Etzioni, 1996; Lau et al., 2004; Kumaran, 2007), and *probabilistic* approaches (e.g., Bui, 2003; Charniak & Goldman, 1991, 1993; Geib & Goldman, 2009; Goldman, Geib & Miller, 1999; Pynadath & Wellman, 2000). The former include hypothesize and revise algorithms, version space techniques, and other closed-world reasoning algorithms, while probabilistic algorithms include those that use stochastic grammars and probabilistic relational models. Both these approaches are sensitive to (1) an incomplete plan library and (2) missing or misclassified actions in the input observations. There have been few attempts to address this issues within these frameworks (e.g., using background goals (Lesh, 1996) or focus stacks (Rich et al., 2001)), but current solutions are usually problem-specific and lack generalizable qualities.

A lesser known approach to the single-agent keyhole plan recognition problem is the case-based approach as exemplified by Cox & Kerkez (2006) and Tecuci & Porter (2009). These investigations focus on the issues of incomplete plan library, incrementally learning the plan library, and responding to novel inputs. But they do not address the issue of error-prone input action sequences. Cox & Kerkez (2006) proposed a novel representation for storing and organizing plans in the plan library based on action-state pairs and abstract states, which counts the number of instances of each type of generalized state predicate. In our work, we start with a similar action-state representation, but process it using a graph representation and store our cases as graphs. As a result, our similarity metrics also operate on graphs.

Most other research on single-agent keyhole CBPR has an application focus. For instance Fagan and Cunningham (2003) acquire cases (state-action sequences) to predict a human's next action while playing SPACE INVADERS. Cheng and Thawonmas (2004) propose a CBPR system for assisting players with low-level management tasks in WARGUS. Lee et al. (2008) integrate Kerkez and Cox's technique with a reinforcement learner to predict opponent actions on a simplified WARGUS task. Similarly, Molineaux et al. (2009) integrate a plan recognition system with a case-based reinforcement learner for an adversarial action selection task involving an American football simulator. In contrast to an application thrust, the focus of our work is to produce a more general single-agent keyhole CBPR approach that is tolerant to uncertainty in observed actions.

In other related work, user traces have been used in a variety of case-based reasoning systems. In CBR systems that learn by observation, user traces are used to automatically extract knowledge and cases so that the system can learn the expert's behavior. These cases typically store state-action pairs (Rubin & Watson, 2010) or state-plan pairs (Ontañón et al., 2007), and retrieval is based on a single state rather than, like our approach, an entire trace. Temporal Backtracking (Floyd & Esfandiari, 2011) has been used in learning by observation to include additional states and actions from a trace during retrieval such that traces are dynamically resized as necessary. Similarly, trace-based reasoning (Zarka et al., 2013) and episode-based reasoning (Sánchez-Marré, 2005) store fixed-length traces in cases and compare the entire trace during retrieval. The primary difference between these approaches and our own is that they represent traces as linear sequences whereas we represent them as graphs.

3   Case-Based Plan Recognition

We now introduce notation that formalizes the general problem of CBPR. A *planning problem* is a 3-tuple $\Pi = (O, s_0, g)$, where O is a set of planning operators, s_0 is the initial state, and g is the goal specification (Ghallab, Nau & Traverso, 2004).

An *action-state sequence* is a sequence $\mathbb{s} = \langle (a_0, s_0), (a_1, s_1), ..., (a_n, s_n) \rangle$ where an action $a_i \in A = \{$all ground instances of operators in $O\}$, and s_i is the state resulting from executing action a_i in state s_{i-1}. A *plan* is a special action-state sequence $\pi = \langle (null, s_0), (a_1, s_1), ..., (a_g, s_g) \rangle$ where s_0 is the initial state of Π and $s_g \in \{s | s$ satisfies $g\}$ is a goal state of Π, where *satisfies* has the same semantics as originally laid out in Ghallab, Nau & Traverso (2004). Action a_0 is *null* because the action preceding the initial state does not need to be specified. While most plan recognition systems represent plans as a sequence of actions, this kind of representation (augmenting action information with the following state information) was proposed originally by Cox and Kerkez (2006). Our rationale for choosing this representation is to offset the inaccuracies of missing or misclassified actions by including information about states.

A *case* is a tuple $c = (\Pi_0, \pi_0)$, where Π_0 is a planning problem and π_0 is a corresponding plan for solving it. A *plan library* (or *case base*) is a set of cases $C = \{(\Pi_i, \pi_i) | 1 \leq i \leq m\}$.

Definition: A *CBPR problem* is represented by a tuple (C, s^{target}), where C is a plan library and $s^{target} = \langle (null, s_0), ..., (a_k, s_k) | k \geq 1 \rangle$ is a target action-state sequence, a subsequence of a plan, that has to be recognized. A solution to this problem is a plan π^{sol} that is predicted using the following CBR process:

- *Plan retrieval*: Retrieve cases $\{(\Pi_i, \pi_i)\}$ from the plan library such that π_i is similar to s^{target} according to some similarity metric
- *Plan evaluation*: Evaluate the retrieved cases according to some evaluation criteria to select a source case $(\Pi^{source}, \pi^{source})$
- *Plan adaptation*: Adapt π^{source} to increase its similarity with s^{target} by resolving the differences between the two, resulting in π^{sol} (and corresponding Π^{sol})
- *Plan repair*: Test and revise π^{sol} to remove inconsistencies

For the sake of simplicity, we assume that the steps of plan adaptation and repair are not required, i.e., $(\Pi^{source}, \pi^{source})$ is equal to (Π^{sol}, π^{sol}). Table 1 captures the differences between the case-based plan synthesis and recognition problems expressed in this notation.

Table 1. Contrasting the general case-based plan synthesis and case-based plan recognition problems

	Case-Based Plan Synthesis	**Case-Based Plan Recognition**
Case Base	$C = \{(\Pi_i, \pi_i)\}$	$C = \{(\Pi_i, \pi_i)\}$
Input	$\langle C, \Pi^{target} \rangle$	$\langle C, s^{target} \rangle$
Output	π^{sol}	π^{sol}
Retrieval	Match Π^{target} with Π_is from cases in C to get Π^{source} (and its π^{source})	Match s^{target} with π_is from cases in C to get π^{source}
Adaptation	Adapt π^{source} to resolve differences between Π^{target} and Π^{source} to get π^{sol}	Adapt π^{source} to resolve differences between s^{target} and π^{source} to get π^{sol}

In online or dynamic CBPR, the above CBR process is employed in recognizing an ongoing plan, before it has ended, by executing the process each time a new set of observations arrive, obtaining an any-time hypothesized plan in each observation cycle.

4 Case Representation

We define a graph data structure called *action sequence graphs* to represent action-state sequences in our cases. Graphs provide a rich means for modelling structured objects and they are widely used in real-life applications to model many objects and processes, from molecules and images to buildings and entire ecosystems. Graphs have also been widely used in planning to represent task networks, goal networks, and plan graphs. The *planning encoding graph* (Serina, 2010) representation, used to encode planning problems, has been particularly influential in the design of our representation. Although there are syntactic similarities between planning encoding graphs and action sequence graphs, important semantic differences exist because the former encodes a problem while the latter encodes a solution (plan).

There are at least two advantages of using a graph representation for cases. First, graphs are well understood. We can rely on a solid theoretical foundation of graph theory to analyze and manipulate our representation. We can also leverage a huge body of work to identify reliable metrics and efficient algorithms. Second, our representation is domain-general; it can be applied to plan recognition problems for a wide range of domains.

4.1 Preliminaries

A *labeled directed graph* G is a 3-tuple $G = (V, E, \lambda)$ where V is the set of vertices, $E \subseteq V \times V$ is the set of directed edges or arcs, and $\lambda : V \cup E \rightarrow 2^L$ assigns labels to vertices and edges. Here, L is a finite set of symbolic labels and 2^L is a set of all the *multisets* on L; this labeling scheme permits multiple non-unique labels for a node or an arc.

An arc $e = [v, u] \in E$ is considered to be directed from v to u, where v is called the *source* node and u is called the *target* node of the arc. Also, u is a *direct successor* of v, v is a *direct predecessor* of u, and v is *adjacent* to vertex u and vice versa.

The *union* of two graphs $G_1 = (V_1, E_1, \lambda_1)$ and $G_2 = (V_2, E_2, \lambda_2)$, denoted by $G_1 \cup G_2$, is the graph $G = (V, E, \lambda)$ where $V = V_1 \cup V_2$, $E = E_1 \cup E_2$, and

$$\lambda(x) = \begin{cases} \lambda_1(x), if \ x \in (V_1 \setminus V_2) \vee x \in (E_1 \setminus E_2) \\ \lambda_2(x), if \ x \in (V_2 \setminus V_1) \vee x \in (E_2 \setminus E_1) \\ \lambda_1(x) \cup \lambda_2(x), otherwise \end{cases}$$

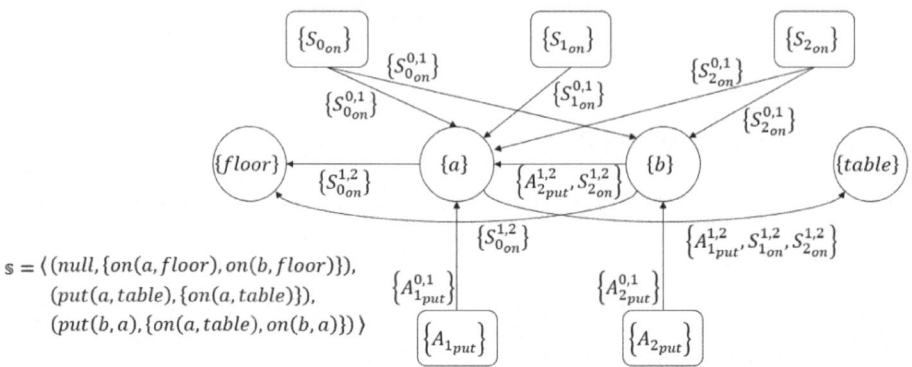

Fig. 1. An example action-state sequence $\mathbb{s}$ and its corresponding action sequence graph $\mathcal{E}^{\mathbb{s}}$

4.2 Action Sequence Graphs

An *action sequence graph* is an order-preserving encoding of an action-state sequence $\mathbb{s} = \langle (null, s_0), (a_1, s_1), \dots \rangle$. Two action-state sequences can be matched by matching their corresponding action sequence graphs. Figure 1 shows an example of $\mathbb{s}$ and its corresponding action sequence graph.

An action-state sequence $\mathbb{s}$ is defined over a planning language $\mathcal{L}$ consisting of P, a set of finitely many *predicate* symbols, and O, a finite set of typed constants representing distinct *objects* in the planning domain. An action a in $(a, s) \in \mathbb{s}$ is represented as a ground atom $p(c_1: t_1, \dots, c_n: t_n)$, where $p \in P$ and represents an action, $c_i \in O$, and t_i is an instance of c_i. Similarly a state s in $(a, s) \in \mathbb{s}$ is represented as a set of ground atoms $\{p(c_1: t_1, \dots, c_n: t_n) | p \in P, c_i \in O\}$.

Definition: Given a ground atom $p = p(c_1: t_1, \dots, c_n: t_n)$ representing either an action a or a single fact of state s in the kth action-state pair $(a, s)_k \in \mathbb{s}$, a *predicate encoding graph* is a labeled directed graph $\mathcal{E}^p(p) = (V_p, E_p, \lambda_p)$ such that:

- $V_p = \begin{cases} \{A_{k_p}, c_1, \dots, c_n\}, \text{if } p \text{ is an action} \\ \{S_{k_p}, c_1, \dots, c_n\}, \text{if } p \text{ is a state fact} \end{cases}$

- $E_p = \begin{cases} [A_{k_p}, c_1] \cup \bigcup_{i=1,n-1; j=i+1,n} [c_i, c_j], \text{if } p \text{ is an action} \\ [S_{k_p}, c_1] \cup \bigcup_{i=1,n-1; j=i+1,n} [c_i, c_j], \text{if } p \text{ is a state fact} \end{cases}$

- $\lambda_p(A_{k_p}) = \{A_{k_p}\}; \lambda_p(S_{k_p}) = \{S_{k_p}\}; \lambda_p(c_i) = \{t_i\}$ for $i = 1, \dots, n$

- $\lambda_p([A_{k_p}, c_1]) = \{A_{k_p}^{0,1}\}; \lambda_p([S_{k_p}, c_1]) = \{S_{k_p}^{0,1}\};$

  $\forall [c_i, c_j] \in E_p, \lambda_p([c_i, c_j]) = \begin{cases} \{A_{k_p}^{i,j}\}, \text{if } p \text{ is an action} \\ \{S_{k_p}^{i,j}\}, \text{if } p \text{ is a state fact} \end{cases}$

Here is an interpretation of this definition. Suppose we have a predicate $p = p(c_1:t_1, \ldots, c_n:t_n)$. Depending on whether p represents an action or a state fact, the first node of the predicate encoding graph $\mathcal{E}^p(p)$ is either A_{k_p} or S_{k_p} (labeled $\{A_{k_p}\}$ or $\{S_{k_p}\}$). Let us assume that it is an action predicate. A_{k_p} is then connected to the second node of this graph, the object node c_1 (labeled $\{t_1\}$), through the edge $[A_{k_p}, c_1]$ (labeled $\{A_{k_p}^{0,1}\}$). Next, c_1 is connected to the third node c_2 (labeled $\{t_2\}$) through the edge $[c_1, c_2]$ (labeled $\{A_{k_p}^{1,2}\}$), then to the fourth node c_3 (labeled $\{t_3\}$) through the edge $[c_1, c_3]$ (labeled $\{A_{k_p}^{1,3}\}$), and so on. Next, the third node c_2 is connected to c_3 through $A_{k_p}^{2,3}$, to c_4 through $A_{k_p}^{2,4}$, with appropriate labels, and so on.

Example: Suppose predicate $p = put(block:a, block:b, table:t)$ appears in the fifth $(k = 5)$ action-state pair of an observed sequence of actions. The nodes of this predicate are $\{A_{5_{put}}\}$, $\{a\}$, $\{b\}$, and $\{t\}$. The edges are $[A_{5_{put}}, a]$, $[a, b]$, $[a, t]$, and $[b, t]$, with respective labels $\{A_{5_{put}}^{0,1}\}$, $\{A_{5_{put}}^{1,2}\}$, $\{A_{5_{put}}^{1,3}\}$, and $\{A_{5_{put}}^{2,3}\}$. The predicate encoding graph for p is shown in Figure 2.

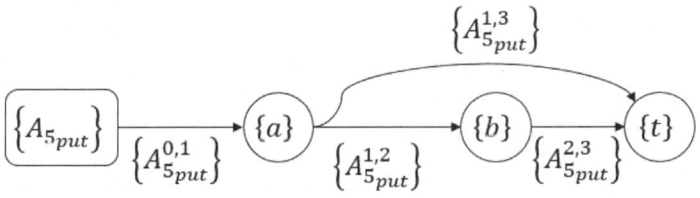

Fig. 2. An example predicate encoding graph $\mathcal{E}^p(p)$ corresponding to $p = put(block: (a, block: b, table: t)$

Definition: An *action sequence graph* of an action-state sequence s is a labeled directed graph $\mathcal{E}^s = \bigcup_{(a,s) \in s}(\mathcal{E}(a) \cup_{p \in s} \mathcal{E}(p))$, a union of the predicate encoding graphs of all the action and state predicates in s. (See Figure 1 for an example of a complete action sequence graph.)

Recall that a case is a tuple $c = (\Pi_0, \pi_0)$, where π_0 is an instance of the action-state sequence where s_0 is the initial state and s_n is the goal state of Π_0. Therefore, π_0 can be represented by an action sequence graph $\mathcal{E}^{\pi_0}$, Π_0 is implicit in π_0 (i.e., π_0 already contains the initial and goal states of Π_0), and it need not be captured explicitly in c. Given this, $\mathcal{E}^{\pi_0}$ can serve as a representation of case c.

5   Similarity and Retrieval

The first step in the CBR process is case retrieval. Given $\langle C, s^{target} \rangle$, where C is a plan library and s^{target} is a target action-state sequence (also a subsequence of a plan), retrieval involves identifying those cases from the plan library that are similar to s^{target}. This requires a similarity metric.

Cases in the plan library are represented as action sequence graphs, as is $\mathfrak{s}^{target}$. Thus, we compute their *structural* similarity using graph matching. One reliable way to match two graphs is to find their *maximum common subgraph* (MCS) (Raymond & Willett, 2002), where similarity is an increasing function of the size of the MCS between G_1 and G_2. Given an action sequence graph $\mathcal{E}^{\mathfrak{s}}$ for $\mathfrak{s}^{target}$, the retrieval step would then compute the MCS between $\mathcal{E}^{\mathfrak{s}}$ and each $\mathcal{E}^{\pi}$ in the case base, selecting the top k $\mathcal{E}^{\pi}$s based on the size of their MCS.

Unfortunately, computing the MCS between two or more graphs is an NP-Complete problem (Raymond & Willett, 2002). The worst-case time requirement of this retrieval step increases exponentially with the size of $\mathcal{E}^{\mathfrak{s}}$, restricting its applicability to small plan recognition problems only. Thus, we seek an efficient approximation of the MCS similarity metric.

5.1   Degree Sequences Similarity Metric

We hypothesize that Johnson's (1985) similarity coefficient, based on graph *degree sequences*, can be used to derive an efficient approximation of the MCS similarity metric. The degree sequence of a graph is the non-increasing sequence of its vertex degrees. The degree sequence is a graph invariant, so isomorphic graphs have the same degree sequences. However, the degree sequence does not uniquely identify a graph. This similarity coefficient has been used for rapid graph matching in several applications, including in case-based planning with planning encoding graphs as a quick filter to *reject* unpromising cases during retrieval (Serina, 2010). In our approach, we will also use this similarity coefficient, but for the opposite purpose - we propose to use it to *select* promising cases.

Johnson's similarity coefficient for two graphs is calculated as follows. First, the set of vertices in each graph is divided into l partitions by label type, and then sorted in a non-increasing total order by degree[1]. Let L_1^i and L_2^i denote the sorted degree sequences of a partition i in the action sequence graphs G_1 and G_2, respectively. An upper bound on the number of vertices $V(G_1, G_2)$ and edges $E(G_1, G_2)$ of the MCS of these two graphs, $G_{1,2}$, can then be computed as:

$$V(G_1, G_2) = \sum_{i=1}^{l} min(|L_1^i|, |L_2^i|)$$

$$E(G_1, G_2) = \left| \sum_{i=1}^{l} \sum_{j=1}^{min(|L_1^i|, |L_2^i|)} \frac{min(|E(v_1^{i,j})|, |E(v_2^{i,j})|)}{2} \right|$$

where $v_1^{i,j}$ denotes the jth vertex of the L_1^i sorted degree sequence, and $E(v_1^{i,j})$ denotes the set of edges connected to vertex $v_1^{i,j}$. Then the structural similarity coefficient can be computed as follows:

$$sim_{str}(G_1, G_2) = \frac{(V(G_1, G_2) + E(G_1, G_2))^2}{(|V(G_1)| + |E(G_1)|) \cdot (|V(G_2)| + |E(G_2)|)}$$

---

[1] The degree (or *valence*) of a vertex v of a graph G is the number of edges that touch v.

This computation can be performed in $O(n \cdot \log n)$ time, where $n = max_i(|L_1^i|, |L_2^i|)$ (Raymond & Willett, 2002).

5.2   Example

Consider the two graphs G_1 and G_2 in Figure 3. The degree sequences of the partitions of G_1 and G_2 (partitioned by label type) are also tabulated in Figure 3. We can compute their (graph) similarity as follows:

$$V(G_1, G_2) = 3 + 1 + 1 + 1 + 1 + 1 = 8$$

$$E(G_1, G_2) = \left| \frac{(7 + 7 + 6)}{2} + \frac{2}{2} + \frac{1}{2} + \frac{2}{2} + \frac{2}{2} + \frac{1}{2} \right| = 14$$

$$sim_{str}(G_1, G_2) = \frac{(8 + 14)^2}{(8 + 12) \cdot (9 + 19)} = 0.8643$$

5.3   Combined Similarity Metric

We use a combination of structural and object similarity to compute the overall similarity of an input subsequence to a plan in the plan library:

$$sim(s^{target}, \pi_i) = sim_{str}(G_1, G_2) + sim_{obj}(s^{target}, \pi_i)/2,$$

where G_1 is the action sequence graph of s^{target}, G_2 is the action sequence graph of π_i, $sim_{str}(G_1, G_2)$ is the degree sequences similarity function described above, and

$$sim_{obj}(s^{target}, \pi_i) = \frac{O_s \cap O_{\pi_i}}{O_s \cup O_{\pi_i}}$$

where O_s is the set of (grounded) objects in s^{target} and O_{π_i} are objects in π_i.

6   Evaluation

In this section we evaluate our claim that plan recognition in SET-PR is robust to two kinds of input errors: misclassified actions and missing actions.

6.1   Empirical Method

We conducted our experiments in the blocks world domain, which we chose because of its simplicity and affordance to quick automatic generation of the plan library with desired characteristics. We used a hierarchal task network (HTN) planner to generate plans for our plan library. Planning problems were created by randomly selecting an initial state and goal state (under the constraint that it is possible to reach the goal state from the initial state), and given as input to the HTN planner. The generated plan (actions and corresponding states) was converted into an action sequence graph and stored, along with the planning problem, as a case. In total, we used this method to generate 100 (error-free) cases for our plan library. We created 4 separate plan libraries that vary in their average plan length (8.94, 12.48, 16.36, and 19.46 actions respectively).

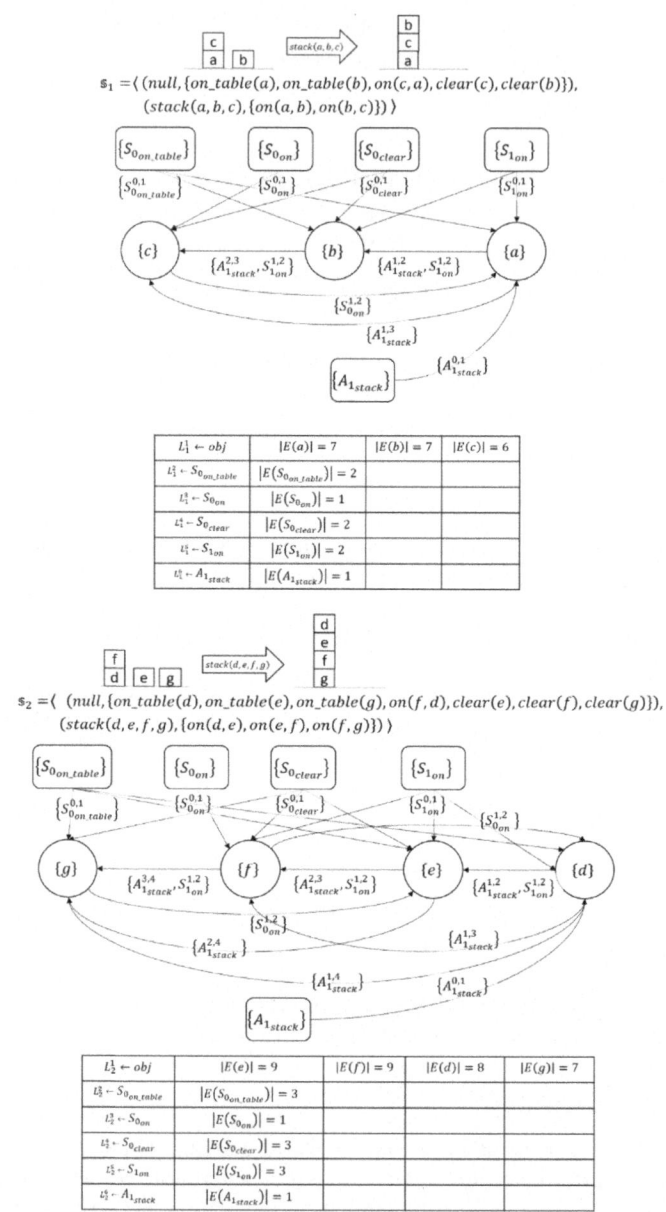

Fig. 3. Two graphs G_1 and G_2 and the degree sequences of their partitions

We varied the following variables in our evaluation of SET-PR: (1) length of the plans in the plan library; (2) percentage of absent ⟨action, state⟩ pairs; and (3) percentage of misclassified ⟨action, state⟩ pairs. Our metrics were convergence, convergence point, and precision. *Convergence* is defined as a boolean value that, for a given query q with n ⟨action, state⟩ pairs, is true if the plan retrieved on the last ⟨action, state⟩ pair matches q. *Convergence point* is defined only for those queries that converge; if they converge after action k, it is defined as k/n. *Precision* is the total number of plans associated with the query's ⟨action, state⟩ pairs that are correct, divided by n.

We evaluated SET-PR's performance for each plan library L using the following method. For each randomly selected case $c = (\Pi, \pi)$ in L, we copied plan π, randomly distorted its action sequence ⟨$(null, s_0), (a_1, s_1), ..., (a_g, s_g)$⟩ (to introduce a fixed amount of error), and used it as an incremental query to SET-PR (i.e., initially with only its first ⟨action, state⟩ pair, and then repeatedly adding the next such pair in its sequence). The error levels tested, separately for both error types (missing and misclassified actions), were {10%, 20%, 30%, 40%, 50%}. For each case c, we also recorded the convergence, convergence point, and precision. Finally, we averaged these metrics across L.

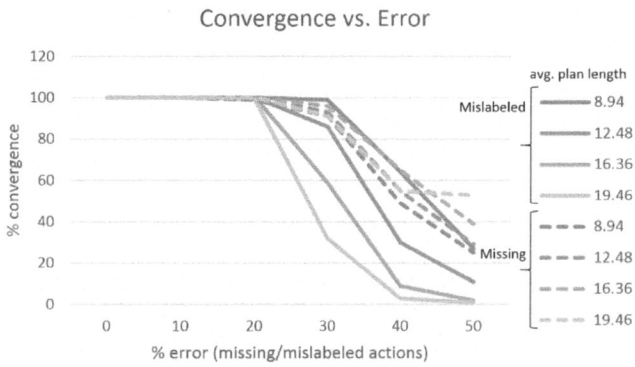

Fig. 4. Convergence rate vs. % error results

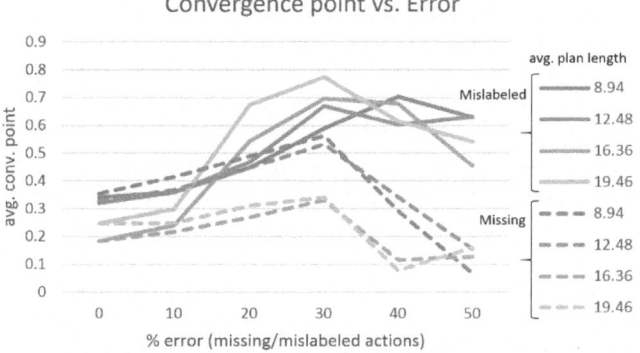

Fig. 5. Average convergence point vs. % error results

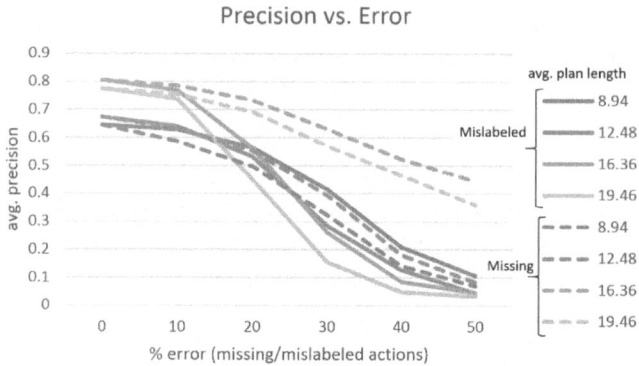

Fig. 6. Average precision vs. % error results

6.2   Trends and Discussion

Figures 4-6 summarize our results. The primary trends, for a given plan length, are as follows. As % error increases (for both error types):

- Convergence rate remains high and near constant until some (yield) point is reached and then sharply decreases
- Average convergence point increases until the yield point is reached, beyond which its value becomes unpredictable
- Average precision decreases steadily and approaches zero

The existence of yield point was surprising; we expected a gradual decrease in convergence rate with a gradual increase in % error. For the blocks world domain, the yield point was around 20% for misclassified and 30% for missing actions. The convergence rate was more than 90% until this yield point.

No clear trends emerged that distinguished the results by average plan length (across the four libraries), although for longer plan lengths, misclassified actions degraded performance of the algorithm more severely than did missing actions. More fine grained analysis is required to explain this.

After the yield point is reached, average convergence point is unreliable; the number of data points in the converged set were too few to indicate any clear trend.

Based on this analysis we conclude that SET-PR is robust to the presence input error until the yield point is reached, which in our experiments was 20%-30% of the input error (depending on the error type). To what extent the yield point is domain dependent and how we can push the yield point further to the right remain open research questions.

We used a variant of SET-PR for our baseline comparison. SET-PR$_{baseline}$ differs from SET-PR only with respect to the representation of action sequences in plan library and queries. While SET-PR uses ⟨action, state⟩ sequences, SET-PR$_{baseline}$ uses ⟨action⟩ sequences. SET-PR$_{baseline}$ better models the representations of classical plan recognition techniques which do not use explicit information about states to infer plans. Recall that the rationale for choosing to include state information in SET-PR was to offset the inaccuracies introduced by missing or misclassified actions. Therefore, we expected the performance of SET-PR$_{baseline}$ to deteriorate more than SET-PR in the presence of input errors.

Table 2. Sample results for SET-PR$_{baseline}$ for one plan library for one type of error (misclassified)

	0%	10%	20%	30%	40%	50%
Conv. Rate (%)	76	14	9	7	1	1
Avg. conv. point	0.21	0.24	0.23	0.43	0.37	0.16
Avg. precision	0.72	0.13	0.11	0.08	0.04	0.02

Table 2 captures the convergence rate, average convergence point, and average precision for different % error levels for one of the plan libraries (average plan length = 12.4) and one type of error (misclassified actions). The results for other plan libraries were similar and we see the following trends in the case of SET-PR$_{baseline}$.

First, when the input error was 0%, we noted that the rate of convergence, average convergence point, and average precision were comparable to SET-PR (rate of convergence and average precision were better in SET-PR, but the convergence point was better in SET-PR$_{baseline}$).

Second, we noted the existence of yield point in SET-PR$_{baseline}$ as well, but it was reached even before the 10% input error level. This suggests that error increments smaller than 10% are required to find out exactly how much error SET-PR$_{baseline}$ can tolerate.

Third, after the yield point is reached, the rate of convergence and average precision in SET-PR$_{baseline}$ dropped sharply to levels seen for 50% error in SET-PR. This suggests that SET-PR$_{baseline}$ is extremely sensitive to input errors.

Fourth, the average convergence point in SET-PR$_{baseline}$ was higher compared to SET-PR, but this is not a meaningful statistic because the data points in the converged set were too few beyond the 0% error level.

Finally, we conclude that SET-PR compares favorably to SET-PR$_{baseline}$ for error tolerance.

7   Summary

We developed a case-based approach to the problem of plan recognition that can tolerate missing and misclassified actions in the input action sequences. We introduced a novel graph representation, called *action sequence graphs*, to represent plans in the plan library. We described a similarity function that uses a combination of graph degree sequences and object similarity for matching action sequence graphs, and used it to retrieve relevant stored plans from the plan library. We also presented an initial empirical investigation of our approach using the blocks world domain. We measured the extent to which our approach is tolerant to missing and misclassified actions, and found that its performance was robust to the presence of up to 20%-30% of error in the input actions (depending on error type). We also found that, in the presence of input errors, our approach compared favorably to the baseline which used more traditional plan recognition representations.

Our study had several limitations. For example, we used a simple paradigmatic domain, the plan library was relatively small, errors were introduced in a systematic manner rather than reflecting a naturally occurring distribution, and while we introduced errors in the query we did not also introduce them in the plan library. As part of our future work, we will address these issues and test SET-PR in different domains, some of which will be real world application domains. One such domain we plan to target is human-robot teams. This will require extending SET-PR to work in domains involving multiple agents and multiple types of plan recognition tasks.

We also plan to compare and test SET-PR's performance using different degree sequence similarity metrics previously reported in literature (e.g., Wallis et al., 2001; Bunke and Shearer, 1998).

Acknowledgements. Thanks to OSD ASD (R&E) for sponsoring this research. Swaroop Vattam performed this work while an NRC post-doctoral research associate located at the Naval Research Laboratory. The views and opinions contained in this paper are those of the authors and should not be interpreted as representing the official views or policies, either expressed or implied, of NRL or OSD.

References

Bui, H.: A general model for online probabilistic plan recognition. In: Proceedings of the Eighteenth International Joint Conference on Artificial Intelligence, pp. 1309–1315. Morgan Kaufmann, Acapulco (2003)

Bunke, H., Shearer, K.: A graph distance metric based on the maximum common subgraph. Pattern Recognition 19(3), 255–259 (1998)

Charniak, E., Goldman, R.: A probabilistic model of plan recognition. In: Proceedings of the Ninth National Conference on Artificial Intelligence, pp. 160–165. AAAI Press, Anaheim (1991)

Charniak, E., Goldman, R.: A Bayesian model of plan recognition. Artificial Intelligence 64, 53–79 (1993)

Cheng, D.C., Thawonmas, R.: Case-based plan recognition for real-time strategy games. In: Proceedings of the Fifth Game-On International Conference, pp. 36–40. University of Wolverhampton Press, Reading (2004)

Cox, M.T., Kerkez, B.: Case-based plan recognition with novel input. Control and Intelligent Systems 34(2), 96–104 (2006)

Fagan, M., Cunningham, P.: Case-based plan recognition in computer games. In: Ashley, K.D., Bridge, D.G. (eds.) ICCBR 2003. LNCS, vol. 2689, pp. 161–170. Springer, Heidelberg (2003)

Floyd, M.W., Esfandiari, B.: Learning state-based behaviour using temporally related cases. In: Petridis, M. (ed.) Proceedings of the Sixteenth UK Workshop on Case-Based Reasoning. Springer, Cambridge (2011)

Geib, C.W., Goldman, R.P.: A probabilistic plan recognition algorithm based on plan tree grammars. Artificial Intelligence 173(11), 1101–1132 (2009)

Ghallab, M., Nau, D., Traverso, P.: Automated planning: Theory and practice. Morgan Kaufmann, San Mateo (2004)

Goldman, R.P., Geib, C.W., Miller, C.A.: A new model of plan recognition. In: Proceedings of the Fifteenth Conference on Uncertainty in Artificial Intelligence, pp. 245–254. Morgan Kaufmann, Bled (1999)

Hong, J.: Goal recognition through goal graph analysis. Journal of Artificial Intelligence Research 15, 1–30 (2001)

Johnson, M.: Relating metrics, lines and variables defined on graphs to problems in medicinal chemistry. John Wiley & Sons, New York (1985)

Kautz, H., Allen, J.: Generalized plan recognition. In: Proceedings of the Fifth National Conference on Artificial Intelligence, pp. 32–37. Morgan Kaufmann, Philadelphia (1986)

Kumaran, V.: Plan recognition as candidate space search (Master's Thesis). North Carolina State University, Department of Computer Science, Raleigh, NC (2007)

Lau, T., Wolfman, S.A., Domingos, P., Weld, D.S.: Programming by demonstration using version space algebra. Machine Learning 53(1-2), 111–156 (2003)

Lee, J., Koo, B., Oh, K.: State space optimization using plan recognition and reinforcement learning on RTS game. In: Proceedings of the International Conference on Artificial Intelligence, Knowledge Engineering, and Data Bases. WSEAS Press, Cambridge (2008)

Lesh, N.: Fast, adaptive, and empirically-tested goal recognition. In: Proceedings of the Fifth International Conference on User Modeling, pp. 231–233 (1996)

Lesh, N., Etzioni, O.: Scaling up goal recognition. In: Proceedings of the Fifth International Conference on Knowledge Representation and Reasoning, pp. 178–189 (1996)

Molineaux, M., Aha, D.W., Sukthankar, G.: Beating the defense: Using plan recognition to inform learning agents. In: Proceedings of the Twenty-Second International FLAIRS Conference, pp. 337–343. AAAI Press, Sanibel Island (2009)

Ontañón, S., Mishra, K., Sugandh, N., Ram, A.: Case-based planning and execution for real-time strategy games. In: Weber, R.O., Richter, M.M. (eds.) ICCBR 2007. LNCS (LNAI), vol. 4626, pp. 164–178. Springer, Heidelberg (2007)

Pynadath, D.V., Wellman, M.P.: Probabilistic state-dependent grammars for plan recognition. In: Proceedings of the Conference on Uncertainty in Artificial Intelligence, pp. 507–514. Morgan Kaufmann, San Francisco (2000)

Raymond, J.W., Willett, P.: Maximum common subgraph isomorphism algorithms for the matching of chemical structures. Journal of Computer-Aided Molecular Design 16, 521–533 (2002)

Rich, C., Sidner, C.L., Lesh, N.: Collagen: Applying collaborative discourse theory to human-computer interaction. AI Magazine 22(4), 15–26 (2001)

Rubin, J., Watson, I.: Similarity-based retrieval and solution re-use policies in the game of Texas Hold'em. In: Bichindaritz, I., Montani, S. (eds.) ICCBR 2010. LNCS, vol. 6176, pp. 465–479. Springer, Heidelberg (2010)

Sánchez-Marré, M., Cortés, U., Martínez, M., Comas, J., Rodríguez-Roda, I.: An approach for temporal case-based reasoning: Episode-based reasoning. In: Muñoz-Ávila, H., Ricci, F. (eds.) ICCBR 2005. LNCS (LNAI), vol. 3620, pp. 465–476. Springer, Heidelberg (2005)

Serina, I.: Kernel functions for case-based planning. Artificial Intelligence 174(16), 1369–1406 (2010)

Sukthankar, G., Goldman, R., Geib, C., Pynadath, D., Bui, H.: An introduction to plan, activity, and intent recognition. In: Sukthankar, G., Goldman, R., Geib, C., Pynadath, D., Bui, H. (eds.) Plan, Activity, and Intent Recognition. Elsevier, Philadelphia (2014)

Tecuci, D., Porter, B.W.: Memory based goal schema recognition. In: Proceedings of the Twenty-Second International Florida Artificial Intelligence Research Society Conference. AAAI Press, Sanibel Island (2009)

Wallis, W.D., Shoubridge, P., Kraetz, M., Ray, D.: Graph distances using graph union. Pattern Recognition Letters 22, 701–704 (2001)

Zarka, R., Cordier, A., Egyed-Zsigmond, E., Lamontagne, L., Mille, A.: Similarity measures to compare episodes in modeled traces. In: Delany, S.J., Ontañón, S. (eds.) ICCBR 2013. LNCS, vol. 7969, pp. 358–372. Springer, Heidelberg (2013)

Combining Case-Based Reasoning and Reinforcement Learning for Unit Navigation in Real-Time Strategy Game AI

Stefan Wender and Ian Watson

The University of Auckland, Auckland, New Zealand
{s.wender,ian}@cs.auckland.ac.nz

Abstract. This paper presents a navigation component based on a hybrid case-based reasoning (CBR) and reinforcement learning (RL) approach for an AI agent in a real-time strategy (RTS) game. Spatial environment information is abstracted into a number of influence maps. These influence maps are then combined into cases that are managed by the CBR component. RL is used to update the case solutions which are composed of unit actions with associated fitness values. We present a detailed account of the architecture and underlying model. Our model accounts for all relevant environment influences with a focus on two main subgoals: damage avoidance and target approximation. For each of these subgoals, we create scenarios in the StarCraft RTS game and look at the performance of our approach given different similarity thresholds for the CBR part. The results show, that our navigation component manages to learn how to fulfill both sub-goals given the choice of a suitable similarity threshold. Finally, we combine both subgoals for the overall navigation component and show a comparison between the integrated approach, a random action selection, and a target-selection-only agent. The results show that the CBR/RL approach manages to successfully learn how to navigate towards goal positions while at the same time avoiding enemy attacks.

Keywords: CBR, Reinforcement Learning, Game AI, Unit Navigation.

1 Introduction

RTS games are becoming more and more popular as challenging test beds that offer a complex environment for AI research. Commercial RTS game such as StarCraft in particular are a polished environment that offer complex test settings. These games have been identified as including numerous properties that are interesting for AI research, such as incomplete information, spatial and temporal reasoning as well as learning and opponent modeling [5]. For these reasons we chose StarCraft as a test bed for a machine learning (ML) approach that tries to learn how to manage combat units on a tactical level ("micromanagement").

Micromanagement requires a large number of actions over a short amount of time. It requires very exact and prompt reactions to changes in the game environment. The micromanagement problem involves things like damage avoidance,

L. Lamontagne and E. Plaza (Eds.): ICCBR 2014, LNCS 8765, pp. 511–525, 2014.
© Springer International Publishing Switzerland 2014

target selection and, on a higher, more tactical level, squad-level actions and unit formations [18]. Micromanagement is used to maximize the utility of both individual units and unit groups, as well as to increase the lifespan of a player's units. A core component of micromanagement is unit navigation. There are numerous influences a unit has to take into account when navigating the game world, including static surroundings, opposing units and other dynamic influences. Navigation and pathfinding is not a problem unique to video games, but a topic that is of great importance in other areas of research such as autonomous robotic navigation [9]. The appeal of researching navigation in video games lies in having a complex environment with various different influences that can be precisely controlled. Navigation is also integrated as a supplementary task in the much broader task of performing well at defeating other players and can therefore be used and evaluated in numerous ways.

The navigation component presented in this paper is also not a standalone conception. We are currently developing a hierarchical architecture for the entire reactive and tactical layers in an RTS game. The AI agent based on this architecture uses CBR for its memory management, RL for fitness adaption and influence maps(IMs)/potential fields [8] for the abstraction of spatial information. This approach is based on previous research [19] where we used RL in an agent in a simplified combat scenario. While StarCraft offers a very large and complex set of problems, our previous RL agent focuses on a small subset of the overall problem space and is limited to one simple scenario.

Our high-level aim is to create an adaptive AI agent for RTS games that minimizes the amount of knowledge that is stored in deterministic sub-functions, but instead learns how to play the game. We aim to do this by using an approach that manages large amounts of obtained knowledge in the shape of cases in a case-base on the one hand, and has the means to acquire and update this knowledge through playing the game itself on the other hand.

2   Related Work

Unit navigation in an RTS game is closely related to autonomous robotic navigation, especially when looking at robotic navigation in a simulator without the added difficulties of managing external sensor input [10]. Case-based reasoning, sometimes combined with RL, has been used in various approaches for autonomous robotic navigation. [13] describes the *self-improving robotic navigation system* (SINS) which operates in continuous environment. One aim of SINS is, similar to our approach, to avoid hand-coded, high-level domain knowledge. Instead, the system learns how to navigate in a continuous environment by using reinforcement learning to update the cases.

Kruusmaa [9] develops a CBR-based system to choose the least risky routes in a grid-based map, similar to our infleucen map abstraction, and lead faster to a goal position. However, the approach only works properly when there are few, large obstacles and the state space is not too big.

Both of these approaches are non-adversarial however, and thus do not account for a central feature of RTS games that raise the complexity of the problem space considerably. We address the presence of opposing units by using influence maps to abstract the battlefield which can contain dozens of units on either side at any one time.

The A* algorithm [7] and its many variations are the most common exponents of search-based pathfinding algorithms that are used in video games while also being successfully applied in other areas. However, the performance of A* depends heavily on a suitable heuristic and is also very computation-intensive as they frequently require complex pre-computation to enable real-time performance. A* variations, such as the real-time heuristic search algorithm kNN-LRTA* [4], which uses CBR to store and retrieve subgoals to avoid re-visting known states too many times, have been used extensively in commercial game AI and game AI research.

Different forms of both RL and CBR are popular techniques in game AI research, either separately or as a combination of both approaches. The reinforcement learning component in this paper is based on our previous research in [19] and a subsequent integration of single-pass CBR. While our previous simple RL approach shows good results for the one specific scenario that is tested, the work presented in this paper enables the generalization to a much larger and more complex area of problems through the integration of CBR.

Using only RL for learning diverse actions in a complex environment becomes quickly unfeasible due to the curse of dimensionality [2]. Therefore, the addition of CBR to manage the obtained knowledge and to retain information in the form of cases offers benefits both in terms of the amount of knowledge that is manageable and in terms of knowledge generalization. [11] describes the integration of CBR and RL for navigation in a continuous environment. Both state- and action-space are continuous and the agent has to learn effective movement strategies for units in a RTS game. As a trade-off for working with a non-discretized model, the authors only look at movement component of the game from a meta-level perspective where actions are given to groups of units instead of individuals. As a result, this approach is situated on a tactical level rather than the reactive level in terms of the overall game (see Section 3).

Influence maps or potential fields for the abstraction of spatial information have been used in a number of domains, including game AI, mostly for navigation purposes. Initially IMs were developed for robotics [8].

[16] use influence maps to produce *kiting* in units, also in StarCraft. *Kiting* is a hit-and-run movement that is similar to movement patterns that our agent learns for micromanagement. Their approach is similar to [19] in the behavior they are trying create and similar to our current approach in that they also use influence maps. However, their focus lies solely on the hit-and-run action and not on general maneuverability and the potential to combine pathfinding and tactical reasoning into a higher-level component as in our current approach.

[6] describes the creation of a multi-agent bot that is able to play entire games of StarCraft using a number of different artificial potential fields. However, some of the parts of the bot use non-potential field techniques such as rule-based

systems and scripted behavior, further emphasizing the need to combine different ML techniques to address larger parts of the problem space in an RTS game.

3   Pathfinding and Navigation as Part of a Hierarchical Agent Architecture

RTS games can be split into logical tasks that fall into distinct categories such as strategy, economy, tactics or reactive maneuvers. These tasks can in turn be grouped into layers [18]. Figure 1 shows the subdivision of an RTS into tasks and layers that are involved in playing the game.

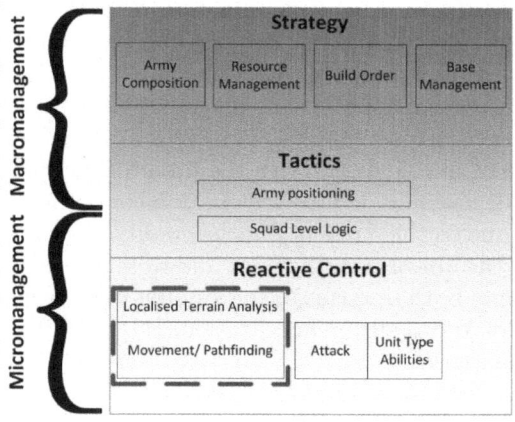

Fig. 1. RTS Game Layers and Tasks

For the creation of an AI agent that learns how to play a game, these tasks and layers can be used to create areas of responsibility for certain parts of an agent architecture. The layering leads to most RTS agents being inherently hierarchical and having a top-down approach when it comes to giving orders or executing actions and plans [12]. A high-level planer will decide on an action, which will subsequently be passed down to subordinate managers. These managers might then pass on parts of these action to managers that are even further down in the hierarchy.

Macromanagement is the umbrella term commonly used for tasks in the strategy layer and for high-level tactical decisions like army positioning in terms of the entire map. *Micromanagement* on the other hand describes the reactive layer as well as squad-level decisions in the tactical layer [12].

The ML agent we are currently working on is based on hierarchical CBR combined with a RL component. The agent learns how to play the entire micromanagement part of a game, from tactical decision making down to reactive control and pathfinding.

Hierarchical CBR is a sub-genre of CBR where several case-bases are organized in an interlinked order. [14] coins the term HCBR to describe a CBR system for software design. The authors use a blackboard based architecture that controls reasoning with a hierarchical case-base. The case-base is structured as a partonomic hierarchy where one case is split into several layers. Finding a solution to a software design problem requires several iterations where later iterations are used to fill lower levels of the case.

Our general architecture is similar, though slightly more complex. The case-bases are not strictly partonomic, since lower-level case-bases do not use solutions from higher level states as the only input for their case descriptions. Furthermore, our case-bases are more autonomous in that they run on different schedules which can result in differing numbers of CBR-cycle executions for different case-bases.

Each level in our architecture is working autonomously while using output from higher levels as part of its problem description. Each level is also represented by different case-bases. The pathfinding component presented here is provided with a target position that is decided on an intermediate level of the overall agent architecture. The goal given by this intermediate level is one of the influences the pathfinding component works with. At the same time, while a unit is trying to reach its goal location, the agent is trying to satisfy two other subordinate goals: minimizing any damage received and avoiding obstacles in the environment. The model that was created to reflect these requirements is described in detail in the next section.

4   CBR/RL Integration and Model

The core part of the pathfinding component is a case-base containing cases based on the environment surrounding a single unit. Navigation cases use both information from the game world and the target position given by a higher-level reasoner on the architecture layer above. Case descriptions are mostly abstracted excerpts of the overall state of the game environment. Case solutions are movement actions of a unit. Units acting on this level are completely autonomous, i.e. there is only one unit per case. As a result, the information taken from the game world is not concerned with the entire battlefield, but only with these immediate surroundings of the unit in question.

4.1   Case Description

The case description consists of an aggregate of the following information:

- Static information about the environment.
- Dynamic information about opposing units.
- Dynamic information about allied units.
- Dynamic information about the unit itself.
- Target specification from the level above.

Most of this information is encoded in the form of influence maps (IMs) [8]. We are using three distinct influence maps: One for the influence of allied units, one for the influence of enemy units and one for the influence of static map properties such as cliffs and other impassable terrain. This influence map for static obstacles also contains information on unit positions, as collision detection means that a unit can not walk through other units.

While higher levels of reasoning in our architecture use influence maps that can cover the entire map, the relevant information for a single unit navigating the game environment on a micromanagement level is contained in its immediate surroundings. Therefore, the perception of the unit is limited to a fixed 7x7 cutout of the overall IM directly surrounding the unit. An example of the different IMs for varying levels of granularity and other information contained in the environment can be seen in Figure 2.

The red numbers denote influence values. The influence values are based on the damage potential of units. This figure only shows enemy influence in order to not clutter the view. Any square in the two influence maps for enemy and allied units has assigned to it the damage potential for adjacent units. The IM containing information on passable terrain only contains a true/false value for each field. In figure 2, impassable fields are shaded.

The 'local' IM that represents the perception of a single unit it surrounds, is marked by the green lines that form a sub-selection of the overall yellow IM. Only values from this excerpt are used in the case descriptions.

In addition to the spatial information contained in the IMs, three other relevant values are part of a case description: *unit type, previous action* and *target*

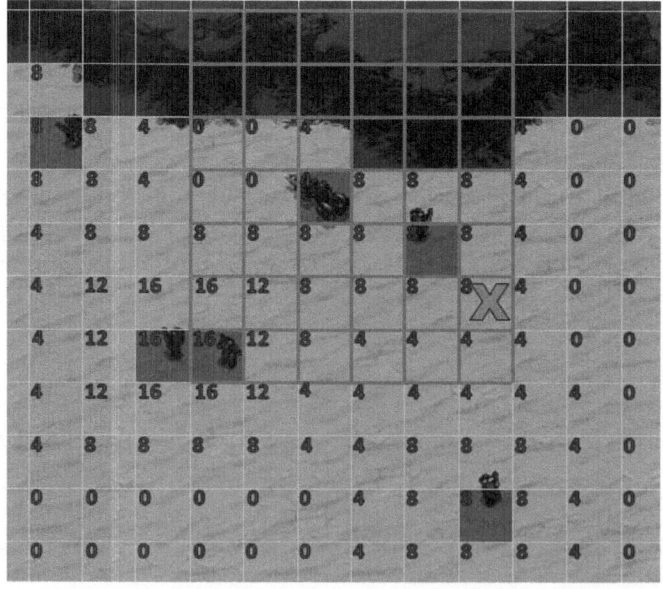

Fig. 2. Game Situation with Influence Map Overlay

location. The *target location* indicates one of the 7x7 fields surrounding a unit and is the result of a decision on the higher levels. In Figure 2, the current target location is marked by the blue **X**. The *unit type* is indicative of a number of related values such as speed, maximum health and ability to fly. The *previous action* is relevant as units are able to carry momentum over from one action to the next.

The case solutions are concrete game actions. We currently use four *Move* actions for the different directions, one for every 90° (see Figure 3).

Fig. 3. Possible Movements for a Unit

The case-base in the CBR part of the pathfinding component reflects a general CBR cycle [1], however does not use all of the cycle's phases. During the **Retrieval** step, the best-matching case is found using a simple kNN algorithm where $k = 1$, given the current state of the environment. If a similar state is found, its solution, made up of all possible movement actions and associated fitness values, is returned. These movement actions can then be **Reused** by executing them. There is no further adaption of the solution apart from RL changing fitness values.

Currently, there is no **Revision** of case descriptions stored in any of the case-bases. Once a case has been stored, the case descriptions, consisting of the abstracted game information and of the solutions from higher-up levels, are not changed anymore. However, the fitness values associated with case solutions are adjusted each time a case is selected. These updates are done using RL (see Section 4.2).

Retention is part of the retrieval process. Whenever a case retrieval does not return a sufficiently similar case, a new one is generated from the current game environment state.

The similarity measure that is used to retrieve the cases plays a central role, both in enabling the retrieval of a similar case, and in deciding when new cases have to be created if no sufficiently similar case exists in the case-base. Part of the novelty of our approach is the use of influence maps as central part of our case descriptions. These IMs are then used in the similarity metric when looking for matching cases.

In all three types of influence maps that are used, the similarity between a map in a problem case and a map stored in the case base is an aggregate of the similarities of the individual map fields. Comparing the map on accessibility of single map plots is comparably easy since each field is identified by a Boolean value, which means that similarity between plot of IMs in a problem case and in a case in the case base is either 0% or 100%. The similarity of a single field describing the damage potential of enemy or own units, is decided by the difference in damage to its counterpart in the stored case. I.e. if one field has the damage potential 20 and the counterpart in the stored case has the damage potential 30, the similarity is decided by the difference of 10. Currently, we are mapping all differences into one of four similarities between 0% and 100%. This also depends both on if unit influence exists in both cases and on how big the difference is. An example similarity computation is shown in Figure 4.

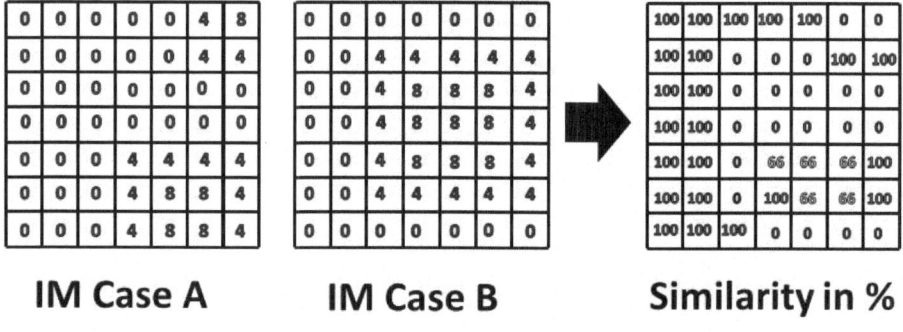

Fig. 4. Example of IM Field Similarity Computation

We decided not to abstract the information stored in the IM fields any further and do a direct field-to-field comparison. However, as a large case-base could slow the search down considerably, we are also investigating methods to reduce the current number of cases. This will start with methods to account for simple geometric transformations on the IMs such as rotation. In higher levels of our architecture (see Section 3), where larger areas of the map are compared and a brute-force nearest-neighbor search would be unfeasible, we are already using more abstract comparisons. Because of the requirement for very precise actions when navigating, using the same methods for the pathfinding component is impractical.

The similarity between *Last Unit Actions* is determined by how similar the movement directions for chosen are, i.e. same direction means a similarity of 1, orthogonal movement 0.5 and opposite direction means 0. *Unit Type* similarity is taken from a table that is based on expert knowledge as it takes a large number of absolute (health,energy, speed) and relative (can the unit fly, special abilities) attributes into account. The similarity between *Target Positions* is computed as a distance between target positions in cases in relation to the maximum possible distance (dimension of the local IM field).

Table 1 summarizes the case description and related similarity metrics.

Table 1. Case-Base Summary

Attribute	Description	Similarity Measure
Agent Unit IM	Map with 7x7 fields containing the damage potential of adjacent allied units.	Normalized aggregate of field similarity.
Enemy Unit IM	Map with 7x7 fields containing the damage potential of adjacent allied units.	Normalized aggregate of field similarity.
Accessibility IM	Map with 7x7 fields containing true/false values about the accessibility.	Normalized aggregate of field similarity.
Unit Type	Type of a unit.	Table of unit similarities between 0 and 1 based on expert knowledge.
Last Unit Action	The last movement action taken.	Value between 0 and 1 depending on the potential to keep momentum between actions.
Target Position	Target position within the local 7x7 map.	Normalized distance.

4.2   Reinforcement Learning for Solution Fitness Adaption

The reinforcement learning component is used to learn fitness values for case solutions. This part of the architecture is based on our previous research [19]. While the complexity of the model here is much higher, the problem domain remains the same, which is why we chose to use a one-step Q-learning algorithm for policy evaluation. One-step Q-learning showed, in this specific setting, the best performance when compared to other tested algorithms by a small margin. Furthermore, its implementation is comparably simple and evaluation using one-step algorithms are computationally inexpensive when compared to algorithms using eligibility traces.

Q-learning is an off-policy temporal-difference (TD) algorithm that does not assign a value to states, but to state-action pairs [17]. Since Q-learning works independent of the policy being followed, the learned action value Q function directly approximates the optimal action-value function Q^*.

In order to use RL, we express the task of updating the fitness of case solutions as a *Markov Decision Process* (*MDP*). A specific MDP is defined by a quadruple $(S, A, \mathcal{P}_{ss'}^a, \mathcal{R}_{ss'}^a)$. In our case the state space S is defined through the case descriptions. The action space A is the case solutions. $\mathcal{P}_{ss'}^a$ are the transition probabilities. While StarCraft has only few minor non-deterministic components, its overall complexity means that the transition rules between states are stochastic. As a result, different subsequent states s_{t+1} can be reached when taking the same action a_t in the same state s_t at different times in the game. $\mathcal{R}_{ss'}^a$ represents the expected reward, given a current state s, an action a and the next state s'.

Reward is computed on a per-unit and per-action basis, similar to our simple RL agent. The reward signal is crucial to enable the learning process of the agent and has to be carefully selected. As the agent tries to satisfy, depending on the given scenario, a number of different goals, we use a composite reward signal. For damage avoidance, the reward signal includes the difference in health h_{unit} of the agent unit in question between time t and time $t + 1$. The reward signal also includes a negative element that is based on the amount of time t_a it takes to finish an action a. This is to encourage fast movement towards the goal position. To encourage target approximation, the other central requirement besides damage avoidance, there is a feedback (positive or negative) for the change in distance d_{target} between the unit's current position and the target position. Finally, to account for inaccessible fields and obstacles, there is a penalty if a unit chooses an action that would access a non-accessible field. This penalty is not part of the regularly computed reward after finishing an action but is attributed whenever such an action is chosen and leads to immediate re-selection. The resulting compound reward that is computed after finishing an action is

$$\mathcal{R}^a_{ss'} = \Delta h_{unit} - t_a + \Delta d_{target}.$$

When looking at scenarios that evaluate the performance of the two subgoals target approximation and damage avoidance, only the parts of the reward signal that are relevant for the respective subgoal are used.

5   Empirical Evaluation and Results

A core parameter to determine in any CBR approach is, given a chosen case representation and similarity metric, a suitable similarity threshold that enables the best possible performance for the learning process. That means on the one hand creating enough cases to distinguish between inherently different situations, while on the other hand not inundating the case-base with unnecessary cases. Keeping the number of cases to a minimum is especially important when using RL. In order for RL to approximate an optimal policy p^*, there has to be a sufficient state- or state-action space coverage [15]. If the case-base contains too many cases, not only will retrieval speeds go down, but a lack of state-action space coverage will diminish the performance or even prevent meaningful learning altogether.

The model described in Section 4 is an aggregate of influences for several subgoals. For this reason, we decided to have two steps in the empirical evaluation. First, the two main parts of our approach, target approximation and damage avoidance, are evaluated separately. This yields suitable similarity thresholds for both parts that can then be integrated for an evaluation of the overall navigation component.

For each of the steps we created suitable scenarios in the StarCraft RTS game. The first capability we evaluate is damage avoidance. The test scenario contains a large number of enemy units that are arbitrarily spread across a map. The scenario contains both mobile and stationary opponents, essentially creating a

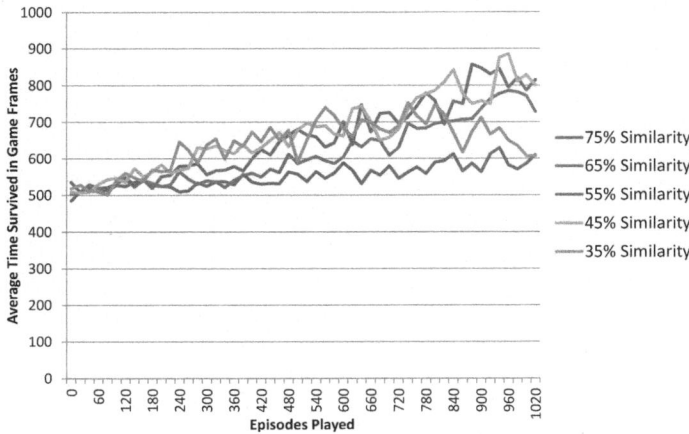

Fig. 5. Results for the Damage Avoidance Scenario

dynamic maze that the agent has to navigate through. The agent's aim is to stay alive as long as possible.

In order to test the agent's ability to find the target position, it controls a unit in a limited-time scenario with no enemies and randomly generated target positions. Therefore, the performance metric is the number of target positions an agent can reach within the given time. Accessibility penalties, *Unit Type* and *Previous Action* computation are part of both scenarios.

Based on initial test runs and experience from previous experiments, we used a learning rate $\alpha = 0.05$ and a discount rate $\alpha = 0.3$ for our Q-learning algorithm. The agent uses a declining ϵ-greedy exploration policy that starts with $\epsilon = 0.9$. This means, that the agent initially selects actions 90% random and 10% greedy, i.e. choosing the best known action. The exploration rate declines linearly towards 0 over the course of the 1000 games for the damage-avoidance scenario and 100 games for the target-approximation respectively. After each full run of 1000 or 100 games, there is another 30 games where the agent uses only greedy selection. The performance in this final phase therefore showcases the best policy the agent learned. The difference in the length of experiments (1000 vs 100 games) is due to the vastly different state-space sizes between the two scenarios. When looking at target approximation, the target can be any one of the fields in a unit's local IM, i.e. one of 49 fields. As a result, there can be a maximum of 49 different states even with 100% similarity threshold. On the other hand, there are n^{49} possible states based on the damage potential of units, where n is the number of different damage values. This also explains why the generalization that CBR provides during case retrieval is essential to enable any learning effect at all. Without CBR, the state-action space coverage would remain negligibly small even for low n-values.

The results for both variants are shown in Figure 5 and Figure 6.

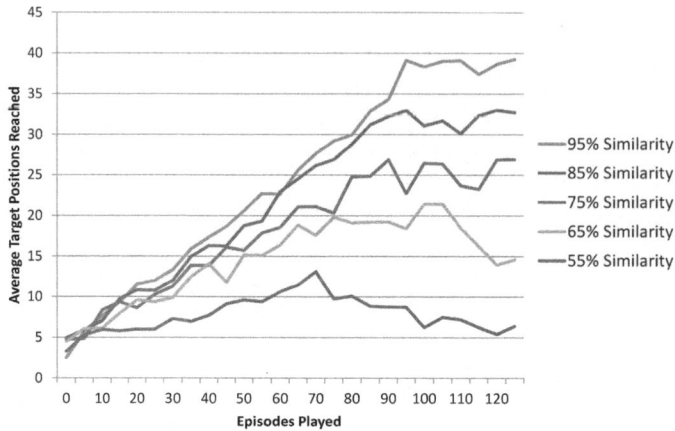

Fig. 6. Results for the Target Approximation Scenario

The results from the two sub-problem evaluation scenarios were used as a guideline to decide on a 80% similarity threshold to evaluate the overall algorithm (see Section 6 for more details). Both subgoals were equally weighted. The scenario for the evaluation of the integrated approach is similar to the one used for the damage avoidance sub-goal. The target metric is no longer the survival time however, but the number of randomly generated positions the agent can reach before it is destroyed. The run length for the experiment was increased to 1600 games to account for the heightened complexity. The resulting performance in the overall scenario can be seen in Figure 7. The figure also shows a random action selection baseline, as well as the outcome for using only target-approximation without damage avoidance.

6   Discussion

The evaluation for the damage avoidance subgoal indicates, that there is only a narrow window in which learning an effective solution is actually achieved in the given 1000 episodes: With a similarity threshold of 35% there is not enough differentiation between inherently different cases/situations and the performance drops of steeply towards the end. On the other end of the spectrum, the worst results with practically no improvement in the overall score is achieved for a 75% similarity threshold. This means that with such a high threshold, the state-action space coverage drops to a level where learning a 'good' strategy is no longer guaranteed: For a similarity threshold of 75% there are 7000 cases in the case-base after 1000 episodes played and less than 1/2 of all state-action pairs have been explored at least once.

Target approximation is slightly different in that an effective solution is achieved for any similarity threshold above 65%. This is due to the fact that

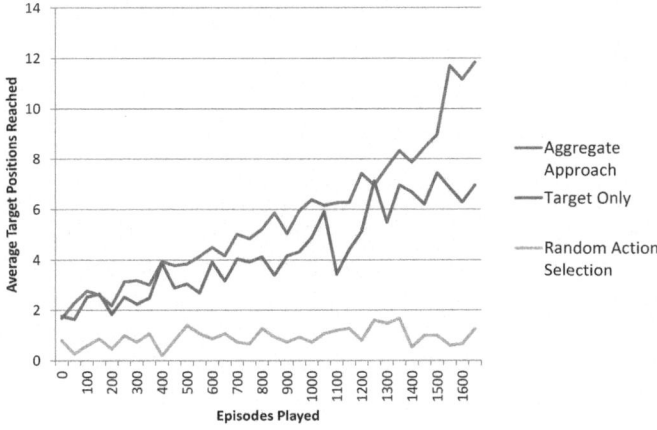

Fig. 7. Results for the Combined Scenario

even at 95% similarity there are still only $49 * 4 = 196$ state-action pairs to explore. The number of cases in the case-base varies however. At 95% similarity threshold, there is 49 different cases, i.e. the maximum possible number. At 85% similarity there exist about 20 different cases and at 75% only around 9 cases. This is important when we combine target approximation with damage avoidance for the overall approach since this also combines the state-space of the two subgoals. Therefore, twice as many cases for only a ∼20% higher reward as for the 75% compared to the 85% threshold or the 85% compared to the 95% threshold is a bad tradeoff. For this reason, we chose to use a 75% similarity threshold for the target approximation sub-goal.

Both subgoal approaches show a performance improvement in their chosen metrics over the run of their experiments. Since the target approximation algorithm manages to exhaustively explore the state-action space, the performance improvement is more visible in this scenario. This also means, that the best possible solution in this scenario has been found, as a 95% similarity threshold with all possible states and a fully explored state-action space guarantee that the optimal policy p^* must have been found. For the damage avoidance scenario on the other hand, there is still a potential for improvement, since significant parts of the state-action space remain unexplored: Even for a 55% threshold only about 3/4 of all possible actions are executed at least once.

The results for the overall algorithm in Figure 7 show, that the agent manages to successfully learn how to navigate towards target positions. The performance of the agent far outperforms the random action selection. However, the results also show that the combined algorithm initially only has a small advantage over the target-approximation-only agent and only towards the end performs better. This is far less of an improvement than expected, especially given the fact that these experiments reached a sufficiently high state-action coverage of about 2/3.

This indicates, that the chosen overall similarity threshold of 80% is too low and there is not enough differentiation between inherently different cases/situations to find the optimal policy.

7 Conclusion and Future Work

This paper presented an approach to unit navigation based on a combination of CBR, RL and influence maps for an RTS game AI agent. We illustrated the general tasks involved in a RTS game, how these tasks can be grouped into logical layers and how navigation is part of the micromanagement problem that we aim to address with an agent based on a hierarchical CBR architecture. The important information was identified and we designed a model that encompasses these different influences that are important for navigation and pathfinding in RTS games.

The empirical evaluation showed how our approach learned how to successfully learn to achieve partial goals of the navigation problem. Using the findings from the partial evaluation we ran an evaluation for the entire approach including all relevant influences and goals. This evaluation shows that our model and approach manage to successfully learns how to navigate a complex game scenario that tests all different sub-goals. Our approach significantly outperforms a random action selection and marginally outperforms the partial, target-approximation-only agent.

In terms of the ML algorithm, a more sophisticated case retrieval method and an evaluation of alternative RL methods are first on our list for future work. We are considering k-d trees to pre-index the cases [20] as a larger number of enemy or own units automatically leads to a more complex state-space, something that could slow down the retrieval speeds considerably.

We are also looking at using a different RL algorithm for our exploration strategy such as Softmax action selection [15] or the R-Max algorithm [3].

The next step in terms of the overall approach will be the evaluation of the integration of the navigation component presented in this paper into the overall HCBR agent architecture.

References

1. Aamodt, A., Plaza, E.: Case-based reasoning: Foundational issues, methodological variations, and system approaches. AI Communications 7(1), 39–59 (1994)
2. Bellman, R., Bellman, R.E., Bellman, R.E., Bellman, R.E.: Adaptive control processes: a guided tour, vol. 4. Princeton University Press, Princeton (1961)
3. Brafman, R.I., Tennenholtz, M.: R-max-a general polynomial time algorithm for near-optimal reinforcement learning. The Journal of Machine Learning Research 3, 213–231 (2003)
4. Bulitko, V., Bjornsson, Y., Lawrence, R.: Case-based subgoaling in real-time heuristic search for video game pathfinding. Journal of Artificial Intelligence Research 39, 269–300 (2010)

5. Buro, M., Furtak, T.: Rts games and real-time ai research. In: Proceedings of the Behavior Representation in Modeling and Simulation Conference (BRIMS), pp. 63–70. Citeseer (2004)
6. Hagelbäck, J.: Multi-Agent Potential Field Based Architectures for Real-Time Strategy Game Bots. Ph.D. thesis, Blekinge Institute of Technology (2012)
7. Hart, P.E., Nilsson, N.J., Raphael, B.: A formal basis for the heuristic determination of minimum cost paths. IEEE Transactions on Systems Science and Cybernetics 4(2), 100–107 (1968)
8. Khatib, O.: Real-time obstacle avoidance for manipulators and mobile robots. The International Journal of Robotics Research 5(1), 90–98 (1986)
9. Kruusmaa, M.: Global navigation in dynamic environments using case-based reasoning. Autonomous Robots 14(1), 71–91 (2003)
10. Laue, T., Spiess, K., Röfer, T.: SimRobot – A general physical robot simulator and its application in roboCup. In: Bredenfeld, A., Jacoff, A., Noda, I., Takahashi, Y. (eds.) RoboCup 2005. LNCS (LNAI), vol. 4020, pp. 173–183. Springer, Heidelberg (2006)
11. Molineaux, M., Aha, D., Moore, P.: Learning continuous action models in a real-time strategy environment. In: Proceedings of the Twenty-First Annual Conference of the Florida Artificial Intelligence Research Society, pp. 257–262 (2008)
12. Ontanón, S., Synnaeve, G., Uriarte, A., Richoux, F., Churchill, D., Preuss, M.: A survey of real-time strategy game ai research and competition in starcraft. IEEE Transactions on Computational Intelligence and AI in Games (2013)
13. Ram, A., Santamaria, J.C.: Continuous case-based reasoning. Artificial Intelligence 90(1), 25–77 (1997)
14. Smyth, B., Cunningham, P.: Déjà vu: A hierarchical case-based reasoning system for software design. In: ECAI, vol. 92, pp. 587–589 (1992)
15. Sutton, R.S., Barto, A.G.: Reinforcement Learning: An Introduction. MIT Press (1998)
16. Uriarte, A., Ontañón, S.: Kiting in rts games using influence maps. In: Workshop Proceedings of the Eighth Artificial Intelligence and Interactive Digital Entertainment Conference (2012)
17. Watkins, C.: Learning from Delayed Rewards. Ph.D. thesis. University of Cambridge, England (1989)
18. Weber, B., Mateas, M., Jhala, A.: Building human-level ai for real-time strategy games. In: 2011 AAAI Fall Symposium Series (2011)
19. Wender, S., Watson, I.: Applying reinforcement learning to small scale combat in the real-time strategy game starcraft: broodwar. In: 2012 IEEE Conference on Computational Intelligence and Games, CIG (2012)
20. Wess, S., Althoff, K., Derwand, G.: Using k-d trees to improve the retrieval step in case-based reasoning. In: Wess, S., Richter, M., Althoff, K.-D. (eds.) EWCBR 1993. LNCS, vol. 837, pp. 167–181. Springer, Heidelberg (1994)

Exploring the Space of Whole-Group Case Retrieval in Making Group Recommendations

David C. Wilson and Nadia A. Najjar

Department of Software and Information Systems
University of North Carolina at Charlotte
Charlotte, North Carolina
{davils,nanajjar}@uncc.edu

Abstract. Case-Based Reasoning has been studied as a methodology to support ratings-based collaborative recommendation, but this predominantly targets the context of an individual end-user. There are, however, many circumstances where several people participating together in a group activity could benefit from recommendations tailored to the group as a whole. Group recommendation has received comparatively little attention overall, and recent research has largely focused on making straightforward individual recommendations for each group member and then aggregating the results. But this examines only the context of the target group, and does not take advantage of other, previous group contexts as a first-class element of the knowledge base. Recent research investigated how case-based reasoning approaches can be applied to retrieve and reuse whole previous groups as a basis for recommendation and showed an advantage over traditional aggregation approaches. In this paper we focus on further exploration of the space. We present our approach for case-based group recommendation, as well as evaluation results across conditions for group size and homogeneity. Results show that foundational group-to-group approaches outperform individual-to-group recommendations across a wide range of group contexts.

Keywords: CBR, group recommenders, collaborative filtering, recommender systems.

1 Introduction

Case-Based Reasoning (CBR) has a long history as a methodology for building recommender systems [1]. In general, case-based recommendation embodies a content-based or knowledge-based approach, with a semantically rich representation of users and/or items (e.g., [2]). However, a number of researchers have also taken a case-based perspective on ratings-based profile representations as employed in Collaborative Filtering recommendation [3–7], and we adopt this perspective.

Overall, research in recommender systems predominantly targets the context of an individual end-user. There are, however, many circumstances where several people participating together in a group activity could benefit from recommendations tailored to the group as a whole. Common examples arise in social

L. Lamontagne and E. Plaza (Eds.): ICCBR 2014, LNCS 8765, pp. 526–540, 2014.
© Springer International Publishing Switzerland 2014

entertainment: finding a movie or a television show for family night, date night, or the like [8]; finding a restaurant for dinner with work colleagues, family, or friends [9]; finding a dish to cook that will satisfy the whole group [10], the book that a book club should read next, the travel destination for the next family vacation [11], or the songs to play for social events / shared public spaces [12].

Though there are a number of such examples, group recommendation has received comparatively little attention overall. Recent research has largely focused on making straightforward individual recommendations for each group member and then aggregating the results. But this considers only the group context of the query or problem specification (on final aggregation), and does not take advantage of other, previous group contexts as a first-class element of the knowledge base. Retrieving and reusing the previous experience of another group — taken as a whole — is a natural application for CBR. Case-based approaches to group recommendation have appeared, but are quite few. To our knowledge, only the seminal work by Quijano-Sanchez et al. [13] has examined retrieval of entire groups as cases, as opposed to the aggregation of individual retrievals.

In this paper we carry out further exploration of the space, investigating the effectiveness of case-based reasoning approaches to retrieve and adapt whole previous groups as a basis for recommendation. This paper presents our foundational approach for case-based group recommendation and evaluation results that investigate a variety of conditions for group size and homogeneity. The paper begins with related work in group modeling and CBR for group recommendation. Section 3 introduces our research question. Sections 4 and 5 formalize the recommendation techniques investigated in our group-based recommender, and experimental outcomes are discussed in Section 7. Our results show that foundational group-to-group approaches can outperform individual-to-group recommendations across a wide range of group contexts.

2   Related Work

Research in group-based recommender systems centers around approaches to model the group as part of the recommendation process. Two primary methods have been proposed: aggregating preferences and aggregating recommendations [14]. These strategies utilize recommendation techniques validated for individual users, and so the aggregation strategy itself has been typically the defining feature for work in group-based recommenders. Group modeling strategies are inspired by Social Choice Theory, and center around modeling the achievement of consensus among the group [15]. Researchers have also looked at group interaction considering personalities and social interactions among group members [16, 17]. A more limited number of researchers have explored using Case-Based Reasoning to model group decisions [13] or as a part of the recommendation technique [2, 18].

McCarthy et al. [2] looked at aggregating individual user models to produce a group model that was used to generate recommendations in a critiquing, case-based recommender. They employed the CATS [19] recommender system to evaluate three different aggregation methods to assess the quality of a case, including similarity to a critiqued case and compatibility to the combined group critique model. To evaluate, they used synthetic groups generated from real-user preference data (34 trial subject profiles converted into a critique-based profile). Four-member groups were randomly generated (3 sets of 100 groups each), with varying similarity levels (*similar, mixed, diverse*). Each test group received three sets of recommendations, with each set containing one recommendation for each group member. Recommended cases were evaluated based on *compatibility* (# shared features) to the known *perfect case* of the individual group members. Group *compatibility* was measured as the average score. They reported an improvement in recommendation quality across the aggregation strategies for the *similar* groups when compared to the individual group member's own case choice, but not for the *diverse* and *mixed* groups, given preference diversity. In this work the CBR perspective was used as part of the recommendation technique in retrieving possible candidate items for the group to critique in a similar fashion to recommendations made to individuals and not adapted for groups.

Recent work from Quijano-Sanchez et al. employed CBR in several aspects of group-based recommendation [13, 18]. In [18] they evaluated a CBR solution to alleviate "cold-start" problems for group members. Cases represent previous movie recommendation events for groups. When a group seeks a recommendation and some group members are considered in cold-start, they find a previous case with users in similar group-roles who are not in cold-start. Ratings are transferred to user(s) in the active group from corresponding users in the retrieved case, and then the updated profile is used in the recommendation process. In this work as well, the CBR perspective was used but not directly for group-based recommendations, as such. It was used to address the cold start problem for individual group members.

Quijano-Sanchez et al. also applied CBR directly in modeling group decisions for group recommendation [13]. Employing user-user similarity, each user in the active group is aligned with exactly one user in the case group. Similarity is measured by comparing group members on their age, gender, personality, ratings and the degrees of trust between members of each group. Group to group similarity is calculated as an average of these one to one similarities. All mapping combinations between the active group and the case groups are checked and the top n cases and mappings are then used in the recommendation process. Using item-item similarity, they map contributions in choosing the selected item from each group member in the case. Predictions for items are based on accumulating the similarities to the selected items in similar cases, weighted by the degree of similarity to those cases. If group and case are of different sizes virtual users are added to that group. An overall improvement in success rate was reported for their CBR approach on a data set of 100 cases, which were individually crafted with review input by a panel of experts. As far as we are aware, this is the only group recommender work to date that investigates holistic retrieval and reuse of entire previous groups as cases. This served as an inspiration for our own exploration of the space.

To help place our investigation in perspective, we here provide a more extensive consideration of [13] as related work in motivating our study. For example, their approach combines many subtle aspects of the group recommendation process all at once (some requiring imputed value support). Since there was not an ablation analysis, this naturally raises questions about the relative contributions of different aspects. Differently from [13], our technique is less complex overall, which enables a clearer understanding of the contribution provided by baseline case retrieval. The evaluation metric (Success@1, etc.) reported in [13] has not been commonly used, and they did not report on significance, which makes comparisons to other approaches more challenging. Here we evaluate with a more widely employed metric, Root Mean Squared Error (RMSE) with reported significance. Their study case-base included only groups of same-age-range for its group members, whereas we examine explicit conditions for intra-group coherence. The case-base is both a strength and limitation of the previous study. Data sets with real group decisions are difficult to create and not readily available, so an expert crafted case-base helps to address validity of the group decision data. At the same time, manual review / creation limits the scope of experimental studies, which has commonly been addressed by generating synthetic groups on a larger scale (e.g., [20]). And so, we expand upon their initial exploration of this interesting space with our own studies. While we look at a general accuracy comparison with [13], the main point here is to better understand the nature and effectiveness of CBR approaches that retrieve and reuse whole previous groups for group recommendation. In particular, to help establish baseline effectiveness for straightforward CBR techniques, as a context for understanding more complex approaches. This includes a streamlined CBR approach, and limiting imputed values for a clearer baseline understanding. It also includes fixing the group size between the active group and the cases considered in the recommendation process eliminating the need to use virtual users. We examine different conditions than [13], such as adaptation from explicit top-1 vs. top-N perspectives, as well as explicit experimental conditions for group homogeneity. Our experiments examine a larger overall case-base size of 1200 groups in comparison to the 100 groups for the cases used in [13].

3   Exploring Whole-Group Case-Based Reasoning

Our main research question is to examine under what conditions will taking advantage of existing group contexts in the knowledge base — in addition to the group context of the target / active group (as query) — improve group recommender performance. That is, if we take a CBR perspective on group recommendation, retrieving whole previous groups as the starting point for predictions instead of directly aggregating on individuals, when will there be an overall benefit for the system? To investigate this issue, we integrate a foundational CBR component (Section 5) with a common group recommender technique, evaluating across a variety of conditions. In this study, we focus on understanding how much traction is really possible with a straightforward whole-group retrieval

CBR approach. And so we embrace the limits of the process — limiting retrieval to groups of the same size, allowing for retrieval failure, and so on. The group recommender component provides a standard aggregation of individual recommendations for group members. The case-based reasoning component retrieves similar groups from the case-base and adapts retrieved group preferences to the target group query. For this study, there is no guarantee of complete coverage by the case base, and so the system acts as a "switching" hybrid recommender [21]: if the CBR component can not make a recommendation with sufficient confidence, then the baseline aggregation method is used. This is similar in spirit to integrations of CBR with generative planners in case-based planning [22, 23].

The standard aggregation technique also serves as the baseline for system evaluation, comparing the CBR approach (with switching as needed) to using only the baseline for each query. Our evaluation examines three main hypotheses: (H1) The foundational CBR approach will be able to respond to a substantial number of queries across a range of conditions; (H2) The hybrid CBR approach will provide significantly better accuracy than the baseline; (H3) Accuracy results for the foundational CBR approach will show similar trends to those reported in [13]. Significance testing is considered for $p < 0.01$.

4 Baseline Group Recommender

Our baseline group recommender employs a common neighborhood-based CF algorithm [24, 25]. Predictions for each individual user in the active group are calculated and then aggregated for an overall group prediction [14]. This applies Pearson Correlation to compute similarity (sim_{xy}) between individual users u_x and u_y, as follows.

$$sim_{xy} = \frac{\sum_{i=1}^{n}[(r_{xi} - \overline{r}_x)(r_{yi} - \overline{r}_y)]}{\sqrt{\sum_{i=1}^{n}(r_{xi} - \overline{r}_x)^2 \sum_{i=1}^{n}(r_{yi} - \overline{r}_y)^2}} \tag{1}$$

Where n is the number of items rated by u_x and u_y, and r refers to rating. To generate predictions a subset of the nearest neighbors of the active user are chosen based on their correlation. A weighted aggregate of their ratings is calculated to generate predictions for that user. We use the following formula to calculate the prediction for item i for user u_x as p_{xi}, where σ_y refers to the standard deviation of the ratings given by user u and m is the number of users in the neighborhood:

$$p_{xi} = \overline{r}_x + \sigma_x \frac{\sum_{y=1}^{m}[(r_{yi} - \overline{r}_y) \cdot sim_{xy}]/\sigma_y}{\sum_{y=1}^{n} sim_{xy}} \tag{2}$$

Once the individual predictions are calculated, a group prediction is calculated by aggregating them into a final group prediction. We focus on three common strategies [15]: average, least misery, and most happiness, defined as follows. For g users in a group and ratings r_{xi} of user x for item i, the group rating for item i is calculated as follows:

- Average Strategy: assumes equal influence among group members and calculates the average rating across group members for a given item as the predicted rating.

$$Gr_i = \frac{\sum_{x=1}^{g} r_{xi}}{g} \qquad (3)$$

- Least Misery Strategy: applicable when the recommender system needs to avoid presenting an item that was strongly disliked by any group member. The predicted rating is calculated as the lowest rating among the group members for any given item.

$$Gr_i = \min_x r_{xi} \qquad (4)$$

- Most Happiness Strategy: opposite of the least misery strategy:

$$Gr_i = \max_x r_{xi} \qquad (5)$$

5   Case-Based Group Recommender

Our case-based group recommender employs ratings-based user profile data as the foundation of the case-base. We formalize our approach as follows. Given a matrix of users (U) and items (I) a case is represented as $\{G_{cb}, I_{cb}\}$. G_{cb} is a group of users, u_{cb}, of size n where $u_{cb} \in U$. I_{cb} is a set of items i_{cb} where i_{cb} is an item rated by all the users in G_{cb} and $i_{cb} \in I$. The active group that is seeking the recommendation is represented as G_a.

5.1   Active Group to Case Similarity Metric

In order to retrieve previous group-cases that are relevant to the active (query) group, we define a similarity metric that considers the correlation between each user in the active group and every user in the case group. Similarity between G_a and G_{cb} is measured by considering the *cartesian product* of these two sets. Which is the set of all ordered pairs (u_a, u_{cb}) where $u_a \in G_a$ and $u_{cb} \in G_{cb}$. Let this set be represented as $CProd_G$ and contain the correlations that need to be calculated to measure the similarity between the active group and a case from the case base. Since we only consider cases with the same size as the active group's size ($|G_a| = |G_{cb}|$) then the size of the set $CProd_G$ is equal to $|G_a|^2$. This is the number of correlations that we need to consider to calculate the group to case similarity. In the next step, we calculate the correlation for all elements of $CProd_G$ using equation 1. The resulting correlations form the group-to-case correlation set $PCorr_G$ where only the possible correlations are stored. The final case similarity can then be calculated as the average of the $PCorr_G$ set. We note here that the similarity metric we use is based only on the user's ratings, differently from [13] where the similarity, in addition to ratings, includes trust, age, gender and personality.

$$GG_{sim} = \frac{\sum_{g=1}^{n} PCorr_{Gg}}{|PCorr_{Gg}|} \tag{6}$$

5.2   Case Retrieval

Using the defined similarity metric, cases are retrieved using the following constraints. First, only cases of the same size as the active (query) group are considered. Second, only cases that meet a defined similarity threshold θ are considered. Subject to these constraints, the top-N set of identified cases is represented as GG_{CB}. For this study, we do not consider a limit on the number of cases that meet the specified constraints, and we examine prediction strategies based on the single best case and on the entire retrieved set. In the circumstance that no case meets the selection criteria, the null set is returned to indicate failure, which serves as a trigger for the hybrid switching mechanism.

5.3   Adaptation for Recommendation

In order to make a prediction on an item rating for the active (query) group, we adapt the retrieved case(s) on the item in question. We also consider several different adaptation variants: either the most similar retrieved case or a set of retrieved cases, as well as scaling. Formally, we examine the items in I_{cb} and the item for which we are calculating a prediction. Let us refer to that item as the active item i_a. Since we are trying to model the group decision for a certain item and attempting to minimize variables in this approach we choose to base our model on cases where the active item considered for recommendation is present in the item set for the case-base group ($i_{cb} == i_a$). Applying this condition to the set of possible cases, GG_{CB} is reduced to $\widehat{GG_{CB}}$.

We consider two main adaptation variants. The first adapts only the single, most-similar group-case that contains i_a. The second adapts the entire set of retrieved cases that contain i_a ($\widehat{GG_{CB}}$). Within each of these two variants, we consider whether group ratings should be normalized between the active group and the retrieved case(s).

Adaptation — Best Case / Multi-Case. In the first variant, we consider only the case with the highest similarity to the active group. We refer to this approach as *CBR_single*. Here the case group rating's are used as the basis for the prediction. The prediction for the active group is modeled as the average of the individual rating for the active item i_a given by the members of the case group G_{cb} formalized as:

$$pG_a i_a = \frac{\sum_{u_{cb}=1}^{n} r_{u_{cb}}}{|G_{cb}|} \tag{7}$$

Where $r_{u_{cb}}$ is the rating for i_a by user u_{cb} and $u_{cb} \in G_{cb}$.

In the second variant, we consider the entire set of retrieved cases. We calculate a prediction for each participating case, as in equation 7, and average the results, formalized as:

$$pG_a i_a = \frac{\sum \left(\frac{\sum_{u_{cb}=1}^{n} r_{u_{cb}}}{|G_{cb}|} \right)}{|\widehat{GG}_{CB}|} \tag{8}$$

Adaptation Scaling. We also consider a scaling condition for each variant, which normalizes the rating scale between the active group and the retrieved case(s). This is formalized as:

$$\hat{p}G_a i_a = \overline{G_a} + \left(\sigma_{G_a} * \frac{\sum \left(\left(\frac{pG_a i_a - \overline{u}_{cb}}{\sigma_{u_{cb}}} \right) * GG_{sim} \right)}{(GG_{sim} * |G_a|)} \right) \tag{9}$$

where σ_{G_a} is the average standard deviation of ratings for the users in G_a, $\sigma_{u_{cb}}$ is the standard deviation of ratings for the users in G_{cb} and $\overline{G}_a$ is defined as:

$$\overline{G}_a = \frac{\sum \overline{u}_a}{|G_a|} \tag{10}$$

where $\overline{u}_a$ is the average rating for a user u_a in G_a. In other words, $\overline{G}_a$ is the average of the average ratings for the users in G_a. The value $pG_a i_a$ is calculated using either CBR_single or CBR_multi.

6 Experimental Setup

6.1 Recommendation Parameters

Based on the recommendations of [24] where a neighborhood of 20 to 50 neighbors provides enough neighbors to average out extremes we set a neighborhood size of 50 for the baseline recommendation.

6.2 Accuracy Measurement

We measure the accuracy of a predicted rating computed for a group across different test conditions using root-mean-square error (RMSE) [26]. To measure the differences between values predicted by a model and the actual values, we compare the group-predicted rating calculated for the test items, using the aggregation approaches described (Average, Least Misery, Most Happiness), to a model of the actual rating (average) across the different group sizes and inner-group similarity levels.

Another evaluation strategy is to compare a recommended list of items to an actual preferences list. A variant of this strategy, *success@n*, was employed in [13] to measure the rate of having at least one recommended item in the top n positions of the actual preferences list. For example, given an ordered set of recommended items *recList* of size n and an ordered set of the actual preferences *actList* of the same size, *success@3* would return 1 if at least one of the items in the top 3 positions of *recList* appears in the top 3 positions of *actList* and 0 otherwise. We use the *success@n=1,2,3* metric to make a general comparison of our results to the approach in [13].

6.3   Case Base

In this experiment we adopt the strategy of generating synthetic groups for evaluating our proposed approach to group-based recommendation. Researchers utilize this approach by generating groups from single-user data sets, to evaluate various approaches to group recommendations [20, 27–30]. We use the MovieLens dataset of 1 million ratings from 6040 users on 3882 movies.

Group Generation. We are interested in evaluating the performance of our approach across various group characteristics. To do so we varied the group size and degree of similarity among group members. The group sizes were varied between 2, 3, 4 and 5. The inner similarity correlation between any two users x, y belonging to group G is calculated using the Pearson Correlation, as defined in equation 1. We defined three inner group similarity levels with the following thresholds: *high* (0.5 <= inner group similarity <= 1), *medium* (0 <= inner group similarity < 0.5) and *low* (-1 <= inner group similarity < 0).

For each group criterion (group size / similarity level) we randomly create 100 groups for a total of 1200 unique groups. We placed an additional constraint on group formation that requires a valid group to have at least 3 items that were rated by all of the members of the group. This constraint provides for a minimal group evaluation baseline across those items. From these items we identify test items for each group.

To generate training and testing sets and to deal with the issue of disparity in profile sizes between group members, we employed a training/test set approach to split based on individual profile sizes within groups. For each group, we identified the commonly rated items among the group members. Then we checked if that set is larger than 40% of the smallest group member's profile size. If it was smaller, then those items would be the testing set for that group. If it was larger then we randomly select items from that set, not exceeding 40% of the smallest group member's profile size, to compose the testing set for that group. We do this to ensure that for each group member we have a majority of their original profile as part of the training set with as many test points as possible.

Once the test items for each group were identified, we created a training and testing set for each group. This ensures that the same training set for each group is used to generate all the predictions for that group. Table 1 outlines the number of test items identified for each group testing category. For the 1200 groups created, we identified 10,543 group/test item pairs. The training set for each group is created by taking the original MovieLens dataset and then removing the ratings of the test items identified for that group from each of the group members' profiles. In other words, the training set for each group is the

Table 1. Number of test items across group sizes and similarity levels

	2	3	4	5
High	1367	493	419	539
Medium	2156	1116	756	496
Low	1896	572	389	344

original MovieLens dataset minus the ratings for the test items for that group, for each of the group members.

Case Retrieval. To retrieve cases that may be used to model group recommendations, we define the threshold θ for group to case similarity (GG_{CB}) with a value of 0.5. This similarity threshold ensures a highly similar group is used as the case-based group. Taking each group at a time as the active group (G_a), we calculate the similarity between that group and all the other groups. The results would be the case groups (G_{cb}) that have the same size as the active group regardless of the groups' inner coherency. Cases that have a group-to-case similarity level higher than the threshold with the active group are then considered for the prediction calculation phase. We note that similarity threshold dependency could also be analyzed to provide a perspective on results for different levels. In this study, we select a representative level to understand initial results, leaving threshold dependency analysis for future work.

We explore outcomes of prediction accuracy for recommendation aggregation using the most commonly used group modeling strategies (Average, Least Misery, Most Happiness) as outlined in section 4. We examine these outcomes with respect to group size and inner group similarity. We contrast this approach, using the baseline neighborhood approach to the *Case-based* model approach, by comparing the prediction accuracy of the predicted group rating of each test item to the average of the actual ratings of the individual group members for that test item.

7   Results

For each created group and test item we calculate a predicted rating using the baseline group recommender (Section 4) and the case-based group recommender (Section 5).

The first hypothesis (H1) we examine is that the CBR approach will be able to respond to a substantial number of queries. Given the constraints, we would consider a baseline 10% response rate for the generated groups to provide an indication of reasonable traction. To some degree, this is a function of the case-base generation model, but it is critical to understand the context of accuracy. It is useful to know for the given similarity threshold in the experiments, the degree to which the approach is able to respond, as this provides context for interpreting the accuracy results. We also note that the model has some influence on intra-group similarity, but not on inter-group similarity, within the generated case-base. To test this hypothesis, we inspect the number of items and groups where the case-based approach is applied for each group category. Tables 7 and 7 show these numbers for the different groups. We compare these numbers to Table 1, which represents the overall number of testing items for each of the 100 groups evaluated in each category. This shows the categories in which the CBR approach was best able to respond to prediction queries. We can see that the CBR approach was most applicable for groups with high inner similarity levels.

Table 2. Number of case-based groups for the different similarity level and size groups

	High	Medium	Low
Size 2	63	81	81
Size 3	40	36	9
Size 4	52	11	1
Size 5	90	8	1

Table 3. Number of case-based items for the different similarity level and size groups

	High	Medium	Low
Size 2	181	326	298
Size 3	68	51	14
Size 4	108	14	1
Size 5	367	9	2

For these groups the CBR approach was overall in effect for 25% of the test items in 60% of the groups. We also notice that as the group size increased, for the highly similar groups, these numbers increased. For example, in groups of size 5 with high similarity among the group members, the CBR approach was applied for 68% of the test items across 90% of the groups. Thus, if the group is highly coherent and the group members are more similar to each other, then the cases that are used are more likely to be of high coherency as well. This, in turn, increases the possibility for those cases to have more commonly rated items within the case group, which in turn, increases the likelihood of the active group's test items being shared with the case groups. This is not the case for groups with low inner similarity levels, where these numbers go down as the similarity level for the group decreases and the size increases. For example, in groups of size 4 and 5 with low similarity level, the CBR approach responded for 1 and 2 test items respectively in only 1 group. Overall, for the representative similarity threshold, the foundational CBR approach was applicable for a substantial number of test queries (H1 accepted), and the applicability of the approach increases as the size and inner group similarity increases.

Our second hypothesis (H2) is that the hybrid CBR approach will provide significantly better accuracy than the baseline. To examine this, we first compare the results of the two case-based approaches, adaptation and adaption scaling. The performance level between the two case-based variations themselves was not significantly different. Given space constraints, here we present and discuss only the adaption scaling approach (as representative of both CBR variants) to the baseline approaches. Figure 1 shows the RMSE for the baseline group aggregation models and both scenarios of the case-based approach (CBR_{single}, CBR_{multi}) across the various evaluated group conditions. From these figures we

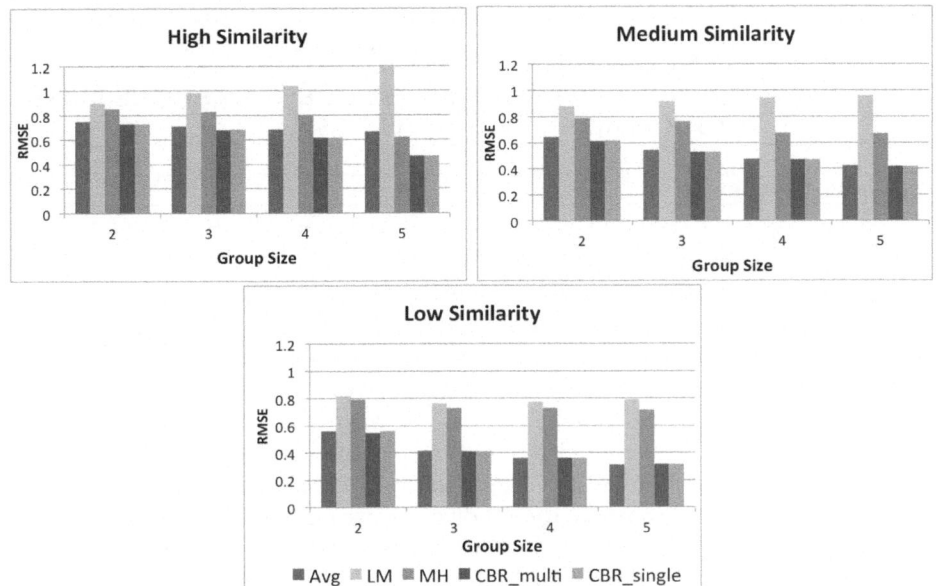

Fig. 1. RMSE for groups across the different inner similarity levels

can see that the case-based group recommender outperforms the baseline recommender for groups with high and medium inner group similarity. Between the two case-based approaches the CBR_{multi} approach is a slightly better technique than the CBR_{single}, but not significantly so. In comparison to the baseline, the CBR approach results in significantly ($p < .01$) higher accuracy predictions for groups with high inner similarity levels (top graph). The performance of this approach also increases as the group size increases for all similarity levels. For groups with low similarity levels no difference is reported between the performance of the CBR group recommender and the average baseline recommender. In this context, the CBR component largely failed to find matching cases. In that circumstance, the group recommender switches to the baseline leading to similar results. We also note for this category (low similarity) of groups of size 2, where 15% of the test items used the case-based recommender in 81 groups, the CBR recommender reported better results. Overall, these results indicate that a significant improvement in accuracy can be achieved by using a case-based group recommender (H2 accepted), where similar cases are considered as part of the prediction calculation.

Our third hypothesis (H3) is that our results would show accuracy trends for a foundational CBR approach to be similar to those reported in [13]. A complete, direct comparison is beyond the scope of the current work, given the overhead for full reimplementation and testing of their approach, as well as the availability of the original expert-crafted case-base[1]. However, we are able to make a general

---

[1] We were very grateful to receive the dataset, but too recently for analysis / inclusion.

Table 4. Success@n rates for our CBR approaches and [13]. * indicates value estimated from figure.

	SUCCESS @1	SUCCESS @2	SUCCESS @3
CBR_multi	54%	83%	97%
CBR_single	69%	89%	100%
Quijano-Sanchez et al. [13]	12%	61%	92%*

comparison, using their *success@n* evaluation metric as a basis. They reported *success@n* evaluation ranking values for $n=1,2,3$. Since it is a ranking metric, we need to be able to predict ratings for more than three items for each group from our case base, and to generate a ranked list of recommendations. To do so, we selected the groups that had a test set with 5 or more items where our case-based approach can be applied to generate predictions for those items. From the 1200 we were able to identify 71 groups that satisfied the condition. For the true ranked list used to compare the predictions, we employed the average rating provided by the group members for those items as the ground truth employing a random selection to resolve equally ranked items. Table 4 shows the success rates at various n. Note that exact numbers were not reported for all levels in [13], and so some are estimated based on the graphs showing their results (as noted). The percentage success rate results show that our approaches can provide comparable (arguably better in some cases) success rates to those previously reported, and a similar trend (increasing as n increases) to those reported in [13] (H3 accepted). While this comparison is limited in comparing outcomes across data sets, it serves to corroborate previous results on the potential benefit of CBR approaches in the context of a different case-base and group conditions. Moreover, it indicates that the benefits of a CBR approach may be realized even in a foundational implementation, without substantial demographic or data imputation requirements.

8   Conclusion

In this paper, we investigated different foundational CBR strategies for retrieving and reusing whole previous groups as a basis for making group recommendations. Inspired by [13], we carried out further exploration of the space, investigating the effectiveness of case-based reasoning approaches to retrieve and adapt whole previous groups as a basis for recommendation. Results showed that even a straightforward CBR approach can be useful across many group recommendation circumstances, providing a significant performance benefit. In comparison with previous work, we examined different similarity and adaptation metrics, larger case-base size, and multiple additional experimental conditions for group size and coherency across the case-base. We confirmed the potential benefit for integrating whole-group retrieval CBR approaches into group recommendation across different case-base and group conditions. We also demonstrated that the benefits of a CBR approach may be found even in straightforward implementations, showing the potential for a broad range of deployments and investigation in the space. Going forward, we plan on conducting larger evaluations for larger group sizes and case-bases; examining other variations in retrieval and adaptation, such as similarity thresholds; and considering the impacts of case storage and case-base maintenance.

References

1. Bridge, D., Göker, M.H., McGinty, L., Smyth, B.: Case-based recommender systems. The Knowledge Engineering Review 20(3) (2005)
2. McCarthy, K., McGinty, L., Smyth, B.: Case-based group recommendation: Compromising for success. In: Weber, R.O., Richter, M.M. (eds.) ICCBR 2007. LNCS (LNAI), vol. 4626, pp. 299–313. Springer, Heidelberg (2007)
3. Burke, R.: A case-based reasoning approach to collaborative filtering. In: Blanzieri, E., Portinale, L. (eds.) EWCBR 2000. LNCS (LNAI), vol. 1898, pp. 370–379. Springer, Heidelberg (2000)
4. Hayes, C., Cunningham, P., Smyth, B.: A case-based reasoning view of automated collaborative filtering. In: Aha, D.W., Watson, I. (eds.) ICCBR 2001. LNCS (LNAI), vol. 2080, pp. 234–248. Springer, Heidelberg (2001)
5. O'Sullivan, D., Wilson, D., Smyth, B.: Using collaborative filtering data in case-based recommendation. In: Proceedings of the 15th International FLAIRS Conference (2002)
6. O'Sullivan, D., Wilson, D.C., Smyth, B.: Improving case-based recommendation: A collaborative filtering approach. In: Craw, S., Preece, A.D. (eds.) ECCBR 2002. LNCS (LNAI), vol. 2416, pp. 278–291. Springer, Heidelberg (2002)
7. Quijano-Sánchez, L., Recio-García, J.A., Díaz-Agudo, B., Jimenez-Diaz, G.: Social factors in group recommender systems. ACM Trans. Intell. Syst. Technol. 4(1) (2013)
8. O'Connor, M., Cosley, D., Konstan, J.A., Riedl, J.: Polylens: A recommender system for groups of users. In: Proceedings of the Seventh European Conference on Computer Supported Cooperative Work (2001)
9. McCarthy, J.F.: Pocket RestaurantFinder: A situated recommender system for groups. In: Proceedings of the ACM Conference on Human Factors in Computer Systems Workshop on Mobile Ad-Hoc Communication (2002)
10. Berkovsky, S., Freyne, J.: Group-based recipe recommendations: analysis of data aggregation strategies. In: Proceedings of the Fourth ACM Conference on Recommender Systems (2010)
11. McCarthy, K., Salamó, M., Coyle, L., McGinty, L., Smyth, B., Nixon, P.: CATS: A synchronous approach to collaborative group recommendation. In: Proceedings of the 19th International FLAIRS Conference (2006)
12. Sprague, D., Wu, F., Tory, M.: Music selection using the PartyVote democratic jukebox. In: Proc. of the Working Conference on Advanced Visual Interfaces (2008)
13. Quijano-Sánchez, L., Bridge, D., Díaz-Agudo, B., Recio-García, J.A.: Case-based aggregation of preferences for group recommenders. In: Díaz Agudo, B., Watson, I. (eds.) ICCBR 2012. LNCS, vol. 7466, pp. 327–341. Springer, Heidelberg (2012)
14. Jameson, A., Smyth, B.: Recommendation to groups. In: Brusilovsky, P., Kobsa, A., Nejdl, W. (eds.) Adaptive Web 2007. LNCS, vol. 4321, pp. 596–627. Springer, Heidelberg (2007)

15. Masthoff, J.: Group modeling: Selecting a sequence of television items to suit a group of viewers. User Modeling and User-Adapted Interaction 14(1) (2004)
16. Gartrell, M., Xing, X., Lv, Q., Beach, A., Han, R., Mishra, S., Seada, K.: Enhancing group recommendation by incorporating social relationship interactions. In: Proceedings of the 16th ACM International Conference on Supporting Group Work (2010)
17. Recio-García, J.A., Jimenez-Diaz, G., Sanchez-Ruiz, A.A., Diaz-Agudo, B.: Personality aware recommendations to groups. In: Proceedings of the Third ACM Conference on Recommender Systems (2009)
18. Quijano-Sánchez, L., Bridge, D., Díaz-Agudo, B., Recio-García, J.A.: A case-based solution to the cold-start problem in group recommenders. In: Díaz Agudo, B., Watson, I. (eds.) ICCBR 2012. LNCS, vol. 7466, pp. 342–356. Springer, Heidelberg (2012)
19. McCarthy, K., McGinty, L., Smyth, B., Salamó, M.: The needs of the many: A case-based group recommender system. In: Roth-Berghofer, T.R., Göker, M.H., Güvenir, H.A. (eds.) ECCBR 2006. LNCS (LNAI), vol. 4106, pp. 196–210. Springer, Heidelberg (2006)
20. Baltrunas, L., Makcinskas, T., Ricci, F.: Group recommendations with rank aggregation and collaborative filtering. In: Proceedings of the Fourth ACM Conference on Recommender Systems (2010)
21. Burke, R.: Hybrid recommender systems: Survey and experiments. User-Modeling and User-Adapted Interaction 12(4) (2002)
22. Cox, M.T., Muñoz-Avila, H., Bergmann, R.: Case-based planning. The Knowledge Engineering Review 20(3) (2005)
23. Spalzzi, L.: A survey on case-based planning. Artificial Intelligence Review 16(1) (2001)
24. Herlocker, J., Konstan, J.A., Riedl, J.: An empirical analysis of design choices in neighborhood-based collaborative filtering algorithms. Inf. Retr. 5(4) (2002)
25. Resnick, P., Iacovou, N., Suchak, M., Bergstrom, P., Riedl, J.: GroupLens: An open architecture for collaborative filtering of netnews. In: Proceedings of the ACM Conference on Computer Supported Cooperative Work (1994)
26. Herlocker, J.L., Konstan, J.A., Terveen, L.G., Riedl, J.: Evaluating collaborative filtering recommender systems. ACM Transactions on Information Systems 22(1) (2004)
27. Salamó, M., McCarthy, K., Smyth, B.: Generating recommendations for consensus negotiation in group personalization services. Personal and Ubiquitous Computing 16(5) (2012)
28. Amer-Yahia, S., Roy, S.B., Chawlat, A., Das, G., Yu, C.: Group recommendation: Semantics and efficiency. Proceedings of the VLDB Endowment 2(1) (2009)
29. Garcia, I., Sebastia, L., Onaindia, E., Guzman, C.: A group recommender system for tourist activities. In: Di Noia, T., Buccafurri, F. (eds.) EC-Web 2009. LNCS, vol. 5692, pp. 26–37. Springer, Heidelberg (2009)
30. Chen, Y.L., Cheng, L.C., Chuang, C.N.: A group recommendation system with consideration of interactions among group members. Expert Syst. Appl. 34 (2008)

Author Index